AF228974

U.S. NAVY ATTACK AIRCRAFT

1920–2020

The quintessential U.S. Navy attack airplane of the 1960s: an A4D Skyhawk of VA-81 prepares to launch from USS *Forrestal*, 14 April 1961. The missile under its centerline is a Bullpup. U.S. NAVAL INSTITUTE PHOTO ARCHIVE

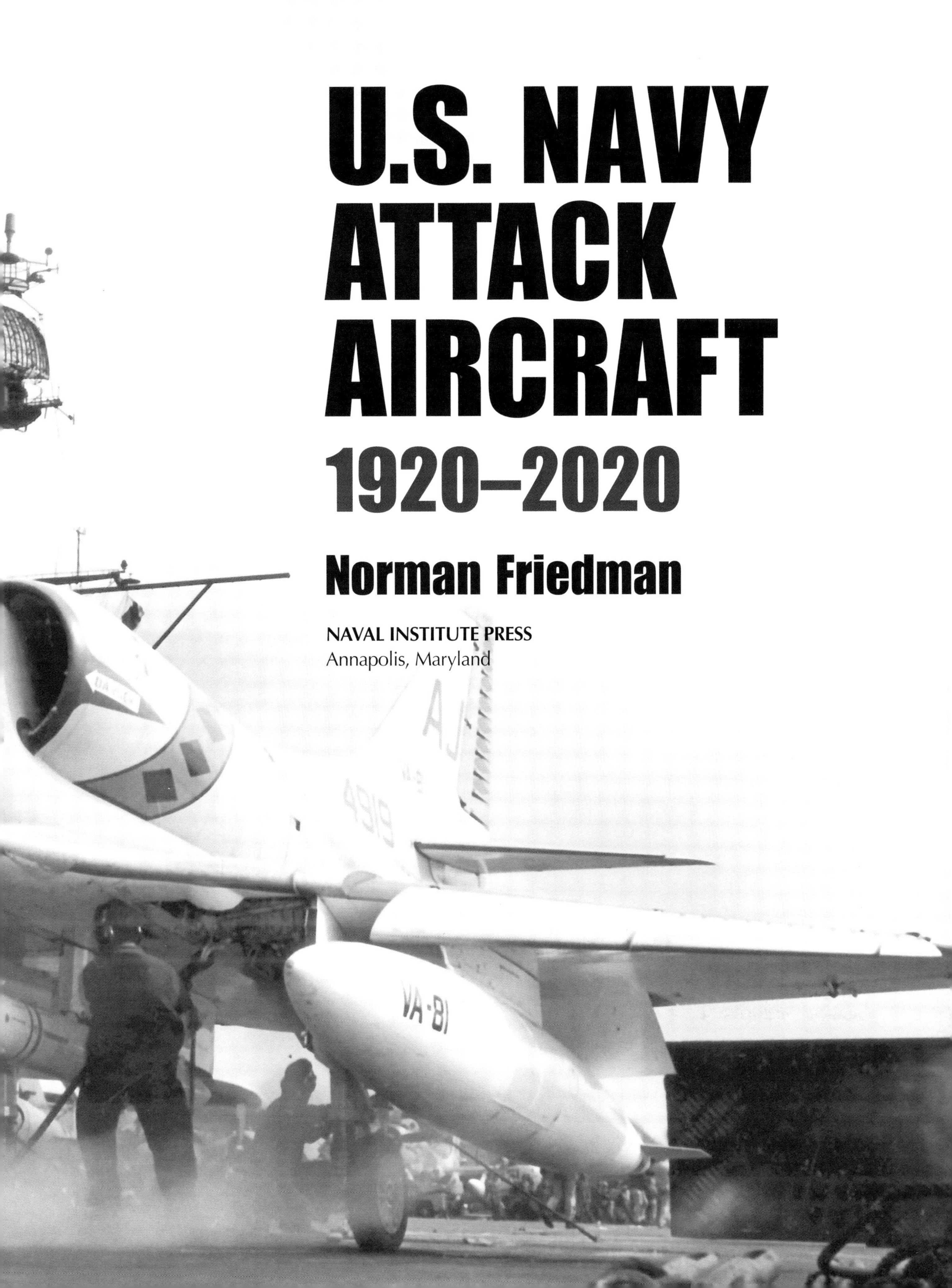

U.S. NAVY ATTACK AIRCRAFT

1920–2020

Norman Friedman

NAVAL INSTITUTE PRESS
Annapolis, Maryland

Naval Institute Press
291 Wood Road
Annapolis, MD 21402

Library of Congress Cataloging-in-Publication Data
Names: Friedman, Norman, 1946–, author.
Title: U.S. Navy attack aircraft, 1920–2020 / Norman Friedman.
Description: Annapolis, Maryland : Naval Institute Press, [2022] | Includes bibliographical
    references and index.
Identifiers: LCCN 2021048265 | ISBN 9781682474174 (hardback)
Subjects: LCSH: Attack planes—United States—History. | Bombers—United States—History. |
    United States. Navy—Aviation—History. | Aeronautics, Military—United States—History.
Classification: LCC UG1242.A28 F75 2022 | DDC 623.74/63—dc23/eng/20211001
LC record available at https://lccn.loc.gov/2021048265

30 29 28 27 26 25 24 23 22   9 8 7 6 5 4 3 2 1
First printing

Contents

Acknowledgments

This book is based mainly on documents held by the U.S. National Archives (both the downtown and College Park, Maryland, branches) and by the old Naval Aviation Division of the Naval History and Heritage Command (NHHC). I also benefited greatly from the former Grumman History Center in Bethpage, Long Island, and its material (including proposals) on that company's aircraft. Some material, particularly on aircraft performance, came from the old Naval Air Systems Command history division at Crystal City in Arlington, Virginia. I am indebted to all their staffs, past and present. I also want to thank Mark Aldrich of *The Hook,* Tommy H. Thomason, Dr. Thomas C. Hone, John Gourley, David C. Isby, Dr. Benjamin Kristy (chief aviation curator of the National Museum of the Marine Corps), Chris Wright (editor of *Warship International*), Richard Worth, and Barrett Tillman for their help, both in deepening my understanding and in helping me avoid error. Errors that remain are my own responsibility. For photographs I would like to thank Janis Jorgensen, photo curator of the U.S. Naval Institute; Lisa Crunch, curator of the photographic archives branch of NHHC; the staff of the still pictures division of the National Archives at College Park; Mark Aldrich of *The Hook*; Bonnie Towne of the National Naval Aviation Museum; John Gourley; and Dr. Raymond Cheung.

Above all I want to thank my wife Rhea for her loving encouragement and her help and her support before—and, even more, during—the strain imposed by the current COVID emergency.

When a crisis erupts overseas, presidents always ask, "Where are the carriers?"—meaning where are their attack aircraft? USS *Nimitz* transits the strategic Strait of Hormuz, 18 September 2020. Apart from an E-2 radar plane, all the airplanes on deck are F/A-18s—attack bombers. U.S. NAVY

1

THE FLEET'S AIRCRAFT

When there are crises, presidents typically ask, "Where are the carriers?" They mean, "Where are the Navy's attack planes?" These aircraft make the carrier the only truly mobile American means of projecting power, either to impress an adversary or to win a local war.

Unlike other books about U.S. naval aircraft, this one is written from the point of view of the Navy rather than of the companies who supplied the aircraft. This is the story of what the Navy thought it needed and of how that changed over time. Almost invariably that was the beginning of an airplane's story. Because this book is oriented toward what the Navy wanted, it includes as "attack" aircraft some types with nonattack designations, such as fighters and even observation aircraft, that come into the attack story.

The Buyers

The civilian Secretary of the Navy buys aircraft, but the uniformed Navy generally decides what it wants. It has two sides of significance in this story, operational and material. The key early development on the operational side was a 1926 reform that created an office of the Assistant Secretary of the Navy for Air, whose holder sometimes made important policy decisions. The operational side of the Navy generally laid out requirements for new airplanes, to be filled, generally under contract, by the material side.

The Navy bought its first airplanes at a time when the operational side was gaining power over the material side. There was considerable political resistance to centralizing Navy operational command. In 1915 Secretary of the Navy Josephus Daniels only reluctantly created the position of Chief of Naval Operations (CNO), with a direct staff known as OpNav. The new structure, its elements designated by numerical "Op-" codes (which were to change in the course of reorganizations), might have withered like earlier attempts to centralize naval leadership, but with the entry of the United States into World War I it became essential, and it grew enormously. The OpNav story matters in the aircraft story because a growing OpNav included an important air component. OpNav operated alongside the General Board, a group of senior admirals who, in theory, advised the Secretary of the Navy. In 1908 the General Board had been given responsibility for the outline requirements—characteristics—of ships, ordnance, and, later, aircraft. The board's influence in the 1920s and 1930s will be seen. It naturally competed with OpNav, and it was eclipsed in the late 1930s. It disappeared in 1950.

After World War I there was ferocious aviation rivalry between the Army and the Navy. Some Army officers, such as Col. William "Billy" Mitchell, tried to unify American military aviation as the British had done with their Royal Air Force (RAF) in 1918.

The Navy successfully resisted. Hearings before the Morrow Board in 1925 led to the important changes in naval structure of 1926. By the early 1930s the U.S. Navy was generally considered the most air-minded in the world. The 1926 reforms convinced men like William F. Halsey and Ernest J. King to gain air credentials. Even before 1926 the Naval War College, in Newport, Rhode Island, integrated naval aircraft and carriers into its war games, giving nonaviators like Raymond Spruance and Chester Nimitz a good idea of what airplanes might do in a future war.

The material side consisted of independent bureaus that, until 1966, answered only to the Secretary of the Navy (the General Board had been created partly to coordinate them). When naval aviation began, the key bureaus were Construction & Repair (C&R), which designed ships, and Steam Engineering (SE), which contracted for engines. Weapons were supplied by the Bureau of Ordnance (BuOrd). It was natural for C&R officers to learn enough about airplanes to take over the aircraft-design role or to contract with the new airplane companies.

Aviation was different from ships, in that the operational and technical sides were intimately related from the outset.[1] In 1921, at the dawn of the attack bomber story, the two sides were united in a new Bureau of Aeronautics (BuAer). It was unique among the bureaus in that it was responsible for all aviation officers and, typically, for the requirements to which airplanes were designed (the General Board had some of that responsibility, however).

BuAer technical personnel—and practices—came from the old bureaus. The C&R engineers who came to BuAer created both a preliminary design section (later called Aircraft Design Research, ADR) and an equivalent to the navy yards, the Naval Aircraft Factory (NAF), at the Philadelphia Navy Yard. ADR designs were used to judge whether stated requirements were feasible and also whether designs offered to the Navy were good enough. For example, BuAer rejected a 1944 Grumman jet fighter proposal because it underperformed an ADR sketch design. In 1946, ADR sketch designs were used to lay out the flight deck of the projected aircraft carrier *United States*. The airplanes that won the ensuing competition bore no resemblance to the sketch designs, however.

Like SE, BuAer never designed engines. Instead, it encouraged civilian builders to produce engines that would suit naval aircraft. Before World War II that meant pressing engine builders to develop the air-cooled engines the Navy much preferred (they could keep running after being damaged, whereas a bullet in the radiator would stop the liquid-cooled engines the Army preferred). The first postwar generation of U.S. naval aircraft, including bombers, was shaped by BuAer's decision to buy the Westinghouse J40 jet—which, unfortunately, failed. So did a contemporary Navy-sponsored turboprop. The Navy found itself buying Air Force–sponsored engines and license-built British engines.

BuAer was divided into divisions. The ones responsible for characteristics were Plans (later Planning) and, after August 1943, Military Requirements. Within Plans, the officer in charge of the Aircraft and Equipment desk initiated and developed proposals, which were reviewed by the Material Branch (particularly its Engineering Division) before being placed before the bureau chief. In effect the bureau chief generally decided what proposals BuAer would pursue. Those proposals generally went up to the Secretary of the Navy for approval. Within the Engineering Division was an Airplane Design Section, which was split into "class desks" for various types of aircraft in 1932. Of these, one designated "D Desk" was responsible for attack aircraft, both torpedo and dive-bomber. Its first product was the TBD Devastator. This desk drafted the type specification for what became the Grumman Avenger.

The Engineering Division developed requests for proposals to reflect the planners' ideas. It also kept track of combat experience. Its various sections all reviewed proposals submitted in design competitions, the division head producing the final report (in effect the conclusion of the competition) for the BuAer chief. This procedure, in which the whole of the technical part of the bureau reviewed all the contestants, was quite different from that used by the Army and its successor the Air Force. The latter two preferred to form an ad hoc group to evaluate each competition. That difference, between a permanent evaluation organization and a temporary one, seems to reflect the Army/Air Force preference for what they hoped would be revolutionary technology; they would argue that the Navy practice was unduly conservative. For its part, the Navy could argue that its far more experienced evaluators had a better chance of spotting what was and was not practical.

Soon after the United States entered World War II, BuAer's Aircraft and Equipment section carried out a formal review of new aircraft requirements based on letters from fleet commanders, reporting on 17 January 1942.[2] This report seems to have resulted in a flurry of requirements for bombers and torpedo bombers, and it was probably responsible for the TB2D project. None of the major carrier battles had yet been fought; the evaluation was based on perceived needs, many derived from reports of the war in Europe. For example, Admiral Nimitz in the Pacific asked for a long-range, land-based dive-bomber because of British reports describing the German Ju-88 (in what turned out to be rather exaggerated terms). His favored airplane never materialized. The following year Military Requirements called for current carrier fighters (Hellcat and Corsair) to be equipped with bombs and torpedoes (4 August 1943).

Soon after this memo was issued, OpNav stepped in. BuAer was an anomalous organization in an increasingly air-oriented Navy. As CNO, Admiral King created a Deputy Chief of Navy Operations (DCNO) for Air, who took over much of the policy role. That left open the question of where the Planning Division should be. BuAer found itself creating a new Military Requirements Division, whose divorce from Planning created some problems. BuAer was still in the policy business: on 30 October 1944 it held a Military Requirements staff conference that among other things discussed the new

single-seat attack aircraft. It took until 1947 for the aircraft characteristics function to be moved to DCNO (Air). Because BuAer was still independent of CNO, typically development contracts were approved by the chief of BuAer and the Assistant Secretary of the Navy for Air. Typical postwar practice was for DCNO (Air) (always referred to as simply "CNO" in official papers) to formulate "operational requirements," as basic requirements were called from 1949 on. BuAer reported back if the basic assumptions involved were unrealistic.[3]

Postwar, both BuAer and BuOrd (Bureau of Ordnance) pursued missile projects, sometimes in competition. In hopes of solving that problem, the two bureaus merged in 1959 to become BuWeps (the Bureau of Naval Weapons). That seems to have killed off the Aircraft Design Research organization. The marriage failed. In 1966 OpNav finally took over control of the bureaus, which became systems commands. As part of this reorganization, the 1959 marriage was dissolved: separate Naval Air Systems Command (NAVAIR) and Naval Ordnance Systems Command (NAVORD) emerged. Fights over responsibility for particular missiles and systems resumed. The Harpoon antiship missile and the Tomahawk cruise missile were probably the main attack examples.

Another important postwar factor was the emergence of a single Department of Defense (1947). Its main early impact on attack aircraft development came from its secretary's decision in April 1949 to cancel the next-generation carrier *United States*, owing to intense fiscal pressure. Secretary of Defense Louis A. Johnson did not cancel the attack bomber it was to have operated, requiring only that it be suitable for the existing *Midway* class. That in turn forced catapult development, culminating in the American adoption of the British steam catapult.

After the outbreak of war in Korea, money was poured into the development of new aircraft and weapons; Korea was widely seen as the precursor of World War III. The new technology began to mature in the late 1950s. Not surprisingly, much of it proved unexpectedly expensive. President Dwight D. Eisenhower felt pressure to control defense spending.

Eisenhower tried to control costs by cutting the expensive mechanized Army, using the nuclear-armed Navy to deter local aggression and the nuclear Air Force to deter all-out war. By this time radical systems like Polaris and the Minuteman intercontinental ballistic missile (ICBM) were on the point of entering production. The grossly underestimated costs of many new systems caused a Navy ship-building-funding crisis in 1959.

Eisenhower bet on U.S. technological superiority to balance out the sheer size of the Soviet and Chinese forces. That expectation made for a severe shock when in 1957 the Soviets orbited the world's first satellite, Sputnik. It appeared to many that a complacent United States had failed to match the Soviets in the new and vital technology of long-range missiles. Reactions included the creation of the Defense Department's interservice Directorate of Research and Engineering (DDR&E), which in turn ran the new Defense Advanced Research Projects Agency (DARPA). DDR&E became interested in cutting-edge projects that might be used by more than one service, such as integrated light attack aircraft avionics under development by BuAer.

Through this period aircraft systems were growing more complex. Until nearly the end of the Korean War, development of such components as electronics and engines were generally (though by no means always) keyed to specific aircraft projects. The BuAer engine section complained that it took longer to develop an engine than an airplane: work on a new aircraft's engine was not authorized until the airplane itself was well along. Electronics ran into similar problems. After about 1953, independent development became the rule. That too caused problems. For example, BuAer sponsored the Integrated Light Attack Avionics System (ILAAS), only to find once it matured that it was not compatible with its only possible platform, the A-7. (The concept involved did prove important and useful, however.)

In 1961 the incoming Kennedy administration adopted the Army's view that it had to be rebuilt both to offer a nonnuclear limited option in Europe and to fight local wars around the Eurasian periphery.[4] Kennedy's secretary of defense, Robert S.

McNamara, had to find the money to do so. His proclaimed goal was greater efficiency, to be gained by seizing control from the uniformed services, so as to eliminate their inefficient and often duplicative practices. To this end McNamara greatly enlarged the Office of the Secretary of Defense (OSD) into something that detractors often described as a parallel defense staff. McNamara's view was epitomized by a book written by one of his staffers: *How Much Is Enough?*[5] McNamara introduced the Planning, Programming, and Budgeting System (PPBS) as a way of controlling defense programs. It seems that the object of PPBS was to stop new programs by requiring that they overcome more and more thresholds before authorizing full funding. His insistence on projecting costs shows in the much more cost-oriented papers of projects of his era. It is debatable whether McNamara's fiscal control really worked, however, and stretching out programs generally made them more expensive. McNamara's centralized practices figured heavily in the attack aircraft programs of the early 1960s.

PPBS entailed a far more standardized approach to procurement as well as the creation of formal five-year defense plans (FYDPs). Before projects could be approved at the OSD level, Tentative Specific Operational Requirements (TSORs) and then final Specific Operational Requirements (SORs) had to be formulated. The Navy had to submit further documents, including Proposed Technical Approaches (PTAs) and Technical Development Plans (TDPs). This was far more documentation than had ever been needed in the past, and it greatly delayed programs.

McNamara and his OSD "whiz kids" often substituted relatively simple analysis for the military judgment the services had previously exercised. That typically meant concentrating on a single scenario or possible situation in which apparently competing weapons could easily be compared. The services' military judgment was based on a broader range of possibilities, but it was difficult to express in numerical form to challenge the "whiz kids." Ultimately the services learned how to do enough analysis to challenge OSD, but since McNamara's time OSD has remained a central factor in aircraft development.

McNamara changed the emphasis in aircraft development. His group kept asking whether a proposed new airplane offered so much that it was worth investing in. As a result, the interval between aircraft projects was considerably stretched. Before McNamara, BuAer had started new projects at roughly two-year intervals, in effect assuming that technology was moving fast enough to justify them. By the 1960s, however, jet aircraft technology was mature and stable enough that no such reasoning was still really valid. Major opportunities, such as variable geometry (swing wings were coming at longer intervals). More and more of the improvement was likely to come from avionics and weapons, which could fit existing airframes—hence the longevity of the A-3, A-4, and A-6. It took stealth—which could not fit into an existing airframe—to justify the A-12 and Joint Strike Fighter (JSF).

Airplanes and engines were generally bought by competition; where possible, the main alternatives are described in this book. First there was a paper competition and then, in some cases, a fly-off. The bureau paid for fly-off prototypes, but in some prewar cases manufacturers built their own airplanes and offered them in competition. Formal competitions stopped during World War II, although in some cases different manufacturers offered designs to much the same sets of requirements. For this book the most prominent example is the single-seat dive- and torpedo bomber, under the VBT designation.

After 1945, given the dramatically rising cost of airplanes, it was generally impossible to buy more than one prototype per requirement. Postwar competition requirements were given numbers, initially as "Outline Specifications" (OS), with numbers above 100, and later as "Type Specifications" (TS), apparently in the same numerical series.

Occasionally the Navy bought airplanes offered unsolicited by industry. Two prominent examples in this book are the Douglas Skyhawk (A4D, later A-4) and the North American Vigilante (A3J, later A-5). The Harrier (AV-8) is a special case. It was offered after it had largely been developed for the British (but partly financed by the United States under the Mutual Weapons Development Program). It is the only airplane in this book that the Marines sponsored and adopted without Navy interest.

Another naval organization very significant for postwar naval attack aircraft was the Naval Ordnance Test Station (NOTS), later the Naval Weapons Center (NWC), at Inyokern, California, better known as China Lake. Although created to test the rockets being developed at the California Institute of Technology (CalTech), after 1945 NOTS began to propose and prototype new weapons, the best-known being the Sidewinder missile. Typically, China Lake conceived and produced prototypes that were turned into production weapons by major defense contractors. In the late 1950s China Lake began to develop nonnuclear air-to-surface weapons, such as Shrike and the "Eye" series.

A few aircraft described here were designed under contract for the Office of Naval Research (ONR). ONR operated largely outside the mainstream Navy; it was responsible for "blue sky" projects and might be considered somewhat similar to the Defense Advanced Research Projects Agency. Like DARPA, ONR let contracts on its own initiative—for example, for the Douglas submarine-launched bomber. These projects seem to have had little or no impact on Navy thinking generally.

The Navy Mission

When the U.S. Navy emerged from World War I, it was assumed that naval warfare meant fleet against fleet. The wartime Royal Navy had invented carriers specifically to support a torpedo attack against the German High Seas Fleet in its sanctuary of Wilhelmshaven. American officers attached to the Grand Fleet were well aware of the planned operation, and British naval constructor Stanley Goodall provided carrier plans and advice to the U.S. Navy. By about 1926 the U.S. Navy had become interested in what was then called "light bombing." Dive-bombing became a dominant theme in U.S. naval aircraft development.

American interwar fleet development was driven by War Plan Orange (meaning Japan). Japan was clearly expansionist and had pressed for "Asia for the Asians" (meaning Japanese domination of Asia) even as an ally of the main European colonial powers during World War I. "Asia for the Asians" would probably mean ejecting the United States from its sole Far Eastern possession, the Philippines. It was generally accepted that the most likely (in fact, almost the only possible) trigger for a war between Japan and the United States would be Japanese seizure of the Philippines.

The endgame was well understood. Japan lived by imports, so it could be defeated by strangulation: blockade. A blockade would have to be enforced by numerous, hence relatively small, warships. The Japanese fleet was there to shield against them, so destroying it was a prerequisite. The only way to enforce a tight enough blockade would be to base those small ships on islands near Japan, so the war plan had to envisage seizing them.[6] If blockade did not convince the Japanese to give up, the alternative, as understood by 1929, was a campaign of air attacks to burn down the rather flammable Japanese cities. Since the air offensive would be based on the same islands needed for blockade, this would not make much difference to naval war planners. Invading Japan was generally ruled out, because it seemed unlikely that the United States could raise or deploy a large enough army.

Through the 1920s and early 1930s the concept of U.S. Navy war planning was the "Through Ticket to Manila"—the fleet would steam west as soon as possible to engage and destroy the Japanese fleet. The "through ticket" was war-gamed at the Naval War College, in a monthlong "big game" each class fought. By 1933 the College had a professional commentator, Capt. Wilbur van Auken, who drew lessons from its games. That year he observed that the fleet's apparent victory in a battle off the Philippines had been illusory. In the game, by the time of the battle the Japanese had taken Manila but not the southern Philippines, where after the battle the surviving U.S. fleet units steamed to a sheltered anchorage. The site had no repair facilities—they had all been around Manila. Ships had suffered under-

water damage. They would sink as soon as their pumps stopped. Damaged Japanese ships, in contrast, would withdraw to well-equipped navy yards. This result was so striking, and so important, that a summary of the game, with van Auken's analysis, went to the Chief of Naval Operations—a unique event. The war plan was completely changed. Instead of charging west, the fleet would move step by step, seizing an island and then using it as the base for the next move.

The Washington Naval Treaty had set ratios for capital-ship strengths. In theory the 60 percent Japanese ratio was justified by the fact that the United States and the United Kingdom had to operate in two oceans, also by an assumption that a fleet lost fighting power the farther it steamed from home. The World War I settlement had given the Japanese "mandates" over chains of islands through which a U.S. fleet using the "through ticket to Manila" would pass. American naval planners understood that the Japanese would seek equalizers, which would include basing naval aircraft on land in the Mandates. American planners came to see a strong carrier force as a prerequisite for the survival of the fleet in this phase of its voyage west. Since its fighters would have to beat off Japanese land-based aircraft, BuAer pressed for the highest possible fighter performance, which in turn led it to invest in powerful lightweight engines—which could power higher-performance bombers. In contrast, the contemporary Royal Navy simply accepted the idea that carrier aircraft had inherently lower performance than their land-based equivalents.

The Mandate problem demanded a land-attack mission to neutralize the islands the fleet would pass. That justified maximum-performance carrier attack aircraft and fighters with enough range to escort them.

The new "step by step" strategy shifted the emphasis to seizing and holding islands. Provision had to be made for Marine attack aircraft to support landings. Given the scarcity of carriers (which were limited by arms-control treaties), these aircraft might have to be flown long distances to reach the

Marines—hence the interest in special Marine versions of dive-bombers, such as the Vindicator and the Helldiver, with temporary float undercarriages so that they could land in lagoons. Pressure for more carrier aircraft seemed to relax. It looked as if long-range seaplane bombers based at seized islands could neutralize Japanese air bases on other islands. The Navy began to build more seaplane tenders, and it reconceived its big seaplanes as bombers. In the event, step-by-step thinking was reflected in the Central Pacific offensive of 1943–44, which brought the U.S. fleet to the western Pacific. The Marines' amphibious preparations served them in good stead at Guadalcanal, in the Solomons, and elsewhere later. The amphibious ships and craft fought effectively in the unexpected war the United States fought in the Mediterranean; and in the Atlantic The Avenger, conceived as a torpedo bomber, proved an effective antisubmarine airplane in the Atlantic.

With the end of World War II, there was no longer an agreed scenario for a future war. The world war had introduced radical new technology that a future conflict would surely feature. The newest submarines seemed to be more than a match for existing antisubmarine forces. It could be argued that only attack at the source—on submarines at their bases—had much chance of success. Carrier bombers thus had a new and vital role, to deliver bombs powerful enough to destroy hardened submarine shelters of the type the Germans had built during the war. This mission (and rationale for carrier bombers) was proposed before the end of the war in Europe; it justified the design of the first big carrier bomber, the North American Savage (AJ). It was soon clear that these aircraft could also carry the massive atomic bomb. President Harry Truman vetoed the Army Air Forces' attempt to gain a monopoly on such weapons. The nuclear mission became a central theme of postwar carrier airpower.

The naval mission shifted from the fight to seize command of the sea to a fight to exploit the command that had been won during World War II. The key (though often unexpressed) question was how carrier bombers could affect a land campaign. One

important advantage of shifting an antisubmarine campaign to attacks on enemy submarine bases ("attack at source") was that once most enemy submarines had been destroyed in this way, carriers could attack targets on the flanks of an enemy advance into Western Europe, perhaps even sensitive targets in Eastern Europe. This idea was advanced in the late 1940s, and it was revived explicitly in the 1970s and 1980s as the Maritime Strategy.

The counterargument was that a future war could be decided by nuclear bombing from land bases. In 1949 it seemed, briefly, that the U.S. nuclear arsenal might deter the Soviets and thus avoid war altogether. The outbreak of war in Korea showed that had been an illusion. Nonnuclear war in which carrier aircraft were extremely important might well be the rule. By the mid-1950s it was widely understood that neither superpower was ready to risk nuclear destruction. Their rivalry shifted to limited warfare in the Third World. Carrier attack aircraft were heavily engaged in Vietnam, and later they were the stick the United States wielded elsewhere—for example, against Libya.

In 1967, in the wake of the Middle East War in June, the Soviets began to build up a Mediterranean force they called the Fifth Eskadra—their fifth fleet. It directly challenged the U.S. Sixth Fleet. In a war, Soviet surface warships armed with antiship missiles would have to be destroyed. After a twenty-year gap, the U.S. Navy became interested once more in what it now called "War at Sea" (WAS). Attack aircraft were its main weapons. By the late 1970s, the U.S. Navy was advocating an offensively oriented Maritime Strategy in which long-range carrier bombers were extremely important.

After the Cold War ended in 1989–91 there was a wave of euphoria in Washington. A senior State Department official wrote that "history is over," meaning that the deep ideological rivalry that had created the Cold War was gone, never to return. As in 1945, it seemed that disarmament was now justified, even mandatory. The Navy found itself giving up its long-range strike aircraft. The post-1945 lesson that there would be considerable turbulence in the wake of war was still understood, however. It seemed reasonable to assume that future wars would be fought in the littorals, on the strip of land directly affected by events offshore (and in the strip of ocean directly affected by events on shore). The former was, after all, where most of the world's people lived. This view certainly seemed borne out by the Gulf War (1991) and the wars in the former Yugoslavia and in Libya. However, it did not apply to the rise of ideological terrorism that led to the war in Afghanistan. Nor did anyone take into account the rise of China as a rival. The move to the littorals justified concentration on relatively short-range attack aircraft; the rise of China has made it necessary to seek greater range, in aircraft such as the Super Hornet and now the F-35.

The Marines

Throughout, there was one other major naval player: the Marine Corps, whose airplanes were bought by BuAer and its successors. Despite their argument that their needs were sometimes quite different from the Navy's, the Marines found themselves compelled to buy standard naval aircraft offered by BuAer. In a few cases their purchases must have made particular airplanes, such as the Helldiver, much more affordable.

In the 1920s, as they fought "small wars" in the Caribbean, the Marines needed mobile precision support, which meant airplanes capable of hitting small targets. They became interested in dive-bombing, which became the U.S. Navy's favored antiship technique in the 1930s. After 1933 the Marines' mission changed to seizing forward bases: amphibious warfare. As they fought their way up the Solomons in 1942–44, the Marines acquired a substantial air force, which they learned to use as mobile artillery. They turned their Corsair (F4U) fighters into tactical bombers. In May 1944, with the Solomons campaign ending, Admiral Nimitz ordered a study of future requirements for naval aviation in the Pacific.[7] Against Navy opposition, the Marines gained the mission of supporting future landing operations from dedicated support carriers: a four-ship division

of escort carriers of the large new *Commencement Bay* (CVE 105) class, each of which would support one squadron of Corsairs (eighteen F4Us) and one of torpedo bombers (twelve Avengers), with 50 percent replacements. Marine dive-bombers would be replaced by fighter-bombers (Corsairs) and torpedo bombers, as the torpedo bombers were considered better suited to ground support thanks to their range, capacity, and versatility in bomb loads. To reflect this change, Commanding General, Marine Air Wings Pacific was redesignated Commanding General, Aircraft, Fleet Marine Force.

Postwar, the Marine Corps' continued existence and strong interest in close air support—which was amply justified in Korea—guaranteed that naval attack aircraft would not follow those of the Air Force to become nuclear bombers. That made it far easier for the Navy to deal with the shift from a nuclear emphasis back toward the conventional war outlook of the 1960s and beyond.

The Marines' aircraft are still tied closely to their troops. Their basic units all combine troops and aircraft in Marine air-ground task forces (MAGTFs). That is a relationship very different from that between the Army and the Air Force or, for that matter, between Marines on the ground and supporting naval aircraft. The current symbol of this relationship is the VSTOL (vertical/short takeoff and landing) attack bomber, the AV-8 and its F-35B successor, which is (in effect) a lineal descendant of the Corsairs on board the Marine escort carriers in the Pacific in 1945.

Technology: Airplanes

Airplane performance is defined by an envelope extending between maximum and minimum speed; its small wing has to suffice at both ends. Carrier aircraft have to land at relatively low speed—there is no vast runway on which to slow down. Limits on arresting gear include how much deceleration a pilot can handle. Typically the assumed limit on deceleration and acceleration (on a catapult) is something less than 3 gs, three times the acceleration of gravity.

An A-6A Intruder of VA-42 on board the carrier *Forrestal* for carrier qualification (25 August 1963) shows its high-lift devices: slats extended from the leading edge of the wing and flaps extended from the trailing edge to increase wing camber and thus lift at low speed. NHHC

Aerodynamics sets a limiting speed on any type of airplane. For example, few biplanes ever exceeded 250 mph, and few piston-engine airplanes exceeded 450. In the 1930s, the advent of lightweight engines in the 1,000-hp range made it possible to build fighters with speeds approaching 400 mph. They were the smallest airplanes that could be wrapped around these most powerful engines. That meant among other things minimum fuel and other payload. The Supermarine Spitfire was an example. To demand more fuel for greater endurance, as in the U.S. Wildcat, cost speed and climb rate. The United States entered World War II with lightweight engines in the 1,700-to-1,800-hp class. Aerodynamics precluded using all that power to gain much more speed. Instead, an airplane like the Hellcat could combine high speed with fuel, armor, and more powerful weapons. Much the same logic applied to carrier attack bombers.

Airplanes fly because the pressure exerted by air flowing over their wings is less than that of the air flowing underneath: an airplane is, in effect, sucked up into the air. The air on top is made to flow faster than that underneath by having to go farther over the curved upper side of the wing. Basic trade-offs come from this physics. Lift per square foot of wing is proportional to the square of airplane speed: more air flowing over the wing increases lift. A thicker wing (more curve on top) increases lift (per square foot at a given speed), but with that lift comes an induced drag. Thus a faster airplane can make do with a thinner (less draggy) and smaller wing. A smaller wing reduces drag also because the air is flowing over a smaller surface. On the other hand, the larger the wing (the lower the "wing loading," or weight per unit wing area) the more maneuverable the airplane.

The difference in pressure above and below the wing causes air to flow not only across the wing but also spanwise along it. That costs lift. The longer the wing, the less this flow and the more efficient the wing.[8] However, greater span costs maneuverability, because it increases the inertia the pilot must fight.

How much wing area an airplane needs depends on its load and its speed. Until the 1930s, the best way to add wing area within limited dimensions was to adopt a biplane arrangement. Up to a point, the drag associated with the struts holding the biplane together was dwarfed by other sources of drag, so they were no real handicap. Moreover, the struts were the main source of wing strength; until designers learned to do better, the internal structures of unsupported monoplane wings (which made them thicker) were a worse handicap. At about 250 mph, though, the drag of struts and wires becomes significant. About the time engines made it possible for tactical aircraft to reach such speeds, materials and structural knowledge were making it possible to build lightweight monoplane wings. This combination of factors explains why biplanes were abandoned.

Lift increases as the wing is tilted more against the direction in which the airplane is moving (i.e., as the "angle of attack" increases). However, at some angle the airflow breaks up: air flowing over the top of the wing no longer smoothly meets air flowing underneath. The wing stalls. It generally loses lift first near its tip, because of the way air flows spanwise around it. This is particularly relevant as a carrier airplane comes in to land, at a steep angle of attack and at low speed. One solution was to change wing geometry by adding flaps at the trailing edge of the wing. When extended, they increase both wing area and wing curvature. Initially they could not be used on takeoff, because they added so much drag. However, slotted flaps were developed that open a slot forward of the flap when it is not fully extended, adding considerable lift for both takeoff and landing. Flaps can be thought of as variable-geometry devices, simpler than swing wings but in the same broad category.

Another such device is the "slat," in effect a small wing carried against the leading edge of the main wing. When deployed it is tipped down, so its angle of attack is less than that of the main wing. When the main wing stalls, airflow over the slat remains smooth and smooths the flow over the rest of the wing. The wing can accordingly reach a higher angle of attack.

Flaps and slats extended the flight envelope, which was particularly important as designers of naval aircraft sought higher maximum performance without an excessive increase in landing speed. Landing speed had to be limited, not only because of constraints in arresting gear but also because (at least in American practice) airplanes were generally coached into correct landings by "landing signal officers" (LSOs), at least before the advent of mirror or Fresnel aids. Coaching was possible because the LSO could see whether an approaching airplane was at the right attitude, altitude, and speed, but the pilot had to have enough time to correct (or to be waved off). This action-observation-decision cycle limited maximum speed of approach.

Even well short of sonic speed, if the airplane is moving fast enough, air over the wing becomes supersonic. It refuses to flow smoothly. Shock waves form. During World War II, the Germans found that they could delay the onset of this problem by sweeping wings back. The results of their research were distributed to American aircraft designers soon after the end of the war, resulting in generations of swept-wing aircraft such as the Douglas Skywarrior (A-3) and delta-wing designs like the Skyhawk. Supersonic flight, as with the Vigilante, demanded further refinements, such as all-flying tails.

Technology: Engines

The greatest driver in aircraft design has probably been engine technology. Both piston and jet engines produce a stream of accelerated air or gas. The equal and opposite reaction to this stream pushes the airplane forward. However, there is a dramatic difference. A propeller creates a stream of air over the inner parts of the wings, so as soon as it begins to turn it generates lift. Until well into World War II, that lift was sufficient for an airplane to take off after accelerating over the few hundred feet available on a carrier deck. The stream of hot gas from a jet engine accelerates the airplane itself down the deck but without any additional stream of high-velocity gas flowing over the wings. The airplane does not gain much lift until it is moving faster than stall speed. Hence the vital importance of catapults for the Navy's first operational jets, and particularly for the Skywarrior.

Until well after World War II, naval aircraft were propeller-driven. The power the propeller can absorb is limited by its area. For high-powered turboprops (in the A2D and A2J) contraprops (two counterrotating, coaxial propellers on the same shaft) made it possible to gain blade area without accepting undue diameter, which would be a problem in an airplane of limited (carrier-suitable) height. That is also why the big TB2D used contraprops.

From a carrier-airplane point of view, the most important development in propellers was "variable pitch," which appeared in the 1930s. A propeller blade is a wing, moving air by generating what amounts to lift. The angle at which air hits the propeller depends on how fast the airplane is moving. As with an airplane wing, the propeller blade can stall out if the angle is acute enough. If pitch is fixed, the engine has to run slower at some airplane speeds, delivering less power unless it is geared down. However, a variable-pitch propeller can operate over a wider airplane speed range; for example, it can provide more power at takeoff speed, and thus reduce takeoff run, and also function efficiently at high speed. Variable pitch was sometimes called "constant speed," because the propeller turning rate did not have to go up or down with airplane speed.

Output per cylinder of a piston engine depended on how hot the combustion inside the cylinder could be, which in turn depended in part on how efficiently cylinder heat could be carried away. At first that meant piping a liquid (initially water) into a radiator and then back to the engine. A hit in the radiator could disable an engine, so BuAer much preferred air cooling, with cooling fins on each cylinder to radiate the heat away. The bureau was well aware also that an air-cooled engine could better survive a long flight back to a carrier. Air cooling also made for a simpler engine, easier to maintain on board ship. In the early 1920s BuAer's goal was to develop air-cooled engines that could produce as much power per cubic inch as the existing liquid-cooled types. If that could be achieved, the air-cooled engine would be lighter and considerably shorter, hence could be mounted in a smaller airplane.

Variable-pitch (constant speed) was the great propeller development of the 1930s. The pitch of the blades was varied by a piston mechanism in the spinner. Its housing is visible in front of the four propeller blades on this Douglas BTD-1, photographed at Patuxent River, 26 June 1944. NHHC

The first high-powered aero engines were liquid-cooled "vees" like this one on board a Martin SC-6. Note the boxy radiator sloping down the side of the fuselage. Because it cooled the engine effectively, the cylinders could be arranged compactly and tightly cowled, for maximum streamlining. However, a single bullet could stop the engine by disabling the radiator. NATIONAL ARCHIVES

Air-cooled engines were far simpler and lighter than liquid-cooled ones, because airflow over the fins of the cylinders cooled them. Nine seems to have been the greatest possible number of cylinders in one row. This T4M-1 torpedo bomber was powered by a nine-cylinder 525-hp Pratt & Whitney Hornet (R-1690). It is unusual in having a three- rather than two-bladed propeller. Note the bombardier's window under the pilot's cockpit. The numbers and letters indicate that this was the tenth aircraft of VT-9. Note the half-axles, originally provided for hooks for longitudinal arresting wires, gone by the time this photograph was taken. NATIONAL ARCHIVES

The key development making high-powered air-cooled engines possible was the NACA cowling. Its airfoil section accelerated air over the cylinders. This is a cowled version of the Vought O2U-1.

The remaining question was whether an air-cooled engine could compete with its liquid-cooled equivalent with respect to drag. Typical liquid-cooled engines were V-shaped, meaning that they presented only two cylinders in frontal area. Many air-cooled engines had nine in cross section. This problem was solved by 1929 by the U.S. National Advisory Committee on Aeronautics (NACA, the predecessor of the current NASA), which had shown that the streamlining difference between liquid- and air-cooled engines could be minimized by wrapping an airfoil-shaped cowling around an air-cooled engine. It created a high-pressure airstream over the cylinders. Typically baffles inside the cowl forced air around the cylinders. An area of low pressure at the rear drew out the air via flaps.

BuAer claimed a key role in encouraging American aircraft engine companies to develop the air-cooled engines it favored.[9] It doubted that such engines would have been developed without its sponsorship, because at least initially they were far too expensive for commercial use. Eventually they became standard American airliner engines, suitable for that role because BuAer had demanded great reliability.

The rush of air created by a cowling made it possible to add a second row of cylinders; a two-row radial was recognizable by the long chord of its cowling. The maximum number of rows in an air-cooled radial seems to have been four, in an R-4360. This fourteen-cylinder Pratt & Whitney R-1535 was installed on the Douglas XFD-1 fighter of 1933. Note the gearbox between the engine and the fixed-pitch propeller. The tank visible to the left is for lubricating oil.

In these engines, the cylinders were clustered around a crankshaft connected to a master connecting rod, which had to withstand the load of all the pistons. That seems to have limited engines to rows of nine cylinders. To increase power, additional rows of cylinders had to be added, the limit being set by the need for sufficient cooling air to reach all the cylinders. For example, the widely used R-2800 was a two-row engine. The limit was four rows, in engines such as the R-4360.[10]

When the process began, the main U.S. aero-engine company was Wright (which merged into Curtiss-Wright in 1929). BuAer found a small-scale prototype radial engine, the Lawrence Aero Engine Company's 200-hp J-1. Wright hired Lawrence to develop a series of engines it called Whirlwinds. Its J-5 powered Lindbergh's *Spirit of St. Louis.* However, Wright's president, Frederick B. Renschler, considered his company's interest in air-cooled engines inadequate. He left with some of his engineers to create an aero-engine division of the Pratt & Whitney machine-tool company. Wright and Pratt & Whitney became the two rival American makers of the air-cooled engines that powered U.S. naval aircraft until jets took over.

Pratt & Whitney designed its first engine, the nine-cylinder R-1340 Wasp, to meet a Navy specification for a 400-hp engine weighing less than 650 pounds. It first ran in December 1925, achieving 425 hp on its third run. The Navy bought two hundred. Later Pratt & Whitey engines were also named after insects. The Hornet (R-1690, 700 hp by 1935) was an enlarged Wasp weighing just half as much as the liquid-cooled Wright engine it replaced in the T4M torpedo bomber. The Wasp Junior (R-985, 450 hp in 1935) was a scaled-down version. BuAer asked for a twin-row engine: the fourteen-cylinder Twin Wasp Junior (R-1535), a doubled-up R-985. The hope was that reduced frontal area would offset limited displacement. Design began in May 1931, and the engine was being tested by November. It powered some interwar fighters, as well as the Vought Vindicator (SB2U),

but was overshadowed by the two-row fourteen-cylinder Twin Wasp (R-1830). The Twin Wasp was one of the most significant World War II engines; about 178,000 were made.

The Twin Wasp well illustrates how rapidly engines grew in the 1930s and during World War II. Initially it was rated at 750 hp, but by 1935 it was rated at 800, the most powerful available fighter engine at the time. In June 1935 the Navy contracted for a 900-hp version (known as the "-64," read "dash 64") for the PBY-1 Catalina flying boat. A -70 version with a two-stage mechanical supercharger (below) developed 800 hp at 15,000 feet and powered most Wildcat fighters. Ultimately this engine developed 1,200 hp. At least initially, about 1940, the Navy considered two-row technology too sensitive to sell overseas, so the export version of the Wildcat had the single-row Wright R-1820 instead of an R-1830. Single-row design meant a shorter engine but larger, hence draggier, cylinders.

The next step was the Double Wasp (R-2800), with eighteen rather than fourteen somewhat larger cylinders.[11] It powered Hellcat and Corsair fighters, but not the big attack aircraft ordered in 1938–39 (they had Wright R-2600s). In 1938 it was rated at 1,800 hp on takeoff. The two-stage super-charged -8 version (B series), produced under a June 1941 contract, was rated at 2,000 hp for takeoff. By 1936 the Navy wanted a 2,300-hp engine, which BuAer thought would have to be liquid-cooled. However, in 1939 Pratt & Whitney offered a four-row air-cooled X-Wasp, or Wasp Major, the R-4360, in effect two R-2800s in tandem. By August 1945 it was producing 3,500 hp.

Wright was initially less successful. Its P-1 air-cooled radial failed when installed in the DT-2 torpedo bomber, and it lost the Navy competition for a 350-pound engine weighing less than 650 to Pratt & Whitney. Then Wright began work on the nine-cylinder R-1750 Cyclone. Initially rated at 500 hp, it passed its type test in 1927. In 1932, the Cyclone was upgraded by increasing cylinder diameter ("bore") to become the R-1820. Like the R-1830, it was widely produced during World War II. The initial F-series was rated at up to 890 hp, but the follow-on G series could produce 1,200 hp. One reason it did was much better cylinder cooling thanks to much finer machining: late (Series G) R-1820 engines had 2,800 square inches of fins, compared to a thousand on earlier ones. By 1945, turbocharged (below) versions of the R-1820 could produce 1,350 hp.

Despite its difficulties in building a two-row engine, in November 1935 Wright began work on a two-row Cyclone 14 (R-2600). Its May 1937 prototype produced 1,500 hp on takeoff, and the standard initial wartime version delivered 1,700 hp on takeoff. The R-2600 powered the two main wartime attack aircraft. Its -20 and -22 versions developed 1,900 hp. The Curtiss-Wright equivalent of the R-4360, for which design work began in January 1936, was the double-row R-3350 Duplex Cyclone, or Cyclone 18, with eighteen cylinders of the same bore and stroke as the fourteen of the R-2600. It was developed in wartime mainly for the B-29 bomber. Rated World War II output was 2,200 hp, with a war emergency rating of 2,500 hp. The Navy's BB series (2,300 hp) was important postwar.

Despite its preference for air-cooled engines, BuAer maintained an interest in liquid-cooled types. That is evident, for example, in late-prewar attack bomber competitions. Liquid-cooled entrants never offered sufficient advantages to win out over the air-cooled types.

Wartime engines gained extra power through "supercharging," compressing the intake air. That provides more oxygen in which to burn fuel. It can also keep an engine running at higher altitude, up to "critical altitude," the highest at which an aircraft's superchargers can deliver standard (sea-level) pressure. Air was compressed by a supercharger driven either directly by the engine (i.e., mechanically) or by a turbine in the engine exhaust (known as "turbocharging"). Compression also heats the air: by World War II, it was necessary to cool the air before letting it into the cylinders, typically using a radiator ("intercooler"). A cooling mixture of water and

methanol could be injected into the supercharger. Such water injection became common in fighters (and, presumably, in single-engine attack aircraft) about 1944. It considerably increased output but could not be sustained; aircraft typically had a ten-to-fifteen-minute supply.

Inside the engine, the air sucked into cylinders is mixed with fuel, and the mixture is compressed before it is ignited to produce power. The greater the compression the higher the output; however, again, compressing the fuel-air mixture heats it. If it gets hot enough, it will explode before it is ignited by the spark plug or other device; this premature ignition is called "knocking" or "predetonation." During World War II U.S. and Allied air forces enjoyed an important advantage in that their higher-octane fuel resisted knocking. The benefits of high octane had been realized during the interwar period. For example, in 1931 tests on the R-1340 engine doubled its output (from 450 to 900 hp) using 98-octane fuel that made it possible to use higher supercharger pressure without knocking. The RAF began World War II with 87-octane fuel, but it fought the Battle of Britain with 100-octane fuel provided by the United States. The octane scale ends at 100, but during World War II fuels with higher "performance numbers" (PNs, corresponding to octane) were produced, ultimately as high as 150 PN. With water injection, 145 PN offered the performance associated with 175 PN.[12]

Jet engines are very different. They compress ingested air with a rotating fan before burning it in combustion chambers. The heating accelerates the air that goes out the tailpipe. Before it gets to the tailpipe, however, the heated air (gas) spins a turbine that in turn powers the compressor. What finally leaves the tailpipe is the "mass flow" against which the airplane reacts by moving forward. The amount of force (thrust) the engine produces depends both on how much hot air it ejects (mass flow) and on how much acceleration its heating imparts. There is no obvious limit to the available power, given a large enough engine. To gain thrust, the engine can use the air it ingests more efficiently by compressing

Piston engines had to be warmed up before takeoff. Initially that limited the number of aircraft that could be launched in a single operation, since the mass of aircraft aft warming up had to leave enough deck for takeoffs. Once the U.S. Navy adopted open hangars, airplanes could warm up in the hangar and could be launched immediately after reaching the flight deck. These SBD-3 Dauntlesses warm up on the flight deck of the carrier *Enterprise* before taking off for Wake, 24 February 1942. The letters *GC* on the second airplane indicate "Group Commander." NHHC

it more. To do that it needs more turbine power, which generally means higher exhaust temperature. Fuel can be fed into the hot stream of air abaft the turbine and burned there as well: "afterburning."

The compressor can be either axial (straight-through airflow) or centrifugal (air thrown out from the center of the engine). An axial compressor is more efficient, but the first successful British and American jet engines had centrifugal compressors, the type used in superchargers. BuAer opted for axial compressors in its initial engines.

The combination of compressor and turbine constitutes a "spool." Some engines have two concentric spools, in which one turbine turns a low-pressure and the other a high-pressure compressor. Splitting the compressor into two components made it possible to turn each at optimum speed. A jet can also accelerate cool air without heating it, using a ducted fan ("turbofan" below). Adding denser, cooler air to the hot exhaust (which is not too dense) adds air mass and therefore thrust.

U.S. jet engines were designated by a sequence number, as in J48, beginning with thirty. Turboprops formed a separate *T* series and much later turbofans were designated in yet another *TF* series and later in an *F* (fan) series. In principle the Navy designated jet engines with even numbers, the Army (and then the Air Force) with odd numbers, albeit the Navy used several Air Force–developed engines after its own J40 program (below) collapsed. As with piston engines, versions were indicated by dash numbers.

When during World War II the U.S. government decided to develop jet engines, an important consideration was not interfering with the massive piston-engine programs of Allison, Pratt & Whitney, and Curtiss-Wright. At this time General Electric (GE) was the world's leading developer of centrifugal-flow superchargers, which were essentially jet engines without combustion chambers. It received Army engine contracts. BuAer considered such engines obsolescent, unsuited to high speeds. It preferred axial flow, and it let its initial contract to Westinghouse, which was a leading manufacturer of steam turbines, which seemed to be a related technology. Since neither General Electric nor Westinghouse had enough spare manufacturing capacity, Allison and Pratt & Whitney were assigned to build engines to their designs. Neither GE nor Westinghouse had designed a jet engine.

Westinghouse's first engine worked well enough in the Navy's first jet fighter, the McDonnell Phantom (FH-1). Immediately after the war, Westinghouse, which BuAer saw as the world's leading maker of axial-flow engines, won a competition to build a series of what BuAer hoped would be its standard engines, under the J40 designation. In December 1948 a J40 prototype produced 7,110 pounds of thrust, which was more than any other engine. Unfortunately, Westinghouse could not meet its delivery schedules, and its engines failed to produce their rated power. The Navy scrambled to find alternatives; for a while, whether some of its aircraft programs lived or died depended on how well they could be adapted to other engines. Westinghouse was forced out of the jet business in 1960.

General Electric mass-produced a centrifugal-flow engine (J33) but also designed the axial-flow J35 (TG-180). Allison took it over and scaled it up to become the J71, one of the alternatives to the J40. GE's developed J35 was its very successful J47, which powered the F-86 fighter and its naval derivative the FJ-2 Fury. In 1952 GE began work on an engine that would give good fuel economy at Mach 0.9 but would function at Mach 2. This J79 engine powered the A-5. Its distinctive feature was that

Douglas' Edward Heinemann evaded the disaster caused by the J40 series engines. His pylon mountings were easily adapted to the successful J57. Two A3Ds on board the carrier *Forrestal* show the extent to which these big aircraft had to be adapted to carrier operation. Not only their wings are folded but also their tails. Although *Forrestal* had hangars with 25-foot clear height, the majority of carriers, built during World War II, had only 17 feet 6 inches clear. A3D height was 22 feet 9½ inches. NATIONAL ARCHIVES

Jet engines drink fuel at a rate that makes air-to-air refueling vital. This A-4E was acting as a "buddy tanker," with three fuel tanks and no weapons. The center tank contained the pump, which was driven by the air turbine visible in its nose. This was the first airplane launched that day, 10 August 1966, from the carrier *Franklin D. Roosevelt.* NHHC

the last seven of its seventeen rows of compressor blades could vary their angles of incidence under electronic control.

Pratt & Whitney realized that it too had to make jet engines. After producing British Nene centrifugal-flow engines under license, it shifted to axial flow, and produced an early two-spool engine, the J57, under Air Force sponsorship. It was the first engine to produce 10,000 pounds of thrust (eventually 18,000). The company went on to produce an early turbofan, in which a turbine stage powered fan blades outside the flow through the burners. (In the first such engines the fan was at the after end, but in most commercial engines it is at the fore end.) The company's TF30 was the first U.S. military turbofan and the first operational turbofan to have an afterburner. It powered the F-111, the A-7, and the F-14. Another version of the turbofan, one that figures in this book, was the Rolls-Royce Pegasus engine of the United Kingdom's Harrier, hence of the American AV-8. The U.S. Mutual Weapons Development Program (MWDP) provided 75 percent of the engine's initial development money.

Until nearly the turn of the twenty-first century, jet engines generally cruised at subsonic speed. Combat aircraft were typically designed for a short supersonic dash, measured in minutes (the supersonic Concorde was a very different proposition). There was, however, considerable interest in "supercruise"—the ability to cruise faster than Mach 1—for self-protection. The F-35, which despite its designation is a light bomber, offers supercruise performance, owing to both its shape and its engine.

Designations

The first post–World War I naval aircraft were given letter designations in which the manufacturer came first, followed by the function—for example, *T* for torpedo bomber. Thus the first important Navy torpedo bomber was the Douglas (D) DT. Versions were indicated by numerical suffixes, like DT-1 and DT-2. A 29 March 1922 directive formalized this system but reversed the order of manufacturer and function. Aircraft (in the same class) developed by the same manufacturer were differentiated by numbers after the type letter, so that the second Douglas torpedo bomber became the T2D-1 (the first in a manufacturer's series did not receive the number 1).

For this book the important type designators were *F* (fighter), *B* (bomber), *O* (observation), *S* (scout), and *T* (torpedo); *A* (attack) was added in 1945. Sometimes function letters were combined, as in *SB*, scout bomber, which generally meant dive-bomber; or *TB*, torpedo bomber with precision bombing capability. After World War II, antisubmarine warfare (ASW) aircraft were initially designated in the *A* (attack) series, but in the 1950s the letter *S* was adopted for them, first with the Grumman Tracker (S2F).

The important attack airplane manufacturers using this system were Brewster (*A*), Boeing (*B*), Curtiss (*C*), Douglas (*D*), Grumman (*F*), Fairchild of Canada (*F*), Great Lakes (*G*), North American (*J*), Martin (*M*), Eastern Aircraft (an arm of General Motors that produced aircraft under license during World War II, also *M*), the Naval Aircraft Factory (*N*), Northrop (*T*), Vought (*U*, because it was an

Jet engines made vertical takeoff and landing (VSTOL) possible. This AV-8B attached to Marine Medium Tiltrotor Squadron 163 (Reinforced) is shown landing on USS *Boxer*, 26 August 2019. The tilted pipe forward directed turbofan air down to provide upward thrust. Smoke marks the tilted exhaust, which directed exhaust gas downward. A single mechanism coordinated all the pipes. Because both were connected to the same single engine, the AV-8B did not risk unbalanced lift, as would be the case with separate lift and cruise engines. The radar nose marks this as an AV-8B+. USMC PHOTO BY LCPL. DALTON S. SWANBECK

Folding enabled large bombers to fit inside carrier hangars. These T3M-2s of VT-1B are on board the new carrier *Lexington*, 10 March 1928. NHHC

Despite BuAer fears that folding might unduly weaken the wings of highly stressed monoplanes, Vought managed to design successful folding wings for its SB2U, shown on 18 January 1938. The distinctive "hay-rake" dive brake can be seen folded down on the underside of the wing. By way of contrast, Douglas adopted nonfolding wings for its SBD. NATIONAL ARCHIVES

Grumman found an elegant way to fold the long wings of its TBF Avenger without excessive height. The wing was turned ninety degrees and then folded back along the fuselage. Grumman-designed aircraft on board an escort carrier all display the distinctive folding method. In the foreground is a Wildcat fighter; all of the aircraft visible farther aft are Avengers. These aircraft are in standard "North Atlantic ASW II" colors: upper surfaces in nonspecular gray, sides in nonspecular white, and the bottom in glossy insignia white. In an alternative "ASW I" scheme, topsides were to be in nonspecular dark gull gray, sides nonspecular light gull gray, and undersides glossy insignia white. Leading edges and the inside of the cowling were to be painted nonspecular white. The choice of scheme depended on likely cloudiness. U.S. NAVAL INSTITUTE PHOTO ARCHIVE

On board the fleet carrier *Essex* (as indicated by the tail marking), the Curtiss Helldiver in the foreground contrasts with the Grumman Avengers farther aft. The conventional wing-folding mechanism of the Helldiver precluded unfolding its wings on the hangar deck, although when folded the airplane fit. These airplanes are painted in the three-color scheme introduced in January 1943, the upper part (including wing upper surfaces) being semigloss sea blue, below which was a broad band of intermediate blue, with undersurfaces painted insignia white. The Avengers are TBF- or TBM-1Cs. The Helldiver appears to be an SB2C-3, as indicated by its narrow spinner and its unperforated flaps. U.S. NAVAL INSTITUTE PHOTO ARCHIVE

arm of United Aircraft), Canadian Car & Foundry (*W*), and Consolidated (*Y*). The Northrop company, founded in the 1930s, became the Douglas El Segundo division; as such, it was separate from the original Douglas Santa Monica company and had its own design staff, but the Navy designated products from both arms as *D* for Douglas.

Versions of a design were indicated by dash numbers, such as SB2C-5 as the fifth version of Curtiss' second scout (dive-) bomber, the Helldiver. Modifications to a version were indicated by a suffix letter, important ones being *A* (Army version or obtained from the Army), *B* (bomber), *C* (cannon), *E* (electronics, usually electronic warfare), *N* (night), *NL* (night winterized), *S* (ASW attack), and *W* (airborne early warning and, in some cases, an ASW search airplane in a hunter-killer combination). Many letters had multiple meanings.

The 1922 order also set up squadron designations, in a system that has survived. It divided aircraft into heavier-than-air (*V*) and lighter-than-air (*Z*) categories. Thus a fighter squadron would be designated *VF*, with a number; a bomber squadron would be *VB*, and so on. Later, helicopters (*H*) were added. Both airships and helicopters used their category letters as prefixes, so that a transport (*R*) helicopter was an *HR*. Marine squadrons were designated by *M*, as in *VMB*.

The Army (later, Air Force) system was completely different, indicating only function and a sequence number, as in the B-52. In 1962 Secretary of Defense McNamara ordered that service designations be unified, the Navy adopting the Air Force system (*A* for attack, *F* for fighter, *S* for ASW) with a single sequence number not related to the manufacturer. That was in line with his stated further intention of forcing the services to adopt each other's aircraft to reduce duplication. However, it is often said that the real driving factor was McNamara's embarrassment at not having realized that the Air Force was considering the Navy's F4H Phantom to be its F-110. Current aircraft are designated far out of sequence. Thus the Joint Strike Fighter became the F-35 because its aerodynamic prototype was the X-35 experimental airplane, not because it succeeded an F-34.

Carriers set the limits within which attack aircraft are designed. With their big flight decks, the two ex–battle cruisers could accommodate the big airplanes needed to lift torpedoes. Seen on her maiden cruise (4 April 1928), *Saratoga* shows T3M torpedo bombers plus two F6C fighters, the latter the first U.S. Navy dive-bombers. The torpedo bombers show horizontal struts near their wheels, intended for hooks to engage lengthwise deck wires initially used. These hooks are evident on the foreground F6C (5-F-22). NHHC

2
CARRIERS

Aircraft carriers inevitably limit the aircraft they can operate. It happened that the first U.S. carriers imposed far less constraint on naval aircraft than might have been expected, given their experimental character and the generally primitive state of aircraft development when they were conceived about 1921. That was largely a historical accident, a consequence of the Washington Treaty on naval arms control. The treaty allowed each of the major navies to convert two existing (or building) capital ships into carriers. At the time, the U.S. Navy was building six 43,500-ton battle cruisers. Two were selected for conversion into the carriers *Lexington* and *Saratoga*. As they were completed to the new design, the two were strongly criticized as white elephants. It took operational experience for the Navy to discover that their huge size was very useful. For one thing, their size made it possible to design large attack aircraft to equip them.

The Navy already wanted carriers. Its aviators had seen the first British carriers operating with the Grand Fleet, and C&R had welcomed the wartime service in the United States of British naval constructor Stanley Goodall, who was later the Royal Navy's Director of Naval Construction. He brought plans of the British carriers, including those then under construction, with him, and he advised the General Board on appropriate characteristics for U.S. carriers.[1] The aviators, for their part, made a strong case

that the postwar American fleet had to have sea-based aircraft, including carrier-based attack bombers.

In 1918 the General Board framed the first formal characteristics for a U.S. carrier. Its air group should comprise twenty-four light aircraft for combat and reconnaissance and six heavy bombers (torpedo planes). War experience showed that the fleet badly needed a carrier to extend its scouting area; construction should "be not delayed." Although other ships should carry their own reconnaissance aircraft and fighters, it took a carrier to operate heavy bombers. Scouting was a particularly sore point, because until 1916 Congress had generally resisted calls for fast scout cruisers and for battle cruisers to back them up. That year's huge naval bill included six battle cruisers and ten scout cruisers, of which the battle cruisers were canceled under the Washington Treaty.

Carriers were included in the tentative U.S. fiscal year 1920 (FY20) program, framed in 1918–19, but were not built. Instead, the General Board supported conversion of the collier *Jupiter* into the experimental carrier *Langley*. The CNO, Adm. William S. Benson, thought even this step premature, but he was overridden. Like the British carrier prototype *Argus*, *Langley* had a flight deck large enough to operate torpedo bombers. Nevertheless, it had limitations that turned out to have important consequences for U.S. naval aviation.

The General Board did not give up on carriers. In February 1921 it submitted characteristics for a carrier to be included in the FY22 program (for the fiscal year beginning 1 July 1921). No air wing was specified, but the ship would stow seventy-two aviation torpedoes. Speed was set at thirty-four knots, and the flight deck would be eight hundred by one hundred feet. C&R offered two alternative carrier designs, one of about 25,000 tons and thirty knots, the other of about 35,000 tons and thirty-five knots. The board chose the larger ship, because the development of naval aviation demanded the largest deck and the highest speed attainable. The larger ship would also be steadier and would offer more stowage space. Anything less would be unwise, given the rapid development of aviation. The Secretary of the Navy approved the characteristics on 12 March 1921.

C&R produced sketch designs for both "flush-deck" and "island" configurations—that is, respectively, with the ship's pilothouse under a single-level flight deck well forward and its uptakes outboard or aft, leaving the flight deck unobstructed, or both in a tower structure at an edge of the flight deck. The General Board liked the former, because it would not restrict airplane size. C&R preferred the island, because it would greatly simplify design without greatly complicating air operations. This question, flush deck versus island, would recur in the 1920s and again after World War II. The main problem in a flush-decker is how to dispose of funnel gas without interfering with flight deck operations. The greater the size of the ship, the higher its power, and the more gas that it would generate. C&R also argued that the flush-decker would need an open stern, which would complicate landing by creating undesirable eddies abaft the ship. Hangar capacity of the island design was nineteen bombers (with folded wings) and thirty-eight fully assembled fighters. C&R forwarded a detailed design in October 1921.

Although C&R's island design was not based on that of the battle cruisers, it was about the same size. A comparable ship could be produced by converting an incomplete battle-cruiser hull. There were already calls in Congress to cut the large ongoing building program. There was already a proposal to cancel four of the six big battle cruisers under construction, converting two of them into carriers. Largely to outflank the disarmament proponents in the Senate, the Warren G. Harding administration called a general naval disarmament conference. It met in Washington in November 1921.

Carriers were very much an experiment, so the U.S. delegation wanted only the experimental *Langley* and two converted battle cruisers. The British argued for much more carrier tonnage. Because the capacity of any one carrier was very limited, to provide their fleet with enough aircraft they wanted multiple carriers. That considerably boosted the total carrier tonnage allowed for each signatory country: the treaty allowed the United States and the United Kingdom 135,000 tons each, a huge figure for an entirely experimental technology (U.S. battleship tonnage was set at 525,850).[2] Maximum individual carrier tonnage was set at 27,000. However, each signatory was allowed to convert existing capital ships into carriers, which could displace up to 33,000 tons. Another treaty clause allowed additions of up to 3,000 tons to existing capital ships to provide them with more air and underwater protection, the idea being that ships designed before World War I might otherwise be outdated, tempting signatories to replace them. The American position was that the two ex–battle cruisers fell into the category of ships that could be given this extra protection; *Lexington* and *Saratoga* were accordingly designed at 36,000 tons.[3] During the interwar period, the sheer size of the two ex–battle cruisers *Lexington* and *Saratoga* defined the sort of aircraft, particularly bombers, that the U.S. fleet could wield.

After the two 33,000-tonners, the United States had 69,000 tons of carrier tonnage left, including *Langley*. It appeared that a 27,000-tonner could carry the same seventy-two airplanes as the ex–battle cruisers. However, three 23,000-tonners could be built within the available tonnage, so in 1925 the General Board asked for a sketch design of such a ship. Other alternative divisions of the available tonnage were four ships of 17,250 tons or five of 13,800 tons.

The fleet certainly wanted carriers. In his FY26 annual report, Commander in Chief U.S. Fleet urged that all the available carrier tonnage be built as quickly as possible. He wanted more carriers so badly that in April 1927 he asked for a sister ship of the slow *Langley*. The General Board turned him down; another second-line carrier would waste valuable treaty tonnage. By this time the General Board was framing a five-year building plan that would include new carriers.

C&R produced a series of sketch designs to define alternatives, but it had very little operational experience to draw on. The British had invented not only the carrier but carrier operating procedure. A senior officer on board a British carrier explained that pilots could not be expected to land until the flight deck (like a runway ashore) was clear, meaning when the airplane that had landed before had been struck below. It followed that the size of the hangar, not the flight deck, determined how many airplanes a carrier could operate. Required hangar headroom would help decide how deep the ship's hull had to be. BuAer wanted seventeen feet for bombers but twelve feet six inches for fighters.

BuAer pointed out that airplanes had to warm up their engines in the open before they could be launched. The number a carrier could launch in one operation, then, was defined by the flight deck. A deck eighty feet wide could accommodate three rows of aircraft. Anything narrower would lose a third of the potential attack force. Slightly greater width would buy very little. An island would squeeze the flight deck.

Most importantly, deck area is not proportional to displacement: on a given total displacement, more aircraft could be packed onto a larger number of smaller ships.[4] With little operating experience of its own by which to test that of the British and these assumptions, the Navy relied heavily on gaming at the Naval War College—where battles often included large carrier forces. Although a carrier might be as difficult to sink as any other large warship, one bomb on its flight deck could eliminate its air capability. Games showed that it would be very difficult to defend a carrier against enemy air attack: effort had to be concentrated on a carrier's own first strike. The more carriers, the more a fleet's airpower could be dispersed to survive. The General Board chose the 13,800-ton carrier. Five of them offered slightly more aircraft. A maximum air effort could be attained more quickly (each flight deck would function about as fast), with greater mobility and less vulnerability. The board included one carrier in each year of a five-year program it proposed in September 1927. It submitted proposed characteristics in November.[5]

As BuAer had pointed out, to get the most out of them, small carriers had to be made flush-deckers. Without an island, uptakes and down-takes (air intakes for the boilers) were difficult to arrange: the less powerful the machinery—the slower the carrier—the better. C&R offered either a 29.4-knot carrier with sufficient space to stow 108 aircraft (770-ft flight deck) or a 32.5-knot ship (700-ft deck, eighty-one aircraft). The choice was probably preordained by the smoke problem, as the slower ship needed only 53,000 shaft horsepower (shp) versus 107,000 for the faster one. Uptakes would be in roughly the same position (at the port edge of the flight deck, aft) as those in *Langley*, a ship that had been successful (but was much slower). To obtain the most aircraft on the available tonnage, the board accepted a nearly unarmored ship, its protection limited to decking over the steering gear and a conning tower. It lacked the sort of sophisticated underwater protection that had been built into the two big carriers. The board's characteristics were formally approved on 7 November 1927. Built under the FY29 program, this ship became USS *Ranger*.

To save weight, the flight deck was built atop the hull rather than being made integral with it, as in the two ex–battle cruisers. The flight deck itself was a relatively light wooden structure laid on thin plating. The hangar was protected from the elements, if at all, by thin roller shutters (some parts could not be closed off at all, which was a problem in cold weather). Airplanes could now warm up inside the open hangar. The carrier could launch many more airplanes in quick succession, as quickly as the warmed-up planes were sent up to the flight deck. Also, any explosive air-gasoline mixture, which could rip apart the hull of a carrier with an integral hangar, would vent through the sides of the open hangar. It might still wreck the flight deck above it but with only limited consequences for the hull.

As the new carrier was being built, American carrier operating technique developed dramatically. At the Naval War College, Cdr. J. M. Reeves Jr. found in a game that given enough fighters he could defeat an air attack; he had to refuel them quickly enough and so keep them in the air. He was struck by the number of fighters he lost because he could not land them quickly enough to fuel them.[6] Reeves left Newport to take the aviation course at Pensacola and then to join *Langley* as Commander Aircraft Squadrons Battle Fleet. At this point *Langley* had only nine aircraft on board, because she had a small hangar.

The single most important U.S. Navy carrier innovation of the interwar period was the deck park: as they landed, airplanes were moved forward, not struck below. They were protected by a cross-deck wire barrier. Off Maui, 2 March 1932, USS *Saratoga* recovers her T4M-1 torpedo bombers, one of which has been waved off by her LSO. Another taxies forward to join the deck park alongside the superstructure. The smaller aircraft in the bow are F4B-2 fighter-bombers and two-seat SU scouts. NHHC

Reeves challenged his pilots to come up with a way to recover aircraft much more quickly. They and he soon realized that after landing, aircraft could be moved to the ship's bow, where they could be protected by a barrier. That was far quicker than waiting for the elevator to take an airplane below. Many of the carrier's airplanes might spend most of their time on the flight deck rather than in the hangar, moving back and forth as they were "recovered" on board and then respotted aft. Reeves' innovation made it possible for U.S. carriers to operate considerably more aircraft than those that, like the Japanese, followed British practice.[7] That is why the three U.S. carriers at Midway operated about as many aircraft as the four Japanese carriers they faced.

To enable airplanes to land at a much faster rate, they were controlled by a landing signal officer.[8] The LSO commanded the pilot into position and then ordered him to cut his engine. With power off, the airplane stalled out and fell onto the deck—in what was often described as a "controlled crash." After hitting the flight deck an airplane still had considerable forward momentum. Normally its tailhook caught an "arresting wire," which absorbed the energy involved.[9] Thus at any particular time the capabilities of existing arresting gear set upper limits on the combination of weight at which an airplane landed and the speed at which it stalled out (energy is proportional to weight and to the square of the speed). Flight-deck design took into account the expected pullout of the arresting wire when the airplane landed.

If an airplane failed to catch a wire, it had to be stopped before it hit airplanes parked forward.[10] Hence the heavy wire barrier, which was intended to wrap itself around the propeller. Airplanes sometimes jumped the barrier and caused dramatic flight-deck fires.

In theory, aircraft parked forward were protected by the wire barrier visible here. The wires were expected to wrap themselves around the propeller, stopping the airplane. Coming in too high on board the escort carrier *Guadalcanal* in 1944, this approaching Avenger is probably going to miss the barrier.
NATIONAL ARCHIVES

It became too common for jets to miss arresting gear. Wire barriers rode up over the smooth nose of a jet and could decapitate a pilot. The successor nylon barrier is visible here, on the deck of the carrier *Oriskany*, 8 February 1955. The airplane is an F2H-2P (photo Banshee). NHHC

While building, *Ranger* was modified to reflect the new way of operating. She was given arresting gear forward and aft, so that she could continue landing aircraft even if one end of its flight deck was destroyed. The two elevators were brought much closer together, between the runs of arresting gear. A third elevator was added right aft, to bring airplanes up to the flight deck after having warmed up below.

By this time British and Japanese carriers were showing secondary "flying-off decks" forward, from which they could launch smaller aircraft while larger ones landed on their main flight decks. As the next best thing, *Ranger* was given space for two cross-deck catapults on the hangar deck (the catapults were never, however, installed). Like the double-ended arresting gear, this was meant as a combat survivability feature: it could continue to launch aircraft even if the whole flight deck was destroyed. While *Ranger* was being designed, the two big carriers entered service. It became obvious that an island was a vital feature; *Ranger* had one added outboard of her flight deck.

Carrier function, which meant the function of carrier aircraft, was reconsidered in 1930, looking forward to the next ship to be built. Initially there was strong pressure to build a "battle-line carrier," intended mainly to provide spotters for the battle line and fighters to protect them. However, the president of the Naval War College argued successfully on the basis of extensive gaming that carriers were valuable mainly as offensive weapons, using dive- and torpedo bombers. Their main defense against being put out of action by air attack was evasion. Any carrier tied to the (rather conspicuous) battleships would be put out of action. The larger the carrier, the greater its air firepower. The General Board dropped the four planned follow-on *Ranger*s in favor of the largest carriers it could get within the available tonnage. That turned out to be two big *Yorktown*s, as well as a smaller carrier (*Wasp*) to use up the remaining tonnage. The two *Yorktown*s were, in effect, the prototypes of the very successful wartime U.S. *Essex* class.

The Naval War College's emphasis on offensive carriers challenged designers to make the ships more survivable. On the 20,000 tons available carriers could have battleship-style torpedo protection and enough armor to deal with the cruisers they might encounter. The difficult question was how to keep them fighting *as carriers* despite likely bomb hits. The lightweight flight deck adopted in *Ranger* turned out to be the solution. A carrier with an integral-hull flight deck—that is, whose flight deck was structurally an integral part of the hull—had to return to a shipyard if the deck was wrecked by a bomb. However, a light wooden deck laid on top of thin steel plating could easily be repaired by a ship's own crew, even during a battle. At Midway, for example, the carrier *Yorktown*'s flight deck was hit repeatedly. The crew was able to repair it and resume flight operations a few hours later. No other navy's carriers could have done the same. Japanese carriers that suffered serious flight-deck damage, such as *Shokaku* at Coral Sea, had to go into a yard. Even had the four Japanese carriers sunk at Midway not suffered fatal internal damage, they would have had to withdraw from operation because of the damage to their flight decks.

Catapults could launch even large aircraft from small escort carriers. This TBM on board the escort carrier *Core* is attached to a catapult bridle. U.S. NAVAL INSTITUTE PHOTO ARCHIVE

Aircraft warmed up in open hangars could be catapulted from the hangar deck, as here (a Hellcat from *Yorktown* in 1943). An airplane launched this way would not interfere with the deck park. This capability was little used, and the cross-deck catapults were eliminated in favor of additional antiaircraft guns at the edge of the hangar deck. U.S. NAVAL INSTITUTE PHOTO ARCHIVE

The *Yorktown*s set a pattern of American carrier designers opting for maximum numbers of airplanes. Their flight decks and other facilities shaped the attack bombers with which the U.S. Navy fought the Pacific War. However, these same airplanes could operate from much smaller decks. When, from 1941 on, the Navy badly needed numbers quickly, it was able to adapt its standard aircraft to converted light cruisers (light carriers, CVLs) and to converted merchant ships (CVEs). It helped considerably that BuAer successfully developed catapults, which made it possible to launch heavy aircraft from these short flight decks. Catapults made their first appearances on the *Yorktown*s. Because they slowed the launch cycle, however, they were not much used on board large carriers until late in the war.

At the end of World War II, the U.S. Navy was completing three much larger *Midway*-class carriers, inspired by the British success in armoring their *Illustrious*-class (and later) carriers' flight decks. Unlike the British ships, the *Midway*s carried their armored flight decks as superstructures, like the unarmored decks of earlier U.S. carriers. That required much larger hulls, with consequences for attack aircraft: the sheer size of the new carriers seemed to justify significantly larger aircraft. After the war, their size suited the *Midway*s to a strategic-attack role. When the supercarrier *United States* was canceled, Secretary of Defense Johnson permitted work to proceed on a Navy strategic bomber—*if* it could be launched from a *Midway*. That in turn made it urgent that a suitable catapult be ready as soon as the bomber became available.

With the advent of jets, existing carriers became inadequate. It was no longer possible for an LSO to coach a pilot into the desired "controlled crash." Unlike a piston engine, a jet engine continues to produce thrust even after its throttle has cut off its fuel, because it retains so much heat. The pilot cannot cause the desired controlled crash by cutting his throttle. Also, the plane was generally moving too fast for the cycle of reactions between LSO and pilot.

Photographed from a blimp of ZP-14 at the mouth of Chesapeake Bay on 15 October 1943, the escort carrier *Block Island* (CVE 21) shows just how small her flight deck was. She had an all-Grumman air group: Wildcat fighters forward and Avengers farther aft. NHHC

BuAer initially seems to have assumed that jets could adapt to the existing system as experience was gained. The British were far less ready to accept that. By 1951 both their aircraft research establishment (Farnborough) and the Royal Navy were interested in having the pilot judge for himself whether he was making a good approach. In 1952 the test carrier HMS *Illustrious* tested a "no cut, no flare" landing method without an LSO. Using it, Vampires landed successfully at high speed, their pilots relying on an innovative "mirror sight." The stabilized mirror reflected a bar of lights. When the pilot approached correctly, he saw the bar centered in the mirror. Otherwise he corrected up or down depending on where he saw the bar. The mirror (later, a Fresnel lens) made the pilot-controlled approach possible, although LSOs are still quite important.

The British did accept, however, that a jet pilot would be more likely to miss a wire and have to go around again. They therefore wanted to revert to their prewar (pre-LSO) practice of landing with power on. That became practicable when the British-invented angled deck was introduced. The British

disclosed this new flight-deck configuration to the U.S. Navy in September 1951. Its landing area was angled out from the axis of the deck, so that even if an airplane missed the wires and had to go around again, it could not collide with aircraft parked forward or in the area opened up by angling the landing zone. The angled deck ended the perceived need for flush decks to operate jets. The first big postwar carrier USS *Forrestal* was designed and laid down with a flush deck but was redesigned on the ways with an angled deck that made it possible to install a conventional island.

The combination of angled decks and mirror sights made jet landings safer, but through the 1960s the accident rate remained unacceptably high. It even brought into question the viability of the jet Navy vis-à-vis the U.S. Air Force. The solution included a much more rigid and detailed approach to air operations, as reflected in the voluminous naval aviation flight manuals (of a series known as "NATOPS") of the late 1960s and later.

Operating heavy jet bombers required much more powerful catapults. The issue was forced by the

The success of the steam catapult led the Navy to investigate another British innovation, the angled deck. This is USS *Essex*, 12 January 1960. Because the landing deck is angled off the centerline of the carrier, the landing aircraft (in this case, presumably a TF-1 for carrier on-board delivery [later designated C-1A]) is not faced with a deck park into which it might crash. Aircraft were parked in the triangular area to the right and at the bow, to the right of the landing path. The after elevator, which would have interfered with the landing path, has been moved out to a starboard position. The original deck-edge elevator remains, at the fore end of the angled deck. The original forward elevator also remained, typically somewhat enlarged so that it could feed strike aircraft to the two catapults in the bow. NHHC

Associated with the angled deck was a shift of responsibility for landing control from LSO to pilot, using another British innovation, the mirror landing aid. The lights made it evident to the pilot whether he was high or low or to one side or the other. This mirror was on board USS *Saratoga*, May 1959. Mirrors were later replaced by Fresnel lenses, but the concept survived. NHHC

advent of the first heavy Navy nuclear bomber, the A3D (later A-3) Skywarrior. An attempt by BuAer to produce a suitable catapult failed, but fortunately the British had developed a viable alternative, the steam catapult. It appears in retrospect that it was having been rescued by the British in this way that made BuAer receptive to the two other British carrier innovations, the angled deck and the mirror sight.

The capacity of the steam catapult set the limitations within which later jet attack aircraft were designed.

The Carrier as a Weapon

Just as strike aircraft constitute the main battery of a carrier, the strike-planning system can be considered broadly equivalent to the fire-control system on board, say, a battleship. Prewar and World War II offensive air planning was based on an air plot separate from the usual surface and flag (the embarked admiral) plots. The air plot showed not only where an enemy force had been spotted but also (at least in theory) where it was going. To aim an air strike at a moving enemy force the carrier's planners had to take into account the course and speed of the enemy and the time it would take the strike aircraft to reach him. At Midway, Admiral Spruance ordered his strike on the basis of a plot that gave him a sense that his aircraft would reach enemy carriers as they recovered and rearmed their own aircraft. Air Plot was physically separate from Air Control, which handled the launch and recovery of aircraft. In the two *Yorktown*s, the first U.S. carriers designed after real carrier experience, the two spaces were adjacent, Air Control having a view of the flight deck.

Initially Air Plot controlled not only strike aircraft but also the carrier's fighters. It was therefore the natural location for a radar plot intended for fighter control. The two functions soon split, fighter control being exercised from the ship's combat information center. As completed, *Essex* had her Air Plot in the island adjacent to Air Control, but CICs needed much more space (it was also important to limit waveguide length from the ship's radars). They were therefore installed in the gallery deck under the flight deck. Typically a wartime Air Plot was adjacent to and not far from the pilots' ready rooms.

Because carriers were essentially so fragile, it was vital that the enemy's carriers be spotted first. When World War II began, that mission was assigned to the carrier's scouts, typically dive-bombers carrying reduced bomb loads (a five-hundred- rather than thousand-pound bomb) and additional fuel. Armed scouts generally flew out in pairs to cover a search pattern. In 1941 the size of the pattern, and hence maximum search range, was defined by visibility: at the end of its search segment the crew of a scout had to be able to see out to the edge of the sector beside theirs to be flown by the next pair of airplanes. That set range at about two hundred nautical miles. With the advent of radar, the horizon receded somewhat (the radar horizon is four-thirds as far away as the visual horizon). Alternatively, the enemy could be spotted by long-range patrol aircraft. In that case much depended on accurate navigation by both the flying boat and the carrier strike aircraft. Moreover, both carrier scouts and patrol planes often wrongly estimated enemy course and speed.

Although American naval war planners always intended that carrier aircraft would strike land targets—for instance, in the Mandates—there was apparently little prewar attempt to gain the necessary intelligence for doing so. Shortly before the war, a version of the Wildcat fighter (F4F-7) was designed specifically for very-long-range photo-reconnaissance. That would be essential in planning strikes. After 1942 this planning/intelligence function was concentrated in an air intelligence room. Carrier air groups included specialized photo aircraft; they could also receive photos taken by long-range patrol aircraft.[11]

Air Intelligence and Targeting

The *Essex* class may have been the first to incorporate an air intelligence room, on the galley deck next to elevator machinery and a squadron office. Its advent may have coincided with the first appointments of air combat intelligence officers, soon after the outbreak of war. The first of them reached the fleet in time for Guadalcanal. Available accounts emphasize their role in collating information about enemy aircraft, but there must also have been photo interpreters to select targets (and to deal with defenses) when attacking Japanese-held islands. Carrier pilots attacking such targets needed timely information about them and their defenses. As the Japanese adopted radar, electronic reconnaissance became more important; typically it involved multiseat Avengers (TBM-3Qs). They could carry the dedicated crewman needed to operate an intercept receiver.

World War II carrier land targets were generally contingent: they were chosen in the context of a particular operation that itself had not been expected more than a few months beforehand. That changed after World War II, when the primary carrier strike objectives were set in the context of an overall nuclear attack plan. Carrier intelligence officers collated intelligence on specific targets the carrier was likely to be ordered to strike. Typically their product was "air objectives folders" for those targets, including navigational aids and target data, as well as the photos and charts a pilot would need to navigate within the target area and to identify it.[12] The folder also included target information needed by commands for operational planning and aircrew briefing. Typically one airplane was assigned to each target, the pilot planning his own flight based on target folder information.

Postwar massed attacks against conventional targets (as in Korea and Vietnam) were planned much as during World War II. However, with the advent of the Grumman Intruder, other tactics were tried. They included one- or two-airplane night strikes. Pilots planned their low-altitude flights based on factors such as enemy radar cover, antiaircraft weapons, and the terrain itself—a pilot should not, for example, plan to fly up a valley if there was no way out—and even likely fuel consumption.

When carriers were assigned nuclear weapons, they deployed with target folders for potential nuclear targets in their safes. Pilots were personally responsible for flight planning. If a strike was approved, the pilot received a special briefing, which might include an assigned "time on target" to ensure that he did not interfere with other nuclear strikes. The intelligence information reflected in the folders was collected by what might now be called "national means." In a presatellite era, that generally meant information obtained from outside Soviet airspace, such as positions and characteristics of radars at or near the Soviet border, and possibly also photographs taken obliquely from outside the Soviet Union. Carriers began to need more up-to-date information once the Soviets deployed mobile radars. They gained an organic means of collecting signals intelligence, initially in the form of AD-5Qs using TASES (Tactical Signal Exploitation System).

This need of carriers for current intelligence, mostly gathered by their own specialized aircraft, was understood; and requirements for a new-generation carrier reconnaissance airplane were framed in 1957.[13] It would combine the previously separate roles of photographic and electronic intelligence gathering. It might also have side-looking radar for night and bad-weather work. As this idea evolved, it became clear that automation would be needed to handle the flood of data involved. The Office of Naval Intelligence (ONI), which was responsible for handling reconnaissance data, was developing a Naval Intelligence Processing System (NIPS). The system was intended to assist an analyst to associate bits of information to create something usable. Without automation, association would rely on an analyst's memory and his ability to find earlier photographs for comparison. Beginning in 1962, ONI developed an Integrated Operational Intelligence System (IOIS) consisting of the new multisensor package plus an automated Integrated Operational Intelligence Center (IOIC) on board a carrier. Given automation, an analyst could quickly find previous photographs of a place of current interest. Digitization was still far in the future—photographs were still stored in safes—but images could be retrieved much more quickly, because their indexing system was now automated. The most dramatic new technology involved was a secret high-speed film processor, a one-hour film lab, which made it possible to develop and quickly file photographs from a returned airplane. (Later this technology was the basis for one-hour commercial film labs.)

IOIS was so important that ONI formed a special task force, including representatives of ONI, NPIC (National Photographic Intelligence Center), BuWeps, and North American Aviation. Project priority may have been raised by Navy experience in the Cuban Missile Crisis, when carrier reconnaissance proved vital. The first system went on board the carrier *Saratoga* in 1963; by 1967 it was on board all the large-deck carriers.

IOIC was associated with a new reconnaissance airplane. The 1953 Development Characteristic was completely rewritten, as CA-03501-1, for a supersonic system and issued on 3 January 1958 (a revised version appeared on 3 July). By this time the Navy assumed that the primary aircraft mission in the 1961–75 era, when the airplane would be operational, would be limited warfare, which demanded much more timely and detailed information than nuclear strikes. In Korea, photo aircraft had provided 85 to 95 percent of all fleet intelligence. The current split between VFP (fighters with cameras), VAP (attack aircraft with cameras), and VAQ (electronic reconnaissance) was undesirable; as far as possible, all sensors should be on board a single airplane. Ideally an existing fleet airplane would be modified for the new integrated intelligence mission. It would operate from *Midway*-, *Forrestal*-, and newer-class carriers. Radius of action should be significantly greater than that of the attack aircraft supported, and the new reconnaissance airplane should have "speed, altitude, and electronic defense capabilities that will permit maximum practicable chance of survival in enemy-controlled airspace." Radius should be two thousand nautical miles and speed Mach 2.0 or better, with a combat ceiling of 100,000 feet. The new airplane would replace others just entering service: the F8U-1P (Crusader) and the A3D-2P and -2Q (Skywarrior versions), in a ratio of about three to four.

The Soviets were beginning to deploy antiaircraft missiles, so it would be difficult to operate safely at optimum photo altitude (2,000 to 30,000 feet). The new reconnaissance airplane should carry vastly improved integrated reconnaissance sensor systems at speeds and altitudes beyond those of current photo and electronic reconnaissance aircraft.

A carrier and her aircraft form a weapon system, the long-range sensor of which is the carrier's organic reconnaissance aircraft. The most elaborate carrier reconnaissance airplane was the RA-5 Vigilante. An RA-5C launches from USS *Ranger* on the Yankee Station in the Tonkin Gulf, March 1966, during Operation Jackstay. These aircraft could be distinguished by their ventral "canoes" containing cameras and other sensors and by the dorsal hump carrying extra fuel. NHHC

Launching from a carrier, an RA-5C of RVAH-1 shows its "canoe" containing its reconnaissance sensors, the aperture for its forward-looking camera visible just behind the strut of the forward landing gear. U.S. NAVAL INSTITUTE PHOTO ARCHIVE

The Development Characteristic emphasized the need to provide current information to fleet and air group commanders: "Where appropriate and by means of data handling [i.e., computers], scanning techniques, and inflight processing, transmission of critically needed photographic, electronic, and infrared data by radio link to surface-based interpreters for preliminary analysis."[14] Digital photography was far off, but some form of television transmission may have been envisaged. In a revised version the airborne system would reduce selected portions of the photography to a transmittable form. The new airplane should be available in 1961.

BuWeps considered much of the Development Characteristic unrealistic. The only naval aircraft that could meet anything like the requirement were the Phantom and the Vigilante. Neither airplane could meet either the desired ceiling (100,000 feet) or radius (2,000 nautical miles). Both had about the same catapult requirements. With ten minutes at Mach 2, the Vigilante's radius was 440 nautical miles, but it could refuel in the air. Phantom radius was 345 nautical miles, but with two 370-gallon drop tanks that figure could stretch to 1,000 nautical miles.

The Phantom could not carry all of the sensors at one time, but it could be argued that the infrared

(IR) and low-light-level television would be used mainly at night, when the conventional cameras would not be needed. (However, it was also suggested that IR equipment would be wanted in daytime.) The Phantom could not record weather information for later analysis. Neither airplane could transmit IR, television, and side-looking airborne radar (SLAR) data other than target coordinates, and they had no capacity for photo transmission. The Vigilante could automatically transmit electronic intelligence (ELINT) information. Work already under way might allow it to transmit other information in future, but only on a highly selective and delayed-time basis.

The "photo/electro reconnaissance system" was briefed to the director of Defense Research & Engineering at the end of July 1958. He concluded that it should be funded in FY60, subject to further briefings on survivability, compatibility of system output with requirements, and funding. Unfortunately, projected Navy funding in FY60 and FY61 was inadequate, so in October DDR&E suggested merging it with the reconnaissance needs of the other services.[15]

That was not enough: even if it were bought jointly with the Air Force, the Vigilante would cost at least a quarter more than the Phantom. However, it had a far more accurate navigational system than the Phantom, which would be important if its data had to be used for strike planning. It would provide most of the desired data-gathering and recording capabilities, and by 1961 it would be able to transmit ELINT data automatically. The Phantom would fall well short of the data-gathering and recording requirements. The Marines tried and failed to have the Phantom specified as the new reconnaissance airplane, to ensure operational and logistical compatibility with their amphibious force.

The Development Characteristic requirements were cut back to make them attainable. Radius was dropped to a thousand nautical miles and combat ceiling to 58,000 feet. The low-light-level television was eliminated in favor of night photography using artificial illumination. The SLAR requirement survived, with photographic recording added. The ELINT requirement survived, but the data link did not. The target date was set back to 1962.

The Navy bought the Vigilante.[16] To improve its radius, it was given additional fuel in a hump behind its cockpit. Its linear bomb bay also stowed fuel (some Vigilantes suffered fires there). By the time it entered service, it had been redesignated the RA-5, the main version being the RA-5C. It had cameras, a side-looking radar, and a powerful ELINT collection system (ALQ-61), but at least in Vietnam it seems to have been valued mostly for its imaging capability.[17]

The RA-5C fed its product into the IOIC. Given automation, information collected by an RA-5C mission was routinely used to plan the next strike, which was remarkable, considering how much the airplane could collect. Strikes were sometimes diverted in flight based on what an RA-5C had just collected.

Vigilantes suffered the highest attrition rate of all U.S. naval aircraft in Vietnam. Vigilante operations included mapping all of North Vietnam, to solve bombing problems arising from erroneous maps. Losses were met by reopening production. Typical Vigilante strength on carriers in 1964 was six aircraft, but by 1971 it was four or five, and it fell to

only three in 1973. After that carriers sometimes deployed with none at all on board; the last Vigilante deployment ended in September 1979.

After the RA-5C was retired, its capability could not be duplicated on board an adapted fighter. However, the F-14 and F/A-18 could carry pods containing digital cameras, whose imagery could be beamed back, in part, to the carrier—in much the way the RA-5C was to have done. An important development since the 1991 Gulf War is that carriers have benefited more and more from national sources of information, such as satellites, and therefore from high-capacity connections back to the United States. The visible indication of this change is the multiplication of satellite communications radomes on carriers.

Tactics

Tactically, the great theme up through the Vietnam War was concentration to saturate target defenses. Attackers were intended to fly out together. They might split up near a target to confuse antiaircraft gunners, but many accounts of World War II attacks emphasize massing up to the point of dropping weapons.[18] Before that war, "saturation" had included the idea that an entire air group should attack together, different kinds of aircraft stressing a target's defenders in a coordinated way. For example, fighters could disrupt antiaircraft fire by strafing while dive- and torpedo bombers attacked at the same time—hence BuAer's determination in the late 1930s to produce dive- and torpedo bombers with roughly the same performance as the fighters that would work with them. The fighters were seen largely as an additional type of attack airplane (a strafer, which would disrupt enemy antiaircraft gunner and fire control, which were considered extremely vulnerable to such attacks) rather than as a means of countering enemy defensive fighters.

Because dive-bombers were used for both scouting and attack, in prewar carriers the same type of aircraft were assigned to both VS and dive-bombing (VB) squadrons. War experience led to their amalgamation into thirty-six-plane VSB squadrons. The normal bombing unit was the six-plane division. As before the war, the guiding principle was that once the enemy had been located, and was within striking distance, all enemy carriers had to be put out of action—their flight decks disabled—as soon as possible. The November 1944 issue of carrier aircraft orders (USF 74B) pointed out that this might seem to be a departure from the earlier idea of concentration to sink rather than disable, but "during the initial phase of an action the results may well be disastrous if even one enemy carrier is left undamaged. If the attack is against a shore based installation, enemy aircraft must be destroyed first and the fields rendered inoperable."[19] Once the enemy had lost air capability, strikes could be concentrated to sink or completely disable ships. Ideally, aircraft would approach at 20,000 feet, losing altitude gradually and pushing over into their dives at 12,000. For maximum accuracy, pilots should drop their bombs at the lowest altitude consistent with recovering above the danger space of the bombs employed. Torpedo bombers were to attack as simultaneously as possible from different sides of a target ship so as to prevent it from evading by "combing" the torpedo tracks.

After World War II, tactics split between those for nuclear and those for conventional attacks. Nuclear attack was solitary, one airplane per target. Conventional attack tactics emphasized saturation, as they had during World War II. Their Vietnam-era symbol was the full-deck-load ("Alfa") strike.

By about 1970 the process of strike planning began by the carrier group flag officer or the carrier commanding officer listing targets and the level of damage or probability of destruction desired. Typically that went to a "strike board" (representatives of Strike Operations, Weapons, Air Intelligence, Navigation, CIC, and the "Flag" (the admiral's staff). The board chose tactics, which at the time meant either an Alfa strike or a series of smaller strikes. Its members took terrain and enemy defenses into account, and they used standard weapons-effect manuals to decide how many weapons of which type were needed. The flights out and back were planned in segments between waypoints, accounting for such factors as enemy defenses and fuel consumption. The chosen plan was reviewed by the strike board as a whole and by the officer in tactical command. Pilots received track charts they could mount on their kneeboards. Flight plans were developed manually, which meant laboriously, typically about one plan per attack group per day. About 1974, ONR began to sponsor studies of computer strike planning, but they attracted little fleet interest.

The disaster of the Bekaa Valley raid in 1983 changed minds. The raid proved unexpectedly difficult, due in large part to delays dictated by a high-level determination to warn Americans in Lebanon beforehand. Instead of coming at dawn, with the light blinding antiaircraft gunners, the raid came at midday, when gunners could see clearly. An A-6E and an A-7 were lost, and a captured pilot was paraded through a street. Secretary of the Navy Dr. John Lehman, who was a reserve A-6E bombardier/navigator, considered the disaster proof that the Navy's chief offensive arm, comprising its carrier bombers, was not fully efficient. Problems in fighter combat during Vietnam had spawned the "Top Gun" fighter school. Lehman ordered a comparable attack school ("Strike U") stood up. In the Bekaa Valley strike it had been impossible to redo the plan quickly enough after the delays had ruined its basis, an attack out of the sun. Strike U became interested in off-axis attacks (not flying directly from the ship to the target) and in dispersed attacks. There was already interest in time-on-target (TOT) attacks, in which strike aircraft would suddenly appear over targets from many different directions.[20]

Strike planning had to be automated. The Navy was already using automated centers to plan Tomahawk cruise-missile missions; Tomahawk was, in effect, a one-way (albeit unmanned) naval attack airplane. McDonnell Douglas, which had already developed the Tomahawk mission-planning system, won the competition for what the Navy called TAMPS (Tactical Aircraft Mission Planning System). It entered service in 1986. Initially it replicated the earlier manual system, producing the sort of paper kneeboard maps pilots already used. The computers on board airplanes were still set up manually. However, the new automated system could feed into a mission rehearsal system, TOPSCENE (Tactical Operational Scene), by which pilots in their ready rooms could see what they would likely see as they flew toward their targets. Simply seeing the target as they would in flight made for much better accuracy.

The next step was a digital connection between the strike planning system and the computer on board the airplane: from Lot 12 on, a DSU (digital storage unit) holding the strike plan that could be plugged into F/A-18s. DSUs became common about 1995. Their data included weapon programming and such settings as Link 16 "slot timing" and IFF codes. Link 16 made it possible to redirect an airplane in flight, inserting a new mission plan into its central computer.

The potential for fully flexible targeting was demonstrated during the NATO war in Bosnia in 1996–97. Strike aircraft could be rerouted and GPS (Global Positioning System)–guided bombs reset to hit new targets whose coordinates were known. The new technique was called "Real-Time [or Retargeting] in the Cockpit," or RTIC. With the advent of RTIC, aircraft mission planning systems (such as TAMPS), Link 16, and the airplane's own mission computer constituted a real-time strike system well adapted to the sort of pop-up mobile targets seen in the 1991 Gulf War and beyond.

Tactics changed radically with the introduction of GPS-guided bombs. Airplanes could attack accurately from well beyond antiaircraft range, and a single airplane could hit multiple targets (with multiple bombs) during a single mission. Discussions of carrier capability shifted from the number of sorties the ship could "generate" against a few targets to the number of targets the ship could hit per day. That number depended on such factors as how rapidly aircraft could be rearmed. This change is reflected in the new flight-deck configuration introduced in the *Gerald R. Ford* (CVN 78) class.

Currently, carrier strike planning is done in the "carrier intelligence center" (CVIC, successor to IOIC) and squadron ready rooms. The CVIC is both an intelligence-fusion and strike-control center, as it contains the mission planning center and the strike plot; it is responsible for both reconnaissance and strike planning. It also contains the debriefing area that provides strike feedback, as well as a photo lab, an SCI (sensitive intelligence) center, an intelligence library, and a multisensor interpretation center. It has access via wideband satellite links to intelligence data stored elsewhere. It maintains and creates strike folders, the modern equivalents of the much earlier nuclear target folders, which bring together all relevant information on the target and on enemy defenses en route.

The TAMPS and other planning workstations are typically located in CVIC—as of 1996, generally six of them. CVIC also contains TOPSCENE terminals. Owing to the increased need for rapid planning to support DSUs, TAMPS was declared a squadron resource in 1995, and systems were placed in squadron ready rooms: at that time, four per F/A-18 and EA-6B squadron, two per F-14 and E-2C. New local-area networks (LANs) were installed to connect these terminals with CVIC. Initially TAMPS was a single-airplane planning tool, but there were plans to extend it to plan integrated and coordinated strikes. Secure closed-circuit television was installed to provide premission intelligence briefings to squadrons. As part of the automation process, the comprehensive hard-copy guide to weapons effects that was used to plan strikes was recast digitally. TAMPS functionality was incorporated in a new Joint Mission Planning System (JMPS), which has now evolved several times into an increasingly powerful platform.

The Gulf War (1991) was the first in which the air effort was unified, strikes being planned under a joint "air tasking order" (ATO). Typically a carrier received an ATO between twelve and eighteen hours before a strike. During the Gulf War, the carriers had insufficient connectivity to receive them electronically and got them instead in paper form, but afterward they gained sufficient satellite bandwidth to receive the orders electronically. On receipt, a strike team convened for rough planning, which would be the basis of individual aircraft plans.

Missions were passed to aircrews, who were individually responsible for detailed flight plans. The strike leader coordinated and reviewed these plans, presenting a single briefing to all strike participants. The pilots planned their individual flights, briefed them to the strike leader, and downloaded data to their aircraft. Through the 1990s strike planning moved from large computers (TAMPS) to laptops. The greater availability of planning devices made it possible to automate the whole process, where previously TAMPS had been reserved for detailed route planning. From about 2004, TAMPS gave way to JMPS, which soon grew to include a force-level component.

Briefing and Ready Rooms

Pilots are briefed in squadron ready rooms, from which they man their aircraft. The need for such spaces was apparently not well understood when the two *Lexington*s and *Ranger* were designed. In the *Lexington*s the squadron briefing spaces were chart rooms, the pilots standing around a chart table. That caused considerable fatigue, as the pilots were already wearing their parachutes. In *Ranger*, the ready-room benches all faced portholes, light from which could blind pilots. The rooms were also too close to "heads," which mattered because ventilations systems were secured when the ships were at general quarters.

Pilots on the carrier *Hornet* prepare for a mission in January 1945 in the South China Sea. The amount of gear they carried and their warm clothing (for altitude) explain why air conditioning was so important. The boards were for navigational information. Although the air group normally flew out together to a target, airplanes generally returned individually. Pilots were (and are) individually responsible for their navigation. NHHC

The ready-room arrangement in the *Yorktowns*, the first modern U.S. carriers, were developed by a pilot officer assigned to *Yorktown* when still under construction. Because, as noted, pilots and aircrew wore their parachutes and other gear to their ready rooms, these spaces were as close to the flight deck as possible, in the gallery deck immediately below (in the smaller *Wasp*, some were in the lowest level in the island). As built the ships did not have enough ready rooms, forcing two or more squadrons to occupy the same space and thereby causing confusion. The heat built up—owing to the lack of ventilation when the ship was completely closed up, the sun beating down on the flight deck immediately overhead—to the point that pilots had to remove their flight clothing and gear. In 1939 the captain of *Yorktown* insisted on and obtained enough air-conditioned ready rooms for all his squadrons. It appears that the *Essex* class was the first for which a separate ready room for each squadron was specified at the design stage, though by 1940 that was also the case in *Yorktown*s and *Wasp*.[21] The *Essex*es split their ready rooms into groups fore and aft for survivability, with an air group commander ready room adjacent to the forward pair. The ready rooms communicated via a teletype with Air Plot, whose output could be displayed at the front of the space.

Movies of U.S. carrier air operations often show an enlarged teletype in a ready room indicating "Point Option," the carrier's expected future position based on its predicted course and average speed, to which it would return after a strike.[22]

Late in the war kamikazes sometimes wiped out spaces in the gallery deck. In late (and modified) *Essex*-class carriers ready rooms (and CIC) were moved under the armored hangar deck. When ships were rebuilt postwar, they were given escalators to bring pilots up to the flight deck; the escalator housings, outboard on the starboard side, are quite visible on the rebuilt ships.

Aircraft Organization

By 1938 carriers had designated carrier air group (CAG) commanders to lead multiple squadrons in battle, implying integrated tactics at least for strikes. The prewar Navy contemplated multicarrier operations in which squadrons of the same type but from different carriers would work together as "wings." They would be governed by doctrine and (preflight) orders, not by tactical signaling in flight.[23] About 1934 the then-existing four-carrier force was split into two divisions, one for the big, fast *Lexington* and *Saratoga* and one for the slower (but incompatible) *Langley* and *Ranger*.[24]

Ready rooms were moved below carriers' armored hangar decks. A kamikaze smashing through the wooden flight deck and its light steel supports could well wipe out the ready room beneath it on the gallery deck. When *Essex*-class carriers were rebuilt postwar, they were given escalators to bring pilots from their ready rooms to the flight deck. The diagonal trunk visible below the island of the carrier *Shangri-La* (25 November 1960) covers her escalator. Most of the aircraft visible on deck are Skyraiders of VA-176. NHHC

By the time the Pacific War broke out, the multicarrier idea had been dropped, on the ground that carriers operating in close proximity might be disabled together. At about the same time the Japanese were developing multicarrier operations, but only for use against land targets (their doctrine, like that of the U.S. Navy, was to disperse carriers facing enemy carriers). The Japanese, however, adopted more closely integrated operating practices than did the Americans, two carriers always working together and their aircraft working as a single integrated unit. The Japanese associated particular aircraft with particular ships and considered a carrier out of action if many of its aircraft had been lost. That is why *Zuikaku* was not reequipped with fresh aircraft after Coral Sea and hence was not present at Midway.

At Midway, the carriers *Yorktown* and *Enterprise* operated more or less as a pair, sharing one "fighter direction officer," but they were too far separated to make this arrangement efficient. Through 1943–44 the U.S. Navy revived the multicarrier operation, forming multicarrier task groups. However, the air groups of the ships of these groups were not integrated. For example, during the October 1944 battle of the Sibuyan Sea, the carrier air groups attacked the Japanese Center Force separately. Without overall coordination, attacks were not spread effectively over the targets. One Japanese battleship was sunk, but two others (and heavy cruisers) were hardly touched, surviving to fight off Samar the next day. When the *Yamato* task group was attacked in April 1945, however, an overall coordinator distributed efforts effectively.

Prewar games and exercises strongly suggested that aircraft and their crews would be lost at a horrific rate and that squadron replacements would be needed. Thus the 4 June 1942 edition of *Location of U.S. Naval Aircraft* showed a "Carrier Replacement Group 10," equipped mainly with training aircraft. By late July two more replacement groups were organizing, and what was now Carrier Air Group (CVG) 10 was being fleshed out with operational aircraft for its four squadrons—VF-10, VB-10, VS-10 (antisubmarine), and VT-10 (torpedo).

By mid-October, three carrier air replacement groups had been equipped completely with combat aircraft. In mid-November, CVG 10 was on board the carrier *Enterprise*, and Carrier Air Group 11 (no longer a replacement group) had departed San Diego for the combat zone. From then on, particular carrier air groups were no longer associated with particular carriers beyond a given cruise.

CVGs were redesignated carrier air wings (CVWs) in 1962 in line with Defense Department policy unifying terminology in the air services, but their commanders are still called CAGs.

Navigation

U.S. Navy pilots were and are responsible for their own navigation.[25] A strike might fly out together, but after attacking individual pilots were generally on their own. Accounts of the battle of the Philippine Sea, for example, have pilots trying to find other airplanes to follow home but often relying on their own efforts. At that time preflight briefings generally designated Point Option, but in wartime a carrier might have to maneuver unexpectedly. Thus the effective combat radius of carrier aircraft took into account time that might be spent searching for the carrier, finding it, and working through the landing circuit (including wave-offs). Until the late 1930s it was assumed that the carrier would not provide them assistance in the form of radio signals; after that there were beacons that developed into the current TACAN (plus GPS and netted navigation aids). Note that World War II beacons were a means of homing on a carrier but not of navigating toward an enemy.

As an illustration of how difficult things were before radio aids, in the postgame discussion of the Naval War College Operations II game of September 1931, Rear Adm. John Halligan, who had come from the fleet as senior aviator, said that the fleet did not like to send planes more than 25 miles from surface ships under conditions of radio silence. However, during the postgame discussion of Operations IV (January 1934) it was pointed out that during Fleet Problem XIII aircraft were sent out in all directions from the two big carriers, out to 125 miles in daylight; all of them got back. That was sustained for six days. At this time Naval War College games assumed much longer ranges based on fuel capacity.

Pilots relied on dead reckoning corrected for "drift" (an airplane was said to drift sideways because of wind). They had small plotting boards and chart boards, the latter carrying essential data such as ships' identification marks and radio frequencies and calls. To estimate wind, aircraft could drop float lights, using a drift sight to judge how quickly they seemed to move. Carriers often launched balloons to measure local wind, but a pilot could not know how wind would vary as he flew. Often he relied on the "seaman's eye," based on the appearance of the surface of the water and an assumption that wind at altitude would be about the same as at the surface. Two-seaters had controls in both seats so that the back-seater could temporarily take over control of the airplane to allow the pilot to concentrate on the plot (autopilots did not appear until about 1939). Prewar aircraft had radio direction finders, but in wartime carriers would observe radio silence.

A viable carrier radio beacon became possible with the advent of very-high-frequency (VHF) radio, which operated on a "line of sight" basis (i.e., recipients had to be visible from the transmitter) and was therefore thought to be nearly safe from enemy interception. Just how seriously this immunity was taken varied through World War II. Work on a beacon began in 1936, and the designation "YE" was assigned in 1937. The YE beam turned and radiated in fifteen-degree sectors, using a different

The mast of the escort carrier *Manila Bay*, photographed on 28 April 1945, is topped by the standard YE aircraft homing beacon. Below it are YJ and CPN-6 homing beacons. YJ worked with the airplanes' ASB radar; CPN-6 (formerly YM) worked with the newer APS-4. Farther down are the ship's surface-search radar (SG) and vertical dipoles used with two radios: the ship–air ARC-1 and the general-service TDQ. Both operated on the same band (100–156 MHz), which was somewhat higher than the single-channel TBS (60–80 MHz). Below are the ship's SC air-search radar and her SM fighter-control radar. IFF interrogator-responders associated with the radars were designated in a *B* series (BM, BO); the transponder for airborne radar was ABK. NATIONAL ARCHIVES

Morse code for each sector. The complete set of codes was called a "pie plate." Every ten cycles the beacon sent out a signal identifying the carrier. Line-of-sight transmission meant that the range at which an airplane could receive YE depended on the airplane's altitude. Conversely, a pilot could estimate the distance to the carrier on the basis of his altitude when he first detected YE signals. YE's rotation could be stopped to transmit keyed or voice messages to airplanes. The airplane receiver was the "ZB," and there was a one-way voice transmitter (ZBDM) whose messages ZB could pick up. ZBDM could be used to control a combat air patrol when radio silence was otherwise enforced.[26] In May 1938 the prototype was installed on board the carrier *Saratoga*, flagship of Commander Air Battle Force, Admiral King. In a 29 August 1938 report King strongly recommended widespread installation, and YE was standard during World War II. It was widely installed ashore, and a version was adopted by the Army Air Forces. "YG" was a portable backup. For an airplane at ten thousand feet, YE range was probably about 160 nautical miles.

Usage varied through World War II. At Coral Sea, the carriers did not use their beacons, because the Japanese were said to be aware of the system and to have captured receivers and codes. Later, however, use was commonplace. Because Japanese aircraft tried to follow American aircraft "home" to attack from within the stream of "friendlies," U.S. aircraft were assigned particular sectors from which to return, using the beacon to find the proper courses. This was standard practice by the time of the Philippine Sea, June 1944. Accounts vary as to the extent that pilots could or would use the beacons to get home; apparently many of them failed to do so on the evening after the Philippine Sea.

In the 1950s YE was superseded by TACAN. After a refit at Hunters Point in November 1957, the carrier *Bennington* shows the big masthead dome of the TACAN antenna. The radar antenna partway down the mast is for an SPS-6 air-search set, with that of an SPS-8 height finder level with the top of the funnel. On the other side of the mast is the antenna of a World War II SC-2 radar. This and similar radars were installed on carriers when it was discovered that later shorter-wave radars sometimes could not detect streamlined jets—an early example of how shaping could avert radar detection. NHHC

Aircraft with "ASB" radars could use the "YJ" radar beacons (YJ) on board carriers. That may be one reason the obsolescent ASB radar survived as long as it did on carrier bombers. YJ was credited with a range of about 120 miles.

By 1947 NRL was developing a successor that became TACAN, operating at 1 MHz (in the ultra-high-frequency, or UHF, spectrum and thus line of sight). It radiated only when "interrogated" by an airplane. A production version was tested in 1951. Maximum range, set by aircraft altitude, was two hundred miles. TACAN automatically gave the pilot bearing (one-degree accuracy) and range (to within a quarter mile plus 0.2 percent of distance). Unlike YE, TACAN could direct a pilot on the initial leg of his outward flight (but targets themselves were generally well beyond its horizon). More than fifty years later, TACAN is still in service in modified form, though now being replaced by a GPS-based system.

Pilots still badly needed some means of navigating over greater distances. BuAer first asked the Naval Research Laboratory (NRL) to develop a single integrated navigational device, a "Ground Position Indicator and Automatic Navigator," in 1933. Only in 1951 did it announce tests of the first such device, the APN-67.[27] Although so massive that it was suited only to patrol planes (in which it was operated by a navigator), it pointed toward future attack bomber systems. It used a continuous-wave (CW) Doppler radar to measure speed over ground or sea directly. Other sensors were a compass (accurate to within a degree) and a vertical gyro. Readouts were position, true ground track, velocity, and heading. By November 1952 APN-67 had flown successfully on board a P2V-3 Neptune patrol bomber and had exceeded the performance of any alternative self-contained navigator.[28]

When Grumman designed the A-6 Intruder in 1957, BuAer wanted a self-contained nonradiating navigational system, meaning something inertial. The Intruder was too small for that, so Grumman incorporated a Doppler radar (APN-153) in its integrated ASQ-57 CNI (communications/navigation/identification) package. Within a few years, shrinking electronics made it possible to incorporate a Doppler system in the A-7A Corsair II.

Doppler navigation was presumably first accepted because it operated with such short waves that its narrow beams could not be detected at any great distance. After many Doppler aircraft had been shot down in Vietnam, the Soviets realized that U.S. presence was often being betrayed by such radiation and deployed numerous passive detection devices.

For small attack aircraft, in 1953 BuAer announced a flight track recorder that would automatically trace out an airplane's past track on a chart, based on compass headings and manual inputs of wind velocity and direction.[29] The first naval attack bomber to use such a device seems to have been the upgraded A4D-2, the A-4B version of the Skyhawk, which introduced an ASN-19 Tapeline automatic dead-reckoning navigational computer. It also had a TACAN terminal and a UHF direction finder. The last production version of the Skyhawk (A-4M) was equipped with an ASN-41 navigation computer supported by an APN-153 Doppler navigational radar.

Because they were intended to attack at high altitude, early heavy nuclear carrier-based bombers used the same device as World War II heavy bombers: ground-mapping radar, with which they could identify the target area. The ASB-1 bomb director of the AJ-2 incorporated an APS-31 radar. A-3 Skywarriors had the ASB-1 or -7 bomb director system, which incorporated a ground-mapping radar, but no separately designated navigation aids. The Vigilante had a nonradiating alternative: missile-type inertial navigation.

Both inertial and Doppler navigation have now largely been superseded by GPS, which became operational about 1990. Thirty years later, the supremacy of GPS is in question, because of both the rise of specialized jammers and the possibility that GPS satellites may be destroyed.

Martin T4M-1s of VT-1 on deck on the carrier *Lexington*. For nearly a decade the big T4M-1 was the standard U.S. Navy torpedo bomber. PRATT & WHITNEY VIA U.S. NAVAL INSTITUTE PHOTO ARCHIVE

3

TORPEDO BOMBERS

The U.S. Navy became interested in torpedo bombers before the country entered World War I.[1] However, interest in carrier-based torpedo bombers stemmed from British efforts to create a carrier strike force to attack the German High Seas Fleet.

The Navy's first torpedo bomber was a Curtiss R-6-L floatplane.[2] On 3 May 1919 two successfully dropped dummy "Type D" torpedoes in Hampton Roads, near Norfolk, Virginia. Others dropped real eighteen-inch Mk VII Type D torpedoes at Pensacola.[3] Range was two thousand yards, which BuAer, BuOrd, and the commander of the Atlantic Fleet all considered inadequate. BuOrd modified about a hundred eighteen-inch torpedoes to reach a four-thousand-yard range.[4] In October 1922 VT-1, the first American torpedo and bombing squadron, asked for four twenty-one-inch, two-thousand-pound weapons, but they never materialized.[5] As a corollary to greater range, the torpedo bombers needed torpedo directors, to guide pilots in "leading" their moving targets. As fire-control computers ships do, they would solve the triangle of own, torpedo, and target movement. American shipboard torpedoes were capable of "curved fire," meaning that they could turn after firing to an intended heading, using a preset gyro. The pilots wanted the same capability.

The aircraft in use, the R-6-L, was clearly a makeshift. To supplement it, the Naval Aircraft Factory produced a hybrid torpedo bomber/trainer under the designation "PT" (patrol torpedo). It used existing parts: the Liberty engine, the Curtiss R-6 fuselage, and the wings of an HS-series flying boat (HS-1L in PT-1, HS-2L in PT-2). Aircraft dropping torpedoes in early Navy tests were often PTs described as R-6-Ls. Neither the R-6-L nor the PT ever operated from a carrier. However, the PT carried out an important demonstration. On 27 September 1922, Atlantic Fleet PT-1 and -2 torpedo bombers "attacked" a moving battleship formation, seventy miles from the ships' Norfolk base, concentrating on the battleship *Arkansas*, which was steaming at full speed. Despite its maneuvers, the attackers made at least seven "hits." Another torpedo hit the battleship *North Dakota*, but it was not counted a success because that was not the selected target.[6]

The First Carrier Torpedo Bomber

By late 1919 the General Board was working on outline requirements (characteristics) for carrier aircraft: a shipboard fighter (conceptually designated *A*), a spotter (*B*), a torpedo bomber (*C*), and a long-range reconnaissance and patrol aircraft (*D*).[7] In December 1919 BuOrd wrote that the torpedo bomber should carry a torpedo weighing not less than 1,650 pounds (a Mk VIII) or an equal weight of bombs. It should also have the most complete machine-gun armament possible within available weight.

Torpedo-bomber characteristics were formally submitted (and approved) on 19 March 1920. The airplane would have a two-man crew; it would carry BuOrd's torpedo (or bombs) plus a flexible machine gun. Speed would be at least ninety knots at sea level; endurance at full speed would be at least three hours, but no specific range was given. Twin floats were specified as the landing gear (wheels were not mentioned); the airplane could be launched either by catapult or from the deck of a carrier. Wings should either fold or be quickly detachable for below-decks stowage; the fuselage might also be detachable. These details seem to have come from the future BuAer (in the form of interested parties from OpNav, C&R, and SE).

OpNav emphasized the need to carry the airplane on board ship so that it would always be available to the fleet for immediate use. It had to be able to take off from a ship—a carrier—under practically any weather condition that would make a fleet action possible and also that "it is most desirable that it be capable of alighting on a carrier's deck." It should be small enough to be carried on board ship in sufficient numbers "to assure a fair chance of success when delivering an attack" yet have sufficient radius of action to reach the enemy. It would use the British tactic of a sudden dive to low attack altitude, hence it had to be strong. It also had to be fast enough to make surprise attacks and survive in the air, with good maneuverability to avoid antiaircraft fire as it attacked. Good visibility ahead and to either side of the nose was important. It was not clear whether the future lay with torpedo bombing or with horizontal bombing using a similar weight of bombs.

C&R studies showed that with a 500-hp engine a two-man airplane could make ninety knots, with an endurance (with a torpedo) of three hours (250 nautical miles) or five hours with a thousand pounds of bombs (400 nautical miles). As a scout, it could remain aloft for eight hours at ninety knots (700 nautical miles) or for fourteen hours at seventy knots (950 nautical miles). It would have a single machine gun in the torpedo-plane role, two in the other roles.

The two competitors were the Curtiss CT and Stout ST.[8] Each was powered by a pair of 300-hp Hispano-Suiza engines. Bids were received on 22 June 1920. C&R ordered nine CTs and six STs, with a proviso that either order might be reduced to three if that aircraft did not meet certain specifications. Curtiss' airplane may have been described as a "government" (i.e., in-house) design, and initially it seems to have been favored. It was an exotic twin-boom airplane, almost a flying wing, and it was much heavier than Stout's ST (11,208 versus 9,187 pounds). Span was sixty-five feet. The ST employed William Stout's favored all-metal construction. Like the CT, it had a low monoplane wing, with a span of sixty feet. Neither manufacturer managed to complete its airplane on schedule, and neither aircraft was compact enough for carrier operation. After unsuccessful initial tests, eight of the nine CTs on order were canceled. The ST seems to have been well liked, but it was canceled after it stalled on takeoff on the fifteenth flight.

In its 1922 annual report, BuAer emphasized the difficulty of developing an airplane that was compact but could nonetheless lift a torpedo. Apparently the chief of the C&R Aviation Division, Jerome C. Hunsaker, thought that Donald Douglas, his former student and assistant at MIT, might be innovative enough to solve the problem. Douglas had moved to southern California in the spring of 1920 to enter the aviation business. He wrote to Hunsaker that he hoped to come to him for business. Hunsaker sent Douglas the request for a proposal for a twin-float seaplane, which, he wrote, was "just the job for him."[9] Presumably he thought Douglas' MIT background uniquely suited him to the difficult problem of designing a light enough structure. Douglas had already done just that in designing the Cloudster, conceived as a single-seater to make the first nonstop transcontinental flight, with a design range of 2,800 miles. In effect Douglas' torpedo bomber substituted a torpedo for much of the fuel weight. Unlike the all-wood Cloudster, it had a welded steel-tube fuselage, its fore and center sections covered with aluminum and its after section with fabric; the horizontal tail surfaces were metal tubing covered with fabric. The wooden wings folded for shipboard stowage. Like the PT, Douglas' airplane was powered by a 400-hp Liberty engine. It carried the same payload as the CT and ST with two-thirds the power.[10] It had interchangeable float or wheel undercarriages. Douglas submitted a formal proposal on 1 February 1921.[11] The Navy ordered three prototype DT-1s.

C&R ordered two foreign aircraft for comparison, the British Blackburn Swift F and a Fokker torpedo bomber.[12] Neither seems to have been considered satisfactory. The Swift, which had been conceived as a carrier airplane, was compact and had performance similar to that of the DT, but it had only a single seat at a time when the Navy demanded at least two: by late 1921 naval aviators wanted a second seat for a defensive gunner.

The single-seat prototype Douglas DT, the first carrier-capable U.S. Navy torpedo bomber, photographed 9 November 1921. Note the semisubmerged torpedo stowage. U.S. NAVAL INSTITUTE PHOTO ARCHIVE

The DT prototype with wings folded for shipboard stowage. U.S. NAVAL INSTITUTE PHOTO ARCHIVE

A standard two-seat DT, with the pilot's seat moved forward and a gunner's position added. This one was built under license by Lowe, Willard, & Fowler (LWF). When not on wheels, the DT-2 had twin floats. NHHC

The first DT-1 was also a single-seater, with the pilot abaft the wing. The second and third prototypes were modified as DT-2s: the pilot was moved forward to a position under the leading edge of the wing, and a gunner's cockpit (with one 0.30-calibre machine gun) placed abaft the wing. To offset the additional weight, the standard 400-hp Liberty engine was replaced by a higher-compression version developing 450 hp.

The DT-1 and the first DT-2 had their radiators on the sides of the fuselage. The third aircraft was rearranged, the radiator moving to the conventional position in the nose. To solve oil cooling problems, an external oil cooler was installed on the left-hand walkway. This version was ordered into production, which was considerable by contemporary standards: thirty-eight DT-2s from Douglas Santa Monica, eleven from Dayton-Wright, twenty by Lowe, Willard, & Fowler (LWF, in New York City), and six by the Naval Aircraft Factory. The DT-2s were delivered in 1922–24. Production aircraft had modified vertical tails with "balanced" rudders (the axis not

The DT became BuAer's engine test bed. The DT-4 (BuAer A-6423) was powered by a 650-hp Wright T-2 engine. Note the radiator alongside the pilot's cockpit. The engine was light enough to be substituted for the original Liberty without structural change, adding 250 hp. Two DT-4s were later given geared T-2Bs as DT-5s. NHHC

The Wright DT-6, shown on 1 May 1925, was a DT-2 reengined with an air-cooled Wright P-1 engine. U.S. NAVAL INSTITUTE PHOTO ARCHIVE

at the leading edge but somewhat farther aft). The Army bought four of a modified version as its World Cruiser, for its round-the-world flight.[13] Douglas proposed, but the Navy did not order, a modified DT-3.

The DT was considered unsatisfactory as a bomber because it did not offer a good position for a bombsight. Moreover, it could not carry a single heavy bomb: at most it could carry four of the obsolete 250-pound demolition bombs. An abortive DT-6 was offered as an alternative to the multipurpose Curtiss CS.[14] With the Curtiss Scout (CS) in production, three DT-2s were modified by Dayton-Wright as long-range scouts for comparison, designated SDW-1.[15] They had additional fuel in a deepened fuselage.

Several DT-2s were modified to test new engines.[16]

The "Three-Purpose" Airplane

During 1921 the BuAer Design Division produced a series of preliminary designs, of which No. 8 showed a two-seat torpedo bomber powered by a Wright engine of 500–550 hp. It was considered a reasonable solution to the torpedo-bomber problem. Plans

Division preferred to mount this engine in a modified DT, but Design argued that the result would be overweight, too heavy for its floats. Only an excessive sacrifice of fuel (i.e., range) would solve that problem. In March 1922, then, Design 8 was chosen for continued development; the merits of a possible DT version would be pursued separately. More importantly, the ongoing study of a long-range scout showed a close resemblance to Preliminary Design 8, to the point that it seemed the two types might be merged, or at least made convertible into each other.[17]

This perception soon led BuAer to advocate a "three-purpose airplane" suitable for scouting as well as attack.[18] It would need a more powerful engine: BuAer was sponsoring development of a lighter-weight engine of 500 to 600 (rather than 450) hp.

Curtiss' CS scout (its Model 31) was often described as a derivative of a government design, which suggests that its source was Preliminary Design 8.[19] As the Design Division suggested, it could lift a torpedo or a thousand-pound bomb, hence it was

Curtiss' CS was the progenitor of the big U.S. Navy biplane torpedo bombers that served through the 1930s. The bombardier or observer sat forward of the pilot. The third crewman (gunner / radio operator) was aft, with large windows on either side of the fuselage and a small cockpit with a flexible Lewis gun. This airplane, the third of the first batch of six CS-1s (A-6500 through -6505), has been modified as the prototype CS-2 long-range scout, with a range of three thousand nautical miles, powered by a 585-hp Wright T-3 engine instead of the usual 550-hp Wright T-2. It has a third (jettisonable) float, to improve water performance. In November 1924 it was fitted with a geared T-3 engine as the sole CS-3. NHHC

usually described as a three-purpose airplane (scout, torpedo bomber, and level bomber). The extra capability came out of additional power: it had a 530-hp Wright T-2 engine. An unusual feature was an underbody channel into which the upper part of the torpedo fit, so that it was carried semisubmerged. Unlike DT, CS had a three-man crew, the bombardier or observer sitting forward of the pilot, and a gunner/radio operator abaft him. The separate radio operator was essential for scouting. This third crew member had a small cockpit, from which he operated a flexible Lewis gun. He could see out through large windows on either side of the fuselage. Like DT, CS could operate either as a wheeled airplane or as a floatplane.

The Navy ordered six CS-1s in June 1922. The first flew that November.

One was modified as the prototype CS-2, with a more powerful engine (a 585-hp Wright T-3) and Curtiss wing-surface radiators on its upper wing. Two more CS-2s were ordered as such. CS-2 was considered a long-range scout, hence did not have to carry a torpedo between two floats. To improve its water performance, it was given a third (droppable) float between the other two. The first CS-2 was fitted with a geared T-3 engine in November 1924 as the sole CS-3. In June 1925 the first prototype CS-1 was modified for landing tests on board the carrier *Langley*. The main problem, poor visibility for the pilot, was solved by moving the cockpit forward one structural "bay" (thirty-eight inches) into the former location of the main gas tank. This 185-gallon tank was replaced by a 60-gallon tank under the floor of the cockpit. Three additional gas tanks (30 gallons each) were fitted in the center section of the upper wing, supplementing the 28-gallon tank already there. The cowling was modified—dished in and cut back—to provide maximum visibility. Both radiators were moved forward ten inches. That reduced, but did not eliminate, their interference with the pilot's vision. Axle hooks and a tailhook were fitted. Special wheels with thicker tires and a heavier tail skeg were also fitted.[20]

In March 1924 BuAer tentatively decided to order twenty or more additional CS aircraft. Before doing so it convened a conference to discuss recent experience with the airplane.[21] Lt. Cdr. Charles P. Mason, who had commanded VT-1, the experimental CS-1 squadron, liked the airplane very much. Many of its problems had already been fixed, and it handled "very nicely" in the air. With a full load it was slightly tail heavy, a condition particularly noticeable in the CS-2, which always had to fly with the stabilizer all the way up. On the other hand, once the horizontal stabilizer had been set, it did not have to be reset for landing.[22] On the water it seemed slightly more difficult to handle than an R-6, probably because its floats were farther apart. In the air, Mason preferred it to the R-6. Overall, Mason thought, the CS could be made as good a torpedo bomber as the DT or the R-6. The bombardier would be in the rear cockpit, with a visibility from the vertical to about 70 degrees forward. Mason thought that would be reasonable from two thousand feet up to the airplane's ceiling with a heavy bomb, about six thousand feet.[23]

Capt. Emory S. Land, BuAer's chief technical expert, pointed out that the airplane would not be worth buying unless it could bomb and torpedo better than any twin-float airplane currently in service.

BuAer's Marc Mitscher, now a commander, said he knew it could bomb better and that he believed it could torpedo better. It was the nearest thing yet to a combined torpedo, bombing, and scouting plane. "Otherwise we will have to admit to the General Board we have fallen down on our mission. . . . I think we can [say it is]." Flight data showed that the airplane performed somewhat better than expected, for example in fuel consumption. Effective radius of action—flight out with a torpedo and back without it—was two hundred miles at cruising speed (70–75 mph) burning aviation gasoline. Compared to the DT, the CS-1 could take off with a far heavier load. It could fly farther. The DT was badly underpowered; it could not take off from the water with the same factor of safety as the CS-1. The consensus in BuAer was that a CS-1 was 30 percent more valuable than a DT-2. On the other hand, officers on *Langley* said that pilot visibility was so bad that they would not land the airplane on board without cutting away some of the engine cowling.

The conference proposed buying twenty CS-1s.

The Navy preferred to award production contracts to companies that had won the initial design competitions, on the theory that they were best suited to building their own designs. Companies tended to underbid for the prototype, in the expectation that they could recoup their initial investments on the production run.[24] In principle, however, the Navy had to award production competitively once it had chosen a successful prototype. BuAer felt that Curtiss' initial bid was too high to deserve that special allowance, but it expected Curtiss to cut its cost enough to make an award reasonable.[25]

When the Navy told Curtiss it planned to order forty CS aircraft, Curtiss bid $32,000 each—sure enough, as it later said, to recoup overruns on the design contract. BuAer estimated that the unit cost should have been $28,000. Curtiss refused to build the airplanes on that basis, so the Navy competed production. It considered only Boeing, Martin, Douglas, and Curtiss sufficiently capable. Douglas, the low bidder, withdrew when it found that the type of construction required was not one it was prepared to do at that price. Boeing could not deliver within a reasonable time, owing to the press of other business. Curtiss bid very high. Martin won, with an average unit price of $23,190. Its bid was based on a set of basic plans. Instead of building to the Curtiss plans (which it did not have), it would redesign the airplane for lower-cost production.[26] BuAer furnished a CS to use as a pattern. Curtiss later argued that this SC-1 (as it was redesignated) was inferior, but it proved successful in service. CS-1/SC-1 seems to have been a unique case of a shift from the design company to a different production company.

Martin built thirty-five SC-1s (its Design 69), which were issued to both torpedo squadrons (VT-1 and -2) and to a scouting squadron (VS-1). These aircraft operated mainly as floatplanes; *Langley* was considered too small to operate them. According to a 1926 report, the SC was a very good torpedo bomber, a fair bomber, and a fair patrol plane.[27] Like the CS, it was underpowered.

In July 1924 BuAer became interested in a follow-on contract for forty more aircraft. Curtiss made an informal bid of $23,500 each, not including engineering; the latter would bring cost up to about $25,000. That was materially lower than what BuAer had been prepared to offer to Curtiss earlier in the year for the thirty-five initial aircraft. It was not excessive in terms of the size of the airplane, anticipated improvements, and the need for the airplane. No criticism would have been received *if* the price had been the result of competitive bidding. However, that was not the case. Martin had bid $23,200 for the initial batch of thirty-five aircraft. Curtiss claimed that if additional engineering costs were factored in for both, its unit price would be less than Martin's. BuAer's analyst rejected that argument, in which Curtiss in effect merged its own experimental contract with Martin's as though one were a direct continuation of the other. However, the bureau chief, Rear Adm. William A. Moffett, doubted that any manufacturer other than Curtiss could supply the forty airplanes promptly and satisfactorily. He considered it just to award the contract to Curtiss—but acknowledged that such an award could be attacked because it was made without competition or negotiation. Moffett proposed to offer Curtiss $15,000 to modify a CS-1 into a production prototype, plus $23,190 per airplane and $40,000 for engineering.

Moffett explained to the Navy's judge advocate general that the CS (or SC) was the best torpedo

Martin's version of the Curtiss CS was this SC-1. The pilot's cockpit is visible in front of the gunner's. Just abaft it is the flush cockpit for the bombardier/observer; his window is visible just in front of the words "U.S. Navy." The modified SC-2 had a more powerful T-3 engine and larger floats. U.S. NAVAL INSTITUTE PHOTO ARCHIVE

plane yet procured by the Navy.[28] Experience with the prototype showed, nevertheless, that there was still considerable developmental work to be done. As *Langley* officers had insisted, the pilot lacked visibility for landing on a carrier, a serious defect when the big carriers under construction entered service. Bombing visibility was also poor. Curtiss offered solutions to both problems, and the company also promised higher performance. BuAer considered its preliminary drawings quite promising. Having conceived these modifications, Curtiss was in a better position than any other company to implement them—and also in the best position to build the airplanes. Awarding the follow-on contract to Curtiss would also support the important goal of maintaining two sources of supply, with a view toward securing effective competition in the future. The only other possible supplier, Martin, was fully occupied with the thirty-five-plane order.

The situation was further complicated when BuAer included design rights in the contract. Martin had paid heavily because these rights had not been included in the contract for the initial CS aircraft. Curtiss did not want to lose out if some other company bid low for a future order. The company argued that it had designed the CS at its own expense. It claimed that CS was a direct descendant of the CT conceived early in 1920, on which it had lost over $76,000 (given the radical differences between the two designs, this must have seemed rich). The CS design had cost it another $68,000, against a design allowance of about $32,000. More had been spent on the proposed modifications, which would not be covered by the proposed $40,000 engineering allowance. At the end of July Curtiss wrote that it planned to patent some essential features of the modified design. To emphasize how much it had done to improve it, Curtiss designated the new airplane S2C.[29]

Moffett, having successfully convinced the judge advocate general that he should order from Curtiss, found that Curtiss' price proposal unacceptable. Late in August 1924 BuAer rejected the Curtiss proposal, which was formally withdrawn on 4 September. The contract went to Martin.

Martin's SC-2 was powered by the 540-hp T-3 engine tested in the CS-2. The main difference from the SC-1 was its larger floats; overall, it was 327 pounds heavier than the SC-1. VT-1 found the SC-2 underpowered, with insufficient cruising radius and a low service ceiling. It was unsuitable for operations in the open sea.[30] As yet the fleet operated mainly seaplanes, with only a few aircraft on its sole carrier, *Langley*. Admiral Moffett pointed out the Scouting Fleet, which had considerably more experience with the SC, was decidedly more favorable. It did not seem to be widely understood that the three-purpose plane had been recommended by the General Board and approved by the secretary, that it was being developed under an established policy. Too, this type had been conceived for use on board carriers. Most of the criticisms involved comparison with flying boats, which could not operate from carriers and were considerably more expensive. By this time BuAer had under contract two experimental designs with new engines and another with twin engines (the T2D). A builder was developing yet another type, with an air-cooled engine, on his own responsibility.

In October 1924, after the first thirty-five aircraft had been ordered (with another forty contemplated), BuAer realized that they were badly underpowered, with insufficient rate of climb and ceiling for day bombing.[31] BuAer would accept this deficiency to get the aircraft into service, but it wanted experiments undertaken to improve performance. Several SC-1s and -2s were modified with alternative engines.[32]

By July 1924 the BuAer design group was sketching a better torpedo bomber: range at full speed would be 1,020 miles, compared to 800 for CS, and at cruising speed it would be 1,245 versus 927 miles. Maximum speed would be 110 rather than 100 mph, and crew would be three rather than the two carried by the CS as a scout. Useful load would be 6,000 rather than 3,184 pounds.[33] This study seems to have matured late in 1924, by which time the main object was to improve the vision of the pilot and bombardier.

Boeing built BuAer's Design 35 as its XTB-1, shown on 26 April 1927. Note the innovation of placing the pilot and observer/bombardier side by side. The observer had a compartment under the cockpit, hence the unusually deep fuselage forward. The gunner's position well abaft the wing is not visible here. The XTB-1 could be distinguished from T3Ms with side-by-side seating by its cut-back fuselage. The floats were interchangeable with wheels. NATIONAL ARCHIVES

The T2D-1 on *Langley*, 15 March 1927, in her arresting gear (note the lengthwise wires held up by vertical frets). The design seems to have demonstrated that the pilot should be as far forward as possible. In the nose was a gunner/bombardier (the bombardier's window is visible). Three T2D-1s were assigned to the ship's torpedo squadron. Trials showed that it was difficult to coordinate the two engines, causing the airplane to swing. That killed support for twin-engine carrier aircraft for a decade, although by 1937 it was apparently accepted that adopting twin engines was the best way to gain desired performance. In 1938 a twin-engine Lockheed Model 12A (the XJO-3) landed successfully on board *Saratoga*. Interest in twin-engine carrier aircraft revived. NATIONAL ARCHIVES

The design group offered two alternatives: first, a single-engine Design 35 built around the Packard 2500 engine, and, second, a twin-engine design built around two Packard 1500s (Design 30) or two Wright or Curtiss air-cooled engines (Design 36).[34] Design 35 had a bombing compartment beneath and abaft the engine and beneath the two-place cockpit, from which the observer's compartment was accessible. Design 36 had its bombardier in the nose, as in the earlier land-based Martin twin-engine bomber. Both aircraft were all metal. The water-cooled designs were too heavy, and their spans were at the limit set by carrier dimensions. The chief of the design group wrote that "the air-cooled design comes within the prescribed limits and is superior in performance in every way." The single-engine plane was somewhat lighter and smaller, even with a water-cooled engine; the twin-engine plane was faster, with a higher rate of climb and higher ceiling. Both had about the same endurance, and neither showed sufficient advantage to be worth building without the other. Given limited funds, investigating the possibility of authorizing changes in the last two SC-2s in accord with Design 35 was recommended. The chief of the bureau pointed out that "the change in fact amounts to a new design [and] there may be some difficulty with the Bureau of Supplies and Accounts."

BuAer bought three Model 35 aircraft (TB-1s) from Boeing under a May 1925 contract. They never became operational; BuAer pursued the modified SC instead. Initially it was designated SC-9, to indicate its connection with the earlier airplane. It introduced a welded steel-tube fuselage structure and was powered by a modified version of the SC-2 engine (the 575-hp Wright T-3B) rather than the Packard 2500. In accord with Design 35, it was rearranged, the pilot and bombardier sitting side by side in a new cockpit forward of the wing (the gunner's position was unchanged). Like the CS/SC series, it had unequal-span wings. An unusual feature was a large radiator slung from the upper wing. Twenty-four aircraft ordered in October 1925 were designated T3M-1s.[35]

BuAer pursued the twin-engine Design 36, both from the Naval Air Factory (as TN-1) and from Douglas (as T2D-1). The T2D briefly operated from *Langley* but was not tried on board the two larger carriers *Lexington* and *Saratoga*.

Two further T3M-1s were modified to take the desired 710-hp Packard 3A-2500 engine. BuAer would have preferred an air-cooled engine, but none was yet powerful enough. These aircraft became the prototypes of the T3M-2, which had equal-span wings (of greater total area, to lift more) and an arresting hook. Tandem seating was revived for carrier operation. The pilot occupied a single cockpit under the wing, the observer remaining forward of him in the front of the airplane. T3M-2 became a standard Navy torpedo bomber, one hundred being delivered,

A modified SC (originally SC-9) was ordered in preference to the Boeing TB: Martin's T3M-1. Like the CS/SC, it had unequal wings. As in BuAer's Design 35, the pilot and observer sat side by side. Note the unusual radiator suspended from the upper wing. NATIONAL ARCHIVES

The T3M-2 reverted to the earlier cockpit arrangement, in tandem. The pilot sat under the wing, the observer forward of him, the gunner well aft. Once aircraft operated on the *Langley*, it was clear that the pilot should be foremost, but that was not done until the next version (T4M). Wings were of equal span. The underwing radiator gave way to more conventional ones alongside the fuselage. The device above the *2* on the fuselage side was a target-towing reel. This photograph was taken 17 November 1928. NATIONAL ARCHIVES

beginning in mid-1927. Once aircraft began operating on board *Langley*, it became clear that the pilot should sit in front. However, the controls of T3M-2s were not relocated, because the successor T4M soon appeared.

With limited engine power, the torpedo bomber needed huge wings to lift the required load. It was so large that it would be difficult to handle on board even the two big new carriers.[36] The ships each had a small elevator aft and a larger one forward, the small elevator being 30 feet by 35.6. The T3M and the twin-engine bombers did not fit the after elevator. T3Ms would have to be stowed forward, where only twelve could fit, with their wings folded. Once moved to the flight deck by the forward elevator they would have to be manhandled aft to prepare for flight. Normally the fighters, with their shorter takeoff runs, would be spotted forward of them, but it was not clear how that could be done if the fighters were stowed aft, using the after elevator. Flight operations would be awkward and slow, despite the premium on fast handling to launch aircraft before an enemy could attack. Landing too would be difficult. The VT-VS would have to be hauled forward to be struck below, seriously complicating flight-deck operations. Limiting the big planes to one elevator would more than double the time required for handling. If the single usable elevator failed, operations would collapse.

The T4M-1 was more compact than the T3M-2, but it was still large because of the considerable length of the standard torpedo, as is evident here. This VT-1B airplane is on board *Lexington*, 29 November 1928. NHHC

A VT-9 T4M-1 photographed in February 1930 shows twin machine guns on the Scarff ring on the gunner's cockpit. Note the windows aft.
NATIONAL ARCHIVES

The nose of a T4M-1 on the carrier *Lexington*, 4 April 1929, shows the closed bombing window (for the bombsight) below and abaft the engine, with a side window above and abaft it for the bombardier. Some aircraft had round side windows; others had none.
NATIONAL ARCHIVES

By this time air-cooled radial engines were finally adequate, so two T3Ms were reengined with the competing Pratt & Whitney Hornet (producing the T3M-3) and the Wright Cyclone (T3M-4). Given the Hornet (R-1700), Martin solved the problem of T3M size. Its Design 74, which was bought as the T4M-1, could carry the same useful load on a ton less gross weight (7,500 pounds), with a maximum speed of 115 mph and a service ceiling of about 12,000 feet, both better than what the T3M-2 offered. Unlike the T3M, it could use the after elevators of the big carriers. The T4M made the three-purpose airplane far more attractive.[37] The test report praised the prototype for its liveliness and controllability, which seemed to make it particularly well suited to carrier operation. Visibility was better than that of any naval airplane other than the small Vought UO series. The Navy ordered the T4M-1. T4M-1 was credited with a maximum speed of 114 mph, and it took 14.1 minutes to reach 5,000 feet. By way of comparison, the twin-engine T2D could reach 128 mph and climb to 5,000 feet in 7.4 minutes. However, initial trials showed that the T2D was not carrier suitable.

The Great Lakes Company continued T4M-1 production under the designation TG-1. This one was photographed on 5 October 1931. Note the tail wheel instead of a skid.
NATIONAL ARCHIVES

The Great Lakes Company acquired Martin's Cleveland factory in October 1928 and there continued production of a slightly modified T4M under the designation TG-1. It had a tail wheel instead of a skid, additional wing struts, a new bomb rack, a new radio loop, exhaust-collector rings (to direct cylinder exhaust safely away from the airframe and crew), and improved wing folding.[38] The successor TG-2 had the more-powerful Wright Cyclone (R-1820-86) and was more streamlined, so it attained higher speed (127 mph).[39]

Still, in his 1930 report CinC Battle Fleet pointed out that the sheer size of the torpedo bombers cluttered flight decks. A carrier maintaining fighter patrols could not afford to keep them on deck, although by 1930 there was some hope that procedures could change. Moreover, the big torpedo bombers could not land safely with a torpedo or a heavy bomb on board: a torpedo sent into the air on board a bomber was a torpedo lost. Each of the big carriers had only sixty-four torpedoes on board and only 120 thousand-pound bombs. In exercises, the big bombers were typically launched with simulated thousand-pound bombs, which the ship could more easily afford to lose them. Torpedoes would be loaded only for a second strike.

Looking ahead in January 1927, the BuAer chief summarized requirements.[40] The three-purpose plane (VS-VT) should be able to deliver a (1,650-pound) torpedo two hours away at full speed. As a bomber, it would carry a thousand-pound bomb to ten thousand feet and have a six-hour endurance at economical speed (dropping the bombs at the halfway point). It should have maximum gun protection: three flexible guns for a single-engine type, five for twin engines. As a scout, it should have an endurance of about twelve hours at economical speed, with a maximum speed of no less than 100 mph and a cruising speed of no less than eighty. This was essentially the T4M-1. In November 1927, the only torpedo bomber planned for near-term procurement was a version of BuAer's Design 70 float-type biplane, an all-metal airplane with a novel float design and a Pratt & Whitney R-1300. Apparently, it was never bought.

By the late 1920s, when the big carriers and their torpedo-bomber squadrons entered service, the standard torpedo-bomber organization was the nine-plane division, which had to maintain tactical concentration or visual contact until it deployed to attack; whenever possible it should have fighter escorts. When the bombers attacked a fleet steaming in circular formation (as did the U.S. Navy), the outer screens would often report them in time for carriers to launch fighters to oppose them. Often there was a layer of haze at the surface: objects on the surface could easily be seen from high altitude, but aircraft above the haze would be nearly invisible to surface ships. In an attempted mass torpedo attack on battleships off Guantanamo Bay, Cuba, in March 1930, the high-flying torpedo-wing commander enjoyed excellent visibility and could not understand why his lower-flying torpedo bombers were not following his orders to strike the battleships. The torpedo bombers could not see the battleships until they were less than three thousand yards out. A few years later, doctrine called for the torpedo bombers to approach at high altitude until they found their objective, deploying outside antiaircraft range and then diving into the haze. Alternative attack tactics were a concentrated strike from one flank of the target formation, a wave attack from one flank, or a divided attack (i.e., "hammer and anvil": half the attack on either bow). The first two tactics were to be used against an enemy battle line unable to maneuver freely because it was already in a gun action, trying to maintain course to make its gunnery effective. A torpedo attack might also be mounted in coordination with a destroyer torpedo attack, one from each side. Throughout the interwar period it was emphasized that torpedo bombers had their best chance when attacking in coordination with a surface torpedo attack. A 1936 lecturer at the Naval War College argued that "most torpedo planes have a very limited endurance, and it may be that they will not be able to wait until the destroyers are ready. . . . Torpedo planes must have protective smoke to be successful, as they are highly vulnerable to gunfire while approaching close to the surface."[41]

Not surprisingly, CinC Battle Fleet. writing in 1930, wanted a heavy dive-bomber, not a new torpedo bomber. It would be far less vulnerable to antiaircraft fire and substantially smaller than a T4M or TG. Torpedo bombers were worth retaining only until such aircraft were developed. By 1932 the fleet had only a single torpedo-bomber squadron, which was also its sole horizontal-bombing squadron.

Where to Go?

Through the late 1920s the Navy sought a way forward for attack bombers. The internal discussion began in September 1925 with General Board hearings on the characteristics of aircraft for the two big carriers, which were then expected (too optimistically) to be completed in 1925–26.[42] BuAer hoped to go ahead with contracts by 1 February 1926; in the event, the project stretched out considerably (the carriers were not completed until 1928). To support this project, in February 1926 the Naval War College summarized war game lessons.[43] The emphasis, the college wrote, should be on torpedo and bombing planes, which could sink the enemy's ships, rather than on fighters, "which may or may not protect own planes and destroy the enemy's." What the college called "VT-S" aircraft had to be able to protect themselves. Fighter endurance was about a third of that of a VT: to require fighter escort would cut striking range by two-thirds. It would be absurd to limit the effective endurance of a fifteen-hour scout to that of a four-hour fighter. Fighter escort was impractical also because fighters were already so much faster than bombers. The Naval War College interpreted all of this as an argument for twin-engine bombers, which alone it seemed could accommodate effective defensive armament (flexible guns) forward as well as aft. The nose was also the natural position for a bombsight.[44] The failure of the T2D in initial carrier tests, however, apparently convinced the fleet that twin-engine aircraft should not operate from carriers. BuAer tried to revive the idea several times, succeeding during World War II.

Laying down a five-year aircraft building program in 1927, BuAer emphasized carrier aircraft, substituting the three-purpose carrier plane for flying-boat patrol types. That January the head of the BuAer Design Section summarized the situation for Admiral Moffett.[45] In 1925, the DT had been in service and the CS had been experimental. In January 1927 the service VT-S aircraft were the SC (production CS) and the T3M-1. The experimental types were the twin-engine T2D and TN-1 and Martin's Design 74 (which became the T4M). VT-S development was lagging that of fighters and spotters, because the three-function role was so difficult. Worse, VT-S were used for patrol, which further distorted their design. Making the airplanes interchangeable between land and seaplane and requiring for them enough strength to be catapulted (and deck-landed) had not helped. Fighters could be divided into four classes, but in these heavier aircraft a single design had four functions. Concentrating these tasks in the largest and heaviest airplane jeopardized carrier operation. This policy had been dictated largely by the limited space on board carriers. No one yet knew whether it worked, because none of the airplanes had yet operated from a carrier—the two big new carriers were not yet in service.

The BuAer preliminary designers argued that the three- or four-role airplane should be abandoned. The current three-role aircraft were the slowest of the carrier planes, whereas speed was a fundamental requirement for scouting, particularly if the airplane encountered headwinds. Patrol ought to be a specialized function, leaving bombing and torpedo attack to other types. Should the primary function be torpedo attack or bomb attack? A torpedo bomber could have a much narrower fuselage. As a guide to decision, BuAer's designers sketched four alternative torpedo-bomber designs.[46]

The BuAer designers argued that no torpedo bomber could attack the enemy's main body until its light forces had been eliminated. Light (dive-) bombers could do that. They might be fighters with a secondary light-bombing function; fighters were already considered useful to harass enemy antiaircraft batteries and fire control points. The Design Section preferred to develop dive-bombers as a special type, intermediate between fighter and heavy bomber.[47]

The Design Section proposed an experimental bomber program:

(a) A single-engine combined torpedo and bombing airplane designed around a large air-cooled engine
(b) A twin-engine combined torpedo and bombing airplane designed around a large air-cooled engine
(c) A two-seat, air-cooled fighter with provision for bomb strafing (dive-bombing)
(d) A carrier observation-two seat fighter / light bomber designed around a large air-cooled engine.

To help develop a revised aeronautic policy, in April 1927 the General Board circulated a questionnaire asking how the functions of the existing three-purpose airplane (V-TBS) should be distributed and what their relative priorities should be.[48] The alternatives were the current combination, a bomber-torpedo plane (VB-VT), a scout torpedo plane (VT-VS), a bomber scout (VB-VS), and pure torpedo plane (VT), bomber (VB), and scout (VS). What priorities should be assigned to various functions, and how many different types should there be? The board also asked about possible combinations of observation (spotting) and other functions: VO-VF, VO-VB, VO-VS, and VO-FB (fighter-bomber).

The Secretary of the Navy created a special board on aeronautic policy, headed by Rear Adm. M. M. Taylor and including as a member Rear Admiral Moffett. J. M. Reeves—by then a captain and, as commander of the air squadrons of the Battle Fleet, conducting innovative experiments in carrier operations—was a member. Lieutenant Commander Mitscher, a future famous carrier admiral (and, before that, BuAer chief) was member and recorder.

Opinions differed enormously, because there was so little concrete experience of carrier air operations. However, the board was impressed by the weight of opinion calling for more carriers but a minimum number of types of aircraft. The board agreed that the triple-function airplane was impractical. The torpedo and heavy-bomber functions could be combined, bombing being secondary. Experiments should be conducted to see whether this airplane could be fitted as a long-range scout, but it was not to be used for tactical scouting.

The two-seat fighter should be developed so that it could be used for spotting, for tactical scouting, or for dive-bombing. Dive-bombing offered greater accuracy than any other method, and it could probably be pushed home where high-altitude level bombing could not. Dive-bombers would probably also suffer less from antiaircraft fire. OpNav's War Plans Division proposed to equip all bombers as dive-bombers. However, dive-bombing was still experimental. To date only small bombs (twenty-five pounds) had been used; how well dive-bombing would work with five-hundred- or thousand-pound bombs was not certain. The Taylor Board concluded that the primary function of observation aircraft operating from carriers should be dive-bombing; secondary functions should be fighting and tactical scouting. The only two-seat fighter to be developed should be a combined fighter and light (dive-) bomber. BuAer was already equipping current fighters and also VO-VS types (the O2U Corsair) to dive with five-hundred-pound bombs. Two experimental designs were being built to dive with thousand-pound bombs. War Plans assumed that bombers would often have to fight beyond fighter range: they needed maximum performance and defensive armament. The need for more forward-firing guns was enough to justify continued interest in twin-engine bombers, which could accommodate them more easily.

Dive-bombers could not penetrate the armored decks of battleships, which aerial torpedoes could sink or disable. BuAer therefore argued that torpedo bombing should survive. Developing a new aerial torpedo might be as important as current work on improved destroyer torpedoes.

The most important new idea promoted by this Taylor Board was that a heavy dive-bomber should be built. It became the T5M and later the BM-1 described in the next chapter. The new policy was formally approved by the Secretary of the Navy on 14 November 1927.

This discussion fed into a General Board study of a new five-year naval aircraft building program (1930–34).[49] The board dismissed current claims that fighters alone could seize control of the air. In a fleet engagement, that could be done only by disabling the enemy's flight decks by bombing. Light (dive-) bombers were more likely to be used than heavy bombers, so the board proposed a ratio of three light bombers to one heavy for new carriers. Even that might be excessive; experience might show that a carrier should have no more than one heavy squadron, either VB or VT.[50] Torpedo bombers were too unwieldy, but "a fair comparison between the torpedo and the bombing form of attack has not yet been made and the neglect of either form of attack is not justified."

The board suggested a new two-seat general service (VS) category. Observation (VO) hardly described the uses of the current O2U Corsair: spotting, observation, tactical scouting, and most recently dive-bombing. VS would indicate combined functions: spotting for gunnery (VO), tactical scouting (VS), and the new dive-bombing. It should be as small as possible consistent with its primary spotting mission and capable of operating either with wheels (from a carrier) or with floats. It should be able to carry a load of up to five hundred pounds of bombs or smoke canisters and be armed with one fixed machine gun forward and one free (flexible) gun firing aft. In some ways the existing O2U-2 was already a two-seat fighter, one that might effectively defend spotters on the "standby circuit" (waiting to take their place to spot for fleet guns, vulnerable to attack by enemy fighters). The proposed redesignation died, however, because it would have caused considerable administrative complication.

Two-seat fighters, which had appeared during World War I, muddied the situation. Advocates claimed that war experience showed that a two-seater was much more difficult than a single-seater to surprise (in the General Board copy the question, "What about fighting?," is penciled in). Advocates of single-seaters argued that maneuverability was likely to be decisive; the free gun in the after cockpit could not be handled effectively during dogfights. It was undesirable to add more types to the carrier. A comparison between the best current single-seater, the F4B (two of which had been built), and the projected two-seat XF8C-2 showed that the single-seater was faster (168.8 versus 149 mph) and had a greater service ceiling (26,900 versus 20,700 feet); it could climb faster (5.3 versus 7 minutes to 10,000, with an initial rate of climb of 2,920 versus 2,070 ft/min.). The General Board recommending assigning one squadron of two-seaters to a carrier in competition with the usual single-seaters. BuAer argued that the two-seat fighter should be limited to fighting and dive-bombing and that all single-seat fighters carry light bombs.

The General Board urged that one carrier alone should have the heavy VTB (with torpedoes) on board; others might have bomb-carrying VTBs. This seems to have been the basis of the decision not to provide torpedo stowage to the one carrier then under construction, USS *Ranger* (CV 4). The General Board's aircraft program left out the Marines, who complained that they alone of the U.S. services had had post-1918 air-combat experience, in both Nicaragua and China. They feared that the Navy saw Marine airmen and airplanes only as a reserve for carrier operations and that their needs could too easily be forgotten.[51]

The FY30 experimental program, formulated in July 1929, included an experimental VBT, which seems to have been the Martin T5M (which became the BM-1), a heavy dive-bomber that briefly had a torpedo-bomber designation.[52] The BuAer Plans Division recommended that a board be convened to determine desired characteristics, to report on the XVBT (and an XVP) by 1 September 1929; other types were to be reported on by 1 December. In his 1930 annual report, for the period ending 30 April 1930, CinC Battle Fleet urged development of a heavy dive-bomber (to dive with a thousand-pound bomb) and a light, fast torpedo plane, priority going to the heavy dive-bomber. Unless something distracted the enemy the torpedo bombers had little chance of success: such distractions would likely be supplied by simultaneous attacks by heavy gunfire or by destroyers, bombers, or a smoke screen.

BuAer badly wanted a higher-performance torpedo bomber. The T5M was smaller and much more agile than the T4M, but it could not accommodate the long Mk VII torpedo. BuAer pointed out that if the length of the torpedo could be cut back to fourteen feet, the T5M could be modified to carry it. The question of future aerial torpedo size had been raised in October 1928, and in December the General Board recommended developing a special aerial torpedo, weighing no more than 2,200 pounds and no more than fourteen feet long. It should be able to reach four thousand yards at twenty-five knots.[53] It happened that a shorter torpedo with greater diameter would have better hydrodynamics, but that was not a factor in the choice of a shorter future torpedo. The Naval Torpedo Station at Newport, which developed naval torpedoes for BuOrd, resisted any new torpedo project in favor of improving the existing Mk VII. It could now be launched reliably in seventy-five feet of water by an airplane flying at up to ninety knots, using a drogue and dropping from no higher than twenty-five feet. However, it turned out that the best dropping altitude was ten feet, which was impossible for a wheeled airplane. BuAer pressed for a more effective weapon or at the least proof that none could be developed.

During the 1929 discussion of the torpedo, Commander Aircraft Squadrons Battle Fleet was not at all sure that a single-purpose torpedo bomber was wanted. He wanted to know how much could be gained in a single-purpose airplane carrying the existing long torpedo. At about the same time that Martin was designing the T5M, the BuAer preliminary designers sketched a single-purpose torpedo bomber, with fixed and flexible guns (Design No. 75). It was a folding-wing biplane with flaps on both wings and slots on its upper wing. Its wheels had brakes, apparently an important feature at the time. Its engine was an R-1740, and it had a radio—a significant weight.[54] On the basis of this study BuAer ordered Martin's XT6M-1 single-purpose torpedo bomber in mid-1929. It was the first American torpedo bomber with a metal-skinned fuselage and tail. With somewhat less power than the BM-1 (from an R-1820), it was slower, little faster than a T4M/TG.[55]

A single-purpose horizontal bomber, the bureau argued, should be designed in parallel. In January 1930, the chief of BuAer pointed out that there was a role for a single-purpose high-altitude bomber.[56] The goal was a substantially higher bombing ceiling than that of the T4M. Without a torpedo, it could be considerably smaller and lighter, hence could climb to 14,000 (rather than 10,000) feet with a thousand-pound bomb. That seemed to be the current limit for a single-engine airplane carrying one bomb and five hours' fuel. As described in February 1930, this three-place airplane would be based on BuAer Design 90, all-metal except for the wing covering.

Given the excessive size of "three-purpose" aircraft, the question was how small a pure torpedo bomber could be. Martin's XT6M-1 was an attempted answer. It is shown on 5 May 1931.
NATIONAL ARCHIVES

The bombardier would lie flat on his belly (with provision for kneeling if necessary). The bombsight would project below the fuselage. Maximum wingspan would be forty-eight feet (the wings would fold). Fuel would be sufficient for 550 miles with full bomb load. Service ceiling would be 15,000 feet. Maximum speed would be 125 mph (using a 575-hp engine), and stall speed would be sixty. The Material Division suggested eliminating the metal tail covering as unnecessary weight and simplifying fuselage construction. It eliminated the prone bombardier's position, which required too much space. There was some question as to practicable range (Design 90 was being revised). CNO considered the project worthwhile; a squadron of such airplanes would replace a torpedo-bomber squadron.

Design 97 (March 1930) was a revised version of Design 90, offering a range of 590 miles at cruising speed (444 at full speed).[57] Maximum speed was 125.4 mph, and weight with a 1,000-pound bomb on board was 7,317 pounds. It was built as the XT3D-1, which was apparently conceived before Commander Aircraft Squadrons made his proposal but became the benchmark three-purpose airplane. Its characteristics suggest that it was intended to test various innovations, particularly a metal wing structure and a drag-reducing Townsend ring surrounding its engine. The XT3D-1 was modified by replacing its engine with an R-1830, removing the forward flexible gun, and providing the pilot with a windshield, becoming the XT3D-2. Neither the T3D nor the single-purpose airplanes were ordered into production.

The Naval Aircraft Factory was tentatively assigned to develop Design 90. It seems to have been revised as Design 98, which was ordered as the BN-1 but canceled before completion.[58]

The XT3D was an experimental next-generation, three-purpose biplane. In contrast to the T4M, the pilot sat in the second cockpit so that the observer/bombardier could easily use the large bombing window under the nose, just abaft the cowling. Presumably this rearrangement reflected rising interest in horizontal bombing associated in part with the development of the Norden bombsight. The airplane was built with a cowled single-row R-1820-E engine, then given a two-row XR-1830-54, which had the long cowling here (the short cowling in the original version was a Townend ring). Other modifications were spats (over the wheels) and a canopy over the two rear cockpits. The sole XT3D-2 became the bombing-trials airplane of the Naval Proving Ground, Dahlgren, Va. It is shown on 7 February 1933. NATIONAL ARCHIVES

At about the same time it ordered the XT6M, BuAer bought the XBY-1 single-purpose horizontal bomber. It was a version of a civilian airliner. NATIONAL ARCHIVES

An unrelated small horizontal bomber was ordered from Consolidated Aircraft in April 1931: the XBY-1, converted from the company's Fleetster, a high-wing monoplane passenger plane (five passengers). It was unusual in having internal bomb stowage in its fuselage. It crashed in 1933 after extensive tests, having been rejected as unsuited to dive-bombing and also as too large for a carrier.

At about the same time, work was proceeding on a specification for an experimental twin-engine VBT. The corresponding sketch design was probably the twin-engine Design 122 (29 September 1932), powered by R-985-36 engines (400 brake horsepower, or bhp, each), with a maximum speed of 145 mph (weight 8,660 pounds); some versions of the design attained about 160 mph. There was apparently no attempt to buy a corresponding airplane.

For BuAer, the question was whether to proceed with a new torpedo bomber or to combine the torpedo and dive-bomber roles by having BuOrd develop a short-range torpedo about the size of a thousand-pound bomb. It was not at all clear what sort of torpedo range was needed. An informal board convened by the sole torpedo squadron (VT-1B) argued that the greater the range, the greater the need for a pilot to work out the required lead angle. To do that, he needed to estimate target course and speed and "target angle" (his own relative bearing as seen from the target)—none of which would be easy under heavy enemy fire. Alternatively, formations of torpedo bombers could fire "browning shots" into the mass of an enemy formation.[59] VT-1B pilots were dubious. They expected to press their attacks home, three or six planes attacking an individual enemy ship at close range. Rather than more range, the pilot wanted to drop his torpedo at the greatest possible speed and, given a choice, the greatest possible altitude. The existing Mk VII was too weak to handle the shock of hitting the water from any considerable height. In April 1929 CinC Battle Fleet pointed out that the new torpedo was the difficult part of the package: it would take at least twice as long to develop as would a new torpedo plane.

The smaller the torpedo bomber, the better. It should be substantially faster—a speed thirty knots

faster than the T4M would justify a new airplane. It did not help that the Battle Fleet wanted both a smaller, lighter torpedo plane *and* extended torpedo range (eight to ten thousand yards).

BuAer wanted the new torpedo to be droppable at 145 knots and at a height of up to forty feet. The pilot should be able to angle the torpedo, giving him much greater freedom of maneuver. Torpedo length should be limited to fourteen feet. BuAer's Plans Division pointed out that length could be cut by increasing torpedo diameter, retaining the same volume. The General Board opted for a maximum weight of 1,750 pounds, a 400-pound warhead, a length of twelve feet six inches, and a diameter of twenty-two inches; it should be launchable at a hundred knots from a height of thirty feet, with a range of seven thousand yards and a speed not less than thirty knots. These figures were soon amended to a length of fourteen feet a range of about four thousand yards at thirty-five knots. This was roughly what BuAer wanted, except for dropping speed.[60] The endorsement on the cover sheet pointed out that it would probably be six years before the new torpedo would be ready. "When we get the new design we can build a plane to carry it." In August 1930 the Torpedo Station, responsible for development, designated the new weapon the "Mk 13." It proposed a variety of exotic technologies to get it under the weight limit.[61]

Another reason for a new torpedo was that the existing Mk VII was too slow (twenty-seven knots) and carried too little explosive (240 pounds).[62] Reportedly, current British torpedoes carried 400 pounds of explosive and could make forty-two knots on a total weight of 1,600 pounds, som what less than that of the U.S. Mk VII. BuOrd pointed out that the reported range of the British torpedo was two thousand yards. Its own new torpedo would make thirty-knots for seven thousand yards and would carry as much explosive as the British.

BuOrd found it impossible to stay within the desired weight; in 1934 designed weight was 1,850 pounds.[63] Mk 13 entered service in 1938 weighing 1,949 pounds with a 400-pound warhead. Range was 4,500 yards at twenty-nine knots. The increased diameter (22.4 inches) made it possible to hold length to 13 feet 5 inches (161 inches). The drop limit was slower than desired, 115 knots, but in theory the new torpedo could be dropped from 60 feet.

Meanwhile, on 30 October 1931, BuAer asked BuOrd to develop a torpedo approximating the size and weight of the thousand-pound dive-bomber weapon, carrying the same four-hundred-pound warhead as the evolving Mk 13. Range should be about two thousand yards at 30 knots or more. It should be droppable from fifty feet at 125 knots.[64] BuOrd pointed out that the desired weight, range, and speed were roughly those of the first U.S. aerial torpedo, Type D—which carried a two-hundred-pound warhead and was twelve feet long. Increasing diameter to nineteen inches, as in the bomb, should make it possible to achieve roughly the same characteristics on a shorter length. The Torpedo Station designers pointed out that such a torpedo carrying 300 or 400 pounds of explosive would be too heavy to float. Each 100 pounds of explosive would add about eleven inches of length and about 111 pounds of weight, so that a torpedo carrying a 300-pound charge would be about 117 inches long and would weigh about 1,225 pounds. With 200 pounds of explosive, the torpedo would be 106 inches long (compared to 84 for the bomb) but would make the desired two thousand yards at 30 knots. The fifty-foot drop and 125-knot speed were unlikely to cause problems.

BuAer revived the idea in July 1932 and in February 1933 formally requested development of a torpedo with dimensions no greater than 20 by 110 inches (1,050 pounds) and a warhead of no less than 200 pounds, with a range of at least two thousand yards at no less than 30 knots, and that could be dropped at no less than 125 knots. BuOrd was already finding Mk 13 development difficult, and it considered a torpedo with an even smaller length-to-diameter ratio impractical. The proposed characteristics for warhead, speed, and range offered little chance of success, and BuOrd questioned whether a heavy dive-bomber could get into a satisfactory firing position against capital ships. CNO agreed.

BuAer's hopes for a very light torpedo led it to imagine that the single-purpose horizontal bomber could carry a torpedo as a secondary weapon. This airplane was one of three new types listed in February 1933, and in March the Material Division produced Design 123, a three-seat horizontal bomber capable of carrying either two five-hundred-pound bombs or a thousand-pound bomb or "the newly projected 1000 lb streamline torpedo."[65] To clean up the design and improve performance, it dispensed with wing-folding; span was limited to forty-one feet (as in the BM-2). Compared to the T3D-2, weight would be reduced about eight hundred pounds, span by nine feet, and length by three and a half feet; speed would increase about 20 mph. The engine would be an 800-hp SCR-1830. Speed would be even greater if, as expected, that engine's output increased.

The Plans Division was skeptical. There was considerable doubt in BuAer as to the ultimate value of any torpedo bomber, and the light torpedo was far from ready. Unless it replaced heavier torpedoes altogether, it was not clear that two parallel lines of aerial torpedoes were worth pursuing. If the ultra-lightweight did materialize, it could be carried by heavy dive-bombers. If not, Design 123 would be a single-purpose horizontal bomber. Its speed of 172.6 mph at critical altitude with a supercharged engine was not sufficiently better than that of the XT3D-2: 162.7 mph.

Material replied that Design 123 had been conceived as a horizontal bomber, without special provision for a torpedo.[66] However, on 7 June the assistant chief of BuAer recommended a design study of a modern VB-VT. It turned out that the added torpedo weight had little impact. Now it did not matter that the thousand-pound torpedo was a fantasy or that the new Mk XIII would weigh 1,859 pounds compared to 1,740 for the current Mk VII-B. With slightly revised landing gear, Design 123 could carry either the new torpedo or three five-hundred-pound bombs. Drag could be reduced by careful redesign. Even without folding wings, the airplane could fit the elevator of the new *Ranger* (foldable wings would cost eighty pounds). Material wrote that this was the best that could be done in the way of compact arrangement for such an airplane.

Material offered Design 125, a monoplane torpedo bomber using flaps. It should, Material claimed, achieve considerably greater speed "at the expense of longer span, vision, and folding wing difficulties." Its data sheet was dated 19 August 1933. At a gross weight of 8,322 pounds (including a torpedo) it offered a maximum speed of 183.3 mph at critical altitude (169.2 mph at sea level). Stall speed was an acceptable 63.7 mph. Service ceiling was 18,300 feet. Its SGR-1830 engine was rated at 750 hp. Span was forty-eight feet.

Design 125 was promising enough to become the basis for a design competition. In October, BuAer circulated a request for proposals.[67] It sought "considerably advanced performance. High speed is especially desired as long as the other requirements are met as to stalling speed, range, and ceiling." Minimum high speed with normal bomber load was 180 mph at critical altitude; stall speed should be no more than 65 mph. Normal weight as a torpedo bomber should not exceed 8,500 pounds. Span was not to exceed fifty feet; if wings were fixed, it should not exceed forty-two feet. Folded span should not exceed twenty feet for a biplane or twenty-five for a monoplane. It must be possible to park two folded airplanes on an elevator forty-eight by forty-four feet with six-inch clearance all around and between airplanes and with no overhang over the boundary. Length should not exceed thirty-five feet or height twelve feet six inches (with propeller horizontal). Although designers were free to choose any approved engine, they were informed that geared R-1830 had been favored in BuAer studies. Proposals were due by 15 January 1934. A note on the cover sheet pointed out that the demand for 180 mph practically excluded biplanes.

BuAer concluded that torpedo bombers were too large because standard torpedoes were too long. The shorter Mk 13 was associated with the next-generation monoplane torpedo bomber, the Douglas Devastator (TBD). To improve performance, the Devastator carried its torpedo semisubmerged, as in this 1937 mock-up photograph. The window above the torpedo was for the bomb aimer, who sat abaft the pilot. THE HOOK

BuAer's new torpedo bomber, the XTBD-1, is shown with two heavy bombs (not a torpedo) slung underneath, 14 October 1935. The three-man crew shows that this was really a three-purpose airplane, with a radioman to report the results of scouting. The sparkling performance of the TBD revived Navy interest in torpedo bombing. NATIONAL ARCHIVES

Competitors were Boeing (Model 289), Great Lakes, Douglas, Martin (Model 144 and 144A), and Seversky (Republic). The finalists seem to have been Douglas, Great Lakes, Boeing (biplane), and Seversky 2A, in that order, with Douglas well in front. As expected, the Douglas prototype showed excellent performance. Its successful proposal became the TBD Devastator.[68] It had the U.S. Navy's first hydraulically folded wings.

The Devastator was sufficiently radical that BuAer ordered Great Lakes' TBG biplane as insurance. It was essentially a scaled-up BG (see chapter 4) with retractable landing gear and a bombardier's station under a low canopy forward of the wings. As in several other airplanes of this period, retractable landing gear required a deep belly, in which was set a bomb bay. Before either prototype had been delivered, BuAer decided that unless tests revealed some major problem the TBD would be selected. Flight tests showed that the TBG had poor stability and inferior performance.

Production Devastators duplicated the prototype except that they had uprated R-1830–64 (900-hp) engines. All aircraft were of the same TBD-1 version.[69] The TBD's high performance revived U.S. Navy torpedo bombing. The two big

As insurance against a problem with the monoplane TBD, BuAer bought the TBG, shown on 23 August 1935. As in the T3D, the forward cockpit was for the bombardier; note the bombing windows under and abaft the cowling. In effect the TBG was a scaled-up (by 15 percent) BG dive-bomber with an additional station for the bombardier. Note the telescopic gunsight in the upper cockpit and the long cowling of the double-row engine. The failure of this design killed the Great Lakes Aircraft Company.
NATIONAL ARCHIVES

carriers in the fleet and the two fleet carriers *York-town* and *Enterprise* under construction as of January 1936 were each to be fitted to carry thirty-six torpedoes for an eighteen-plane VT squadron. In 1939 a Naval War College lecturer on naval airpower described aerial torpedoes as the "most effective means of reducing an enemy's speed"—a prime consideration for the U.S. Navy, whose battle line was slower than that of its most likely enemy, the Imperial Japanese Navy.

In November 1933, the BuAer Plans Division asked for a parallel study of two-place torpedo bombers, both single- and twin-engine. With a crew of two rather than three, either the bombsight or the free gun could be manned at one time. No corresponding sketch-design data sheet has survived. Similarly, nothing came of a 1937 BuAer preliminary design for a twin-engine torpedo bomber (Design 145) powered by two XR-1535-92 engines. It was expected to make 275.9 mph at 8,000 feet and to have a service ceiling of 27,400 feet; it would climb to 15,000 feet in eight minutes (initial rate of climb would be 2,070 ft/min.). Endurance at high speed would be 1.55 hours; at 60 percent speed it would be 5.79 hours. The takeoff run would be 554 feet without wind, or an acceptable 230 with twenty-five knots of wind. Maximum range, the bomb being dropped at half-range, would be 1,040 miles.

In June 1941 Rear Admiral Halsey, then commanding Aircraft Battle Force, wrote to BuOrd that "recent developments abroad" had emphasized the effectiveness of determined torpedo attack. He probably meant the British strike on Taranto the previous November and the attacks against the German battleship *Bismarck* that had crippled its steering and in effect doomed it (an anonymous skeptic pointed out that in the face of serious opposition at Matapan four months later, British torpedo attacks failed). Halsey echoed the much-earlier call for torpedoes that could be deployed by all bomb-carrying aircraft in the fleet. Among other requirements, he wanted it droppable at speeds of up to about two hundred knots and from altitudes as high as two hundred feet. He would accept a range of only two thousand yards.[70] The big dive-bombers then being developed, the SB2D and SB3C, could carry the full weight of a Mk 13 torpedo; there was no need for some special developmental project. Because the SB2D and SB3C had tricycle undercarriages (i.e. wheels under each wing and the nose, as opposed to "three-point" gear under the wings and the tail), they could carry torpedoes externally with minimal modification (the change to torpedoes would take about thirty minutes). A BuAer officer made an annotation that "an outstanding advantage to be gained . . . is that the design load factor would permit of executing a dive torpedo attack. This will not be possible in the TBF [Avenger]. Reports from Europe indicate that this form of attack to be more effective and to afford greater security for attacking planes." These advantages outweighed the problems of limited carrier torpedo stowage and the possible difficulty of providing a torpedo director.

The production version of the TBD differed from the prototype mainly in having a roll bar to protect the crew. That pushed up the fore end of the canopy, giving the airplane a hunchback appearance. Just abaft the cowling, above the exhaust pipe, is the muzzle of the forward-firing 0.30-calibre gun on the right side of the cowling, firing through the propeller arc. There was a single flexible 0.30-calibre gun in the after cockpit. The paint scheme indicates that this photograph was taken in 1941; this is the tenth airplane of VT-3. THE HOOK

The next step, which had not yet been taken, was to see whether the two dive-bombers approaching service (the SB2C Helldiver and the SB2A) could also carry the Mk 13. Initially it seemed that extensive redesign would be needed. However, it turned out that the Helldiver could be adapted, albeit initially by a four-hour process. During World War II this adaptation was improved to the point that in 1944 an SB2C-4 could be converted into a torpedo bomber in forty minutes. A thousand adaptation kits were made, but they were not used in combat.

Making Torpedoes Work

Through the interwar period it was understood that torpedo bombers had to attack from higher altitude and at higher speed than had been common. All the aerial-torpedo navies found that torpedoes oscillated (nose up, then nose down) as they dropped. That ruined their accuracy. Torpedoes often went too deep on their initial dives, even hitting the bottom. A torpedo might run as much as a thousand yards before recovering from a dive to reach its set depth. BuAer found that attaching drogues to the noses of its torpedoes solved the diving problem—but made aircraft operation difficult.[71] It ordered the drogue kept secret, but by the early 1930s it wanted an alternative.

After a series of tests in 1926 to develop techniques, the Bureau of Ordnance recommended that torpedoes not be dropped from more than twenty-five feet, the airplane having a slight up-angle.[72] Launch speed could be as great as ninety knots. By this time, however, much faster torpedo bombers were in prospect. To survive enemy fighters and guns, a torpedo bomber should approach and get away at maximum speed, with maximum maneuverability. However, instructions issued in 1929 reduced maximum dropping speed to seventy-five knots, consistent with fleet experience. Given the higher performance of the new T4M, CNO ordered the Torpedo Station in Newport to modify the Mk VII torpedo to make it capable of being dropped from fifty feet at 125 knots. Late in 1929, in pressing for the new, shorter torpedo (which became Mk 13), BuAer pointed out that it was now possible to build a 120-knot (138-mph) torpedo bomber that might well fly in a 25-knot following wind, so torpedoes should be droppable at 145 knots.[73]

By March 1938 Newport was experimenting with "boards" (auxiliary wings) attached to torpedo tails. In the air, the boards could limit or eliminate oscillation and thus make it possible to drop torpedoes from greater altitudes. They would break away on impact with the water. The earliest ones were conceived for patrol planes, which carried their torpedoes underwing. For carrier torpedo bombers, which carried their weapons internally or semi-internally, BuOrd developed torpedo stabilizers with wider-chord (that is, the distance from the leading edge to the trailing edge) but much shorter biplane wings.[74] With their vertical structural members, they became the standard "box tails" of wartime torpedoes. The plywood stabilizers broke away when torpedoes hit the water. The shorter-span stabilizers were being tested by April 1941. Torpedo air stabilizers were finally approved in August 1942.

That was too late for the TBD, which experienced serious problems when dropping torpedoes.[75] Because the TBD dropped it nose down, its torpedo oscillated in flight and could not be dropped at a speed over 115 knots or at an altitude other than between 75 and 150 feet. Battle experience with Mod (Modification) 2 in the Solomons showed that the torpedo could be dropped at 240 knots or more, but there were some duds, which BuOrd thought might have been caused by damage to the exploder when it hit the water.[76]

To make matters worse, pilots could rarely be sure of their attack altitudes.[77] Their barometric altimeters were affected by variation in air pressure from place to place. Pilots found it nearly impossible to estimate altitude over a featureless sea: one pilot mistook his altitude by four hundred feet, dove into the sea, and was killed. Pilots generally dropped high, the question being how high a drop BuOrd should plan for. A January 1941 BuOrd conference concluded that torpedo launching height should be a minimum of two hundred feet, three hundred feet if possible. It turned out that strengthening the torpedo to withstand a 250-foot drop would cost only about fifty pounds in weight. The new torpedo bombers (Avenger and Seawolf) were likely to be capable of 240 mph when loaded, so future torpedoes should be launchable at that speed.

For higher drops, Mk 13 was given strengthened propellers and radial tail braces and a stronger warhead section. In this form it turned out to be more rugged even than expected. As of early February 1942, Army A-20C light bombers were managing drops at 175 mph from two hundred feet using torpedoes with heavy warheads. At the time it seemed that the limit would be 200 mph, which would give a safe maximum of about 150 knots—better than the British or, it seemed, the Japanese could show. The altitude could be anything from fifty to three hundred feet. In April 1942 the strengthened Mk 13 was considered strong enough to withstand impacts when dropped from two hundred feet at 200 knots, but only when it was dropped nearly horizontally.[78]

By July 1942 Mk 13s could be dropped at two hundred knots and 275 feet. The minimum and maximum altitude limits for dropping at two hundred knots were soon set at 235 and 355 feet. A torpedo dropped at 275 feet would run in the direction of flight about 470 yards in the air. It would not fully arm before it had run another 420 yards in the water, so it would have to be dropped at least 890 yards from a static target. If the attack were ideal and the target was converging on the torpedo, dropping range would have to be about 1,200 yards. By this time combat reports showed that pilots were sometimes dropping at much shorter ranges and that in low visibility they might even be able to drop at point-blank range (600 yards). Many pilots thought that at such ranges they could not miss. BuAer therefore wanted torpedoes to arm much more quickly.

In December 1942 BuAer asked for higher release speed, which would greatly increase the average overall speed of a torpedo (since it would travel for much more of its path in the air). A reinforced Mk 13 Mod 2 dropped at 200 knots at 200 feet would make an average of forty knots if dropped at 1,500 yards. If the torpedo could be dropped at 350 knots from 275 feet at 1,500 yards, its average speed would be about 120 knots and its chance

of hitting a fast target would be much greater. In theory, a capital ship making thirty knots could turn ninety degrees in forty-five seconds, so decreasing torpedo running time below that would make hits far more certain. Although current torpedo bombers could not drop torpedoes at 350 knots, the Helldiver currently in service could drop at well over 300. The SB2D, due in service in April 1943, offered a similar speed. The XBTC (October 1944) would considerably exceed 350 knots, and the TB2D (December 1944) would do substantially better than 325. In fact, apart from the Helldiver, the envisaged faster torpedo bombers never entered service, although by the end of the war new single- seat attack bombers, such as the Mauler and the Skyraider, certainly reached the envisaged speeds.

Unfortunately, it seemed that the Mk 13 was nearly at the limit of possible release speed: local failures due to deceleration on impact were becoming more and more common as release speed increased beyond two hundred knots. Could BuOrd begin work on a new higher-speed torpedo? Some of the Army torpedo bombers could already exceed three hundred knots in gliding approaches.

By this time there was another source of weapon research, the National Defense Research Committee (NDRC), which coordinated wartime university-laboratory work. In the latter part of 1942 the Navy asked it to investigate whether aircraft torpedoes could be slowed using rockets. The project was assigned to CalTech. This was a low-priority effort; only in April 1943 did CalTech report, to the effect that a torpedo could be decelerated by about 120 knots using twelve aircraft rockets. Nothing came of this idea, but it was the beginning of what proved a fruitful resource for torpedo work.

In July 1943, BuOrd formally asked NDRC to develop a new aircraft torpedo that could be dropped at speeds of up to 350 knots and altitudes between six hundred and a thousand feet. The launching mechanism CalTech had built for the retrorocket tests would be used. CalTech began by attacking the "hooking" problem—that after hitting the water torpedoes might move as much as two hundred yards to one side before steadying on course, on a track offset from that originally intended. That effect turned out to result from the torpedo's tendency to tumble while decelerating in the water; it took some time for the torpedo's control mechanism to take charge. Caltech found that Mk 13 torpedoes were highly unstable in the water, controllable only within a very narrow range of angles of water entry. It turned out that an airfoil-section ring greatly improved control and cured, among other things, hooking. This device was immediately applied to Mk 13 torpedoes, mounted inside the box stabilizer; it was very successful. At San Diego, seventy-eight torpedoes were dropped at speeds up to 300 knots from 1,100 feet. Only thirteen hooked, nine broached, and five did both. Hooks over 25 feet were rare. The torpedo did tend to dive deeply before reaching its set depth. The rings were first used in combat in August 1944.

At about the same time, the Torpedo Station produced for BuOrd a preliminary solution to the high-speed-drop problem, a wooden drag ring (the "pickle barrel") surrounding the nose of the torpedo.[79] Combined with the box tail, it helped dampen aerodynamic oscillations as the torpedo fell through the air.[80] Because the stabilizer alone gave little damping, before the drag ring appeared airplanes had used different stabilizers for different speed ranges. Developers also discovered, to their surprise, that the pickle barrel cushioned impact with the water. When Torpedo Station Newport gave torpedoes with both drag rings and stabilizers to a training group, it found about a 40 percent reduction in measured deceleration when the torpedo had a drag ring. Without the ring, torpedoes suffered minor damage when dropped at 225 knots, severe damage at 270, and catastrophic

(including extensive internal) damage at 300. By mid-1944 BuOrd considered drag rings mandatory for high-speed drops; without it, permissible dropping speed was seventy-five knots slower. Tests indicated 40 percent less shock on water impact with the drag ring installed. Approved in April 1944, the drag ring made it possible to drop torpedoes at up to three hundred knots and from any altitude that would result in a water-entry angle of between twenty and thirty-three degrees. A torpedo could be dropped at four hundred knots and 2,600 feet. According to a 12 May 1944 BuOrd memo, no other development then under consideration showed any promise of replacing the drag ring.

The higher and faster the torpedo was dropped, the more time it spent in the air, so that its average speed over the entire run was higher. With an average speed of 200 to 250 knots, the torpedo could be fired as if the target were dead in the water, with very little lead angle and hence no need for a director. At 250 knots, the torpedo would travel a thousand yards in only 7.1 seconds.

Torpedo bombers used the pickle barrel and the drag ring for the first time during the 16–17 February 1944 strike on Truk. They made eighteen hits out of thirty-seven drops at an average height of three hundred feet and an average speed of two hundred knots.[81] Complete information on the modifications went to the fleet in June 1944. Soon they were being made and installed both by CalTech and by Naval Air Station San Diego at the rate of two hundred a week for the first 1,600. Naval Torpedo Station Keyport, in Washington state, was installing them on a large scale. No additional pilot training was needed.

It took two years to cure the problems plaguing the Mk 13 aerial torpedo. Torpedoes being readied on board the carrier *Bennington* in March 1945 off Okinawa show two of the three solutions: the wooden box tail and the "pickle barrel" (drag ring) around the nose. The circular shroud ring is hidden by the box tail. NHHC

Soon modified torpedoes were being dropped at up to three hundred knots and from 1,600 feet, although lower figures were recommended.[82] Performance continued to improve, so that in September 1945 BuOrd reported that of 103 torpedoes launched at between 100 and 1,000 feet (mostly at 750) and at 100–290 knots (mostly 250), running 1,000 to 3,500 yards (mostly 1,500), and set for between six and twenty feet, 42 percent made definite hits. Another 5 percent made possible hits, and there were 2 percent duds. These torpedoes had accounted for one battleship (presumably *Yamato*), one light cruiser (*Yahagi*), and two destroyers sunk, plus three light cruisers, six destroyers, and two escort destroyers damaged. Mk 13s had also sunk the other Japanese superbattleship, *Musashi*.[83] During the whole war, 2,487 Mk 13s were launched (presumably including those by PT boats) for 514 hits.[84]

CalTech's new torpedo was Mk 25, which became available in 1946. Only twenty-five were made, because so many Mk 13s were still available.

Horizontal Bombing

The *B* in "TBD" indicated precision horizontal bombing, which the interwar Navy considered a viable alternative to torpedo bombing. By the late 1930s, the Navy believed that pattern bombing from high altitude, which it analogized to salvo firing from guns, could hit moving ships. The Navy's belief in pattern bombing arose from the supposed precision of the Norden (Mk XV) bombsight BuOrd had developed.[85] Assuming that precision, hitting probability could be calculated; such calculations appear in prewar tactical handbooks.[86] Wartime reality was rather different: horizontal precision bombing utterly failed to match its advertising.[87]

Before the war, given problems with torpedoes, the Navy seems to have considered the horizontal-bombing role of the TBD and its TBF successor more important than torpedo attack. When Admiral Halsey wrote that recent experience showed how effective torpedoes could be, a BuAer officer wrote on the route sheet that the British had been forced to rely on torpedoes because they lacked the sort of bombsight BuAer had. Moreover, the British often fought under such low-ceiling conditions that high-altitude bombing was impossible. The assumed threat of high-altitude bombing led prewar officers to argue that they had to have fighter defense.[88]

Owing to the priority accorded horizontal bombing in prewar thinking, the Devastator and its successor the Avenger were both equipped with Norden bombsights, and provision for these sights affected their designs. Their bombardier needed a window on the bottom of the airplane through which he could see the target as the airplane approached; in the Avenger, for example, the window was in the after wall of the bomb bay. Norden's key innovation had been a gyrostabilized platform that enabled the bombardier to measure airplane motion over the ground, both toward the target and across the line of sight (drift due to wind). The bombardier could compare speed over the ground (as observed) with air speed to calculate wind along the direction of flight. A "Pilot Director" (later, an autopilot) caused the pilot to fly a steady course at a steady speed. The SBAE (Stabilized Bombing Approach Equipment) was a related device, an autopilot to guide the airplane toward a target.

War

After the United States entered the war, Devastators made a good showing (as level bombers) in carrier raids on the Mandated Islands in February 1942 and (as torpedo bombers) at the Coral Sea in May. In the latter, the *Lexington* air group made a perfect coordinated attack on the small Japanese carrier *Shoho*, one squadron of SBDs dive-bombing while another, with Dauntlesses, coordinated with it. Wildcats provided effective top cover against the six Japanese fighters present. The *Lexington* torpedo bombers made a classic split, or "anvil," attack. Of nine torpedoes dropped, at least five hit. A second attack claimed at least another two hits. In all, twenty-two torpedoes were launched. However, in the subsequent attack on the larger carrier *Shokaku*, nine torpedoes dropped by Devastators in line abreast all failed to hit. Some later attacks also failed. None of the Devastators was shot down, and several were rugged enough to survive battle damage. The failure against *Shokaku* might be attributed to the lack of coordination with dive-bombers; the post-action

TBDs of VT-6 line up for takeoff on the carrier *Enterprise* on the morning of Midway. Few would return. Note the unfolding wings; the Devastator was the first U.S. Navy monoplane with folding wings. The main problem at Midway seems to have been not poor performance by the TBD but poor coordination between fighters, dive-bombers, and torpedo bombers. All prewar Navy handbooks on aerial tactics had held that without such coordination torpedo attacks would fail. NHHC

report emphasized the value of such tactics. There was some evidence that defending Japanese fighters had concentrated on the dive-bombers, allowing the low-flying torpedo bombers to get through. The post-action report also pointed to the numerous hits needed to sink even the small carrier *Shoho*: aircraft torpedoes needed larger warheads, and the standard thousand-pound bomb was ineffective. It is not clear to what extent this perception led to the adoption of the Mod 2 version of Mk 13.

In contrast to Coral Sea, Midway was disastrous for the Devastators. Of forty-three launched that day, only six made it back to their carriers, two being forced to ditch. Only four of fourteen from *Enterprise* made it back, one so badly damaged that it had to be pushed overboard. *Yorktown* lost ten of twelve Devastators. The entire *Hornet* squadron, VT-8, was wiped out. Admiral Nimitz, Commander in Chief U.S. Pacific Fleet (CinCPacFlt), blamed the losses on the requirement that the Devastators slow to no more than 125 knots to launch their torpedoes, making them vulnerable to Japanese fighters. Future torpedo bombers had to be able to launch at higher speeds.[89] That development was already well under way.

It seems to have been far more important that, in contrast to the Coral Sea, the attacks at Midway were poorly coordinated. *Hornet*'s air group was sent on the wrong heading; its experienced torpedo-bomber leader, Lt. Cdr. John C. Waldron, broke off and headed in what turned out to be the right direction. His inexperienced VT-8 attacked alone. Without any distraction, the twenty-seven Zeroes over the Japanese carriers massacred it, only one man surviving. VT-6 (*Enterprise*) also made an attack uncoordinated with dive-bombers, leaving the Japanese fighters free to concentrate on it. Communication with the carrier's fighters (VF-6) apparently failed. *Yorktown* (VT-3) did manage something approaching a coordinated strike. In its case, by concentrating on the torpedo bombers the Zeroes missed the dive-bombers, which made effective hits on three Japanese carriers.

Midway also demonstrated that high-altitude bombing of ships was pointless. Although the bombers were Army, they had the same Norden sights as Navy bombers.

The greatest of all dive-bombers above the greatest U.S. carrier of World War II: a Dauntless (SBD-5) above the carrier *Enterprise*, March 1944. The hook is down; the bomber is turning into the landing circuit of its carrier. The vertical strut under the wing supports one of two antennas of an ASB radar. Note the channel for one of the two forward-firing guns and the underwing bomb rack. Note too that there is no telescopic sight; the dive-bombing sight was a fighter-type reflector inside the windshield.

4

DIVE-BOMBERS

As early as 1914 pilots discovered that they could hit small targets by diving at them and releasing bombs before pulling out.[1] U.S. Army pilots attached to British squadrons during World War I picked up the technique. Marine pilots supporting the counterinsurgency campaign in Haiti in 1919 also dive-bombed. Marine pilot Maj. (later lieutenant general) Ross E. "Rusty" Rowell adopted the technique, which he learned while attending the Army's Advanced Flying School at Kelly Field near San Antonio, Texas, when he took over VO-1M in the summer of 1924.[2] Rowell's squadron demonstrated the new tactic publicly on the West Coast in the spring of 1925. Dive-bombing would clearly be useful against the small targets the Marines faced in counterinsurgency. Rowell later said that he immediately realized that dive-bombing was also the natural way to hit moving ships. That October Capt. Joseph Reeves became Commander Aircraft Squadrons Battle Fleet. Among the "thousand and one" questions he posed for his pilots was how to repel an enemy force attempting to land on a beach. The pilots of VF-2 discovered that dive-bombing at a steep angle offered both the element of surprise and great accuracy. It was probably significant that VF-2 trained alongside VO-1M during the summer of 1926, both being part of the Battle Fleet aircraft command.

On 22 October 1926, Lt. Cdr. Frank D. Wagner led VF-2 in a simulated attack against a battle fleet emerging from San Pedro.[3] Although the fleet had been warned, he achieved complete surprise, diving from 12,000 feet. All on board the ships agreed that no existing defense could have stopped the attack. The CNO, Adm. Edward W. Eberle, ordered this "light bombing" included in the FY27 gunnery exercises, beginning in the late fall of 1926. Six squadrons participated: three observation squadrons (VO-1, -2, and -4) flying UO-1 floatplanes and three fighter squadrons (VF-1M, VF-2, and VF-5) flying F6C-3s and FB-5s.[4] Rules called for dives from one thousand feet and pullouts at four hundred.[5] The three observation squadrons managed 44.5 percent direct hits on a relatively small static target, two hundred by forty-five feet. The two fighter squadrons made 67 percent hits, diving from 2,500 feet. Some pilots found that their targets disappeared from view as they dove, so Wagner shifted to steeper dives (forty-five to seventy-five degrees) starting much higher (ten thousand feet), the speed gained being about the same. At about the same time, the Scouting Fleet fighter squadron on the East Coast (VF-5) also began dive-bombing; on 4 May it demonstrated seventy-degree dives over Smithfield, Virginia. BuAer considered the new form of attack an extension of the strafing with which fighters were already expected to neutralize ships' antiaircraft batteries. It ordered all fighters and observation planes equipped with two underwing racks to carry five 25-pound fragmentation bombs. Alternatively, each rack could carry a 100-pound bomb (later 116 pounds) intended specifically to disable a carrier's flight deck.

Dawn: the first U.S. Navy dive-bomber, an F6C-3 fighter. It was adapted to glide or dive-bombing by virtue of being strong enough to pull out of a dive. It dropped light bombs from wing racks, not visible here. Note the drop tank abaft the radiator. This photograph was taken in 1932, when the airplane was assigned to utility squadron VJ-4; the pilot has no gunsight and no guns. NATIONAL ARCHIVES

When the outline of a cruiser 400 by 50 feet was drawn over plots of bomb hits and misses, it seemed that 44.6 percent hits would have been made from 500 feet, 38.3 from 1,000 feet, and 45 percent from 1,500 feet. With the target zigzagging, 15 percent hits were made from 1,000 feet. From 3,000 feet, 30 percent hits were made. Commander Air Squadrons Scouting Fleet recommended development of a high-performance dive-bomber to work at low and intermediate altitudes. He expected it to be effective against cargo and troop ships and landing operations.

To test the new technique, the destroyers *Putnam* and *Breck* were ordered to tow targets at high speed (for which the ships were converted specifically) for the F6C-3 fighters of VF-5S (Scouting Force) to dive-bomb.[6] Bombing seemed most effective from one thousand feet; bombing at five hundred feet created a "mental hazard" for the bomber, who would have only a short distance left in which to pull out. Some attackers pulled out before releasing their bombs, causing them to miss. The steeper the dive, the more accurate. It was best to drop bombs either up or down the apparent wind with respect to the ship, that created by its speed and course. If the plane was turning when it released, the bomb was thrown off, because the sideways motion of the airplane was imparted to the bomb. Airplanes diving across the apparent wind had to change dive direction to correct for range errors. Airplanes attacking out of the sun from directly overhead were particularly difficult to counter. Zigzagging did not seem to help the target.

All fighter pilots were to be trained in the new tactics, greatly increasing the carrier's offensive capacity without requiring new airplanes. However, Admiral Moffett wanted the technique expanded to create a new type of ship-killer, one that would deliver a five-hundred-pound bomb. In June 1928, after issuance of a call for a specialized dive-bomber,

BuAer hoped that single-seat fighters could be adapted to dive-bombing, giving a carrier an all-attack air group. This F4B-3 shows a rack for a five-hundred-pound bomb between its wheels. The *B* in the circle indicates a Marine bombing squadron, VMB-4. NATIONAL ARCHIVES

he wrote that a special single-seater to be delivered within a month, almost certainly Boeing's F4B-1, was being considered to carry two 100-pound bombs or one of 500 pounds. It was delivered to Anacostia in August 1928, slightly later than Moffett had expected.

By this time BuAer was interested in two-seat fighters, whose more extreme advocates thought they might be better than single-seaters.[7] During the summer of 1928 BuAer decided to compare single- and two-seat fighters operationally, buying some under the 1929 program. As noted in the previous chapter, the obvious two-seat candidate was the Curtiss Falcon.[8] A conference attended by commanders of aircraft squadrons of the Battle Fleet and of the Scouting Fleet and the chief of BuAer decided to buy twenty-seven two-seaters (in fact only twenty-five were bought). This was a somewhat controversial decision; immediately afterward, Commander Aircraft Squadrons Battle Fleet reversed himself and asked for fifty-four single-seaters and no two-seaters at all (BuAer and CinC Battle Fleet continued to want two-seaters).

In mid-1928 BuAer was negotiating with Curtiss to redesign its two-seat F8C-1 fighter to dive-bomb with a five-hundred-pound bomb. That required a much stronger structure, including a strengthened tail. Unlike the F8C-1, the new airplane had equal-span wings, and it incorporated some of the design details of Curtiss' F7C shipboard single-seat fighter. This F8C-2 was expected to exceed 150 mph when carrying a single 500-pound bomb or three 100-pound bombs or ten 30-pound bombs.[9] The loss of the prototype on 3 December 1928 during a dive at Anacostia prompted a conference with BuAer that laid out requirements for the production version and also for a corresponding Marine Corps observation airplane designated O2C-1.[10] The new prototype was designated the F8C-4 Helldiver. It had a greater "strength factor" (twelve rather than ten) and a cutout at the base of the rudder for its oleo tail skid. Thirty-six were ordered, twenty-seven to equip a squadron (VF-1B on board *Saratoga*) and nine as observation planes for the Marines. The air-plane gained fame when it appeared in the 1931 movie *Helldivers*, starring Wallace Beery and Clark Gable. This movie is sometimes credited with having inspired foreign interest in dive-bombing.

BuAer also planned to fit the new high-performance O2U Corsair observation plane to carry a five-hundred-pound bomb when it operated on wheels from carriers.[11] In 1930 CinC Battle Fleet pointed out that although both the F4B and the O2U had been designed to dive-bomb with five-hundred-pound bombs, neither had ever dive-bombed with such weapons or seemed to be strong enough to do so. BuAer added dive-bombing (with a five-hundred-pound bomb) to standard fighter requirements. BuAer let a contract to Vought for its XF2U, which

BuAer saw Curtiss' two-seat Falcon fighter as a viable basis for a two-seat fighter/dive-bomber. This is the first production F8C-4 Helldiver. Although designated a fighter, it was strengthened as a dive-bomber. Curtiss called it and all its later dive-bombers "Helldivers" rather than using the raptor names it used for fighters. The panels alongside the pilot covered cheek fuel tanks. They were introduced in the Vought O2U and had been incorporated in Curtiss' XF7C-1. Note the cutout, introduced in this model, at the base of the rudder to accommodate a new oleo tail skid. Although not initially fitted with cowlings around their cylinders, many Helldivers later had them. VF-1B (*Saratoga*) had eighteen of these aircraft in 1931, presumably as a test, but in 1932 they were replaced by nineteen F4B-3s and an SU-1. U.S. NAVAL INSTITUTE PHOTO ARCHIVE

embodied many features of the O2U series. It never entered production, apparently because it offered too little advantage over the Helldiver.[12]

Six months of F8C-4 operation convinced pilots that the two-seater had great possibilities both offensively and defensively, but something with greater performance was needed.[13] The fleet commander pointed out that the F8C-4 would not be useful as a scout either, because of its poor carrier-landing qualities, its high landing speed, long takeoff run, and comparatively slow climb rate. The long takeoff run was a particular problem, because carriers had to spot them forward of the rest of the aircraft on the flight deck.

BuAer was interested in an even more powerful weapon, a special bomber that could dive with a thousand-pound bomb. Its speed would be about 150 mph, "with characteristics which will compare favorably with the single-seat fighter, although this machine being somewhat larger must sacrifice some maneuverability." A BuAer design was developed for production under the designations "T2N" and "T5M" (Martin 77), the difference being their engines (respectively, R-1750 and R-1690).[14] The new design had an important indirect effect on American aerial-torpedo development.

All of this turned out to be premature. Bombs sometimes hit the releasing aircraft's own landing gear or propeller. A steep enough dive could avoid the problem, but the average pilot miscalculated his diving angle by fifteen to thirty degrees. For the moment, big centerline bombs were abandoned, and fighter specifications showed only wing racks for much smaller weapons. Thus F8C-4s of VF-1B normally carried five 30-pound bombs or two 100-pound bombs. They were hardly the dive-bomber-fighters envisaged in 1928.

The T5M first flew in March 1930. Its tests convinced BuAer that efficient heavy dive-bombers could be built. They would be pointless, however, until the problem of bombs hitting parts of airplanes could be solved. BuAer developed a device that swung the bomb clear before it was released: a trapeze ("displacing gear"). The trapeze was successfully tested on a T5M on 7 and 8 January 1931. A new heavy dive-bomber designation, VB, was created. The T5M became the BM-1 and was given strengthened wings. BuAer described this specialized dive-bomber as a response to fleet recommendations. The trapeze featured in all U.S. dive-bombers employed in World War II.

With displacement gear demonstrated, a service test became urgent. To maintain a full-strength (nine-airplane) squadron for a year required a total of twelve airplanes. To pay for the test, one of the two torpedo squadrons was cut to nine aircraft. CNO

Despite its designation, Martin's XT5M-1, shown on 22 May 1929, was a heavy dive-bomber, designed to carry the thousand-pound bomb shown. The project faltered as it became clear that a bomb might hit part of the airplane on the way down. The XT5M itself was badly damaged in a terminal-velocity dive on 15 October 1929. Its lower wing caved in. NATIONAL ARCHIVES

The solution to the dive-bombing problem: displacement gear, a trapeze or crutch that pulled a bomb clear of an airplane before releasing it. Not visible here is the fulcrum at the fore end of the trapeze, connecting it to the airplane so that it could swing down. This thousand-pound bomb was on board the prototype Great Lakes XBG-1, 14 June 1933. NATIONAL ARCHIVES

Displacement gear was installed in existing fighters. This 116-pound bomb is under a Boeing F4B-4 fighter, 29 September 1932. NATIONAL ARCHIVES

With the trapeze, the T5M became the BM dive-bomber. This BM-1 carries a drop tank instead of a bomb, but its displacement gear extending aft from the engine is visible. Despite its pure bomber role it was assigned to a torpedo squadron, VT-5 (it is the squadron commander's airplane). U.S. NAVAL INSTITUTE PHOTO ARCHIVE

Many aircraft were given low-drag modifications: ring cowlings and sometimes spats over their wheels. This new BM-1, photographed on 30 September 1931, was from the last production batch of these airplanes. NATIONAL ARCHIVES

and then the Assistant Secretary of the Navy for Aeronautics almost immediately approved; twelve BM-1s (Martin 125) were ordered on 9 April 1931. Another sixteen were ordered in October (FY32) as BM-2s (Martin 129s) and finally five more BM-1s were ordered from Martin in 1932. The heavy dive-bomber was so important that work on further dive-bombers began before the BM-1 entered service.

Displacement gear aside, many airplanes were not sufficiently strong to withstand the stresses of dive-bombing. In November 1929 BuAer circulated a cautionary memo describing the failure of the upper wing of an F3B-1 fighter resulting from a very short pullout from a two-thousand-foot, power-off dive (the pilot had also put the plane into a right turn so that he could see the impact of his bomb). The entire leading edge of the wing collapsed, about half the top fabric and ribs were torn off, and the front spar fractured. Careful examination showed that none of this had been caused by poor maintenance or workmanship.[15]

Carrier Scouts

Scouting was vital both to the fleet as a whole and to carriers, which had to find and hit enemy carriers before they could be hit themselves. In his FY30 annual report, CinC U.S. Fleet reported that the existing O2U was not nearly fast enough, nor did it have sufficient endurance.[16] Carrier scouting was soon tied into dive-bomber development. BuAer argued that the new BM-2 heavy dive-bomber would be good enough for tactical scouting, out to two hundred miles.

The need for something with much greater range seems to have been understood in the fleet. As naval aircraft inspector at Boeing in April 1931, Lt. Cdr. G. S. Gillespie had been discussing the problem with Boeing engineers for two years.[17] Boeing was then working on long-range monoplanes. Gillespie grasped the issue, having commanded the scouting wing of the Battle Fleet during Fleet Problems 9, 10, and 11; its CinC had considered his scouting work excellent. Gillespie had invariably located the enemy despite considerable personal risk. Gillespie envisaged a two- or three-seater, since the pilot should

not occupy himself with navigation, and radio operation (including decoding) would be a nearly full-time job. Boeing proposed its Model 233, a fast low-wing monoplane with retractable landing gear.

Gillespie sent the Boeing sketch to Cdr. C. A. Pownall at BuAer, who also was interested in scouting. Pownall pointed out its excessive span and supercharged engine, which would make for a sluggish takeoff. However, he found the proposed design promising enough to be worth building. Pownall expected an attack group to be accompanied by two scouting (navigating) airplanes, which would orbit at a set rendezvous and lead the strike back to the ship. Expert navigation by those on board the scouts would extend strike range, as "you would then have assurance that the squadrons would not get lost no matter how far you sent them." The scout should be comparatively fast. The cover sheet of the letter shows that a more senior officer (identity not clear from his initials) rejected the low-wing monoplane and argued that carrier scouts should *not* be used for long-range scouting.

The BuAer Plans Division laid out general requirements for a two-man carrier tactical scout in a 22 September 1931 memo.[18] The date suggests that it was prompted by the need to equip the new carrier *Ranger*, although that is not indicated in surviving papers. Scouting required excellent navigation by the pilot. To perform that, he sometimes had to relinquish control. With autopilots not yet in existence, the crewman in the back seat (who, in a fighter, was only a gunner) would sometimes have to take over and so needed another set of controls. The scout also required a longer-range radio, sometimes described as a "four-hundred-mile set," which would be operated by the back-seater. Bombs could be limited to a single 116-pound or two 30-pound bombs under each wing; no trapeze was envisaged. The plane could be powered by either a liquid-cooled V-1570 or a two-row radial. Maximum speed would be the highest possible. Range would be six hundred miles at 80 percent maximum speed. Since the scouts would take off ahead of the rest of the carrier's aircraft, takeoff run had to be limited to two hundred feet in a 30 mph wind. Strength had to be sufficient to pull out of a sixty-degree dive with three-quarters fuel—this was not quite dive-bombing as exercised in the fleet. Span should be no more than forty-two feet, a typical figure. These requirements defined the changes to a two-seat fighter required to turn it into a scout. BuAer's Design 110 (for a heavy dive-bomber powered by an R-1830) fitted with an auxiliary forty-gallon tank as a scout offered much of what BuAer wanted. In this form its maximum speed was 176 mph, which was not considered fast enough. Range at 90 mph was 880 miles.

BuAer Plans Division also wanted a much longer-range (1,500 miles) strategic scout, which would not carry bombs. Given the very long range, hence long flying time, cockpits should be enclosed. As with the tactical scout, the desired engine was either the V-1570 or a two-row radial. Extreme range required a long, narrow wing, so Plans accepted a forty-five-foot span unless the wings could fold. BuAer preliminary designers developed a design using a 700-hp R-1830 engine, with a wingspan of either fifty feet three inches or fifty-one feet seven inches. Maximum speed at sea level would be either 178.5 or 175.6 mph, very respectable figures for the time. Range at 120 mph would be either 1,205 or 1,304 miles, well short of Plans' requirement; maximum range would be either 1,405 or 1,514 miles. Bellanca,

not normally associated with naval aircraft, offered a design that BuAer accepted: the XSE-1, with a 600-hp R-1690-C engine and a lighter gross weight than that of the BuAer design (5,509 rather than 6,050 or 6,320 pounds). The span of its single high wing was forty-nine feet nine inches, and maximum speed at sea level was 171.4 mph. Service ceiling was 19,200 feet. Maximum range was 1,375 miles. On the cover sheet Pownall wrote that the 600-hp, two-row radial with a NACA drag ring should be considered preferable to the liquid-cooled V-1570, "which is generally passing out of the picture." The single XSE-1, delivered in 1932, crashed, and nothing came of a modified SE-2, which was delivered with the same serial number. Presumably the long-range scouting role was taken by flying boats, which had the vital advantage of not being limited by treaty. Anything that could allow a treaty-limited carrier to concentrate on offensive power was to be welcomed.

In January 1932, the assistant bureau chief squelched the idea of a new scout. Sometime in 1931, BuAer ordered Vought to build an up-engined version of its Corsair (600-hp R-1690 Hornet instead of 450-hp R-1340 Wasp) as the O3U-2. It offered a maximum speed of 166 mph and a ceiling of 20,000 feet. It seemed good enough to become a carrier scout. Fourteen were ordered, to equip a

The strategic scout: Bellanca's XSE-1, 15 December 1932. The modified XSE-2 was still being tested in 1934. This mission was transferred to long-range flying boats. NATIONAL ARCHIVES

This SU-1 from the Anacostia test center was photographed on 13 December 1934. The underwing object was a flare holder, which kept a lit flare (for night photography) away from the fabric-covered wing. Attached to the wing struts is an aerograph, which automatically measured atmospheric conditions. As bombers, SUs had wing racks but not displacement gear. NATIONAL ARCHIVES

This SU-4 was flown by VS-1B. The panels alongside the cockpits covered cheek fuel tanks characteristic of the Vought biplane Corsairs and of their UO predecessor. They saved internal space. W. T. LARKINS VIA MALD COLLECTION VIA *THE HOOK*

carrier scouting squadron. A second order, for fifteen, was intended to provide spares and liaison aircraft for the four fighter squadrons of the Battle Fleet. By September 1931 another forty were needed to provide full complements of scouts for the two big carriers. The Marines adopted the O3U-2 for the twenty-five VO planes they wanted by 1 July 1933 (they had to choose a standard Navy model). Only the first twenty-nine O3U-2s were built as such, and all were redesignated SU-1s. The carrier SU-1s were flown by Marines of squadrons VS-14M and VS-15M.

The remaining Hornet-engine Corsairs became O3U-4s. All but one of the sixty-five built were redesignated as scouts, SU-2 and SU-3. Another forty were built as SU-4s. SU-2 and -3 could be distinguished by their vertical tails, with vertical trailing edges. The SU-4 prototype was a modified O3U-4 Corsair with a "greenhouse" (long cockpit canopy) and a long-chord engine cowl (as in the abortive XF2U-1 fighter). However, production aircraft reverted to the short Townend cowling and open cockpits of the SU-1 series. Some were provided with canopies (designed for the Marines' O3U-6) late in their service lives, generally after they had been relegated to utility and staff roles or in Marine Corps service. SU-4s replaced SU-1s in the scouting unit on board *Lexington*. Marines flew those assigned to USS *Saratoga* between 1932 and 1934. The scout pilots were the only Marine aviators assigned to carriers before 1944.

It was still worth developing requirements for scouts, in case new ones were wanted. Takeoff distance should be no more than 150 feet in a 30 mph wind, and range no less than 600 miles (maximum 800). Maximum speed should be at least 175 mph. Ceiling should be 20,000 feet if possible, and at least 15,000. Muffling the scout's engine should help it reach its objective unnoticed. The airplane need be stressed only for shallow and relatively low-speed dives. Something like Grumman's new FF-2 two-seat fighter might be suitable. At about this time BuAer tested the XF10C-1 two-seat fighter prototype as the XS3C-1 (it crashed in February 1932).[19]

The question arose again in the spring of 1933: fifty carrier scouts were required.[20] BuAer hoped to reengine Corsairs with a 600-hp R-1690-12 so that they could reach 190 mph. An SU-2 (O3U-4) modified as the prototype SU-4 was not nearly fast enough: it was rated at only 167 mph at sea level. The only fast two-seater even nearly ready for production was the Grumman FF-1, but it had not yet been through service tests. The question was revisited in October, when BuAer needed thirty-four

Grumman's SF was externally indistinguishable from its FF fighter. The XSF-2 is shown on 14 March 1935. NATIONAL ARCHIVES

Curtiss built the S3C-1 as the private venture "Cyclone Helldiver," named for its R-1820 Cyclone (rather than the 450-hp R-1340-88 of the Helldiver). The Navy bought it as the XF8C-8. It and two duplicates became O2C-2s. With the later R-1820-E (620 hp) in a new airframe, the second airplane became the XF10C-1 and then the XS3C-1 shown here. It achieved 180 mph but crashed in February 1932. Despite its scout designation, the S3C could dive-bomb. NATIONAL ARCHIVES

scouts for the new carrier *Ranger*.[21] By this time Grumman's SF had completed extensive trials. It was the only choice, the alternatives being the SU-4 and two new two-seat fighters that had not yet completed trials (XF3U-1 and XFD-1). BuAer ordered thirty-three SF-1 scout versions of the Grumman FF-1, deliveries beginning in March 1934.

Production Dive-Bombers for *Ranger*

In the summer of 1931 BuAer had to develop new aircraft for *Ranger*. Prototypes would be bought the following year (the ship would commission in 1934).[22] Plans Division argued against single-seat dive-bombers. Radiomen were needed to receive orders and bearings by radio. Also, because they would face determined opposition, they needed rear gunners. It appeared that a two-seater carrying one 1,000-pound bomb could reach medium ranges (400-mile endurance), or 750 miles with a 500-pounder, or 1,000 miles as a scout without bombs (but with a belly tank). Without a belly tank, it could reach 750 miles as a scout, and stalling speed would be considerably reduced. The BuAer sketch designers produced Design 110, which was powered by an R-1830 engine.

The assistant chief called a 24 August 1931 conference to discuss future dive-bombers, initially for *Ranger*. Opinion was almost evenly split as to whether design and procurement of a new two-seater should await BM-1 tests in the fall. However, the question had to be decided sooner, so that a next-generation airplane could be completed the next summer. Plans suggested a two-seater carrying a thousand-pound bomb over medium ranges (400 miles) or a five-hundred-pound bomb over long range (750 miles). It could also function as a 1,000-mile scout without bombs or with a belly tank, or as a 750-mile scout without bombs or belly tank.[23] Desired maximum speed was 175 mph, desired cruising range was 500 miles, and service ceiling should be 15,000 feet. Stall speed (without a bomb) was set at 63 mph. If possible, the landing gear would retract. The Propeller Division was directed to push development of a variable-pitch propeller. The Radio Division described a 250-mile radio (fifty-five pounds) for

the airplane. Frontal area should be minimized, to give the pilot a better view of the target. That made a two-row radial attractive: Pratt & Whitney was developing the R-1830. The single-row supercharged R-1690C was also of interest. Although BuAer was still financing liquid-cooled-engine development, such engines were not at all favored.

BuAer was very much interested in folding-wing monoplanes but feared that foldability would unduly weaken the wing of a dive-bomber. There was considerable interest in carrying the bomb internally. Without sufficient space for a trapeze, it would have to be ejected sideways. Nothing came of this idea.

As a yardstick, in October the Material Division produced Design 110. Span was 36 feet, length 28 feet 3⅝ inches, and height 10 feet 4½ inches, with a gross weight of 5,659 pounds (3,086 empty) as a 1,000-pound dive-bomber. The Material Division described it as the smallest possible design (for maximum performance), kept as aerodynamically "clean" as possible. Plans initially called for a Hornet (R-1690) engine, but in August 1931 BuAer had decided to use the newer R-1830 instead, to achieve a speed of 175 mph. With bomb and gear omitted, the same airplane could be a two-seat fighter (4,577 pounds gross/3,086 empty). As a scout, it would carry a drop tank in place of its bomb. As a dive-bomber, maximum speed at sea level would be 175.5 mph (178 at 6,000 feet), service ceiling would be 18,600 feet, and range at 90 mph would be 546 miles. As a fighter, maximum speed would be 180.4 mph at sea level and 183.5 mph at 6,000 feet. Performance as a scout would be roughly that of the dive-bomber. Estimated takeoff run into a 25-mph wind (with a 1,000-pound bomb) would be 251 feet (217 feet using a variable-pitch propeller). The sketch design was approved, and on 28 November 1931 BuAer circulated it as part of a request for proposals from aircraft manufacturers.

Single-Seat Dive-Bombers

Despite the argument that a back-seater was needed, the August 1931 meeting voted unanimously to proceed at once with the design of an advanced single-seat dive-bomber. The single-seater would carry one thousand-pound bomb (five-hundred-mile endurance). It might also serve as a two-seat scout (same five-hundred-mile endurance). This payload seems to have been too great for a fighter-bomber, and some time in 1932 the fighter-bomber payload was set at five hundred pounds.

The fleet had already asked for a dive-bomber version of its current Boeing F4B fighter. In June 1931 BuAer ordered the XF6B-1 (later designated XBFB-1).[24] Boeing involvement in Army projects delayed its delivery to February 1933.

Some time later in 1931 or very early in 1932 Curtiss offered a version of its long-running Hawk fighter series. It is not clear from surviving documentation whether or not they were considered alternatives to the Boeing project, but they were available much sooner. BuAer decided to order two

With retractable wheels and metal wings the BFC became the BF2C shown. It turned out that the natural frequency of the wings matched the frequency of engine vibration, with unfortunate results. The commander's BF2C-1 of VB-5B (USS *Ranger*) was photographed on 18 March 1935. The solid tail color (willow green) was associated with the squadron, but under the 1 July 1937 reorganization tail colors were associated instead with carriers. This practice of marking airplanes by carrier has continued in different forms since then. The color of the stripe around the fuselage and on the cowling, royal red, indicated the three-plane section, in this case the first, with a complete cowling indicating the first airplane in the section. This was also the color of the wing chevron. Standard practice was to paint the upper surfaces of the upper wings (in monoplanes, the wings) orange-yellow for visibility. Overall finish was light gray (aluminum dope over fabric surfaces). NATIONAL ARCHIVES

Experience with the F4B as a near-dive-bomber convinced BuAer that a fighter could be so adapted. Curtiss' BFC-1 started life as the F11C fighter. This is a BFC-2 of VB-3B from USS *Saratoga*, probably in 1937, when the squadron was renamed from VB-2B (it retained its "high hat" insignia) under a 1 July 1937 reorganization that aligned squadron numbers with ships' hull numbers. The tail color changed from red to white. The following year VB-3 received SB2Us.
U.S. NAVAL INSTITUTE PHOTO ARCHIVE

versions, which became the XF11C-1 and -2. In March the Navy bought Curtiss' Hawk II (the Goshawk, Model 64A) demonstrator, powered by a 700-hp R-1820. It had a welded-steel-tube engine mount on an aluminum alloy monocoque structure—that is, on a fuselage whose skin was load-bearing. In April 1932 it ordered Curtiss' F11C (Curtiss' Model 64), which could carry a five-hundred-pound bomb on its centerline.[25] Both aircraft had the single-strut undercarriage the company had offered on the XF6C-6 and the Army P-6E. Model 64 was powered by the new 600-hp R-1510 engine and had metal-covered tail surfaces. Model 64A, which became the XF11C-2, had a fabric-covered tail and longer undercarriage legs.

In January 1932, Design 110 (the two-seater) or its equivalent was one of the four highest-priority aircraft design and procurement items, the others being the XF11C-1 fighter, Grumman's VJ amphibian, and the XF11C-2 fighter, in that order.[26] A conference headed by the assistant chief of the bureau decided to buy the two XF11C prototypes and to hold informal design competitions for two-seat and single-seat fighters, one of the two-seaters to be powered by a liquid-cooled engine. Further efforts to produce the desired strategic scout would be stopped pending delivery of the Bellanca airplane. There was also interest in creating a single-seat dive-bomber

by removing the gunner and rear gun (and fairing over the rear cockpit) of the XF10C, of the XF2J two-seat fighter (derived from the Army's Y1P-16), and of various BM versions.

Looking toward the fighter-bomber competition, in January 1932 Plans circulated general requirements for carrier-based standard (i.e., single-seat) fighters that could carry two 116-pound bombs in wing racks in lieu of a belly tank. The airplane had to be strong enough to dive at terminal velocity and to pull out without diving more than 1,500 feet with bombs in place and with half fuel on board. The assistant chief wanted at least 200 mph. Preference for air-cooled engines should be explicit. The Armament Division wanted provision for a five-hundred-pound bomb on the centerline, with displacing gear, "to permit carrying out secondary mission after control of the air has been gained." This was accepted. It is probably why the BFC was modified (in the design stage) to carry a five-hundred-pound bomb. In June 1932, BuAer ordered two competitive single- seat fighter-dive-bombers: the Berliner-Joyce XF3J-1 and the Curtiss XF13C-1. The XF13C-1 was delivered as a high-wing monoplane, with the understanding that it could be converted into a biplane. Neither reached production. For FY33 the Navy bought F11C-2s alongside Grumman FF-1s.

The Curtiss light dive-bombers beat out Boeing's dive-bomber redesign of its F4B fighter, the F6B shown here. NATIONAL ARCHIVES

Displacement gear of the F6B, carrying a five-hundred-pound bomb, the standard light dive-bomber load. This Mk 33 bomb rack was photographed on 4 May 1933.
NATIONAL ARCHIVES

Meanwhile BuAer received reports that European air arms were testing very-high-performance fighters. It used the known performance of European aircraft as a reasonable proxy for the unknowable performance of Japanese aircraft.[27] To survive its planned passage through the Japanese-held Mandates, the U.S. fleet would have to defeat land-based Japanese aircraft. To do that it needed maximum-performance ("pure") fighters, aircraft designed solely for air-to-air combat and without any ability to dive-bomb.[28] The F11C was therefore designated a light dive-bomber (VFB: BFC-2 in March 1934, with a partial canopy). One became the XF11C-3 (Model 67), with metal-framed wings and manually retractable landing gear similar to what Grumman had used in the FF-1. Its production version became the BF2C-1. It was taken out of service within a year because of mechanical problems with its undercarriage and vibration in its metal-framed wings; the successful export version reverted to wood-framed wings.[29]

The shift to pure carrier fighters killed pure scouts. The carrier air group needed both scouts and light dive-bombers, which had to be combined in a single type, the dual-purpose scout-bomber—VSB.

Back to Heavy Dive-Bombers

On 28 November 1931 BuAer circulated Design 110 as the basis for a new heavy dive-bomber. (This was the last time BuAer circulated one of its sketch designs for a bomber.) Later an additional condition was added, by which the airplane carried a five-hundred-pound bomb and used additional internal tanks to add fifty more gallons of fuel, for greater range. After the bureau had circulated its design, it decided to substitute a geared Pratt & Whitney R-1535 or Wright R-1510 two-row radial (700 hp at sea level, 625 at 6,000 feet) for the originally specified direct-drive R-1830 (700 hp at sea level). On this basis estimated maximum speed was 178 mph at critical altitude. However, there was good reason to estimate that the geared R-1535 would deliver 700 hp at 9,000 feet, in which case speed at critical altitude would be 190.3 mph and service ceiling would be 23,800 rather than 19,700 feet. Design 110 incorporated a steel-tube (framework) fuselage, but some bidders proposed a monocoque structure instead. There was some difficulty over weight. BuAer expected the airplane to weigh 3,126 pounds empty (gross 5,657 pounds); Vought representatives said that it could not be built to weigh less than 3,343.

A BG-1 of VB-4B at Union Air Terminal, Burbank, Calif., 18 December 1937. Production amounted to sixty aircraft under three contracts, half going to the Marines. Production aircraft like this one had a greenhouse over the two cockpits. Navy BG-1s equipped VB-3B, VB-4B, VB-7 (on formation), and two Marine Corps squadrons. They were replaced by SBUs and SBCs beginning in 1937. This sixth-section airplane has a lemon-yellow fuselage band. Note that color is applied over only half the cowling. The squadron was assigned to USS *Ranger*, so the tail was green.
W. L. SWISHER VIA MALD COLLECTION, VIA *THE HOOK*

The XB2Y lost the competition to the BG-1. It is shown on 24 November 1934 with its bomb in displacement gear. NATIONAL ARCHIVES

In March 1932 BuAer rated as best Consolidated's designs, both its own alternative and its version of Design 110, closely followed by Great Lakes and then by Vought. The other, less successful bidders were Great Lakes (Design 110 with monocoque fuselage), Douglas (monocoque), and Curtiss (far behind the others). Martin could not be considered, because it quoted too high a price and did not quote a delivery time. Although Consolidated came first in both engineering and total merit, BuAer considered it undesirable to award that company both prototype contracts. Contracts were awarded to Consolidated for its alternative design and to Great Lakes for its version of BuAer's Design 110.

Despite the initial rating, Great Lakes won the production contract late in 1933. The production version had a closed greenhouse over both pilot and gunner. It had provision for a long-range radio, and as a scout it could carry additional fuel. Meanwhile the Marines formally requested a dive-bomber for expeditionary service.[30] The BG-1 met the Marines' requirements; half the aircraft went to them.

By this time dive-bombing was considered the best way to disable enemy carriers. Attackers had to be able to deliver bombs that could penetrate flight decks to burst in the hangars, where foreign navies typically stowed their aircraft. Disabling enemy carriers was so important that dive-bombers on scouting missions were authorized to attack on sighting them. A dive-bomber could be loaded with extra fuel (as a scout) but still carry a somewhat lighter bomb—by 1942, typically a 500-pounder (the standard dive-bomber weapon was the 1,000-pound bomb). In this form it could conduct armed reconnaissance.

Exercises showed that pilots should not dive vertically, because they might find it difficult to perceive skidding or "skipping"—away from the direction of the target. They had to avoid even slight turns, which might cause bombs in wing racks to hit their propellers. Instead, the ideal dive angle was seventy-five degrees, which would enable a pilot to compensate for wind and for the movement of the target ship. Pilots had to compensate for a time lag, a function of dive angle, release altitude, and speed of the airplane. The drift due to wind (true wind on a static target or relative wind on a moving one) would tend to counteract lag. Ideally, a number of bombers should approach the same target and from different directions, as nearly simultaneously as possible, maneuvering to confuse antiaircraft fire. They would pull out immediately after dropping their bombs. Tests showed that it was best for them to retire at maximum speed at low altitude in irregular flight, to frustrate antiaircraft fire. If supporting heavier bombers, the dive-bombers should attack first—ideally about three minutes before the torpedo bombers.

Prewar exercises showed that aircraft flying above 12,000 feet were rarely seen by unalerted surface ships. On a clear day against a bright blue sky, a division of light dive-bombers (fighter type) was seen at 13,000 feet, a division of heavy bombers at 15,000 feet. Maneuvering aircraft were often given away by the glint of the sun on dipping wings. Prewar doctrine had aircraft approaching in a loose formation at an altitude of at least 10,000 feet, ideally from out of the sun. World War II tactics typically called for dive-bombers to cruise at 18,000 feet.[31] The attack was made from down-sun and upwind, pushing over at 15,000 feet, beginning with the airplane on top of a "stack" of bombers. Pilots followed the peel-off direction of the leader, rolling into a seventy-degree dive. Bombs were released between 1,500 and 2,500 feet.

BuAer found that high diving speed seemed to degrade dive-bombing accuracy. It limited diving speed to 250 mph in the 1934 dive-bombing competition. That was why the BT/SBD and the SB2U both had dive brakes, which were not in any previous, slower-diving aircraft. BuAer also sought a specialized dive-bombing sight. In 1934 dive-bombers used the same gunsights that fighters had, the Mk III telescopic sight. Its reticle had crosshairs and speed rings to facilitate deflection shots (for 100, 150, and 200 mph). It also had sights like those of a pistol, a blade in front and an after sight to be lined up with it, both on top of the telescope barrel.

This was hardly good enough. The released bomb did not simply carry on in the direction in which the airplane had pointed. Rather, it followed a ballistic (curved) path determined by factors like air resistance; it was also affected by the wind. The height of release was an important factor, but pilots in dives had no means of knowing their altitude at any given moment. Their barometric (pressure) altimeters could not sense rapid changes of altitude. Nor did pilots have any gauge of their actual diving angle. The aim point might be as much as a hundred yards from the target, to allow for the curved trajectory of the released bomb.

BuOrd was aware of the problem. It wanted to increase attack height so that bombs would hit at high enough velocities to penetrate the armored decks of capital ships. To that end, on 2 January 1934 it let a contract to the Carl L. Norden firm for a completely stabilized dive-bomb sight.[32] Assuming that this project would succeed, BuOrd also developed a

1,600-pound armor-piercing bomb.[33] The combination of sight and bomb would remove the one great limit on dive-bombers, their inability to destroy capital ships with heavily armored decks.

At this time Norden was completing development of his superb Mk 15 horizontal bombing sight. Like it, the dive-bombing sight used a gyro to track the target. The pilot could see the target using a periscope protruding under the device. The mechanism pointed a long sighting tube that the pilot used to aim the airplane in the right direction.[34] The shape of the sight was poorly adapted to installation forward of the cockpit of a single-engine airplane. Its body had to be mounted clear of the engine, insulated against vibration, a fundamental problem. It was also extremely complex.

Development was protracted, so BuOrd conceived a simpler interim sight based on the existing Mk III Mod 4 telescopic gunsight. Sperry turned BuOrd's outline concept into the prototype Dive Bombsight Mk I (later Aiming Angle Sight Mk I).[35] The pilot saw both fixed and moving horizontal crosshairs and had to place the target between them. The device calculated the required "sight angle" or "aiming angle," that between the flight path and the line of sight at the moment the bomb was released. It took into account the airplane's "lift angle," between its axis and the path the airplane flew. A gyro measured the angle at which the airplane was diving. The pilot saw a bull's-eye that moved up or down depending on the lift angle, its center displaced from that of the eyepiece according to that angle. The bull's-eye contained concentric circles marked (inside to outside) 100, 150, and 200. The pilot set the device for intended release air speed and altitude. When the pilot was at three times dropping altitude, the moving crosshair would be a third of the way between the two crosshairs (midway between when at twice dropping altitude). They would merge at the dropping point. The pilot would see a green warning signal when he was five hundred feet from the set release altitude. If the pilot tried instead to keep the target on the moving crosshair, his dive would steepen rapidly. The sight enormously simplified the work

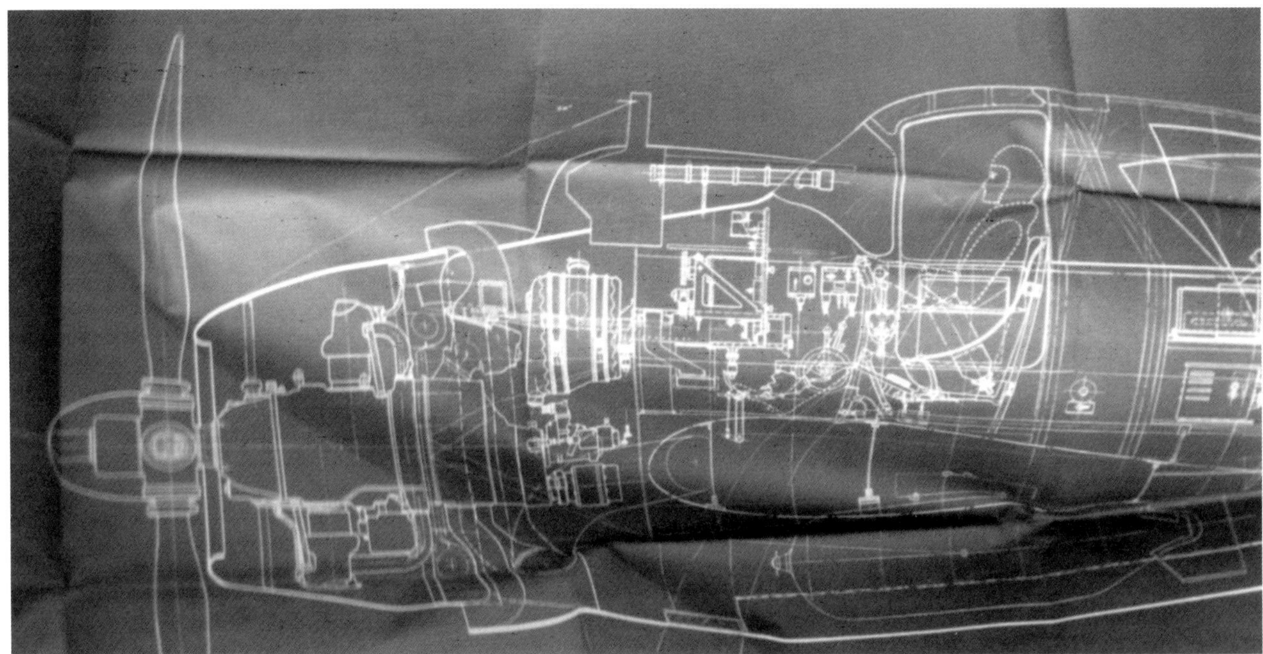

The side view of North American's unsuccessful bid in the 1938 dive-bomber competition shows the Norden dive-bombing sight as it would have been installed. It is the box in front of the pilot. The pilot looked through the telescope. The sight used a periscope projecting above it; its lines of sight are indicated by the prominent angled line pointing down, limited by the engine cowling and by a fainter horizontal line. The original of this blueprint is in the National Archives.
AUTHOR'S COLLECTION

the pilot had to do. In October 1935 Dive Bombsight Mk I was installed on a BG-1 for tests. By August 1937, aiming-angle sights had been made for service testing: eleven for Commander Aircraft Battle Force (ComAirBatFor) and six for the Marines' Commander Aircraft One.

BuOrd, however, much preferred Norden's more sophisticated sight. It first became available for tests, on board the BG-1, in the late summer of 1937. The airplane's big cowling blocked the sight at moderate diving angles, and only fifteen to eighteen bombs had been dropped; still, personnel at the Dahlgren Naval Proving Ground in Virginia were quite optimistic. The project was important enough to rate using the best available bomber, a BT-1. Norden's sight was renamed Aiming Angle Sight Mk II. The Army joined the program. However, tests showed that the sight was not ready for service use. Norden was asked to redesign it, and requirements had to be relaxed.[36]

In April 1941 BuOrd considered reviving the simpler Sperry sight, but the BuAer chief, now Rear Adm. J. H. Towers, vetoed that on the ground that it did not offer enough. The Sperry sight restricted forward vision and limited maneuvers prior to, and methods of going into, the dive and took up space in the crowded area forward of the instrument panel.

BuOrd remained sufficiently enthusiastic about the Norden sight that it set up a production facility for it in the spring of 1942. By that time there was no hope of installing the sight in any existing dive-bomber, the only type for which it was planned being Curtiss' abortive XBTC-1. The sight was redesignated Bombsight Mk 16. Work was completed at the end of 1942, but the sight never entered production.[37] Norden himself shifted to a full radar bombsight for horizontal and glide bombing, one that could hit a maneuvering target.

American dive-bomber pilots were left with standard fighter gunsights, initially the Mk III telescope. That was replaced by a reflector sight (Mk 8), which became the standard dive-bomber sight.[38] November 1944 instructions to Navy pilots pointed out that the sight could be used as a rangefinder:

the distance subtended on the target between the 100-knot (i.e., inner) ring and the bead of the sight could be computed for any desired altitude.[39]

With the dive-bombing sight program stalled, in January 1942 BuAer deferred modification of dive-bombers to drop the 1,600-pound bomb. Interest revived early in 1943, as it appeared that much less modification was needed than had been thought, mainly changes to sway bracing and acceptance of some reduction in safety factors. This new bomb could be effective even without a new bombsight. During the July–August 1944 British Fleet Air Arm attacks on the German battleship *Tirpitz*, a U.S.-supplied 1,600-pound armor-piercing bomb from a dive-bombing Barracuda penetrated the armor deck and reached the ship's inner bottom (lower platform deck). Unfortunately it failed to detonate; had it done so, it would have knocked out the main fire-control and switchboard rooms and flooded spaces probably as far as the forward auxiliary boiler room. The Germans described the damage it might have done as "immeasurable."

Most existing battleships lacked deck armor too thick for dive-bombers to penetrate. In November 1942 dive-bombers contributed to the sinking of the Japanese battleship *Hiei*; in July 1945 dive-bombers alone sank *Haruna*, *Ise*, and *Hyuga*.[40] On the other hand, dive-bombers could not penetrate the deck armor of the more modern *Nagato*, *Yamato*, or *Musashi*, although they contributed to the destruction of the two latter battleships.

An alternative to dive-bombing was "toss-bombing," invented in 1943. The bomb was released during pullout rather than just before, giving it added velocity that compensated for gravity sufficiently that the bomb fell in a straight line and hence could be aimed better. Calculating the appropriate moment for release required a computer. An altimeter reading and the dive angle gave the range to the target, and the rate at which altitude decreased was the criterion for releasing a bomb. A target could be hit accurately from up to 6,000 feet in a fifteen-to-sixteen-degree dive, the pilot dealing only with wind and target motion. Typically the pilot dove at the target, pressing a release when steady on course. About a sixth of the way to the target the bomb director turned on a light indicating that the pilot should pull up. The bomb was released automatically. BuAer ordered five hundred of these Bomb Directors Mk I in August 1944, but few had been delivered by the end of the war. BuAer considered them revolutionary. Bomb Director Mk 3, which equipped Skyraiders, could be used at altitudes as great as 25,000 feet. The last of this series, Bomb Director Mk 5, became the ASB-1 of postwar heavy carrier bombers (AJ and A3D).

Scout Bombers

In the fall of 1931, BuAer required—as it did with its single-seat fighters, prior to the decision to adopt "pure" ones—the new two-seat fighters it bought that year to be able to dive with five-hundred-pound bombs. Design Division had determined that this capability would cost so little in terms of weight and performance that it should be required. It produced sketch designs as benchmarks.[41] Curtiss offered a high-wing monoplane, its XF10C-2. It was strong enough because it had been conceived with all three roles (fighter, dive-bomber, scout) in mind. Douglas offered a high-wing (gull-wing) monoplane similar to, but smaller than, the XA-31 it was offering the Army. Its controllable-pitch propeller would limit takeoff run. Lockheed offered a low-wing monoplane with retractable landing gear. It seemed that neither the Douglas design (BuAer wondered how much strength the cuts for wheel wells cost) nor the FF-1 was strong enough to be a dive-bomber. BuAer favored the Curtiss F10C-2 and would also buy the Douglas airplane if there was enough money.

The XF10C-2 was only loosely related to the earlier XF10C-1 biplane. Curtiss used this design to display a range of new features it thought might interest BuAer: slots and flaps, a variable-pitch propeller, a watertight fuselage for inherent floatation, folding wings, and retractable landing gear (droppable in the event of a water landing). The slots and flaps and the variable-pitch propeller were intended to reduce takeoff run, which might not matter to a fighter but would be vital for a scout.

BuAer was already aware that simply adding power to the existing T5M/BM would cost size and weight for little gain. A study showed that eliminating the rear gunner might gain as much as 30 mph, which might be better protection than the gun. With a thousand-pound bomb and 125 gallons of fuel, a single-seat BM-1 would have an endurance of 1.65 hours at full speed (549 nautical miles at 80 percent speed). As a scout, without a bomb but with a second crewman, the airplane could carry seventy-five gallons of fuel, increasing range to about 880 miles. Limiting the airplane to a five-hundred-pound bomb could also improve performance, even if the backseater were retained.

BuAer became interested in a compromise dive-bomber/scout, for which it opened a competition (it also wanted to buy the XF10C-2).[42] Seven companies submitted designs in April 1932; Douglas (for the XFD-1) and Vought (for the XF3U-1) received contracts on 30 June. Both were powered by two-row R-1535 engines. BuAer also ordered the Curtiss Model 73 (XF12C-1), which seems to have been derived from the XF10C-2 (it was described as a version of the Army's O-40). It too was a "parasol" monoplane (the wing attached above the fuselage, like the upper wing of a biplane), with retractable landing gear and able to carry a five-hundred-pound bomb. Of existing two-seat fighters, the FF-1 was too small and the XF2J too heavily loaded (and too small) to carry enough fuel.

In May 1933 Plans Division proposed to study converting the XF3U-1 or the XFD-1 to a 500-pound-bomb dive-bomber, the bomb to replace the belly gas tank.[43] Douglas' XFD-1, which was designed to carry a pair of 116-pound bombs underwing, had a spreader bar between its main landing-gear struts that would have precluded installation of a trapeze.[44] Vought's airplane had no such obstacle, and it was superior in other ways.

In June, Plans became interested in a scout dive-bomber (VS-VB).[45] Converting a fighter into a scout bomber required increased fuel, a scouting (long-range) radio, a second set of controls in the rear cockpit, and greater oil capacity (twelve rather than eight and a half gallons). The gunner's seat had to

The SBU-1 shown here on 9 February 1940 was a modified version of Vought's XF3U-1 fighter, with larger wings, more fuel, and provision for a five-hundred-pound bomb in displacement gear (not visible here). This one was from one of two scout squadrons on board USS *Ranger*, VS-41; the star on the cowling indicates that it was part of the Neutrality Patrol. Note the fighter-type telescopic sight passing through the windshield; it would remain standard through about 1943. The object between pilot and gunner was the loop of a radio direction finder, seen edge-on. The finish shown was the initial stage of a move toward camouflaging naval aircraft, the bright yellow on wing upper surfaces being eliminated, together with the fuselage band indicating the section (but note that the cowling was still painted). The tail color indicating the carrier survived. Only at the end of 1940 were all naval aircraft ordered painted nonspecular (nonreflecting) light gray. NATIONAL ARCHIVES

be changed so that he could use the controls and the radio. The added oil weight could be bought by eliminating one fixed gun. The scout bomber had to have both a trapeze for a five-hundred-pound bomb and light underwing bomb racks. BuAer would turn one of the three two-seaters under test into a scout fighter. By this time Curtiss was already completing its XF12C-1 as the S4C-1 scout. The S4C-1 had the best range of the three alternatives but unfortunately had poor takeoff characteristics. In November, the XF3U-1 having beaten out the XFD-1, BuAer asked Vought to modify it as a scout bomber; the Navy promised to order at least twenty-seven aircraft (in the event, eighty-four were built, in addition to the prototype). In this form it flew on 29 May 1934, soon entering production as the SBU-1. SBU-2 was a noncarrier version for the reserves.

The XS4C-1 began evaluation in December 1933. It crashed during a 14 June 1934 company test flight, having been redesignated XSBC-1. On 6 July Curtiss proposed building a replacement using thirty-four-foot-span biplane wings instead of the forty-one-foot

folding parasol, with full-span flaps instead of leading-edge slats on the lower wing. The Navy reordered the prototype as Curtiss had proposed in April 1935 as the XSBC-2. It had a new fuselage, a new canopy (faired into the after fuselage), a new tail, and a three-bladed propeller. Its retractable landing gear was similar to that of the XSBC-1 but with less of an underbody bulge. Its only major drawback was its unreliable XR-1510-12 engine. The successful XSBC-3 version had a two-row R-1535-82. A contract for eighty-three aircraft was issued on 29 August 1936, and deliveries of production SBC-3s (R-1535-94 engine) began on 17 July 1937. The final SBC-4 version reverted to a single-row R-1820-34 engine (the cowling was noticeably larger). It was the last combat biplane produced for the U.S. Navy. Fifty were sold to France, diverted from Navy production to allow quicker delivery. By this time two monoplane dive-bombers were in production; many SBC-4s went directly to reserve units. However, two carrier squadrons were still flying SBC biplane dive-bombers in December 1941.

Curtiss offered an alternative in the form of a two-seat fighter that became the S4C-1 and then the XSBC-1. It is shown on 17 February 1934. Its outer wings swung back, pivoting on the big struts (the joint is visible). This prototype crashed during a 14 June 1934 company test flight. On 6 July Curtiss proposed building a biplane replacement. It was ordered in April 1935 as the XSBC-2. The production version, with a more reliable engine, was the SBC-3. NATIONAL ARCHIVES

The prototype SBC-3, shown on 17 April 1936, was essentially the SBC-2 with a more reliable R-1535-82 twin-row engine. Differences from the XSBC-1 were new wings, a new fuselage, a new canopy (faired into the fuselage), a new tail, and a three- rather than two-bladed propeller. NATIONAL ARCHIVES

The final production version of the SBC was the SBC-4. It could be recognized by its short cowling, housing a single-row R-1820-34 Cyclone engine; note its appreciably greater diameter. It was the last combat biplane produced for the U.S. Navy; two carrier squadrons of these aircraft were still in service in December 1941. Note the displacement gear holding the drop tank in place, the wing bomb rack, and the trough atop the cowling for a forward-firing gun. This SBC-4 served at the Naval Reserve Aviation Base, New York. U.S. NAVAL INSTITUTE PHOTO ARCHIVE

Higher Performance

In March 1934 BuAer circulated a request for proposals for a new high-performance scout dive-bomber.[46] Desired maximum speed was about 235 mph, which at the time was roughly fighter speed. Performance would be limited by weight (not to exceed five thousand pounds normal gross), stall speed (65 mph), and size—maximum span was thirty-six feet for a biplane and forty-two for a monoplane. Folding wings were neither necessary nor desirable; BuAer was not sure they could be strong enough to handle the stress of dive-bombing. The airplane would typically dive at a seventy-degree angle and release a bomb at two thousand feet. Speed in a dive—a streamlined airplane diving from ten thousand feet might exceed 350 mph—had to be limited to 250, to maintain accuracy. Excessive diving speed might also make for a dangerously low recovery altitude. By May, designs had been received from Northrop, Bellanca, Brewster, Chance Vought, and Curtiss.

Chance Vought's Design A monoplane received the highest rating, on the strength of its performance.[47] Next was Chance Vought's B, which differed mainly in its engine. Vought's C, which came next, was a modification of the biplane XSBU-1 then undergoing trials; the main changes were retractable landing gear and a means of limiting diving speed. Vought's D, not described in the memo summarizing the aircraft, came fourth. After that, in order of preference, came two Curtiss designs, one by Bellanca, and one by Brewster (which had not previously competed).

The recommendation of the BuAer test report was to order Vought's A design, which became the XSB2U-1. If, as it turned out, the SBU-1 performed as expected, one Vought C should be ordered. The Vought C became the XSB3U-1. Both prototypes were delivered at about the same time. Compared to the biplane, the monoplane XSB2U-1 was 15 mph faster (carrying a thousand-pound rather than a five-hundred-pound bomb) and had the same stalling speed. As built, the SB2U-1 could fold its wings.

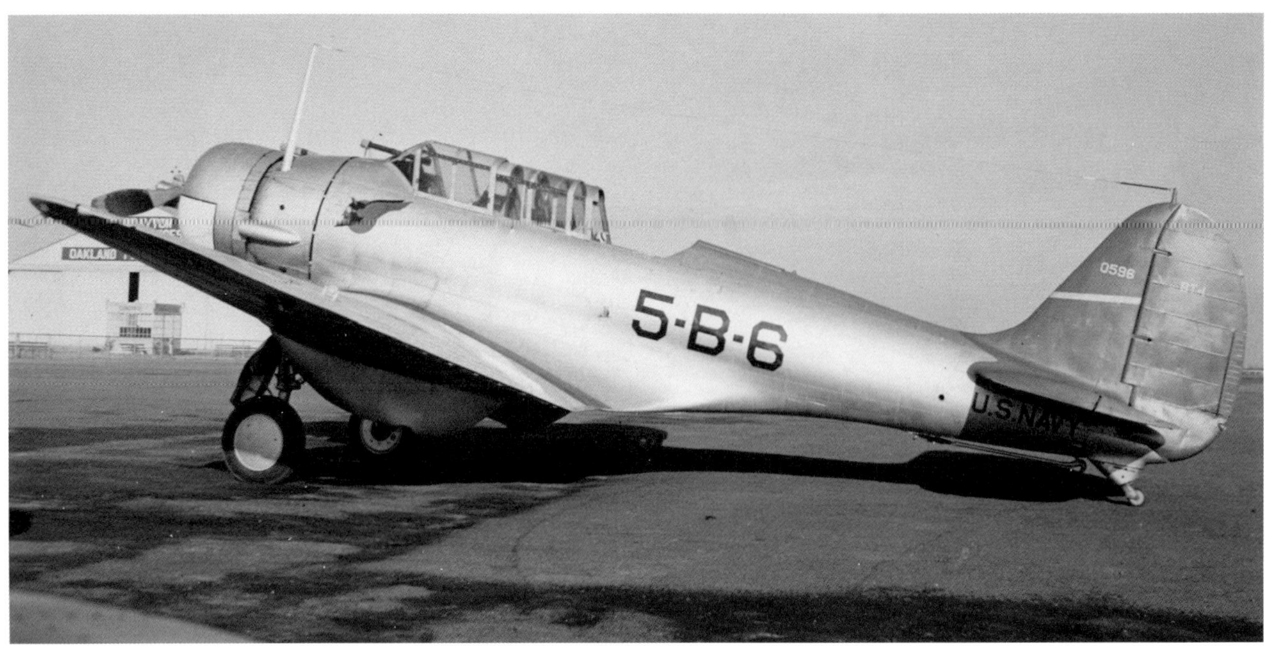

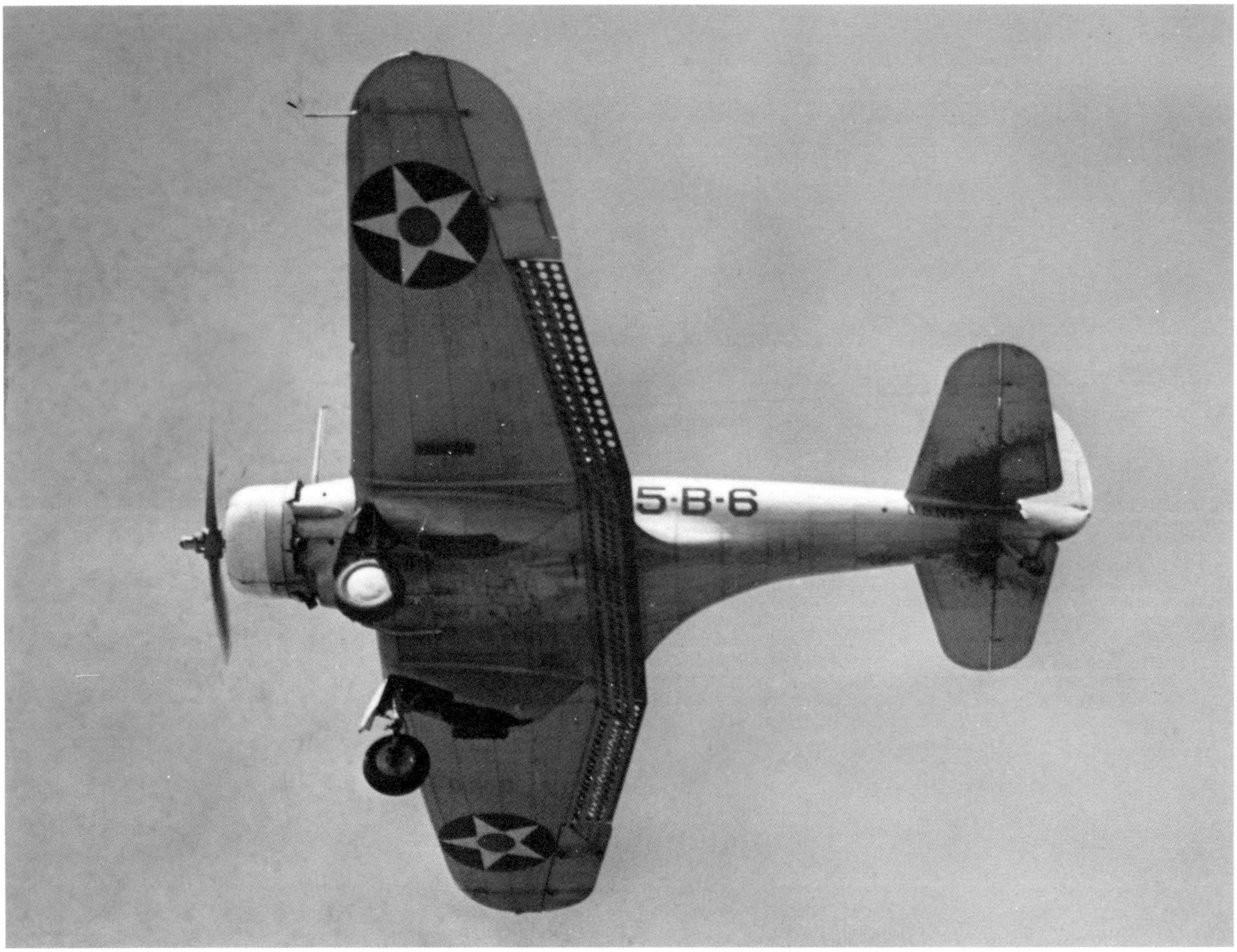

The shape of the future: Northrop's BT-1 heavy dive-bomber. Note the perforated flaps this design introduced. This one is from VB-5 on board *Yorktown* in 1936–37. The stripe on the tail was intended to help the LSO gauge the angle at which the airplane was approaching. The tail was painted red. U.S. NAVAL INSTITUTE PHOTO ARCHIVE

Great Lakes' XB2G was bought as insurance against failure of the monoplane BT. U.S. NAVAL INSTITUTE PHOTO ARCHIVE

Brewster's midwing monoplane was considered intriguing despite its low rating, which could be attributed to the company's inexperience and limited engineering capability. Unusually, the pilot was located immediately abaft the engine, but placing the flexible gun and gunner/radioman well aft would have made coordination (e.g., to report scouting results by radio) more difficult. There were possible structural difficulties in retracting landing gear. The company had apparently misunderstood the NACA variable-density and full-scale wind-tunnel reference data on the wing section used. However, according to the BuAer evaluator, "the general conception of this design is of sufficient interest to warrant further consideration and development." Brewster should be asked to resubmit, correcting obvious errors.

There was a parallel heavy dive-bomber (VB) competition. BuAer planned to ask for a speed of over 200 mph, but Commander Aircraft Battle Force wanted at least 235 mph.[48] The new experimental torpedo/horizontal bomber was expected to make 220 mph, and a dive-bomber should be at least as fast. Dive-bombers needed adjustable vertical stabilizers and ailerons so that they could dive

without extra pressure on either rudder panel, to eliminate the tendency of diving airplanes to skid. Their pilots needed better downward visibility, which meant bombing windows under the airplane.[49]

BuAer circulated its request for proposals on 15 March 1934. These aircraft would deliver thousand-pound rather than five-hundred-pound bombs. Maximum speed should be at least 200 mph, and diving speed in a seventy-degree dive should be no more than 260 mph. Strength should allow a terminal velocity dive with a pullout not to exceed 9 *gs*, carrying 75 percent of the fuel used with a thousand-pound bomb (but no bomb). Stall speed with normal bomb load should not exceed 65 mph. Maximum takeoff run into a twenty-five-knot wind was 350 feet. Service ceiling should not be less than 22,000 feet. Range should be 500 miles (435 nautical miles). Normal gross weight should not exceed six thousand pounds. Size should be minimized for carrier stowage. The after cockpit should have auxiliary controls, and oxygen should be provided for both pilot and gunner. Gun armament was two 0.30-calibre machine guns, one fixed (synchronized) and one flexible (soon the fixed gun was made 0.50-calibre).

Proposals were opened on 1 May. On the 22nd the BuAer Material Division recommended Northrop's, which was practically unanimously preferred within the division.[50] It was the first major design by Edward Heincmann, who became a classic naval aircraft designer. The Material Division was so impressed that it recommended that BuAer exercise an option for twenty-seven production aircraft. Northrop had entered the same airplane in the VSB competition, but it was not rated in that competition.

Second and third places were taken by the Great Lakes B2G and the Consolidated B3Y. Great Lakes' B2G (Design 66-A) was a modified BG with Grumman-type retracting landing gear. That required a deep belly, which became an internal bomb bay. The Material Division proposed giving Great Lakes a twenty-seven-airplane option (which was not exercised). Consolidated claimed a maximum speed of 210 mph, which had not been checked. The division proposed buying one of its airplanes if the company could meet price and performance requirements, but the B3Y was not built.

It might seem that Northrop's victory in the 1934 competition proved that the VB and VSB classes were really the same, but as late as September 1935 BuAer circulated an internal memorandum deny-

The fully redesigned XBT-2 became the prototype of the SBD Dauntless. The change in designation reflected both Douglas' absorption of its Northrop division (*D* rather than *T*) and a change in concept, amalgamating the scout (*S*) and bomber (*B*) roles.
U.S. NAVAL INSTITUTE PHOTO ARCHIVE

Vought's SB2U won the VSB competition.
NATIONAL ARCHIVES

A VB-3 (*Saratoga*, white tail) SB2U in flight over the San Joaquin Valley, 29 June 1938, shows its large wings. It had hay-rake dive brakes rather than the perforated flaps of the Northrop/Douglas dive-bombers. NATIONAL ARCHIVES

ing that.[51] The difference was in useful load: about 1,700 pounds for a VSB and about 2,500 for a VB. Given the same strength and stalling requirements, a VSB could weigh about 8,600 pounds, a VB about 9,500. Key VSB requirements were limited stall speed (60 mph with normal useful load on board) and scouting range (not less than 750 miles at 75 percent maximum speed). At 75 percent speed VB range (as a bomber) should be no less than 500 miles. With a VB load, a VSB would have a higher stall speed. However, a surviving copy carries a handwritten comment by a member of Plans Division: "There is nothing sacred about our requirements for VSB and VB classes. I am still of the opinion that one type of plane can be developed to do everything and *more* [emphasis in original] than both types are supposed to do at present. It is immaterial whether we call them VBs or VSBs." All a VB needed to qualify it as a scout was a long-range radio.

In June 1935, while their airplanes were being built, Chance Vought and Curtiss proposed modifying the SB2U and the SBC to carry thousand-pound bombs. In the case of Curtiss, that meant replacing the R-1820-F with an R-1525-82. The extra load raised stall speed to 70 mph and required some wind if the aircraft were to be catapulted.

The 1936 Competition

The 1934 competition had selected aircraft worth building as prototypes, but in 1936 BuAer had to decide which to build in quantity. In April 1936 Plans Division laid out the tentative 1936 procurement program, for VB, VSB, and a possible merged type.[52] VSBs from past competitions were Vought's XSB2U-1 monoplane, its XSB3U-1 biplane, and Curtiss' XSBC-3. Brewster's XSBA-1 had just been delivered to Anacostia. The only new type was Grumman's SBF-1, a scout-bomber version of its

FF-1/SF-1 two-seater. Four airplanes were offered as VBs: Northrop's XBT-1, Vought's XSB2U-1, Curtiss' XSBC-3, and Great Lakes' XB2G-1. Production bids were opened on 29 May 1936.

Vought's entry offered outstanding performance but suffered penalties in other areas; overall it came in somewhat behind the BT-1 (78.2 versus 79.9 points).[53] The BT-1 was rated very high under general characteristics, "due principally to the many very favorable reports received during trials." It had "reasonable performance and by far the most satisfactory general characteristics." The SB2U-1 had superior performance but needed work, which would delay it. Curtiss "suffers because of its low performance, particularly as a bomber"; it trailed, with 71.6. Great Lakes was far behind (59.2). The 108-plane VB order was split evenly between the BT-1 and the SB2U-1. The BT-1 developed into the classic SBD Dauntless, which was built in large numbers. The SB2U appeared in much smaller quantity.[54]

In the VSB category, Brewster's XSBA-1 came first (81.3), followed closely by the XSB2U-1 (79.0), the XSBC-3 (77.6), the XSBF-1 (73.3), and then the XSB3U-1 (63.5). Grumman's XSBF-1 offered reasonable general characteristics but was too slow. The XSBC-3 also had good general characteristics, but it was 34 mph slower than the Brewster as a bomber and 30 slower than the SB2U as a scout. That left the two monoplanes, which were about equal in performance (Brewster was 6.6 mph faster as a bomber, but Vought was 4 mph faster as a scout). Under general characteristics, Brewster was rated higher than Vought. Vought's biplane was too slow and had unsatisfactory carrier and scouting characteristics. The SB2U-1 was already being bought as a VB. That narrowed the choice to the SBA-1 and the SBC-3. Probably because Brewster was a new company, BuAer bought eighty-three SBC-3s alongside the SBA-1. Brewster had insufficient plant capacity to produce its thirty airplanes, so they were built by the Naval Aircraft Factory, as the SBN-1.[55]

The BT-1 and the Dauntless

The BT-1 had excellent flying characteristics and "seems to be a rather successful product."[56] It had split flaps in its wings to limit its terminal dive to 250 mph. The lower half of the flap also provided extra lift on landing to limit stall speed. Early tests showed severe tail flutter occurred when the flaps were extended, owing to vortices they created. The solution, which characterized both the BT and its offspring the SBD, was to cut circular holes in the flaps to break up the vortices. Fears that reducing flap area would increase diving speed proved unfounded.[57]

With a more powerful R-1535-94 engine and other modifications, the BT-1 entered production, the first due for delivery on 1 June 1937. It showed a tendency to drop one wing and the nose, presumably near the stall, as it landed. Modifications and better pilot training seemed to cure the problem; BT-1s modified for better lateral control and stability seem to have been successful on board the two *Yorktown*-class carriers.

Heinemann and his Navy project officer decided to redesign the airplane (as the BT-2) with a new fuselage, center section, outer wing, canopy, trapeze, and arresting gear. Instead of retracting backward into the wing (leaving the wheels half exposed), the landing gear retracted fully. The engine was replaced by a 1,000-hp R-1820-G133 driving a constant-speed propeller. Cockpit equipment, controls, and control surfaces were all replaced. Virtually the only surviving part of the BT-1 was the distinctive pierced dive brakes. The more powerful engine added about 35 mph. The first BT-2, the twenty-seventh production airplane, flew in April 1938. Trials showed engine problems at high altitude and high angles of attack. After tests in the full-scale NACA wind tunnel, wingtip slats were installed to improve lateral stability and control at or near the stall—that is, when landing. Slight longitudinal and lateral instability at maximum weight was cured by increasing dihedral (the angle between the wing and the horizontal) by two degrees and by changing the tail.

Brewster's XSBA-1, 20 May 1937. Note the considerable distance between the pilot and the gunner/radio operator. NATIONAL ARCHIVES

In 1938 BuAer had to decide between the BT-2 and a proposed upgraded SB2U-4. The SB2U airframe could not take a much larger-diameter engine, so the -4 would have been powered by an R-1535–96 (825 hp on takeoff, 750 up to 9,500 feet). With its much less powerful engine, the SB2U-4, at 245 mph, could not match the 262 mph (260 with extra gas) of the BT-2. Its range at economical speed would be 955 miles versus 975 (1,685 miles in over-load versus 1,630, with extra gas, for the BT-2, however). Takeoff distance as a scout with overload would be 315 feet versus 235 feet, an extremely important point for scouts. The SB2U had already suffered some problems, such as failures of wing-center sections

The SBD-1, photographed 2 August 1940, was assigned to the Marines because its wings could not fold, hence took up more hangar deck space than an SB2U Vindicator. This one is in the full prewar color scheme, including colors indicating the section and yellow wings with chevrons; another photo of this particular airplane shows that the aircraft number within the squadron (in this case *1*) was repeated on each wing between the chevron and the fuselage. The three vertical stripes on the rudder were a standard feature on prewar Marine Corps aircraft. Note the Marine Corps insignia below the after end of the greenhouse. Despite its shore assignment, the airplane had an arrester hook. NATIONAL ARCHIVES

and wing flutter (in two cases) in dives at 350 mph indicated airspeed (IAS). It is not clear whether the company's "hay-rake" dive brakes, near the wing's leading edge, were a source of the flutter problem.

BuAer chose the BT-2, which was redesignated SBD-1 (soon named Dauntless) when Douglas Aircraft absorbed Northrop as its El Segundo Division (a new Northrop company emerged). It became the most successful American dive-bomber of World War II, famous for its achievements at Midway. Production deliveries began in June 1940. Probably because its wings did not fold, the SBD-1 was initially assigned to the Marines. In 1941 it went to sea on board *Enterprise* (VB-6) and *Lexington* (VB-2).

In June 1939 the inspector of naval aircraft described major desired changes.[58] The R-1820-32 would be replaced by an R-1820-251 rated at 1,000 hp from sea level to 4,500 feet, at 1,200 bhp for takeoff. Greater engine weight would be balanced by replacing the current main and auxiliary tanks with an integral tank in the center section. At the same time, fuel capacity would increase from 210 to 300 gallons. All external surfaces would be flush riveted. Speed as a scout would increase from 262 to 279 mph, and endurance at 60 percent maximum speed with full fuel from 7.3 hours (1,150 miles) to 10.3 hours (1,710 miles) without any increase in weight. Airplanes so modified were designated SBD-2. They could be distinguished from SBD-1s by the elimination of a prominent carburetor air scoop atop the engine cowling and of the gun. The Army evaluated an SBD-2, deciding to adopt it as the A-24.

SBD-3s. The underwing radar antennas (posts with trainable Yagi antennas) served the ASB radar, a metric-wave (60-cm wavelength) set adapted from a British prototype by RCA. There was a similar antenna under the port wing. It could detect a capital ship forty miles away, within one degree in azimuth. Each of the two antennas had a sixty-degree beam width or search over water; the antennas were set perpendicular to the line of flight, or for homing, parallel to it. The operator had a five-inch scope, signals from the two antennas being presented separately, to left and right. ASB operated with the YJ beacon on board carriers. The star-and-bar national insignia was adopted on 28 June 1943, with a red border changed to dark blue on 4 September. Note the short wire antennas above and below the airplane body, for, respectively, IFF and YE homing. NHHC

An SBD is prepared for a strike against Tulagi, 7 August 1942: ordnancemen are about to mount a five-hundred-pound demolition bomb on the displacement gear. Note the connection to the airplane at its fore end. NHHC

The prototype arrived at Anacostia on 20 October 1939 to recheck changes made in response to BuAer's Trial Board. Curing its longitudinal and lateral stability problems had added a small amount of directional instability; a new section of vertical tail was added. BuAer wanted further changes: a Sperry autopilot was installed, the left-hand fixed 0.50-calibre machine gun was eliminated (to keep the center of gravity in its previous position), and an optical drift sight (for navigation) was added in the rear cockpit. Reports from the European war led BuAer to require self-sealing tanks. Because Douglas could not make the big (three-hundred-gallon) tank self-sealing, it reverted to smaller ones. It also began work on armor for pilot and gunner.

The fleet wanted more forward-firing firepower, as it appeared that strafing had proven very effective in Europe. It seemed to demoralize shipboard topside personnel completely.[59] In December 1940 ComAirBatFor (Rear Admiral Halsey) asked for four 0.50-calibre machine guns, which he thought would be effective against destroyers, motor torpedo boats, and other light surface craft. BuAer offered more gun power in the new dive-bombers (SB2C and SB2A).[60] The second 0.50 was restored during the production run of the next Dauntless version, the SBD-3.

Delivery of a new SBD-3 version began in February 1941. By that time the international situation was so serious that delivery was not delayed for trials.

In January 1941, 174 aircraft were on order, 74 of them for the Army as A-24s (Navy designation SBD-3A). Total Navy production, including 175 originally ordered for France, amounted to 585 aircraft, compared to 87 SBD-2s. This version had pressurized, self-sealing tanks, cockpit armor (production designs released September 1941), and a new R-1820-52 engine. During the production run the single 0.30-calibre flexible gun was replaced by a twin gun with continuous belt feed; beginning in mid-1942 these improvements were retrofitted to many earlier SBD-3s.

By mid-1942, SBDs were being flown badly overloaded. With normal fuel and a five-hundred-pound bomb, the takeoff run was 355 feet. It grew to 420 feet with a thousand-pound bomb and 150 gallons and to 495 feet with a thousand-pound bomb and 254 gallons, the maximum fuel load. The cure was a new, more powerful 1,200-hp R-1820-40 engine. The extra 200 hp solved the takeoff-run problem, but it did not much improve performance at altitude. Increasing electrical demands, particularly installation of the ASB radar (generally at forward bases) were met by switching from a twelve- to a twenty-four-volt system.

Because engine deliveries lagged, the SBD-4 had the new electrical system but the earlier engine (with an electric boost fuel pump). SBD-4 production was 870 aircraft, delivered between October 1942 and April 1943.

The seventy-fifth SBD-4 became the SBD-5 prototype, with the -40 engine. The cowling was reshaped to accommodate the new engine. The SBD-5 also had provision for two fifty-eight-gallon drop tanks.[61] It was delivered with the Mk 8 reflector sight. Because it was not pierced by a telescopic sight, the windshield could be heated to prevent it from misting over as the airplane dived. A total of 2,965 aircraft were delivered between February 1943 and April 1944.

Late in 1942 BuAer Plans Division ordered a further SBD-6 version, to be powered by a 1,350-hp R-1820-56 engine, which would offer sufficient power at altitude to add considerably to speed and rate of climb. Maximum weapon load increased to two thousand pounds. In December, however, Plans decided to use all -56 engines in Wildcat fighters, so the planned 1,000 airplanes were produced as repeat SBD-5s. The SBD-6 program reappeared in mid-1943, the prototype flying on 2 August. At this time, 3,000 more SBDs were projected: 1,550 SBD-5s and 1,450 SBD-6s. Douglas was already working on the follow-on BTD, so to transfer personnel to the new project it decelerated SBD production late in 1943. SBD-6 prodution began in

An SBD-6, the final version of the Dauntless, at Patuxent River, 25 May 1945. NATIONAL ARCHIVES

March 1944. With little capacity available, orders were cut by 1,000 airplanes, the last being delivered on 22 June 1944. When the fleet asked for Dauntlesses to replace its Helldivers after the Philippine Sea (June 1944), nothing could be done.

The fleet by this time was demanding more ground-attack firepower. As of January 1944, aircraft were being fitted with two Mk 42 bomb racks under their center sections, to carry up to six 100-pound bombs. The wing bomb racks could also carry two 0.50-calibre "package guns" (in a package that could be hung from a bomb rack) on each side. Alternatively, six rocket rails were fitted to the outer wing panels.

During July 1942, Douglas proposed building a version of the SBD as a torpedo bomber. BuAer rejected the idea because the airplane would have insufficient range.

By the end of World War II, having overcome considerable problems, the SB2C Helldiver was the standard U.S. Navy dive-bomber. These show ASB radar antennas, 20-mm wing cannon, and the big spinners of SB2C-1Cs. These Helldivers are flying over Kaneohe Bay, Oahu. The small *K* on the rudder probably indicates CAG-1, which was based at Kaneohe before coming on board the carrier *Yorktown* before the battle of the Philippine Sea, probably about the end of May 1944. By this time carrier air groups were being replaced periodically. Aircraft on board *Yorktown* were already distinguished by their diagonal tail stripes, but in a few cases they retained the earlier *K* rudder marking, at least for a time. CAG-1 received its Helldivers about the end of March 1944. When it returned to the United States, it exchanged them for Dauntlesses for training. It is not clear to what extent CAGs *not* assigned to carrier had their own markings; tail markings generally indicated the carrier. U.S. NAVAL INSTITUTE PHOTO ARCHIVE

5

DIVE-BOMBERS OF WORLD WAR II

The 1938 Dive-Bombers: SB2C and SB2A

The key requirement for the 1938 dive-bomber was higher speed, so that bombers could operate closely with new high-performance fighters such as the Grumman Wildcat and the Brewster Buffalo. From 1934 on, BuAer had found it difficult to gain really high performance without using twin engines, but in 1938 the big R-2600 finally offered enough power (1,700 hp).[1] In February, BuAer's preliminary designers produced Design 171, a 10,948-pound dive-bomber that the R-2600 powered.[2] Work on specifications began about March.[3] Presumably on the basis of the preliminary design, the BuAer Plans Division called for a maximum speed of 283.7 mph at 17,500 feet and 243 mph at scouting altitude (2,000 feet), with a 1,000-mile range (carrying a five-hundred-pound bomb, fuel for 1,500 miles in overload condition); four fixed guns for strafing surface and shore targets; and an autopilot (now required on all new VSB).[4] This was considerably better than the BT-1 and the SBD. To gain higher performance, stalling speed could be 70 rather than 65 mph, as in new fighters. This higher limit was justified on the ground that lateral control at and below stall speed could be much improved. Takeoff run in a twenty-five-knot wind should be no more than 250 feet, and service ceiling not less than 20,000 feet. The diving speed limit was raised to be more than the maximum level speed of the airplane. By the time the aircraft appeared all carriers could be expected to have catapults, that could make overload takeoffs routine. A 50 percent overload was the maximum additional fuel that could be accommodated without exacting a penalty in wing structure, weight, and arrangement.

Overall weight was a problem. The limit imposed by existing arresting gear seems to have been 9,000 pounds (requiring some modification to Design 171). Anything beyond 11,000 pounds would require new arresting gear. It seemed that would be the case to accommodate the R-2600. Plans soon increased maximum weight to 9,500 pounds.

Plans' draft specification (27 April 1938) called for the bomb to be carried internally ("desired" was written in) to limit drag.[5] Maximum diving speed was cut back to the earlier 250 mph. With the decision to merge the VB and VSB classes, the airplane was soon required to carry a thousand-pound bomb as well as two underwing hundred-pound bombs. As a scout *and* a dive-bomber, the airplane needed maximum performance both at optimum bombing approach altitude and at the much lower optimum scouting altitude (one to two thousand feet).

Reviewing the draft requirement, the chief of the Engineering Division emphasized the possibility of handling two airplanes on a standard forty-one-by-forty-eight-foot elevator, with twelve-inch clearance all around. To limit the complexity of the cockpit, the rear gunner would control wing folding.

Bidders' letters went out on 6 August.

Meanwhile, Commander Aircraft Battle Force (Rear Adm. E. J. King, formerly chief of BuAer) questioned the bureau's policy of seeking higher and higher performance at the cost, he thought, of reliability and effectiveness. In a 20 September letter he complained that practically every modern airplane in his command, except the F3F, was operating under restrictions. He preferred biplanes to unsafe folding-wing dive-bombers. He also preferred what he considered safer fixed landing gear. The criterion should be what the fleet's aircraft could reliably do, not their maximum performance. The fleet's aircraft should offer maximum effective power for the limited deck areas of the carriers. On the cover sheet was written "believe Chief of Bureau should see this letter (received yesterday) without delay."

The head of Engineering thought King's view absurd. The biplane SBC-3 and the monoplane SB2U-1 both used the same engine (except for its gearing ratio), had the same stalling speed, and carried the same military load. The monoplane was over 11 percent faster, had 12 percent greater range on thirty pounds less fuel, and 10 percent higher ceiling. SBCs occupied 17 percent more deck than folded SB2Us. King's preferred fixed landing gear cost performance. The complexity of hydraulic operation was justified because hand operation involved too much time and work. Folding wings were no weaker than fixed ones.

To get its desired higher performance, BuAer retreated on weight; on 8 October it raised the limit to 10,250 pounds.[6] Probably reacting to reports of air action during the ongoing Spanish Civil War, BuAer asked bidders to consider the effect of adding armor for both pilot and gunner. It slipped its deadline to December 1938.

Only Curtiss and Brewster used the R-2600 engine, with its promise of really high performance; theirs were the only serious proposals. BuAer wanted more alternatives, because the inexperienced Brewster company might well be unable to produce airplanes quickly enough.

Curtiss offered two alternatives, "A" with a retracting undercarriage and "B" with a fixed one ("spatted," with streamlined fairings); both were dated 6 December 1938. Each was powered by an R-2600 Wright (i.e., within the Curtiss-Wright group) engine. To remain within the weight limit the company used integral fuel tanks, which it disliked. If allowed another seventy-five pounds it would adopt separately mounted tanks. Another eighty gallons would bring gross weight as an overload scout to the limit set by catapult capacity and design stall speed. The bomb was housed internally, the wing spar passing around the bomb bay. Curtiss offered enough space for two 500-pound bombs side by side, or one 500-pound and an extra fuel tank, or a very large fuel tank for maximum range. No other bidder could offer two 500-pound bombs.

The Helldiver prototype, 11 January 1941, in prewar finish. The bumps on the cowling were for the projected pair of forward-firing guns. Note the absence of the usual telescopic gunsight.
NATIONAL ARCHIVES

To maintain control and stability, Curtiss used automatic wingtip slots, a thinner airfoil at the tip, 2 percent wing twist, and six-degree dihedral. Automatic tip slots ahead of the ailerons would help prevent stalling of the wingtips and hence help maintain lateral control near the stall—that is, as the airplane approached to land on a carrier. To control this heavy airplane, Curtiss proposed to substitute a yoke (carrying a control wheel) for the usual stick and rudder pedals. In theory the pilot could pull back on the yoke to bring the airplane out of a dive.[7] The yoke did not survive into prototypes or production aircraft.

Curtiss claimed that the combination of high-lift devices and folding wings made it possible to meet BuAer's request that two airplanes be able to share one elevator. The company argued that folded width, because it made for quicker movement between hangar and flight deck, was far more important than maximum height while wings were being folded. Although the height of the airplane with folded wings fit within hangar height, in the process of folding the wings stretched higher, which precluded folding or unfolding Helldiver wings inside a hangar deck. Curtiss limited folded width to twenty-two feet six inches. Wings would fold hydraulically. Curtiss claimed, on the other hand, that overall length (thirty-five feet three inches) had been chosen to gain proper balance, tail length, and satisfactory vision down and forward, not to fit two airplanes onto an elevator. In any case, the BuAer evaluator pointed out, the weight of the Curtiss and Brewster designs made it impossible to place two on an elevator at the same time.

Dive speed would be limited by opening large double split flaps and by holding bomb-bay doors open during a dive. To prevent tail buffeting, the wing was as low, and the horizontal tail as high, as possible. Split-flap chord was limited to reduce the width of the wake from the wing, which would hit the tail. Curtiss wanted BuAer to accept a higher diving speed (266 versus 250 mph) to avoid any need for perforated flaps, which would reduce maximum speed and increase minimum speed.

To provide power for several different systems, Curtiss connected a hydraulic pump to the engine.[8] This centralization was later a vulnerability, as leaks or holes anywhere in the hydraulic system could put it out of action.

Gun armament was either one 0.50-calibre or two 0.30-calibre synchronized machine guns plus one gun in each wing, either a 0.50- or a 23-mm Madsen cannon (then being advocated for many aircraft). The was also one 0.50 (200 rounds) or one 0.30 (600 rounds) in an enclosed turret.

The only important deviation from BuAer requirements was a 265-mph diving speed.

Curtiss estimated that with one 1,000-pound bomb Proposal A would have a gross weight of 10,697 pounds and a sea-level speed of 292 mph (324 at 15,000 feet). Service ceiling would be 33,450 feet. In a twenty-five-knot wind, the airplane would need a 258-foot deck run. When evaluating designs, BuAer made its own weight estimates. On that basis it estimated that with a five-hundred-pound bomb Curtiss A would make 315.5 mph at 15,000 feet or 283.6 at sea level. Service ceiling would be 30,500 feet, and takeoff run with twenty-five knots of wind over deck would be 248 feet. Range would be 1,030 miles.

Curtiss estimated that with one 1,000-pound bomb, its Proposal B would have the same gross weight. Sea-level speed would be reduced to 276 mph (307 mph at 15,000 feet). Service ceiling would be 32,300 feet. In a twenty-five-knot wind, this airplane would need a 252-foot deck run. BuAer estimated that fixed landing gear for Curtiss B would cost 16.3 mph at 15,000 feet (500-pound bomb), although there would be a slight gain in range (1,048 miles). Curtiss argued that eliminating gear-folding would eliminate the most frequent type of service failure requiring a major overhaul. Weight saved went into the additional fuel and oil required to reach the specified range.

Brewster offered an enlarged SBA-1. Unlike Curtiss, in its own Proposal A it offered a two-stage supercharger (with two compressor stages, for optimal performance at high and low altitudes) for its R-2600, which gave it better high-altitude performance than airplanes with two-speed superchargers (one stage, geared to operate at two speeds as required). Powered by an R-2600-A83, the Brewster A was expected to make 340.5 mph at 24,000 feet as a scout (1,340 bhp at 22,000 feet) or 339 mph as a bomber with a 500-pound bomb. Takeoff run with a twenty-five-knot wind over deck would be 250 feet. Brewster's data sheet showed a lower gross weight with a 1,000-pound bomb than with one of 500 pounds, hence a shorter takeoff (243 feet), because the company assumed that when carrying this bomb the airplane would carry much less fuel. Brewster expected the bomber to make 281 mph at sea level and 314 at 15,000 feet (with either a 500- or 1,000-pound bomb). Calculated stall speed was somewhat high, 70 mph. Range was given as 475 miles at economical speed, compared to 1,000 with the 500-pound bomb. Curtiss A was faster below 15,400 feet—and probably up to about 17,000 feet as well, thanks to the ram effect (air pressure generated by the aircraft's motion) on its carburetor. BuAer estimated that Brewster's design was 319 pounds over the specified gross weight limit and 1.1 mph over the stalling-speed limit as a result of the addition of 225 pounds empty weight and 94 pounds of fuel and oil to reach the specified thousand-mile range. Brewster B had no supercharger. It was at least as fast as Brewster A from 15,000 feet down, taking the ram effect into account. Maximum speed as a scout would be reduced to 311 mph at 15,000 feet, and gross weight would be 9,840 pounds; service ceiling would be reduced to 28,000 feet. Given the overweight of Brewster A, BuAer preferred Brewster B.

Compared to Curtiss A, Brewster B offered lower performance, because of its lower aspect ratio (5.8 versus 6.5) and for other aerodynamic reasons. Because it was less efficient aerodynamically, it needed more fuel and oil. It was also heavier, because its wings had less taper and were thinner at the root. On the other hand, its similarity to existing Brewster airplanes (XF2A-1 and XSBA-1) seemed to assure satisfactory control and stability. Still, the folded width of thirty feet seven inches would preclude placing

two airplanes on an elevator. BuAer wanted the airplane's wing to assume an of attack of about fifteen degrees on the ground, so that its wing could develop 90 percent of its lift for takeoff in the three-point position. Brewster offered an 11½-degree angle. The airplane was sixty-three pounds over the weight limit, 0.3 mph over the 70-mph stall limit, and 11 feet over the 250-foot takeoff limit.

Brewster used split flaps and extended landing gear to limit diving speed. BuAer did not like using landing gear for that purpose, because it would take time to retract (and to adjust elevator tabs for balance and control) and thus slow getaway after dropping a bomb. It preferred Curtiss' bomb-door solution.

BuAer also disliked Brewster's crew arrangement, in which pilot and gunner / radio operator were separated by a solid bulkhead, the gunner's forward vision further limited by a metal covering over part of the greenhouse. After the competition, Brewster moved the oil tank from between pilot and gunner to between pilot and engine. It tipped the wing up two degrees (it would be tipped down a degree in level flight); the new 13½-degree three-point angle was good enough. The wing hinge was moved inboard to reduce folded width to twenty-one feet six inches.

Curtiss A was rated highest, with 83 points; Brewster A was rated at 78.0.[9] In January 1939 the chief of BuAer decided to award contracts for Curtiss A and Brewster B, based on a 30 December engineering evaluation. They became the XSB2C-1 and the XSB2A-1. They gave BuAer the high maximum speed it wanted.[10] Excessive diving speed was accepted, because BuAer had not yet managed to reduce diving speed in any other airplane to the desired extent.[11]

The SB2C Helldiver

The SB2C Helldiver was often called, without affection, "the beast"—and worse. In 1943 it was declared unfit for carrier service until it was heavily modified. After the battle of the Philippine Sea in June 1944 there was some fleet sentiment in favor of replacing it with the earlier SBD. The SB2C probably deserved

A pair of SB2C-1s from VB-8, assigned to the new carrier *Intrepid*, fly near a convoy in the Hampton Roads area, December 1943. These early Helldivers had 0.50-calibre wing guns, distinguishable from the later 20-mm cannon. Note the ASB antenna post and underwing bomb rack. NHHC

better: it delivered half again as much bomb weight as the Dauntless, and its loss rate to Japanese anti-aircraft fire and fighters was comparable. At the Philippine Sea, Helldivers did not suffer badly until they were forced to ditch as they ran out of fuel on their return. During that night recovery, thirty-nine of forty-seven returning aircraft were lost. Only four had been lost in combat, out of a total of eighteen aircraft shot down.[12]

The most serious problem seems to have been longitudinal strength. Apparently a design defect in the tailwheel strut or in its connection with the fuselage transmitted too great a stress each time an airplane landed. An underlying problem may have been poor workmanship, which in turn can be traced to the intense mobilization of American industry during World War II.

By the end of the war, the problems seem to have been cured. In the late summer of 1945 the Helldiver was the only precision bomber the fleet had. Production was ordered ramped up (and Avenger production reduced), presumably to meet the expected demands of the invasion of Japan. Helldivers continued to serve on board U.S. carriers until 1949.

They were transferred to France, Greece, Italy, Portugal, and Thailand, seeing service as dive-bombers in French, Greek, and Thai service against communist guerillas.

As the SB2C was being developed from prototype to production, the Germans overran Western Europe, their attacks spearheaded by dive-bombers. Not only did the Navy suddenly need many more dive-bombers to fill newly authorized carriers but the Army wanted them too. It adopted the Helldiver as its A-25. At this time Curtiss, mass-producing the P-40 fighter, was the only American aircraft company mass-producing aircraft of any type. It understood that every aspect of the SB2C program had to take future mass production into account. The prototype itself would be built using production tooling; mass-producible structures (such as forgings and die castings) would be used as much as possible. (In the case of the P-40, prototypes had been built by hand: there had been plenty of time to cure problems before production.) Curtiss understood that it would take longer to engineer the airplane in this form but considered that justified because production would be much quicker. The disadvantage was that the closer the prototype Helldiver was to the planned production version, the more difficult it would be to change the production version to reflect experience with it. This was a common wartime problem. The American solution was generally to keep producing a standard design and send completed aircraft to modification centers, which would bring aircraft up to the desired standard until the production line could be stopped, changed, and restarted.

To complicate matters further, in March 1941 Curtiss decided that Helldivers would not be built at its existing plant in Buffalo but rather in a new government-financed plant at Columbus, Ohio. That separated the production plant from Curtiss' design team. It also required Curtiss to train new production workers.

When the SB2C was ordered into production in May 1939, its prototype was to have been delivered on 15 October 1940. But while it was being built, BuAer asked for armor and self-sealing tanks in response to reports about combat in Europe. Also, it became interested in a Marine Corps (expeditionary) version, analogous to the SB2U-3. Finally, in July 1940 wind-tunnel and full-scale tests showed that the lift coefficient of the wing had been overestimated: stall speed would be unacceptably high. The wing area had to be increased from 385 to 422 square feet, at a cost in speed and in production time.[13] The prototype first flew on 18 December 1940.

Curtiss' test pilot attained 323 mph at 16,500 feet. He reported that the airplane was extremely satisfactory as regarded vision, control near the stall, and all-round flying qualities. Unfortunately, on 9 February 1941, before delivery to the Navy, the prototype crashed at Buffalo Airport. The crash, however, did not reveal any fundamental problem. One wing, the landing gear on both sides, and the tail had to be replaced. Flight tests resumed on 6 May 1941. This delay gave enough time to develop hydraulic control for the turret carrying a 0.50-calibre gun. Curtiss was also able to revise its wing design to provide more forward-shooting firepower. Wing and fuselage structures were modified, because static testing had shown some weakness.

A larger, redesigned vertical and horizontal tail was fitted in September to cure longitudinal instability in the carrier-approach condition. A production-type engine mount was installed. Delivery of the modified prototype slipped from 1 October to November 1941. But on 21 December 1941 the prototype crashed again and this time was completely destroyed (the test pilot got out). It had apparently disintegrated in the air while pulling out of a high-speed dive. The Helldiver was later generally required to dive using dive brakes and with bomb-bay doors open to limit its speed.

Curtiss was told to expedite the first four production aircraft so they could be used for formal Navy trials. By this time American industry was mobilizing, so demands on forgings and dies grew and supplies became tight. In another program they would not have been needed until mass production began, but Curtiss' decision to build all aircraft with production parts delayed the construction of what amounted to further prototypes. These aircraft were due in May 1942, but the first production airplane flew on 30 June. Overweight was now estimated at seven hundred pounds. Estimated takeoff runs were 300 feet with a five-hundred-pound bomb and normal fuel, 335 feet with a thousand-pound bomb and normal fuel, and 380 feet with a thousand-pound bomb and 290 gallons of fuel. In August the first air-plane was ready to begin diving tests. By 7 September the preliminary demonstration was about two-thirds complete, and an airplane had dived at 360 knots and had withstood 8 *gs*. The worst problem was that the ailerons had proved very heavy at high speed.

As of January 1942, 865 aircraft were on order (471 for the Navy, 294 for the Marines, and 100 for the Army). In March that figure exploded to 6,000 aircraft, half to be built in Columbus for the Navy, half in St. Louis for the Army. Another 1,000 aircraft would be obtained from Canadian Car & Foundry (CCF) at Fort William, Ontario, under the designation SBW-1. Of these, 450 were earmarked for the British (later 1,000 were to have gone to the British). Late in April, orders stood at 3,571 SB2C-1s, 294 Marine SB2C-2s, and 3,100 A-25s, plus the projected SBWs. The fifty-first production airplane would be the first SB2C-2, after which 1 of each 3 would be the Marine version.[14] Later a contract for three hundred more was let to Fairchild of Canada (SBF-1s). Curtiss had jigs in place, but it could not begin assembling fuselages until it had seven large, critical forgings—but such forgings were in short supply.

In November the BuAer Bomber Desk reported that "the airplane appears to be good except for the complicated hydraulic system." Landing characteristics were satisfactory, and the Helldiver had bettered the expected takeoff run. It had shown "surprisingly good" characteristics under overload. Some desirable changes were being made on a not-to-delay basis; there had been relatively few mandatory changes. The fifteenth airplane would go to Canadian Car & Foundry as a pattern for its SBW-1 program.

The SB2C was modified so that it could carry a torpedo instead of a bomb. The thirteenth airplane went to Naval Air Station Quonset Point, in Rhode Island, for torpedo tests. They were made at higher speeds than had been believed possible when the airplane first arrived, as high as two hundred knots. Unfortunately, it took about four hours to install or remove the torpedo fittings. However, in the fall of 1944 a new adapter for the SB2C-3 could be installed or removed in forty to sixty minutes; a

thousand were ordered. (They were never used in combat, presumably mainly because so few torpedo targets remained by the time they were available in quantity.) With both forms of adapter, the torpedo was carried semi-exposed (as was the two-thousand-pound bomb, another overload item), and the "side area" (i.e., surface exposed to air) added by its nose caused some control problems.

Early in January 1943 the BuAer Bomber Desk reported that

> the airplane is showing considerable promise as the trials progress. It has very good landing characteristics, handles overloads well, and as a torpedo plane has exceeded our expectations. There have been considerable maintenance troubles and a number of parts have had to be strengthened as a result of arresting, catapulting, and accelerated service test trials. Considering the fact that the XSB2C-1 crashed to complete destruction and that the production articles were the first airplanes available for Navy pilots to fly, the airplanes have done very well[;] . . . the airplane has bettered the anticipated performance.

It had been impossible to incorporate some important armament changes in early airplanes before delivery. The airplane seemed to be ready for service.

This SB2C-4 (BuNo 20411) photographed at Patuxent River with torpedo and fairing in place, 16 March 1945, shows the final torpedo-carrying installation. It was never used in combat. NATIONAL ARCHIVES

The first twelve fleet airplanes went to Norfolk in January 1943 to join Carrier Group 9 on board the new carrier *Essex*. The carrier's personnel found maintenance extremely difficult, largely because of poor workmanship. By early April 1943 the Bomber Desk was referring "many deficiencies in design, workmanship, and inspection" and demanding an intensive modification program at Columbus; two hundred airplanes would be involved, about a hundred of them coming directly from the production line. This was the first of three modification programs.[15] The airplanes were used intensively, and new deficiencies kept appearing. Most disturbingly, after the first carrier-landing qualification exercise; the mechanism of the tailhook had to be changed. Tail wheels were collapsing. A second round of carrier qualifications revealed yet further deficiencies.

The 201st airplane received 20-mm wing guns. In October 1944 airplanes with two 20-mm cannon and all-hydraulic (as opposed to screw-jack) flaps were designated SB2C-1Cs (the -1As were Army A-25s). The 979th airplane was scheduled to have the R-2600-20 (rather than -8) engine with a four-bladed propeller, as the first SB2C-3. It also had the first redesigned turret with a new hydraulic unit (CH-150-2). The new engine was also intended for the Marines' SB2C-2.

By 1 June 1943, Carrier Air Group 17 (on board USS *Bunker Hill*) had practically completed training. It had found three major needs: a lightweight gun installation, a self-sealing fuselage tank, and reinforcement of the fuselage near the tail wheel, which was locked in the extended position. The wing-folding mechanism was modified. The 0.50-calibre turret (CH-150-1) was removed from the first 978 airplanes, replaced by a twin 0.30-calibre flexible gun on a Douglas mount.

Two SB2C-1Cs from *Yorktown*, probably in 1944. They had the usual two wing 20-mm guns and a twin unpowered 0.30. The diagonal tail stripe indicated the carrier. She may have been the first to adopt such markings. They became necessary once carriers were grouped together tactically, late in 1943. Markings were initially unofficial, but on 7 October 1944 large symbols were assigned to each fleet carrier (most carriers did not apply them until January 1945). On 27 July 1945 they were ordered replaced by letters. NHHC

Curtiss set up a modification center at its plant in Louisville, Kentucky. Three aircraft were allocated to Curtiss Columbus to serve as mock-ups on which to incorporate changes as they arose, in hopes of speeding modifications. Nevertheless, the modification program lagged badly; as of October 1943 Curtiss Louisville was more than a month behind, and Curtiss Buffalo "modification production" had been reduced to twenty aircraft. A month later five modification centers were operating: Curtiss Columbus, -Louisville, and -Buffalo, plus Fairchild (Canada) at Montreal and Consolidated-Vultee at Allentown, Pennsylvania. Columbus and Fairchild seemed capable of modifying aircraft in sufficient quantity.

The 601st was the first Helldiver delivered embodying all modifications planned as of the date of its production.

As of September 1943, it was hoped to have a thousand SB2C-1s ready for combat by 1 January 1944. Planned production was now 978 SB2C-1s, 294 SB2C-2s, and 2,593 SB2C-3s (Fairchild SBW-3s). Planned Canadian Car & Foundry production for Britain was cut, initially to 410 aircraft (SBW-1B and -3B, the latter the CCF version of the SB2C-3). In November the 1,000 British Helldivers were cut to 26; CCF would now be building Helldivers mainly for the U.S. Navy. By August 1944 it had been contracted for 1,320: the original 40 SBW-1s, the 26 SBW-1Bs for the British, 414 SBW-3s, 580 SBW-4s, and now 260 SBW-5s (of which 220 were canceled about September 1944 because of reduced demand). Fairchild of Canada had had 300 on order (50 SBF-1s, 150 SBF-3s, 100 SBF-4s) and a letter of intent for another 60 SBF-4s (which were canceled). Unfortunately, Fairchild (and Curtiss Columbus) depended on Chrysler for SB2C-4 center sections, whereas CCF made its own. Chrysler lagged and was unable to shift to making SB2C-5 center sections. To keep production running, 220 SBFs were transferred to CCF, to become SBW-5s.

The Marines' SB2C-2 was deferred in January 1944 to the end of the year, then killed altogether at a 31 March 1944 BuAer production planning conference. The Marines no longer needed a special airplane to be delivered to them via lagoons, and production manpower was very tight. They received 410 ex-Army SB2C-1As.[16] The planned -2s were reordered as SB2C-4s.

When an SB2C-3 was tested at Naval Air Station Patuxent River, Maryland, it turned out that the turret pushed the center of gravity too far aft. A pair of synchronized 0.50s in the nose would push it back forward, but their added weight had to be balanced by eliminating the two 20-mm wing guns.

An SB2C-3 from the carrier *Hancock* flies over Formosa en route to attack shipping, 13 October 1944. Note the underwing package gun suspended from its bomb rack. This version had a four-bladed propeller instead of the three-bladed propeller of the SB2C-1 and, typically, a small spinner. Having a name written below the cockpit (in this case "Satan's Angel") was unusual among carrier aircraft. SB2C-3s began to replace -1s early in 1944. NHHC

Provision for these guns, however, was retained; force commanders could fit the 20-mm cannon at their discretion.

About November 1943, Curtiss proposed using R-2800 or even R-3350 engines, Emerson Dumbo and Curtiss turrets, and a variety of single- and two-seat versions with a lengthened fuselage and increased protected fuel capacity. BuAer decided to ask for a cost quotation for the R-2800 version, including production tooling.

Meanwhile, the first action report (by VB-17 in January 1944) criticized mainly the weight of the airplane, which lengthened its takeoff run. That, it was felt, should be improved greatly by the extra 200 hp offered by the R2600-20 engine in the SB2C-3. Curtiss was seeking "with highest priority" a cure for malfunctioning flaps in the SB2C-1C. One of the early production airplanes was tested at Patux-ent River with perforated dive flaps, as in the Dauntless. Opened to thirty-five degrees, they cut diving speed to 270 knots.

The next version (SB2C-4) would incorporate a new power turret, either Norge's modified one-gun turret (CH-150-3) or Emerson's two-gun Dumbo (the Norge turret was chosen in March, because the Dumbo was too heavy). This version had a new hook strut, intended to cure a hook-bounce problem that was causing carrier crashes, and the new perforated flaps. In March the 2,091st airplane was designated tentatively as the first SB2C-4.

An SB2C-4 shows important improvements: launching stubs for up to eight 5-inch rockets and antennas for the APN-1 radar altimeter inboard of them. Racks were reinforced to carry a thousand pounds of bombs underwing. Some late -3s apparently also had the stubs. The finish is the overall glossy blue ordered in June 1944 (effective that October), with nonspecular blue when needed to reduce glare for the pilot (it is visible forward of the cockpit). U.S. NAVAL INSTITUTE PHOTO ARCHIVE

An SB2C-4 from the carrier *Shangri-La* at the end of the war, functioning as a photo plane (note the camera in the rear cockpit). It sports the end-of-war dark-blue color scheme and the carrier identification letter on the tail, introduced on 27 July 1945. Note the two whip antennas, one atop the canopies (presumably for IFF), the other below the bomb bay (for homing). There are no ASB antennas, because the airplane can be adapted to carry the APS-4 radar "bomb." This -4 can be distinguished from a -5 because its cockpit lacks a frameless canopy. It has the fat spinner, like an SB2C-1 but a four- rather than three-bladed propeller. NHHC

An SBW-4E at Patuxent River on 12 December 1945 shows its APS-4 radar "bomb" on the starboard bomb rack, inboard of the rocket stubs. The "bomb" housed the whole radar apart from the control box, indicators (scopes), and their amplifiers. The B-type scope provided a maplike picture. A large freighter could be detected at a fifty-mile range, and a submarine at eight. The "bomb" could be released quickly in flight if necessary. Compared to ASB, it was more resistant to jamming, and it was pressurized, so that it could be used up to 20,000 feet. It also generated less drag. Low-altitude bombing devices (APA-5 and -16) could be attached to the scope. There was a 150-mile sweep for beacon detection. The radar "bomb" was five feet long and seventeen inches in diameter. NHHC

During February 1944, a trial rocket-launcher installation was completed, as COMINCH had directed that all carrier aircraft be fitted for such. Trials were successful enough for contractors to be authorized to install these "zero-length" launchers (the weapon leaving the rail immediately or nearly so). Later the fleet became interested in diving (braked) rocket attacks, so about October Curtiss was asked for a means by which the pilot could select only center-section (rather than all) dive brakes. Otherwise the rocket blast would damage the outer panel flaps. Production was deferred pending completion of design work for the change. Also, about February, the existing single-bomb underwing racks were ordered replaced by double-suspension racks. They could handle a wider range of ordnance: bombs, fragmentation clusters, or mines, as well as droppable fuel tanks, smoke tanks, etc. The ex-Army SB2C-1As had the new racks. By this time Douglas had developed a 0.50-calibre package-gun installation, which could be mounted on the bomb rack.

In March 1944, BuAer production representatives concluded that for efficiency of production planning and material procurement, Curtiss should be authorized to build another 900 Helldivers, which would be SB2C-4s with revised cockpits. BuAer issued a letter of intent for 894 of a new version it designated SB2C-5, with more powerful R-2800-28 engines and twin 0.50-calibre turrets. It wanted the fuselage lengthened to improve longitudinal stability. When bureau, Curtiss, and NACA representatives reviewed wind-tunnel data on 22 February 1944, they decided that the engine should be moved forward about fourteen inches, in which case the tail should be about twenty inches farther aft to keep stability characteristics about the same as the SB2C-1C's. For production reasons, Curtiss proposed to make most of the planned changes in a -4C (cannon) version, leaving only the engine change and the lengthened fuselage to the -5. The arrangement was set at a 24–25 March 1944 conference. To maintain balance, the centerline bomb rack would be moved forward 6.6 inches and would be able to handle a two-thousand-pound bomb. The airplane would be modified to use package guns. This version would incorporate the new cockpit, including provision for an APS-4 radar and APN-1 radio. It would be strong enough to pull 7 *gs* in flight at 17,000 pounds gross weight and be able to land and be catapulted at that weight. Two additional fifty-gallon wing tanks would be installed. This version would have the double-suspension wing bomb rack, which could take a fifty-gallon drop tank.

A new cockpit would be installed in the three-thousandth airplane or later. Compared to the existing one, it was considerably cleaned up, with a hydraulic console on the left side, an electrical and electronics console (including radar and radio controls) on the right, and armament switches and controls on a horizontal strip below the lower instrument panel. A Mk 30 torpedo director could be rotated out of sight when not in use.

The SB2C-4C proposal was so attractive that this version was redesignated SB2C-5 in April 1944. The airplane with the new engine and lengthened fuselage would be the SB2C-6. The letter of intent for SB2C-4s and -4Cs (ex-SB2C-5s) was superseded in April 1944 by one for 1,794 SB2C-5s, to be delivered from December 1944 through June 1945. All would have the revised cockpit. The first SB2C-6 aerodynamic prototype would fly in June 1944 and the second in July.

There were still problems, including an increasing backlog of airplanes on the West Coast awaiting modification. In May 1944 Patuxent River reported that an SB2C-3 with a thousand-pound bomb had displayed inadequate longitudinal stability in carrier-approach, level-cruising, and high-power-climb conditions. It also had inadequate directional control in the carrier-approach condition. A 15–16 May conference at Columbus proposed corrections. Tilting the engine up five degrees would improve stability by increasing the distance between the airplane center of gravity and the thrust line. Doubling the number of ribs in the rudder moved the neutral point of the airplane aft as much as 1.5 percent. Redesign of the entire control system, using ball bearings and larger sheaves whenever possible, was considered. The stability investigation proceeded through June at high priority. Possibilities included redesigning the fin and rudder and modifying the trailing edge of the elevator. The stability problem explains why SB2C-4s retained their twin 0.30s and did not get the CH-150-2 turret.

The production changeover from SB2C-3 to -4 was completed early in July 1944 without delaying deliveries. Accelerated service tests confirmed better

An SB2C-5 at Patuxent River, 11 January 1946, showing its characteristic frameless cockpit canopy and its small spinner. This version had a modified tailhook. The main difference from the -4 was increased fuel capacity. NHHC

maneuverability in the dive, better stability, and improved workmanship. Diving speed was reduced in the split-flap condition. Curtiss continued experiments with a larger rudder and an improved control system with reduced friction.

In August 1944, it was expected that the XSB2C-5 would be completed on 15 October, with the rearranged cockpit. A separate armament prototype had 20-mm (T-31) guns, Mk 5-1 rocket launchers, Mk 5-7 wing bomb racks, an improved Mk 8 sight installation, and an improved gun-camera installation and could accommodate a two-thousand-pound bomb. Its centerline bomb rack had been moved forward slightly. Fuselage fuel-tank capacity was reduced from 110 to 55 gallons, two new 44-gallon wing fuel tanks being installed accordingly. That increased protected internal fuel by 33 gallons. The change was intended to keep the center of gravity as far forward as possible to keep performance, including takeoff, as nearly as possible that of the SB2C-4. The extra fuel would increase combat radius (on internal protected fuel) to three hundred nautical miles. The center of gravity location of the -4 would be retained without any need for propeller ballast weights, underengine armor, or a "spinner" (streamlined fairing over the propeller hub).

The first XSB2C-6 flew on 24 August. BuAer, which must have been nervous about delays with the SB2C-5, told Curtiss not to continue production engineering of the -6 since it was pushing as hard as possible on the -5; the production version could not be defined until the prototype had been delivered and evaluated or production had been ordered, or both.

By working seven days a week and twenty-four hours a day, Curtiss completed all static tests of the SB2C-5 except the final drop. In March 1945, Columbus shifted production to the SB2C-5, after delivery of the last SB2C-4E. Curtiss managed the shift without losing a single airplane, and by 26 April 1945 202 SB2C-5s had been delivered. They and the last 270 SB2C-4s had a new improved-vision canopy, which was well liked. To increase striking power, Curtiss was asked to begin work on adapting the aircraft to accept to fire the 11.75-inch Tiny Tim aircraft rocket.

As of 15 April, the two SB2C-5s assigned to service testing at Patuxent River had accumulated, respectively, 113 and 58 hours of flight time. Only minor maintenance problems had been reported. By 25 May, 370 SB2C-5s had been delivered, and the two service-test aircraft had accumulated about 260 hours each.

The SB2C-6, prototype of the planned final production version. THE HOOK

In the fall of 1944, with Avenger production not meeting fleet requirements, it had been proposed that Helldivers replace some Avengers, on board both fleet and escort carriers. In November 1944 an SB2C-3 equipped with the "special ASW weapon" (presumably the Mk 24 FIDO homing torpedo) was catapult- and arresting gear–tested at the Naval Air Material Center at Philadelphia. SB2C-3 take-off-performance curves (showing how much deck was needed at various weights and engine settings) were forwarded to all carriers, including escort carriers, on 8 November. The big escort carrier *Santee* had successfully operated SB2C-1s in April 1943. It reported that the airplane compared favorably with the Avenger and was much better than the SBD-3 Dauntless. The Helldiver landed in a shorter distance than the Avenger, and with better longitudinal control. Except for the SBD-3, it was more rugged than other types operated from the ship; in barrier crashes only its propeller was damaged, and it had no tendency to nose-up. It did require about 25 percent more takeoff run than an Avenger with the same bomb loading but about 20 percent less than an SBD-3. Contemporary Pacific Fleet evaluation was less favorable. The Helldiver was much larger than the Dauntless and faster in a dive but less maneuverable.

During the run-up to the invasion of Japan, the Helldiver became increasingly important, presumably because it was the only available precision bomber. Plans called for increasing production to about 325 per month, despite Curtiss' continuing problems; production of aircraft already under contract would continue through November (or, at the accelerated rate, through September 1945). The same plans called for a reduction in the rate of Avenger production.

The Alternative: The SB2A Buccaneer

It is not clear from the original documents whether the SB2A was considered a backup for the Helldiver, but it was never ordered in the same numbers and was never operational as a U.S. Navy dive-bomber. Unlike the SB2C, Brewster's dive-bomber was released for export. The British and the Dutch each ordered about a hundred airplanes. Like the Helldiver, the export version could carry two 500-pound bombs side by side instead of a single 1,000-pound bomb. It had a 0.50-calibre turret gun, armor, and self-sealing gas tanks. The protected tanks gave a range of one thousand miles (plus another five hundred with unprotected tanks), but the oil tank was unprotected. To accommodate the additional weight of the armor (gross weight increased from 10,180 to

Brewster's SB2A won the 1938 dive-bomber competition alongside the Helldiver, but it was far less successful. The British called the aircraft that were sold to them "Bermudas," and no U.S. name was ever applied. Bermudas never saw action. U.S. NAVAL INSTITUTE PHOTO ARCHIVE

11,850 pounds), the engine mount was lengthened by about a foot. Stall speed rose to 75 mph from 70. These figures gave some idea of the performance cost of the armor BuAer soon wanted.

In April 1940 estimated delivery slipped to 24 October. Eager to get a prototype into the air, in August BuAer decided to eliminate some ground tests. A mock-up inspection in April was not entirely satisfactory, but the need for dive-bombers was urgent. During the week of 19 August BuAer negotiated to buy 409 aircraft. As with the Helldiver, it was buying an airplane that had never flown. Moreover, Brewster was far less experienced than Curtiss. In December 1940 BuAer ordered 140 more aircraft. By about that time Brewster had sold 912 to the British (as the Bermuda) and the Dutch (as Model 340D).

The Brewster firm was badly overloaded. It would build its airplanes at an as-yet uncompleted plant at Hatboro, outside Philadelphia. It planned to get more than half its parts from subcontractors. Although in January BuAer expected the prototype to fly about 1 April 1941, it flew only on 17 June. As of March 1941, 203 aircraft were on order, to be delivered beginning in June, which was wildly unrealistic. The contract called for a production rate of twenty per month by February 1942. By this time five hundred were on order for the British and the Dutch. To speed delivery, BuAer and Brewster eliminated some equipment, saving 180 pounds and about $1,000.

A frustrated BuAer Bomber Desk staffer wrote in the March 1941 status report that the prototype contract was being carried out in a "most extraordinary" way, that "none of the contract requirements have been completed or probably ever will be completed." Given the airplane's structure, large tanks had to be replaced by smaller protected ones, roughly halving fuel capacity. An additional tank was added in the fuselage. It and the armor moved the center of gravity too far aft, so the engine had to be moved forward ten inches. As yet the subcontracted gun turret was incomplete. The XSB2A differed so greatly from the planned production version, owing mostly

to its lack of armor and protected fuel tanks, that Brewster wanted BuAer to acccpt it as an experimental type.

The BuAer representative decided that the program was in such trouble that he would avoid making any special arrangements with Brewster; he described the company's condition as "confused."

Brewster was desperate enough to offer to pay for any modification of the prototype into a true representation of the production version. It also offered to deliver stripped-down production aircraft at reduced cost. When the prototype finally flew with a turret and a modified after fuselage, the top of the turret blew off at only 200 mph. In October 1941 it seemed that the first airplane would not be delivered until January or February 1942, and the first ten not until March or April. A final mock-up was inspected on 30 December 1941, but the turret had not yet flown with a real (instead of metal) dome. By this time flight tests had shown a need for a larger tail. At this point estimated overweight in the production version was 1,100 pounds.

In March 1942 the Bomber Desk reported that it had been impossible to dive the prototype at high speed, because its wings had been proof-tested only to about 6.5 gs. The bomber was now up to 1,250 pounds overweight. At about this time the BuAer Plans Division decided to abandon heavy powered turrets in both the Brewster and the Curtiss dive-bombers in favor of twin flexible 0.30-calibre guns. That would save some weight, but in a twenty-five-knot wind the bomber would still need 360 feet to take off with a 500-pound bomb (normal fuel), 410 feet with a 1,000-pound bomb (normal fuel), and 430 feet as a scout (362 gallons).

No Navy delivery schedule could be set, although Brewster had already delivered aircraft to the British. Recently a British Bermuda had experienced aileron overbalance at only 360 mph. This dangerous condition had to be corrected before the SB2A could be accepted for U.S. service.

In June 1942 BuAer split the type, the more quickly to get the airplane into production. The SB2A-1 was the full Navy prototype with the long engine mount, turret, folding wings, and carrier provisions. SB2A-2 had the British free gun, fixed wings, carrier provisions, Navy bomb installations, Navy fixed-wing 0.50-calibre guns, and a British synchronized gun installation (0.30s). SB2A-3 was the same but with folding wings. BuAer wanted 93 (later 80) -2s and 119 (later 122) -3s. The first SB2A-2 was due to fly about 1 September 1942. It was badly overweight (BuAer estimated 1,100 pounds). The only optimistic note was that in July it seemed that the British were about to accept the Bermuda formally, as most problems had been eliminated. As for the 162 Dutch aircraft—the Netherlands East Indies (for which they had been ordered) having been overrun by the Japanese—the Navy took them over, as SB2A-4s. They differed from British aircraft in bomb installation and radio equipment. The first production airplane flew on 31 July 1942. By September 1942 an equivalent British Bermuda had dived at 400 mph and 6 *g*s.

As of November 1942, one SB2A-2 had been delivered, but no SB2A-3s. BuAer considered these airplanes quite inferior to the Helldiver and planned to allocate all of them to operational training. However, they carried a good load and had a fair range. Operated from land, they were better than the SBD-3. BuAer suggested that the Marines might find them very useful if the spares problem could be solved— but logistics demanded minimizing the number of different planes. Deficiencies included poor directional stability, sloppy control characteristics at low speed, questionable aileron characteristics (borderline for and use and unacceptable on carriers), high landing speed, poor visibility on approach to land, poor brakes, and excessive carbon monoxide concentration. It did not help that Brewster was very slow in adopting requested changes. The 2 January 1943 Bomber Desk status report recommended not buying any more. The SB2A never saw active service.

The 1941 Dive-Bombers

With the 1938 dive-bombers about to enter production, in January 1941 BuAer held a new design competition. It hoped that new engines would make it possible to achieve 350 mph at full military power (100 percent throttle). BuAer would accept a higher stall speed (75 mph, even 77 with the tricycle landing gear it now favored).[17] To accommodate both the heavier engine and protection, BuAer would accept an airplane 40 percent heavier than in 1938: 14,000 pounds gross with a 500-pound bomb (14,250 with a tricycle undercarriage). The most powerful engines contemplated were the R-2600 (1,800 hp), R-2800 (2,000 hp), and R-3350 (2,300 hp for takeoff). Maximum takeoff run was no longer specified. As in 1934 and 1938 dive brakes were needed, in this case to limit diving speed (with bomb-bay doors open) to no more than maximum attainable level speed. They should be extended within ten seconds at maximum attainable level speed without objectionable buffeting or vibration. Vought and Vultee questioned the dive-bombing speed requirement, and in March it was changed to a maximum indicated airspeed of 350 mph with bomb-bay doors open and landing gear retracted. Flaps and bomb-bay doors should close rapidly and automatically once the bomb was dropped and the displacement gear was back in the bomb bay.

As in 1938, required range with a five-hundred-pound bomb was held at 1,000 miles (1,500 for scouting with overload fuel).

Designs with full-span flaps would be accepted if they appeared to make workable provision for lateral control and dive braking. As in the previous competition, good stability and control near the stall was essential. Special consideration would be shown to design features favoring low-speed stability and control.

BuAer still wanted to stow two airplanes on an elevator (forty-four by forty-eight feet or forty-one by fifty-two). That limited weight to 14,250 pounds, as the elevator capacity of the new *Essex* class was 28,000 pounds. According to a handwritten comment on the draft circular to vendors, the two-airplane requirement was a way of holding down size and folded wingspan. Folded height should not exceed seventeen feet (twenty during folding might be accepted). Power wing-folding was mandatory. Designs that could be spotted more compactly would be favored.

The normal bomb load would be one 500-pound bomb carried internally; alternatives would be one 1,000-pound or two 500-pound bombs carried internally, or two 100-pound bombs underwing, or two to four 325-pound depth charges on the wing racks and in the bomb bay. The standard fixed-gun installation was now two 0.50-calibre guns, with two additional guns and more ammunition carried when the airplane operated as a scout. An alternative special loading condition was 20-mm HSB (Hispano-Suiza-Browning) cannon (Commander Aircraft Battle Force wanted at least four 0.50s). The standard flexible-gun installation would be two 0.50s. To achieve the desired flexible-gun coverage, BuAer recommended study of a remote-control turret. It visualized two guns in streamlined blisters on the sides of the airplane, both controlled by a single gunner with a double periscopic sight whose ends would project above and below the fuselage. Companies already working on remote-control guns included GE, Maxon, and Douglas. Since remote control was developmental, these guns were not made mandatory.

Protection should include leakproof gas tanks for the thousand-mile normal range (other tanks would have a purging system or be droppable). Pilot and gunner should have armor.

Since an autopilot was required, there was no need for the gunner to have auxiliary controls.

The letter was circulated on 4 February; proposals were due by 30 April. As in 1938, the highest speed required the most powerful engine, in this case the new R-3350. It was offered by Brewster, Curtiss, Douglas, and the Naval Aircraft Factory. The most powerful engine required the heaviest airplane: the weight limit had to exceed 15,000 pounds. For maximum performance below 18,000 to 20,000 feet, the two-stage supercharged R-3350 was best; it offered about 10 mph better than the R-2800 in a Curtiss design, about five at intermediate altitudes in the Douglas design. Above 20,000 feet the two-stage R-2800 was best. The BuAer evaluator did not

consider that this high-altitude performance outweighed higher speed lower down. Remote-control turrets (Douglas and Vought) cost between £200 and £300 compared to Curtiss' conventional turret, even though the latter's aircraft had wings that were heavier because of their greater span. The penalty would have been worse had Curtiss made do with much the same wing arrangement. Only one bidder met or bettered the £14,250 limit, though Douglas and Vought came close to it.

Curtiss offered a design similar enough to its SB2C to suggest that it could quickly enter production. It used an R-3350 with two-speed supercharger, tricycle landing gear, and a conventional two-gun upper turret. The turret seemed much closer to realization than the more complicated types offered by Douglas and Vought. The power-plant installation was good, and lateral control seemed likely to be good. The pilot would enjoy good vision over the nose (nine degrees). Curtiss enjoyed the highest armament rating, due in part to its flexible bomb/overload-fuel-tank arrangement. It could carry two 1,000-pound bombs 1,400 miles at an overload weight of about 16,800 pounds. BuAer found this capability, which was not required, attractive. The bomb bay was so large that, used for fuel tanks, it would provide a range of 2,540 miles, much increasing the value of the airplane as a scout. Curtiss also offered the best fixed guns and next to the best flexible-gun coverage (assuming that certain undesirable features of the turret design were overcome). It was the lightest of all the designs offered, and it had a relatively short takeoff run of 258 feet. Against that, to limit height Curtiss folded the wingtips as well as the main part of the wings, adding weight. From a control point of view, the design was very close-coupled, so there would probably be trouble turning trim power on and off. Rudder forces might be too high. In "battle ready" condition the airplane would lose 35 mph (probably 25 mph with guns facing astern but with the enclosure open for vision). Although the tricycle landing gear would be new to Curtiss, it was not expected to present any difficulties.

Curtiss' R-3350 design would make 351 mph at 14,000 feet and would stall at 76.8 mph; takeoff run would be 306 feet, service ceiling would be 30,900 feet, empty weight 10,728 pounds. The alternative tail-down design using a two-stage R-2800 would weigh about as much (10,734 pounds empty) and would make 350 mph at 22,800 feet; stalling speed would be 76.8 mph, and deck run would be 353 feet. Another alternative added upper and lower remote-control turrets; BuAer's armament section did not like it as much as the Douglas or Vought offers. In view of a decision to proceed with the Douglas bomber, with its twin turrets, there was no point in proceeding with Curtiss' version, the most expensive of the lot. Adding the twin turrets would, moreover, eliminate Curtiss' weight advantage.

BuAer rejected Douglas' favored contraprops, because they added too much weight and because engine makers could not yet quote details. That left one option powered by a two-stage R-2800 and one powered by an R-3350, both with gull wings—which threatened the sort of lateral stability problems then being experienced with the Corsair (F4U). Douglas' 517M (R-2800) offered the highest speed and would be particularly attractive if its NACA laminar-flow (minimum drag) wing succeeded. Against that, ceiling was slightly lower than Curtiss' and takeoff run slightly longer. Douglas proposed a new type of double-slotted flaps, on which it had done preliminary tests for an Army project but that had to be proven in wind-tunnel tests. BuAer disliked the armament installation: the wing guns were in the outer wing panels and their ammunition in the center section, offering access only after the wings were folded. Douglas pointed out that because space in the engine section of the R-2800 version was limited, it was impossible to provide an intercooler (between the two supercharger stages) to make full use of the second stage climbing during bad weather.

Below 18,000 feet the version with the R-3350 and two-speed supercharger had the highest performance of any design submitted. Maximum speed would be well over the requirement, at 370 to 392 mph. To achieve that Douglas had kept wing area to a minimum consistent with the stall-speed requirement. Ejector exhausts converted the maximum amount of heat energy from the engine into thrust, and the latest type of NACA cowling would minimize cooling drag. The engine installation and maintenance burden with a two-speed supercharger were expected to be better than with a two-stage supercharger. Size had been held down to the point that four airplanes could be spotted abreast on an 82-foot-wide flight deck. The gunner's sighting periscope could also function as a drift sight for navigation.

BuAer considered Douglas' R-3350 proposal next best after Curtiss'. It considered Douglas probably best organized to develop the remote-controlled turret. BuAer also wanted to try the new laminar-flow wing, on which Douglas "probably has done more work . . . than the rest of the bidders." Tests on Army airplanes using similar wings (presumably the P-51) would soon be available. If they were negative, there would be plenty of time for the bureau to order a conventional wing for a Douglas dive-bomber.

Brewster's midwing Model 37-A was also powered by an R-3350. Its only attraction was its low price. Otherwise it was unattractive: too heavy (15,429 pounds gross, 10,556 pounds empty), poor engine installation, poor spin recovery, probably poor stability at high speed (due to stabilizer location), and probably excessive control forces. It had no bombing approach window, spotting was poor, and armament installation was poor (the fixed-wing guns were practically at the inner end of the ailerons). Brewster's other designs were no more appealing. Its Model 37-B used a buried R-2800 to improve sighting angles over the nose, but BuAer foresaw an expensive developmental program for cooling and difficult maintenance even if that worked. BuAer had not yet discovered just how much trouble Brewster would have trying to develop its previous-generation bomber.

Naval Aircraft Factory's R-3350 proposal was ruled out because it required a nonexistent version of the engine (with offset reduction gear) and because it was far too heavy (15,150 pounds with the 500-pound bomb).

Vought used an R-2800 with a two-stage supercharger; it claimed a maximum speed of 350 mph at critical altitude, a stall speed of 77 mph (as required), a service ceiling of 29,450 feet, and a takeoff run of 289 feet. BuAer liked its armament installation (rated third) and its flexible-gun coverage (rated fourth). It had the best-rated powerplant installation. BuAer disliked Vought's approach to bomb displacement—a telescoping device instead of a trapeze—and it rejected altogether the company's use of wing bomb racks inside the propeller arc. The location of the stabilizer seemed to guarantee stability trouble.

Vultee offered its Model 72, which had been designed for the French army and had now been sold to the British and to the U.S. Army as the A-31 Vengeance. Its wing was not large enough to meet the landing speed requirement and would be difficult to fold. To meet the speed requirement, the airplane would need a larger engine, and that in turn would require redesign. Vultee dropped out.

At the end of May 1941 BuAer ordered two R-3350 prototypes, one each from Curtiss (SB3C-1) and Douglas (SB2D-1). Initially it seemed that the only problem would be to obtain R-3350-14 engines in time.[18] A mock-up of the SB2D was inspected in September 1941. Curtiss did not do as well: as of mid-September 1941, its SB3C mock-up was expected in October, but it was not inspected until 8–11 December.

Both airplanes were somewhat overweight from the outset, and they were expected to have excessive takeoff runs. However much it prided itself on building its prototypes essentially as production airplanes, Curtiss found itself machining key parts rather than using the planned aluminum-alloy forgings. Its flexible-gun mounting was badly delayed. Late in April 1942 both prototypes were at about the same stage of shop assembly, but Curtiss was beginning to divert men to speed production of its two major Navy airplanes, the SB2C Helldiver and the SO3C Seamew floatplane scout. Late engine delivery was expected to delay the prototypes to October and November 1942. By September Curtiss was working on a new single-seat dive-bomber (the BTC, for which see chapter 7), and the SB2D promised considerably better performance. Overweight kept increasing (as it did for the SB2D). In November, with the prototypes still far from delivery, the

Douglas' SB2D won the 1941 dive-bomber contest but was never built in quantity. For a time the project was continued specifically to promote development of the remote-controlled turrets evident here. MALD COLLECTION VIA *THE HOOK*

VSB Desk wrote that given its lower performance, its lengthening guaranteed takeoff run, and the fact that no production was contemplated prior to prototype flights, the SB3C could be deferred. To ease the workload at Curtiss Columbus, only one prototype should be built, and that with a low priority. The XSC-1 (floatplane scout) and the XBTC were to be expedited. Late in December the SB3C program was suspended altogether.

The SB2D probably survived, largely because it was Douglas' only new Navy program. By late April 1942 its overweight had been reduced to 175 pounds, the best it ever managed; the takeoff run would be 325 feet with one 500-pound bomb and normal fuel (223 gallons). By mid-June overweight was back up to 250 pounds. It kept growing: 600 pounds at the beginning of September. BuAer's evaluation was that with a typical load (a 1,000-pound bomb and full fuel) the airplane would be "logy" and a disappointment in service. In November, estimated overweight was 1,350 pounds. After a conference, Douglas El Segundo agreed to an aggressive weight-reduction program, concentrating on changes that could be incorporated in production aircraft without delay. At the beginning of April 1943, with a first flight imminent, overweight was 1,340 pounds (the 1 June report gave a figure of 1,375 pounds, nearly 10 percent). Aside from detail changes, Douglas suggested installing a jet engine (Westinghouse 19A or 19B [J30]) in the eighth production SB2D-1. That would appreciably increase speed and also solve the takeoff problem. It would eliminate any need for a gunner and for the remote-controlled turret, dramatcally reducing weight. BuAer was interested; Douglas was authorized to begin engineering. Meanwhile, as completed the prototype was 1,439 pounds ove weight. Flight tests showed that its "picket fence" wing dive brakes were not working properly.

On 7 August 1943 BuAer authorized Douglas to convert all planned SB2Ds to single-seaters, retaining only a few two-seaters to develop and test the Emerson remote-control turret. That project, using the third SB2D, was soon given a very low priority. The single-seater was initially designated STD-1 and then BTD-1. It was hoped that eliminating the gunner and the remotely controlled turrets would nearly cure the gross overweight of the SB2D. It seemed that the main remaining issue was the dive brakes, whose redesign turned out to be quite simple. Unfortunately, the resulting airplane shared the overweight and related problems of the SB2D.

As of 30 October 1943, Douglas had already produced eleven fuselages. Conversion to single-seaters with jet engines would be relatively simple. This airplane was now designated BTD-2.[19] This project was given priority second only to the BTD-1 itself. Against that, in the late fall of 1943 R-3350s were in short supply, and they were vital to the higher-priority B-29 Superfortress program. A pessimistic DCNO (Air) officer wrote that if there were not enough engines, or even enough resources to build them, for B-29s, there was little hope for the BTD; across the bottom of his memo was written in pencil, "The *out* is to push the BTD priority up equal to the B-29." Given the unhappy experience with the BTC, BuAer personnel were to think about what should be done if the BTD turned out to be unsatisfactory.

By early November it seemed likely that the BTD-1 prototype would fly about 20 November (it actually flew on 15 December 1943).[20] It was heartening that the SB2D had succeeded in diving at sixty degrees at a speed of up to 210 mph without buffeting. Space liberated in the after fuselage was given over to fuel tanks. By 1 February 1944, 622 production BTDs were on order, and the prototype had made about twelve flights.

There were problems. As in the other single-engine tricycle aircraft, the nosewheel bay could let carbon monoxide from the engine exhaust into the cockpit (the ventilator intended to cure the problem blew air into the pilot's face). The picket-fence-type wing dive brakes caused buffeting that could throw off the pilot's aim. Increased-area fuselage dive brakes and even the use of the landing gear were expected to help (eventually it might be possible to eliminate the troublesome wing brakes, left over from the SB2D, altogether). The large three-bladed propeller had poor ground clearance and had been damaged by ground contact during taxiing. Replacing with a Curtiss four-bladed, electrically controlled propeller and extending the wingtips (to a span of forty-eight feet) much improved the rate of climb after takeoff and reduced the takeoff roll by about a hundred feet. It also helped that production R-3350s were expected to produce 2,500 rather than 2,300 hp. Aggressive weight-reduction measures included elimination of the internal bomb bay (gaining three hundred pounds directly).[21] Douglas identified nearly a thousand pounds that could be saved; takeoff run was reduced from 558 to 376 feet.

In the late spring of 1945 the Secretary of the Navy agreed to waive flight demonstration tests; the contract was terminated and the SB2D prototype stricken. The Helldiver (SB2C) was considered good enough for the present, and by 1944, when a production decision might have been made, much better single-seat dive-bombers were nearly ready.

The Avenger was the standard U.S. Navy torpedo bomber of World War II. These TBM-3Es were photographed postwar, as indicated by their national insignia with the red bar, adopted January 1947. The band around it indicates that they were operated by the reserves. The lack of any station name indicates an early postwar photograph, about 1950. The *F* denoted Oakland. Like aircraft operated late in World War II, they retained their 0.50-calibre turret guns but not their tunnel (ventral) guns. J. W. HAWKINS COLLECTION COURTESY JOHN GOURLEY

6

TORPEDO BOMBERS OF WORLD WAR II

The 1939 Torpedo-Bomber Competition

In February 1937 the chief of BuAer laid out a proposed FY38 experimental airplane program.[1] BuAer wanted higher bomber speed so that all components of the carrier air group could fly out together for mutual support at the target. The next carrier type should be the VTB. The chief feared that a twin-engine torpedo bomber would be so large that numbers would be cut unacceptably, so he asked the Design Section what it could get with a single R-2180 or R-2600. A speed of 250 mph or more would make a new design worthwhile. The new scout bombers were already that fast. Studies by the design group showed that twin engines were unavoidable but that a torpedo bomber using them might be no larger than the existing TBD.[2] Plans concluded that BuAer would have to accept heavier airplanes if it wanted higher performance. In May the bureau chief wrote that he did not want to start a twin-engine VTB, because of uncertainty about landing performance and probably the need for stronger arresting gear, "coupled with the paucity of compelling reasons for high speeds in this type."[3]

Plans had its Design Section make a performance study using Allison liquid-cooled engines like those the Army had developed (presumably the V-1710).[4] The bureau chief did not find the new designs particularly promising. "This project had best wait over until 1939. On this basis we will have more information resulting from operations of TBDs." Given the recent purchases, no new VTBs would be required for delivery for five and a half years (i.e., through 1943).

Plans Division's preliminary designers returned to the subject late in 1938, looking toward a FY39 prototype award. On 12 January 1939 Plans Division forwarded military requirements. The most important development was that the Ship Installations branch (responsible for arresting gear) was now willing to accept a maximum weight of 13,000 pounds (12,500 desirable). Somewhat later C&R, which designed carriers, found that even larger aircraft could soon safely be handled on a flight deck. Interest in twin-engine designs evaporated when new studies showed that they would weigh about 2,000 pounds more than a single-engine type yet be only 5 or 6 mph faster.

The yardstick was Design 183.[5] BuAer hoped that the new torpedo bomber would have much the same performance as the 1938 dive-bomber, so that the two could operate together. As with the 1938 dive-bombers, high performance demanded an enclosed weapons bay. As with the TBD, this aircraft would be a dual-purpose horizontal and torpedo bomber. Since bombs took up less length than a torpedo there was space left over, which designers used for the bombsight. With the closed bomb bay and the new

R-2600 engine, this torpedo bomber had much the same performance as the 1938 scout bomber using the same engine. It was about 35 mph faster than existing scout bombers with R-1830 two-stage engines. Rate of climb and takeoff were both excellent despite the high gross weight. Stall speed was under the usual 70 mph limit except in torpedo overload condition (when it rose to 71.9 mph).[6] This airplane was 50 percent faster than the current TBD, fast enough to substitute for the 1938 scout bomber. Plans pointed out that meeting the stall-speed requirement (with full fuel and a torpedo) would require greater span, weight, and drag and therefore lower maximum speed.

The bomb bay could enclose an auxiliary fuel tank, which would not affect flying qualities as an external belly tank could be dropped if necessary. It would extend range to at least three thousand miles. Plans argued that great range and endurance would make this an excellent scout, particularly because it had a crew of three. All would be needed because a very long flight would place great demands on a crew. The only shortcoming would be an inability to strafe surface craft. That could be corrected by requiring provision for 20-mm or 0.50-calibre wing guns. Heavy guns would be carried only when the airplane was used as a scout.

Given what Design 183 showed could be done, Plans called for thousand-mile range with a torpedo or with three 500-pound bombs, provision for at least 50 percent more range, and the ability to carry a drop tank inside the bomb bay. Gun armament in normal condition should be one 0.30-calibre wing gun and one 0.50-calibre gun in a power turret (with provision for two 0.50-calibre or 20-mm wing guns). Because it might be used as a scout, the airplane would carry a scouting radio set and also an internal radio direction finder. Maximum speed should be as high as practicable, and the 70 mph stall speed was to be met with half fuel. Maximum takeoff run should be three hundred feet (twenty-five-knot wind over deck), and service ceiling should be 30,000 feet. A strength of 6 *g*s with full normal fuel and a torpedo was specified to enable torpedo

bombers to pass over a distant screen and then dive steeply to attack altitude. It would permit a fifty-degree dive from 10,000 feet with a loss of altitude of 2,000 feet on pullout. Maximum span was set at 60 feet, overall length at 39 feet, with a 17 foot folded height. The weight limit was 13,000 pounds with a torpedo or three 500-pound bombs and overload fuel. In three-point position, the tail of one plane should pass over the unfolded part of the wing of another, to ease stowage. At this time elevator weight limits were 16,000 pounds on the forward elevator of the *Lexington*s, 17,000 pounds for all others.

BuAer circulated requests for bids on 9 February 1939. Missions were torpedo and heavy bombing, as well as scouting and smoke laying.[7] The airplane would be powered by a single mechanically supercharged engine, the air-cooled type being preferable but liquid-cooled allowed. Requirements were much as Plans proposed, but maximum span was cut to 50 feet. The plane had to fit on a 41-by-48-foot elevator with twelve inches of clearance all around. Unlike the 1938 dive-bomber, there was no requirement to fit two on an elevator. Maximum overload weight was not to exceed catapult capacity. Range with additional fuel and a torpedo should be 1,500 miles. Enough oil had to be carried for the range achievable with maximum gas. The airplane should make at least 300 mph with normal fuel load. Takeoff run against the usual twenty-five-knot wind over deck should be no more than 325 feet, carrying a torpedo and fuel for a thousand miles. Service ceiling in this condition should be at least 30,000 feet.

Special consideration would be given to design features favoring low-speed stability and control. If a design did not include wingtip slots, an alternative with slots was wanted. To facilitate torpedo runs, control had to be excellent with flaps down, wheels up, and bomb-bay doors open.

The Norden bombsight (Mk 15) was required only when the airplane carried bombs. If necessary, BuAer would consider a retractable "dust pan" mounting for the bombsight. Specified gun armament was two fixed forward-firing guns, as close as possible to the centerline, and one flexible gun aft (single 0.50, or twin 0.30 if the latter did not require a much larger fuselage), as in a dive-bomber. The airplane was to be able to drop its torpedo when flying at maximum speed. That was hardly possible at the time.

Maximum allowable weight, with the torpedo and fuel for a thousand miles, was 12,500 pounds. Equipment would include an autopilot. The other aircrew were an assistant pilot / bombardier (called the "bomber") and a gunner; the gunner would operate the radio when the assistant pilot was at the bombsight and therefore needed easy access to it from his gun enclosure, facing forward or aft and acting as a lookout. He might be given an auxiliary seat for use when not inside the gun enclosure, though this was not required.

Grumman, Brewster, Vought-Sikorsky, Douglas, Hall-Aluminum, and Vultee all bid.

Grumman's G-40 would be powered by a two-stage (Proposal A) or a two-speed one-stage (Proposal B) supercharged R-2600 engine, developing 1,500 bhp from sea level to 6,700 feet and 1,350 bhp up to 13,000 feet (geared 2:1). Takeoff rating was 1,700 hp. Proposal A was eliminated because of poor performance at high altitude. Proposal B (single-stage supercharged to 22,000 feet) was considered the best all-around design. It offered equal or better speeds at all altitudes up to about 25,500 feet, which covered the usual operating range. Gross weight would be 12,199 pounds. The highest overload weight would be as a scout (12,761 pounds). As a torpedo bomber, G-40 would make 300 mph at critical altitude (14,500 feet) and 272 mph at sea level using a single-stage supercharger. With a two-stage supercharger, it could make 320 mph at 23,500 feet and 268 at sea level. With the single-stage supercharger, it would need 1.9 minutes to climb to 5,000 feet; service ceiling would be 30,000 (36,000) feet. Climb would be slightly slower with a two-stage supercharger (two minutes to 5,000 feet), but service ceiling would increase to 36,000 feet. Takeoff run into a twenty-five-knot wind over deck would be 250 feet (265 feet with the two-stage supercharger). Although Grumman claimed a 1,000-mile range at

economical speed (870 nautical miles), BuAer estimated it would be 937 miles. To meet the requirements Grumman should add about ten more square feet of wing and 149 pounds, bringing it up to 12,648 pounds.

The wing design was generally that of the company's F4F-3 fighter, but it had fixed slots forward of the inboard end of the ailerons. Grumman offered a novel wing-folding arrangement that would simplify carrier stowage (it later applied this arrangement to its fighters). It conformed to a requirement that the center section be removable, but the section was so short that this was pointless and leaving it fixed would save some weight. BuAer considered the five-degree dihedral too small, so it asked that the airplane be arranged so that it could be changed during trials.

Grumman's bombsight looked forward through the bomb bay; it did not have to be removed if the torpedo was in place. BuAer liked that, and it also liked the crew arrangement, which placed each member within reaching distance of another. That also placed the gunner as far forward as possible, minimizing interference from the tail.

Vought offered one fuselage design with alternative wings and engines. It claimed many new aerodynamic refinements and design features, including wings with high "maximum-lift coefficients" (lift-to-drag ratios, in this case 2.29). All its wings used deflector-type flaps, spoilers, Maxwell-type slots, and drooping ailerons. Spoilers worked with the small ailerons to provide lateral control. Slots and a two-degree "washout in incidence" (a way of reducing lift toward the wingtip) were intended to prevent stall near the ailerons. Leading-edge slots were intended to prevent rolling instability (wing dropping) at the stall. The different wings all used fabric-covered outer panels aft of the leading-edge beam and torque box (the structural center of a wing) to save weight; Grumman and Brewster wings were all-metal. Space limitation between the bombsight and the flexible gun made it difficult for the Vought to have the fixed installation BuAer preferred (the bombsight and window had to be folded back out of the way when the torpedo was carried).

Vought Designs A and C used two-speed engines (R-2600-8 and R-2800, respectively) and had high aspect ratios (8.55 and 7.45) to gain a high ceiling (30,000 feet). B and D had more highly supercharged single-stage engines (R-2600 and -2800, respectively) and broader wings (ratio 5.3) that saved wing weight. BuAer eliminated Design A for poor performance at altitude and B for inferiority to Grumman and to a Brewster proposal (see below). Design C offered what BuAer considered outstanding performance on about the same weight as Grumman. At 16,000 feet it was 14 mph faster than the Grumman airplane, though at 24,000 feet it was 28 mph slower. Design D (R-2800 two-stage engine supercharged to 20,000 feet) was slightly inferior to Grumman below 19,500 feet but faster from this altitude up to about 23,500 feet. Its main drawback was its 12,978-pound weight, which would probably increase to well over 13,000. Like Grumman's entry, the favored Design C was heavy: 12,650 pounds with torpedo. Speed at sea level would be 267.3 mph (320.4 at 29,800 feet). It would climb to 5,000 feet in three minutes. Take-off run with twenty-five knots of wind over the deck would be 288 feet. Range would be 1,011 miles.

Douglas's high-wing monoplane was powered by an R-2600-B659 supercharged to 27,000 feet. It was the heaviest of all the submissions with this engine (12,980 pounds), and it was slower than the other three. BuAer credited it with 155 miles more range than required, which would make it possible to reduce wing area by about six square feet and to reduce gross weight to 12,732 pounds. Vision for scouting would be good, but the pilot's view for formation flying would be poor. The pilot would also find it difficult to bail out, because the size of the side doors that let him into and out of the cockpit was limited by the wings' leading edges. The horizontal tail seemed low (Douglas cited contrary wind-tunnel data), the dihedral seemed insufficient, and the plane would probably stall prematurely at the wing root. The wing guns could not be serviced from the top of the wing when it was folded, and servicing from below would require a stool out on the flight deck.

BuAer had not yet concluded that Brewster could not manage another program in addition to its SB2A. The company offered a midwing monoplane powered by either an R-2600 or an R-2800. With the same engine as Grumman's, it offered about the same speed, but general design was not as good, and the takeoff distance was 332 rather than 261 feet. With a two-speed R-2800, Brewster's airplane was slightly slower than Vought's equivalent, and it had a much longer takeoff run (420 versus 298 feet). Contrary to requirements, Brewster offered no alternative design with slots (without which, BuAer thought, its entry would not be stable enough in the stall, landing on a carrier). Nor had it provided the detachable wing center section BuAer wanted (to make shipping easier). There were other problems, such as a poor position for the bombardier. Brewster was rated fifth for carrier adaptability, as a result mainly of the wide tread of its wheels and its large folded span (25 feet 2 inches). The low tail would be difficult to get down for landing and would make for poor spin recovery. Brewster offered a gull-winged version (to shorten landing gear tread), but BuAer was unwilling to try it until the idea was proven in the new F4U Corsair fighter. As in Brewster's SB2A design, the assistant pilot, at one end of the cockpit, was blocked from the pilot, in this case by a solid rollover structure. The cockpit itself was too short, too shallow, and too crowded. The bombardier position was poor.

Vultee's gull-wing airplane would be powered by either the R-2600-B659, supercharged to 27,000 feet, or the R-2800 two-speed engine. BuAer expected a change of wing taper at the knuckle of the gull wing to cause an unsatisfactory stall. The speed with the R-2600 was the lowest of all the entries except for Hall-Aluminum's. With the R-2800 it was slower than both Brewster and Vought. There were no outstanding favorable features.

Hall-Aluminum's midwing monoplane was slowest at all altitudes except near sea level, where it was practically tied with Vultee A for lowest. It was worst for carrier adaptability. Its center section had no dihedral but its outer panels had four degrees, which BuAer considered an unsatisfactory combination. The wing root was mutilated by air intakes and by the landing gear, which BuAer considered would produce poor flow.

In writing the specification BuAer had failed to emphasize the requirement for quick change between torpedo-attack and level-bomber roles. Painfully aware that any of the airplanes would be close to the weight limit, BuAer specified that the bombsight be removable when the airplane carried a torpedo. As a torpedo plane, it would use an autopilot. BuAer soon decided that it would be best to make the bombsight permanent and eliminate the autopilot. That was no problem for Grumman (it saved 110 pounds), but it required modification of the Brewster and Vought designs. In both, the bombsight interfered with the tail of the torpedo. BuOrd was currently interested in a longer torpedo, for which the Grumman design could easily be modified. That was impossible in the Brewster and Vought designs, due to interference between the bomb bay and the flexible gun.

Grumman's B design was the clear winner, based not only on performance but also on arrangement, folding-wing scheme, simplicity, armament provisions, and price. Alternatives for second place were Brewster's Design A and Vought's C.[8] The R-2600 in Grumman's airplane was more readily available than the R-2800 in Vought's, and the latter did not seem to offer much better performance. Vought offered insurance against any problem with the R-2600. BuAer ordered two prototypes (XTBF-1, later named Avenger) from Grumman and one (XTBU-1, later named Seawolf) from Vought.

Just before recommendations were made, the issue of defensive firepower was raised. The latest report on German lessons learned in Spain pointed to a need not only for increased firepower but also for wider gun arcs. An officer returning from the West Coast said that the most frequent topic of conversation among pilots of Aircraft Battle Force was the inadequate gun defense of multiplace aircraft. The fleet found the defenses of the PBY Catalina far

from sufficient.[9] Reports of individual battle practice indicated that VSB and VTB, particularly the latter, rarely brought their fixed guns to bear. It was unlikely that VTBs would be used for strafing. The bureau chief ordered a study of rearranged armament—for example, moving one of the two fixed guns to a tunnel in a retractable enclosure or placing two 0.50s in the single upper turret. A tunnel gun had been included in the original military requirement, then omitted because it seemed difficult to incorporate. By late October, the BuAer Armament Division wanted to move one of the two fixed 0.30s into a tunnel. That was entirely possible in the Grumman design. Early in December, Vought submitted a revised design with a tunnel gun.[10]

Fortunately, the new requirement did not change the order of preference among the designs. Grumman could deliver aircraft more quickly than Vought-Sikorsky and at a much lower unit price: $74,000 versus $90,000. Initially (August 1940) BuAer planned to buy 196 Grumman TBFs and 90 Vought TBUs in FY42 (1941–42), of which 144 were to be delivered by 1 April 1942 and 142 afterward. Grumman could probably deliver 144 aircraft by 1 April. Limiting the FY42 program to Grumman saved $2 million and allowed Vought Sikorsky to concentrate on the OS2U Kingfisher cruiser/scout-plane program.[11]

The TBF/TBM Avenger

Grumman's Avenger became the standard U.S. Navy torpedo bomber. From the outset, its carrier-landing characteristics were considered outstanding, although its controls were considered very heavy. Compared to the Helldiver it was little modified, and it was always well liked. In the Atlantic it was the standard carrier antisubmarine airplane, and it survived for about a decade postwar in that role in the U.S. and allied navies. The contract for two prototypes was let on 8 April 1940, and the first flew on 7 August 1941. As with the Helldiver, the initial production contract (for 286 airplanes) was let before the prototype flew, on 23 December 1940.[12] Between contract and first flight the design was slightly modified: the bomb bay

An early Avenger shows its characteristic 0.30-calibre nose gun. This version could also be recognized by its canted radio mast, anchored in the solid mass of the roll bar between pilot and second (midships) crewman. The midships crewman visible here had a second set of controls, deleted after the fiftieth production airplane. He operated the power turret gun. The third crewman was bombardier and radio operator; he also operated the tunnel gun, not very visible here (note the triangular window just ahead of it, with the bottom of a door above it). After the controls were eliminated from the second seat, the instrument panel there was replaced by more radio equipment, including the transmitter and receiver of an APN-1 radar altimeter. The seat was removed and an additional radio receiver (ARB) installed. The British equivalent of the TBF-1C (Tarpon II) retained the seat amidships. The thirteen alternating red and white tail stripes were ordered by CinCPacFlt on 23 December 1941 as an emergency recognition measure. The two-tone paint scheme employed blue-gray for upper surfaces and light gray below. U.S. NAVAL INSTITUTE PHOTO ARCHIVE

It took big wings to lift a heavy torpedo. A TBF-1 of VT-5 launches from USS *Yorktown* (CV 10) as she works up in 1943. Squadron markings were removed when ships reached the combat area. The national markings are consistent with this date. The red disc was ordered removed on 15 May 1942 because it could be confused with the Japanese insignia; bars (not visible here) were not ordered added until 28 June 1943. NHHC

was enlarged to take four rather than three five-hundred pound bombs, flight controls were eliminated from the second cockpit, the turret motor and ammunition were armored, self-sealing fuel lines and tanks were installed, and flotation gear was replaced by a life raft. To maintain balance, the engine was moved forward twelve inches. The first four production aircraft were delivered in January 1942.

The contract specified the two-stage R-2600-10 engine, but to save time the prototype used the two-speed R-2600-8. In January 1941 BuAer agreed that the first hundred aircrafts would have it, but it wanted production shifted after that to the two-stage R-2600-10, which was still experimental. Grumman argued that production would suffer, and the shift was pushed into the future. By November 1941, BuAer had been forced to accept that all 526 aircraft then under contract would have the -8 engine. The twenty-first production airplane had a -10 engine as the prototype XTBF-2 (it flew on 1 May 1942). BuAer decided that numbers were more important than better performance, so this version never entered production. A version with a new engine (in the TBM-3) did not enter production until 1943.

As with the Helldiver, the demands of a rapidly expanding carrier force drove production. On 23 December 1941, just after Pearl Harbor, 1,141 Avengers were on order, and Grumman was expected to reach a peak rate of 100 per month in January 1943 (all the aircraft on order would be delivered by about August). Around 110 aircraft would probably be delivered before August 1942. The Avenger became the standard airplane on board numerous escort carriers. Grumman soon agreed to build 125 per month. To produce even more Avengers, it set up production at the new General Motors Eastern Division, which used plants at Trenton, Tarrytown, and Baltimore. Eastern agreed to a peak rate of 150 aircraft per month. Avengers produced under this program were designated "TBMs." Grumman and Eastern began to accelerate production to the point where they expected to reach 540 aircraft per month. However, in mid-1942 battle experience in the Pacific showed that fighters were desperately needed. To concentrate on Hellcat fighters, Grumman tapered off TBF production; by the end of 1943 Eastern was sole producer. Eastern, which also took over engineering responsibility, planned to double its peak rate to 300 per month.

A TBF-1 operating out of Quonset Point, 11 September 1943. The prominent black box above the *3* on the side was part of the radio equipment that replaced the after cockpit. Note the early VHF whip protruding down from the radio operator's position, visible to the left of the ASB radar antenna. This is the original TBF, without wing guns. The two whips extending down from the bomb-bay doors are for the ZB homing device (working with a carrier's YE) and for IFF. The ZB antenna was omitted from shore- and CONUS-based aircraft. The T-shaped radar altimeter antenna can just be discerned near the wheel well under the wing. *THE HOOK*

Before that, BuAer had decided to buy TBUs to supplement Avengers: on 12 June 1942 it sent Vought a letter of intent for 1,100 TBUs at a peak rate of 150 a month. However, Vought was fully occupied with the Corsair fighter program. BuAer approached Vultee (which soon merged with Consolidated) to build TBUs under the new designation "TBY"; it received a letter of intent for 1,100 on 31 October 1942.

As of September 1942, Grumman had 1,350 TBF-1s and 1,150 TBF-2s on order. Eastern had a contract for another 1,500 aircraft, of which it was hoped that the 301st would introduce the new -10 engine. Consolidated promised a peak rate of 150 TBYs per month, so that total torpedo-bomber production would reach 450 per month, which seemed to be just enough—BuAer stopped looking for further sources of these aircraft. As it happened, the TBY program failed. By July 1945 Eastern was producing 15 TBMs per day, an average of 376 a month;[13] it had reached 150 a month in the fall of 1943. At that time 11,847 TBMs had been ordered, of which 6,829 had been accepted through 1 July 1945. Only with the end of the European war was production at Eastern finally enough. Painfully aware of the desperate need for these aircraft, BuAer resisted changes the fleet requested. Even at the beginning of 1942 it considered the TBF at the upper limit of acceptable weight for strength and performance.

Glide Bombing

Probably the most important wartime Avenger development was adoption of glide bombing, which employed a shallower dive than dive-bombing. This technique was proposed in 1938; it was incorporated in Design 183, before the 1939 competition.[14] As in dive-bombing, the pilot aimed the airplane, in this case at a maximum angle of from forty-five to fifty degrees; strength to withstand 6 gs was ample for a pullout. No dive brakes were needed, but the fleet wanted dive flaps.

In the fall of 1941 VT-3 became interested in using a steep, high-speed glide to reach torpedo firing position very quickly, getting past screening ships and evading antiaircraft fire. To that end, its commanding officer (CO) asked in November 1941 for dive flaps to slow his airplanes in their glides.[15] During 1942 fleet torpedo squadrons generally became increasingly drawn to the idea of glide bombing as a viable (and preferable) alternative to either torpedo or horizontal bombing. In May 1942 Dahlgren demonstrated that bombs released in a sixty-degree dive would clear the aircraft structure without a trapeze. Airplanes at Dahlgren dropped thousand-pound bombs at 300 knots indicated airspeed, holding the dive until indicated speed reached 315 knots. VT-6 soon reported that it was starting sixty-degree dives at 6,500 feet at 120 to 130 knots indicated speed, releasing bombs at 2,500 feet, and taking 800 to 1,000 feet to pull out (3 gs, with a maximum stress of 3.5 gs). Many pilots lowered their wheels and opened bomb-bay doors to limit diving speed and improve accuracy. BuAer rejected calls for dive flaps.[16]

As the Avenger gained weight, steep diving angles became more and more dangerous. BuAer tried to limit dives to thirty-five degrees, but at such angles the target disappeared below the sight before the bomb had to be released.[17] In August 1944, for example, Fleet Air Force West Coast limited all Avengers to thirty degrees, which could be increased to forty for training provided gross weight did not exceed 14,500 pounds. Airplanes could enter their final forty-degree dive from no more than 6,500 feet with bomb-bay doors open, releasing their bombs or rockets at a slant range of 1,500 yards or at 2,900 feet. Steeper dives made for much greater accuracy. Aircraft were diving slowly enough to be extremely vulnerable to antiaircraft fire.

VT-3 apparently was the first to request more forward-firing firepower, in November 1941. Enemy fighters would probably attack from ahead, and the single 0.30 nose gun was inadequate. Instead the squadron wanted a 0.50 plus two rather than one of those guns in the turret. BuAer had scoffed at demands for ahead fire, doubting that torpedo bombers would ever strafe, but shortly after war broke out someone in BuAer pointed out that airplanes were being used for unexpected purposes and that strafing might

A TBM-1C shows four rockets on underwing rails (Mk 4 rocket launcher) and the post of an underwing ASB radar antenna. Tape covers the gun muzzle near the wing root. TBF- and TBM-1Cs had their radio masts relocated aft slightly and made vertical. The third whip is probably for VHF, replacing an earlier one angled down from the radio operator's station. The white dot on the tail indicates the carrier *Hornet*. U.S. NAVAL INSTITUTE PHOTO ARCHIVE

become important in the Mandates. Another writer admitted that the single gun had been put in the turret only to obtain fore-and-aft balance.[18]

The first major modification to the Avenger, the TBF-1C, traded the nose gun for two 0.50-calibre wing guns.[19] These weapons seem to have been adopted specifically to deal with U-boats that had been fitted with heavier antiaircraft armament.[20] TBF-1Cs and TBM-1Cs began to come off production lines in September 1943.

Provision for two fifty-eight-gallon underwing drop tanks began to appear on production aircraft in August 1943 (TBM-1Cs got the tanks in December). Airplanes delivered early in October had modernized bomb bays.

A TBM-1CP photo version had a Trimetregon camera (actually, a set of three cameras, from which stereoscopic imagery of terrain could be produced).

Radar

Associated with the -1C version was installation of ASB surface-search radar, whose two Yagi (directional, with a distinctive arrangement of elements) antennas were clearly visible underwing. Avengers and Dauntlesses with ASBs were the only radar-equipped aircraft on board Pacific Fleet carriers in 1943. Some Avengers, which were much faster than Dauntlesses, were teamed with pairs of fighters to counter night-flying Japanese attackers.

A much better air-to-air radar, the ASD (APS-3), was being developed specifically for night fighters, to be mounted in a wing nacelle. In the fall of 1943 Commander Fleet Air Atlantic (ComFairLant) Quonset, which had been developing naval night fighters and their tactics, proposed having Avengers fitted with the ASD radar train with night fighter squadrons. DCNO (Air) approved. The first five of these TBM-1Ds were quickly delivered, three of them assigned to VF(N)-76. Another ninety were planned, to provide one for every two night-fighter squadrons; thirty were given the highest priority. The radar nacelle was installed on the starboard wing.

Unfortunately, a TBM 1D required an unusually long takeoff run and climbed very slowly.[21] Once at altitude it cruised at a speed five or six knots slower than a TBM-1C. Airplanes showed some right-wing heaviness and buffeting when landing. Lateral control in landing condition at low speed near the stall was limited: it took nearly full aileron to overcome the heaviness of the right wing and maintain level flight. Lateral stability was deficient and dihedral effect (tending to return the aircraft to level flight) of the starboard wing was lost, particularly in landing condition. Also, the starboard wing stalled prematurely when pilots tried to take off at minimum speed. Since much of the trouble came from the asymmetry, NAS Philadelphia modified at least one Avenger with the radar nacelle atop its greenhouse, on the centerline. Nothing came of that project.

The radar itself required considerable maintenance. When it did work, it was effective both air-to-air and air-to-surface. However, the TBM-1D was too slow to work effectively with Hellcats, and it could not closely follow airplanes making sharp turns. It could help select a ship for a torpedo attack, but as the bomber closed in sea clutter would bury the blip on the operator's scope, so visual contact had to be made. The earlier ASB was somewhat more successful keeping a clear blip at low altitude and close range. For glide bombing, ASD could coach pilots into pushover position at six to twelve thousand feet, but then pilots had to make visual contact with the target. ASD had a far narrower beam than the earlier ASB, but the ASB would home better at minimum range. At least initially it was more reliable by virtue of being simpler. VT-50 found that because it had so narrow a beam, APS-3 (ASD) often lost targets during evasive action prior to dropping torpedoes. Compared to ASD, ASB searched a slightly larger area dead ahead, consistently detecting large ships at about twenty-eight miles (ASD could detect a fleet carrier at fifty). APS-3 was much better for navigation. For its part, ASB was liked because it worked with a YJ shipboard homing beacon, which could easily be picked up at seventy miles by an airplane at three thousand feet.

Overall, ASD was valuable enough that it was also fitted to TBM-3s.[22]

A more compact radar "bomb," the APS-4 (ASH), could be hung from a specially wired underwing bomb shackle. It could be dropped if necessary. Like ASD, it had a much narrower beam than ASB. It was not considered an air-to-air radar, although it could detect air targets. Avengers were not rewired for it

This TBF-1D gives a sense of the size of the radar nacelle. That the radio mast is vertical shows that this airplane was converted from a TBF-1C rather than a TBF-1. Note the omission of the turret gun and of the tunnel gun. The heavy framing of the midships part of the greenhouse appears to support a hatch that could be opened into the slipstream to control a searchlight; note also the unusual equipment amidships. This is presumably a "Night Owl." This photograph was taken in August 1944. NHHC

until about the end of 1944. In contrast to the big APS-3 nacelle, the "radar bomb" apparently had no ill effects on aircraft performance. TBF/TBM-1Cs wired for APS-4 were designated TBF/TBM-1E.

Late in the war, a few aircraft were fitted with the APG-4 (nicknamed "Sniffer") radar intended to support low-level bombing. It could be recognized by its antenna—a pair of Yagis over the wing, similar in configuration to the Yagis of the initial Avenger air-to-surface radar (which were installed underwing).

Searchlights and ASW

Avengers were the primary antisubmarine attack aircraft operating from U.S. escort carriers during the Battle of the Atlantic.[23] The first Avenger-equipped escort carrier assigned to the North Atlantic, USS *Bogue*, began operating in January 1943, as the battle reached its peak. *Bogue* operated in daylight, using a British HF/DF (high-frequency direction finder) fitted in the United Kingdom in May 1943, an important Allied ASW sensor because U-boat operations were coordinated by HF radio. HF/DF picked up the relatively short-range surface wave produced by an HF signal. A destroyer could run down the bearing, but by the time it neared the position where the U-boat had transmitted, the submarine might well be gone. Carrier aircraft—Avengers—were much likelier to get there in time. Late in May 1943 U.S. escort carriers shifted from their initial role in direct support of convoys to open-ocean hunting, following up U-boat positions revealed by code breaking.[24] For example, carrier-based Avengers managed to break up German U-boat refueling operations.

The Avengers were at this point limited to daylight operations.[25] They were armed with depth charges and with the new Mk 24 "mine," a small homing torpedo, and in at least some cases sonobuoys.[26]

Allied successes during the summer of 1943 drove most U-boats out of the North Atlantic, but the remaining boats' exaggerated claims of success in September led the U-boat command to try again in October, with a peak effort.[27] It failed disastrously: only twenty merchant ships were sunk, at the cost of twenty-seven U-boats. In November, then, the Germans adopted new tactics oriented more to self-preservation than to sinking merchant ships. They ceased to operate on the surface in daytime, surfacing only at night to charge batteries or attack convoys.[28] The Avengers were equipped only to find their U-boat targets on the surface, although sonobuoys made it possible to attack them just after they dove. Now they had to find U-boats at night in order to strike them.

Night flying from escort carriers seems to have begun with *Croatan* (CVE 25), which left Norfolk on 14 January 1944.[29] *Croatan* was credited with the first night Avenger attack on a U-boat, on 15 February.

For night search, Avengers were adapted as "Night Owls"—TBM-1Cs whose guns, weapons, and armor had been removed to save weight.[30] The crew was reduced to two, pilot and radioman. Night Owls were intended to cue other aircraft to attack. Night attacks required radar, typically the ASB on board a TBM-1C. As was already well known, target blips dissolved into sea return at low altitude, just as an airplane neared a U-boat. Many long-range patrol aircraft were accordingly fitted with searchlights. Because they had attacked in daylight, Avengers had been last in line to get searchlights. The change in U-boat tactics made a stopgap urgent: flares.

Flare attacks were exemplified by an attack by aircraft from the escort carrier *Block Island* described in the July 1944 issue of the *U.S. Fleet Anti-Submarine Bulletin*. At 0255 on 29 May, an unarmed TBM (presumably a Night Owl) made a radar contact seventy-eight nautical miles from the task group containing the carrier. A coordinated attack was set up, one plane dropping flares and the other attacking. In the light of the flare, the second plane saw the wake of a submarine and glided down to 150 feet; there the pilot saw the dim bulk of a submerging submarine about eight hundred yards away. Then the flare went out, leaving the pilot blinded, unable to continue. All the task group could do the next day was search along the estimated track of the U-boat. The next night the carrier expected to launch six search planes plus a fighter to act as VHF radio relay, but the weather was deteriorating. Instead, two Night Owls were launched late in the afternoon to see how badly the weather would limit operations and to search as well as possible before night. Nothing came of this. *Block Island* was surprised by (and sunk) by an unsuspected U-boat, probably not the one spotted the previous day.

The need for a searchlight to support Avenger attacks on submarines had been established late the year before.[31] At that time prototypes were planned for January or February 1944. It is not clear when operational searchlight aircraft first appeared. Initially the light (L-8C) was mounted in the bomb bay, controlled by an operator who leaned out into the airstream, secured by a gunner's belt, protected to an extent by a hinged hood. The Engineering Section of the A/S (Anti Submarine) Development Detachment of the Atlantic Fleet developed a periscopic sight, the prototype of which was delivered on 24 May 1944.[32] On the periscope was mounted a standard ASW bombsight (Mk 23). The operator remained inside the cockpit. An August 1944 test of a searchlight-equipped TBF-1D showed that, thanks to the searchlight, it could make attacks on 80 percent of its runs.

TBF-1Ds with searchlights became operational about mid-1944.[33] The first such attack was conducted on the early morning of 20 August. The Avenger pilot was coached in by his radar operator. He switched on his searchlight about three-quarters of a mile from his target, which was completely surprised (but, in this case, escaped).

NAS Quonset pressed for an alternative arrangement, in which the L-8Q searchlight (Q for Quonset) was modified so that it could be mounted underwing (on a gas-tank shackle) in a fairing similar to that of the APS-4 radar "bomb." BuAer developed an alternative L-11 using a teardrop fairing; seventy-eight were on order as of November 1944, for delivery beginning in December. Unlike L-8Q, it was designed to be quickly detached or jettisoned. The external arrangement, which became standard, was successfully test-flown in November 1944. Twelve

production L-8Qs were made for use pending availability of the later searchlight. Initial L-11 tests on board a TBM-1D showed excessive drag, but that was corrected by modifying the searchlight pod. The prototype was flown on board a TBM-3 with APS-3 and APG-4 radars.

In addition to ASW searchlights, BuAer sponsored a test of camouflage using lights mounted on an Avenger.[34] A TBF was modified and tested early in 1944, in a joint project by the National Defense Research Committee and the A/S Development Detachment of the Atlantic Fleet. The idea was that an airplane could be lit in a way that made it invisible beyond three thousand yards to the naked eye of an observer approached head-on. That is, an observer normally saw the line represented by the wing against the sky, but with the lights on, that line would seem to disappear (it would reappear if the airplane turned more than four and a half degrees). Normally an Avenger would be visible twelve miles away. The installation included five lights on each wing, four spaced evenly inboard and another farther out toward the wingtip. Four more lights were spaced around the airplane's nose. The whole trial installation weighed ninety pounds, and an operational version would probably weigh seventy. The same technology could be applied to patrol aircraft. Results were sufficiently promising for the A/S Detachment to propose converting a full squadron of twelve Avengers. If the U-boats could not spot the airplanes, they would be forced to use their own radars rather than passive search receivers, giving themselves away to properly equipped aircraft. Against this, the airplane would still be visible through binoculars, which were standard among U-boat lookouts. The squadron installation was deferred pending completion of a new prototype by 1 January 1945. Experiments continued at least through April 1945, but the device was not adopted.

By June 1944 TBM-1Cs had gained considerable weight, which markedly reduced their rate of climb (to no more than 350 feet per minute with two thousand pounds of bombs and full fuel), acceleration,

This Avenger was modified at Quonset for ASDEVLANT, the Atlantic Fleet's antisubmarine-development unit. It combined the APS-3 nacelle radar with the searchlight needed to attack submarines at night. The hinged top of the greenhouse made it possible to erect the periscope used to control the searchlight. The searchlight could be dropped (like an APS-4 radar "bomb"), so the airplane was not redesignated with an *L* suffix. Note the gun muzzles of the TBM-1C and the Mk 4 rocket launchers underwing. The turret and tunnel gun have both been removed. AUTHOR'S COLLECTION

A TBM-3D with a searchlight, presumably the production version with a more compact light. The pair of Yagis atop the wings were for the APG-4 "sniffer" low-altitude attack radar, which used FM techniques (hence required separate send and receive antennas). *THE HOOK*

speed, and maneuverability. Weight also affected endurance. Even before the Philippine Sea, VT-50 wrote that recent operational losses had far exceeded combat losses. By far the greatest percentage had been due to airplanes running out of fuel before they could safely land on board.

BuAer was already well aware that the Avenger needed more power, which it finally got in 1943 with the TBF/TBM-3 (Avenger III), powered by an R-2600-20 (rather than the -14 planned in November 1941).[35] As Grumman had warned it would, changing the engine considerably delayed production, because the cowling had to be reengineered. The prototype XTBF-3 retained the SBAE of the earlier version, but production aircraft had a lighter-weight GE hydraulic autopilot, which was easier to maintain. The new version was being flight-tested in September 1943. Deliveries began in April 1944.

VT-80 liked the new version, which offered a considerably better takeoff than the TBM-1C (500 rather than 550 feet with thirty knots of wind over deck, including a 75-foot safety allowance) and a much faster climb (it could keep station with a Helldiver carrying a thousand-pound bomb).[36] The TBM-1C had labored to remain in position. Instead of slowly climbing during the run in to the target, the -3 could easily climb to between 12,000 and 14,000 feet for a coordinated attack, even one that ended at low altitude. The new version was slightly more maneuverable than earlier ones, thanks to a slight decrease in stick pressure. It had the same gliding characteristics as the -1C and the same excellent landing characteristics. It also needed less fuel. Its modified ASB-7A radar was slightly better than the ASB-7 in a TBM-1C.

A TBM-3 of VC-84 (USS *Makin Island* [CVE 93]) shows rockets on zero-length launchers (stubs), 10 January 1945. Aircraft on this Pacific Fleet carrier were painted in standard fashion, with dark tops and gray sides. The two antennas of the ASB radar point to the sides, for sea search. On the starboard (upper) wing the T-shaped antenna of the radar altimeter can be seen abaft the landing-gear leg. The muzzle of the starboard gun is evident as a darkened area outboard of the landing gear leg, the slot through which cartridge cases were discarded visible farther back. Note the countershading under the wing root and under the horizontal tail. At this time aircraft operated by the ship had a white stripe across the tail; in mid-1945 that was changed to a white tail tip plus a white stripe on the upper side of the upper right wing. These markings were needed because she and other escort carriers were operating in groups in the Pacific. NHHC

Two TBM-3s of VT-82 on board *Bennington* prepare to take off, February 1945. The nacelle change is clearly visible, an oil-cooler intake has been added below the engine. The upper intake was for the carburetor. Additional cowl flaps were added. This version added zero-length rocket launchers (stubs). NHHC

Many TBM-3Es had their tunnel guns removed and the space faired over, the radio operator (moved up to the former second cockpit) handling the radar. The external tailhook was installed only on later production aircraft. U.S. NAVAL INSTITUTE PHOTO ARCHIVE

About this time provision was made to carry a two-thousand-pound bomb. During December 1944 aircraft were given provision for the new APS-4 radar "bomb" (instead of ASB) and also for twin 0.50-calibre gun packages underwing. About January 1945 the commanders of the air forces of the Pacific and Atlantic Fleets (ComAirPac and ComAirLant) approved removal of the tunnel gun in TBM-3s (and in TBY-2s in production).[37] It is not clear whether this change reached the production line before the program was terminated in August 1945.

As with the -1 version, there were specialized variants: -3P with a Trimetregon camera, -3D with an APS-3 radar, and -3E with provision for an APS-4 "bomb." The -3E was the last production version, redesigned in detail to reduce weight gain represented by the APS-4; 1,480 were built.

Some TBM-3Es were modified as electronic countermeasures (ECM) and ELINT aircraft, designated TBM-3Q. Initially a few TBM-3s were modified in forward areas with intercept receivers and jammers (many TBMs carried chaff).[38] Their success led to a more formal program, the *Q* designation (not for "Quonset" this time) being adopted formally in November 1945 (the TBM-3Q was the first airplane so designated). Internally, the main change was to the radio operator's position. He had a receiver, a pulse analyzer, a jammer, and a manual chaff dispenser.[39]

Some aircraft were converted in 1944–45 to serve in night torpedo squadrons, VT(N)s.[40] They were not specially designated, but postwar up to thirty TBM-3s were modified into TBM-3Ns specifically for night attack. Their turrets were removed and their canopies extended. They seem to have operated from 1947 on, both in carrier air groups and in composite night squadrons. Antennas were similar to those of the -3Qs.

Some TBM-3s were converted into TBM-3W early-warning aircraft (fitted with the Cadillac I system) in 1944–45, with massive APS-20 antennas in a big belly radome and two operators in the fuselage, the after end of the greenhouse being covered. The airplane transmitted its radar picture to a surface ship, generally a carrier. This version served postwar. A related version had a significant postwar ASW role, because the big radar could detect a snorkel against sea clutter and pass its radar data to an accompanying TBM-3S attack bomber.

There were also postwar noncombat versions: the TBM-3R transport (the first carrier-on-board delivery airplane) and the TBM-3U (utility type) for target towing.

The big Avenger was the natural choice to carry the APS-20 early-warning radar developed under Project Cadillac. These aircraft were redesignated "TBM-3W." Postwar essentially the same aircraft were used for ASW, the big radar proving useful to detect periscopes and snorkels. U.S. Naval Institute Photo Archive

Further Versions

When the war ended, BuAer was preparing to put a new version of the Avenger into production. This project began in May 1944 with a request from the BuAer Military Requirements Branch to the VTB Branch to investigate using a more powerful engine, such as the R-2800C then being installed in the TBY-2.[41] Nothing came of this, but in October the Engineering Division asked Military Requirements to issue a directive cutting the weight of the TBM by eliminating the turret, some radio and other electronic equipment, and relocating the radar/radio operator's seat. This was the origin of the TBM-5 project.

The alternative XTBM-4 was strengthened to withstand 6 *gs* at 16,000 pounds gross weight. The major and most visible change was a new-design wing-hinge fitting. The wingtip slots were eliminated. Other changes were 75ST material in main spar-cap strips in the wing structure, reinforced wing ribs, and strengthened fuselage/center-section carry-through structure (which joined the two wings).[42] As of May 1945, production was expected to switch to the TBM-4 by August.

XTBM-5 was a redesigned version begun in November 1944.[43] NAMU (Naval Air Modification Unit) Johnsville, Pennsylvania, produced one prototype, Eastern Aircraft another, both from TBM-3Es. Each reflected a different attempt to "clean up" the TBM aerodynamically. The major question was whether to retain the big 0.5-inch machine-gun turret or eliminate it and its operator. At a conference at Eastern Aircraft in April 1945, BuAer representatives were reluctant to eliminate the turret although Japanese air opposition was by now minimal and U.S. naval attack aircraft were being heavily escorted. They wanted to delay improvements indefinitely, because any changes would have slowed production. A prototype two-seat version (with an SBD twin 0.30 mount on a scarf ring) was already being tested at Patuxent River. After it was damaged, elimination of the turret or free gun was considered. As of April 1945 Eastern was testing a prototype with special wheel-well fairings, a faired-in tunnel compartment, jet exhaust stacks (to add exhaust thrust), a faired turret opening, and a new integral, bullet-proof windshield. Eastern was also proceeding with a prototype powered by an R-2600-34 engine. The two-man prototype was built by NAMU Johnsville. It had jet stacks and was generally cleaned up aerodynamically. The fixed-wing slots were eliminated, and span was increased three feet. BuAer estimated that these alterations could save a thousand pounds,

The TBM-5 prototype was accepted with its cockpit completely faired abaft the pilot. It is shown at "Pax River," 18 January 1946. NATIONAL ARCHIVES

adding 15 mph to maximum speed and reducing stall speed by 3 mph. Climb rate would increase by three hundred feet per minute and cruising speed by 18 mph. Takeoff run (twenty-five knots wind over deck) would be reduced by sixty feet.[44] Flights by the two prototypes began in June 1945. As of May the earliest production date was May 1946. BuAer expected to request the TBM-5 designation formally as soon as it decided the final configuration. With the end of TBM production, only three prototypes (one TBM-4 and two TBM-5) were accepted. The NAMU TBM-5 was accepted with its cockpit completely faired abaft the pilot.

TBM production terminated on 14 August 1945.

The TBU/TBY Seawolf

On 23 December 1941 BuAer informed CNO that it was shopping for another TBU builder. After giving up on North American, BuAer turned to Consolidated Vultee. As of 4 June 1942 the War Production Board expected deliveries to begin with 10 aircraft in June 1943, but that was impossible. The best Consolidated could offer was to deliver the first airplane fifteen months after the contract, building up to 150 in the twenty-fifth month. Production required further engineering, partly because the planned production version had an R-2800-20 ("B," i.e., R-2800B)

engine rather than the -6 of the prototype. Testing revealed other required changes. For example, in September 1942 the span of the ailerons had to be increased for better lateral control and the horizontal tail surfaces somewhat modified for better longitudinal control.

Consolidated received a letter of intent for 1,100 aircraft on 31 October (superseding a similar letter sent to Vought in September), at which time it was expected to deliver its first airplane in September 1943 and to reach the desired 150 per month in June 1944. It had no surplus engineering staff, so it agreed to produce the airplanes only if Vought retained full engineering responsibility. Unfortunately, the rationale for having Consolidated in the program in the first place was that Vought was so badly overloaded—by, among other things, engineering responsibility for the Corsair fighter. To make matters worse, the War Production Board demanded that Consolidated use a Mack Truck plant in Allentown, which an Army Ordnance cutback had made available. Consolidated had hoped to use Holmes Airport on Long Island, near both Vought and the best available source of skilled labor, the Greater New York City area.

Even so, in January 1943 Consolidated's chairman said that if Vought completed production engineering by May 1943 and the plant was at least

Avengers were modified postwar for ASW as TBM-3s like this one. The small radome covered one of several search receivers. This one served with VS-24. U.S. NAVAL INSTITUTE PHOTO ARCHIVE

partly ready by then, he could meet the desired schedule. Consolidated signed a separate contract with Vought, which would retain engineering responsibility for the first 300 airplanes. Nothing worked as planned. The plant was not ready for occupancy until October. Due to a shortage of aeronautical engineers and draftsmen and delays caused by the distance between engineers in Stratford, Connecticut, and the Allentown factory, design engineering for production aircraft was not completed until 20 November 1943. The most complex problem was the planned replacement of the current R-2800B by the more powerful -2800C, as required by BuAer. In the spring of 1943 the two companies agreed that the change—and the change of engineering responsibility—would be made with the 301st airplane, which would be designated TBY-2. By that fall, however, it seemed that the -2800B (-20) engine, made in an Army plant, might not be available: all aircraft would be TBY-2s powered by the R-2800C (-22). Consolidated had to accelerate its own design work.

Early in January 1944 the Vought prototype was grounded by engine failure. Given the delay in changing to the new R-2800B engine, various other changes were ordered. An APS-3 (ASD) radar nacelle was installed on the starboard wing as standard equipment (rather than as a special installation, as in the TBF/TBM). An entirely new after fuselage replaced the original. Production radio equipment was fitted. In January 1944 BuAer informed Consolidated that it wanted a larger-diameter oil cooler. In May 1944 the bomb bay had to be modified to accommodate the new "pickle barrel" of the Mk 13 torpedo. It did not help that in May Vought's Allentown powerplant engineer resigned without replacement.

The major changes were incorporated in the original TBU-1 prototype; they were completed in February 1944, and tests were completed by August. By this time TBY-2 design work was, finally, virtually complete. The first airplane was completed that month (and flown on 20 August 1944), but because

The alternative 1939 torpedo bomber was Vought's TBU-1. Its production version, which never entered service, was the Consolidated TBY-2 shown here. U.S. NAVAL INSTITUTE PHOTO ARCHIVE

of minor problems it was not accepted until November. Problems persisted, however, to the point that in January 1945 the company's president personally took charge at Allentown. They were in part a result of Consolidated's decision to rely heavily on subcontractors. It took forward fuselage sections from Tucson and wing center sections and outer panels from Briggs. Later the wing-hinge subcontractor was a problem. Allentown had to make twenty-five forward fuselage sections for initial production aircraft. Most of the eight airplanes accepted in January 1945 had been completed earlier. Production gradually accelerated to thirty-three acceptances in June 1945. By that time 94 aircraft had been accepted out of an initial contract for 250; in August it seemed that the remaining 156 would be delivered on time (to the revised schedule), the last in October 1945.

Through the end of April 1945 the TBY-2 program was still accelerating, for assignment to *Essex*-class carriers as an alternative to Avengers.[45] According to a 30 April program summary, another 600 were planned after the initial 1,100, for production in May–September 1946. At this time aircraft from the 551st on were to be strengthened (like the TBM-4) and designated TBY-3s. This change was expected in January 1946.

In August 1944 the VTB Desk had written that the TBY-2 was necessary because Eastern would be unable to meet torpedo-bomber requirements; "The critical shortage of torpedo planes is one of the main problems of the navy program." Compared to the TBM, the TBY offered better performance, particularly higher speed, but that was offset by equally important and less favorable range and deck-spotting considerations. Moreover, the TBY was not combat proven, and like any other new airplane it would have teething problems. Moreover, having two carrier torpedo bombers in service at the same time was to court logistical and training issues. The TBY was worth sending to the fleet only so long as there was a severe torpedo-bomber shortage. Unfortunately for the TBY, it became available in numbers just as the shortage evaporated with the end of the war in Europe.

The program was canceled on 6 July 1945, cut back to the 180 already built or nearly completed (94 had been accepted as of 1 July 1945, 134 as of 2 August). All 600 planned TBY-3s were canceled. Aircraft already on the West Coast scheduled for assignment to a carrier air group for evaluation were held back.

The Big Torpedo Bomber

In the fall of 1941, BuAer projected a new torpedo-bomber competition for the spring of 1942, but the war intervened. It seemed that a three-seat torpedo-bomber was a natural scout, since it would provide space for a radar operator. This idea was ratified at a 21–22 January 1942 BuAer conference. The scout/torpedo bomber would become a VTSB.

The conference envisaged a two-engine midwing monoplane with tricycle landing gear, designed to work at high altitude (engines would be supercharged to about 20,000 feet). Twin engines would provide better visibility, leaving a clear forward view for glide bombing. Lower-rear gun protection would be essential in the envisaged high-altitude bombing role. The nose could accommodate the projected ASD radar; single-engine airplanes would have to make do with inferior wing or belly installations. Twin engines also meant the desired tricycle landing gear, greater safety, simpler engine installation, and outstanding performance. In a single-engine airplane, really high power seemed to demand complicated contraprops to hold down propeller diameter. The conference report included a comparison between the TBF-2 and the proposed VTSB, both airplanes accommodating a crew of three and one torpedo.[46]

Douglas was apparently already aware of Navy interest. The conference proposed that Douglas take it up immediately, as its engineering staff would become increasingly available as the SB2D project proceeded toward production. A letter soliciting a design went to Douglas on 9 February.[47] The new airplane should fit existing elevators, and its folded width was not to exceed seventeen feet (height during folding should not exceed twenty-one feet). It was desired that eighteen could be parked on a flight deck in the same space as twenty-one TBFs; a tight hangar-deck park could accommodate either twenty-five of the new airplanes or thirty-five TBF.[48] BuAer considered that the requirement could best be met with two R-1820s or R-1830s with two-stage superchargers.

Douglas was receptive, and on 12 March 1942 Heinemann, its chief engineer, who had designed the Dauntless, wrote that engineering was already under way.[49] In April 1942 he showed BuAer four alternatives: one with twin engines and turbochargers, one with twin engines and two-speed superchargers, one with an R-4360 (X-Wasp) and a two-speed supercharger, and one with an Allison V-3420 and a turbocharger.[50] Heinemann himself thought the X-Wasp, with contraprops, most attractive. BuAer estimated that his airplane would make 336 mph at critical altitude, 15,600 feet. Takeoff distance with one torpedo would be 360 feet. The BuAer evaluator considered the design a "definite step forward."[51] The head of BuAer's Engineering Branch wanted a letter of intent issued to Douglas "in view of the significant advantages of this type, and the imperative need for continuation of development of the VTB class." No Navy airplane had yet been ordered using the X-Wasp; BuAer agreed to get two engines, with two-speed reduction gear and two-speed superchargers (both with single rather than contra-rotating propellers) by November 1943.

Douglas was told to submit a formal proposal by the end of May 1942; production engineering should be based on either a total of 600 airplanes (50 per month) or 1,200 (100 per month). The mock-up should be ready late in August, and a production contract should be let by the end of January 1943, well before engines could be delivered and a prototype flown. As of 1 September, BuAer thought the Douglas project was proceeding rather slowly but that it could be accelerated during the month. The company did not submit a formal proposal until 16 October, about four and a half months late.[52] Douglas blamed its large engineering backlog and numerous emergency change requests due to "tactical developments abroad." Both the SBD and the SB2D had higher priorities, but ultimately Heineman

redesigned the airplane to carry its bombs and torpedoes externally under the wings, eliminating bomb-bay doors and mechanism. As in the SB2D, double-slotted flaps could be used as dive brakes. Experience with the SB2D led Heinemann to increase design gross weight from the original 21,800 pounds to 23,000 to provide a 500-pound margin over the 22,500 pounds needed for one torpedo and a thousand-mile range. To hold down weight, he made no provision for catapulting or for barrier crashes (fittings for catapulting would add 60 pounds).[53] Douglas estimated that a wooden mock-up could be completed by March 1943, followed by a metal one (to show armament installation) by June 1944. It complained that it had been unable to maintain a sufficient drafting force because of the needs of the services, the shipyards, and other companies.

Douglas' design completely missed the usual requirement for a limited rolling-takeoff run with twenty-five knots of wind over the deck. However, by 1943 work was under way on the *Midway*-class carriers, which were considerably longer than the *Essex*es, hence could accommodate airplanes with much longer takeoff runs. As conceived in the fall of 1942, they were to accommodate six squadrons, about 120 aircraft. At that time BuAer envisaged three types of future aircraft: an 18,000-pound fighter, a 19,000-pound VSB, and a 23,500-pound VTB, the latter clearly corresponding to the new VTSB. The projected airplanes were so large that the ship would be able to accommodate only a total of 81, compared to 133 of current types. The reduction to 81 was partly due to the longer takeoff runs envisaged, which would push the deck park aft.

A mock-up board—the final step before an airplane was ordered—met at Douglas from 15 to 19 March 1943. It noted that to reduce development time Douglas had limited experimental features to full-span flaps working on both sides of the wing fold (takeoff would still be satisfactory if the full-span flaps had to be replaced by partial-span ones); Douglas-type double-slotted flaps (which had flown successfully in the XA-26); the use of landing flaps

Soon after the outbreak of war, BuAer became interested in long-range torpedo bomber / scouts. It bought Douglas' big TB2D, shown on 26 February 1945. U.S. NAVAL INSTITUTE PHOTO ARCHIVE

as dive brakes, in an alternative position (care had to be taken to avoid excessive disturbance to airflow over the tail); and the XR-4360 Wasp Major driving contraprops ("Opinion on this engine is favorable"). Possible problems included drag due to external weapon stowage, marginal cockpit view during carrier approach, and the blister assembly containing the bombsight and tunnel gun (which could be eliminated if production demanded it). Douglas offered two more fixed machine guns, which the board agreed were necessary. External stowage made it possible for the airplane to carry four torpedoes if a single-point suspension for them could be developed.

Twenty-five TB2Ds were ordered as the pilot phase of a larger program. After the comparable Grumman TB2F (see below) was canceled, this program was reviewed and its production canceled on 28 July 1944.[54] That left only two unarmed prototypes, the first of which was completed on 13 March 1945.

Although it suspended open competitions for the duration of the war, BuAer still wanted alternatives.

Grumman got wind of the Douglas project, and asked BuAer about the upcoming conference. In return, BuAer asked for informal comments on the virtues of twin versus single-engine designs.[55] Grumman strongly preferred twin engines. Two single-row R-1820s (1,350 hp each) would weigh about as much as one R-3350 (2,200 or 2,300 hp) but would offer about 500 more hp. The higher-powered single engine would require either a very large propeller or complex contraprops. An airplane with shorter-diameter twin propellers would be smaller and lighter, with shorter landing gear. BuAer's favored tricycle gear would be easier to provide in a twin-engine airplane. Deck handling would be easier. Pilot vision and safety would be better, and weight would be distributed better over the wing span. Existing twin-engine aircraft accelerated faster and took off in shorter distances. Grumman chief engineer Tom Schwendler added that "these advantages have actually been observed in the twin engine airplanes which we have designed and are not merely theoretical."

BuAer decided to try Grumman as an alternative to Douglas; it probably made a formal request in June 1942.[56] Like Douglas, Grumman moved slowly, because it too was heavily loaded. In October 1942 it promised a preliminary proposal by 23 November, but it did not offer a design until January 1943.[57] Its G-55A could carry two torpedoes using R-2800s; G-55B would carry one torpedo using R-2000s.[58] These were large airplanes: G-55A gross weight was 33,924 pounds. Its powerful engines offered high speed: 360 mph at 25,000 feet with military (as opposed to civil-standard) rating (295 mph at sea level). Range would be 1,000 miles at 5,000 feet. As an experienced supplier of naval aircraft, Grumman was well aware of the need for a short takeoff roll (264 feet with twenty-five-knot wind) and a low stalling speed (70.8 mph with half-fuel and power on).

At a January 1943 conference BuAer asked Grumman to shrink its airplane. One way to do that was to have it carry a single torpedo as normal load, with two as an overload; alternatively, it could carry the torpedo externally, or it could even dispense with a bomb bay altogether. Grumman was to consider the use of a two-speed rather than two-stage engine, an increase in load factor from 4 to 5 *g*s, suitability for carrier elevators, and a reduction in height for stowage below decks.

Grumman finally submitted two twin-engine designs without bomb bays on 19 March 1943: G-55C and -55D.[59] Both had four 0.50-calibre guns (which could be replaced by four 20-mm cannon) in the wing roots, two in a manned power turret (approximately amidships atop the fuselage), and two more in a belly power turret (a "bathtub," in G-55C) abaft the space occupied by the torpedo or bombs. Both versions could glide-bomb at up to fifty degrees (-55C had dive brakes), but if it carried bombs underwing G-55D was limited to forty so that bombs would clear its propellers. The maximum load was four 1,600-pound armor-piercing (AP) bombs or two torpedoes. G-55D carried its weapons under its wings and on the sides of its body. That cleared forward arcs of fire for its lower turret.

Grumman much preferred the -55C. It argued that the external weapons in -55D might cause buffeting or cost lift in landing. Wing racks might interfere with wing structure and fuel tanks. Moreover, although it was conceived as lighter and more compact than -55C, the G-55D would have a shorter range at any but maximum-range speed. G-55C maximum speed at critical altitude would be 330 mph, and service ceiling would be 31,400 feet. Weight in landing condition would be 25,244 pounds (all torpedoes dropped, thirty minutes of fuel and one hundred rounds per gun left). It could take off in 283 feet (twenty-five knots of wind over the deck) and it stalled at 74.8 mph with full power.[60] Both versions were longer than -55A and -55B, but stowage would be easier thanks to their rearranged tails (Grumman argued that a longer tail would be valuable if the airplane was flying on one engine). Both designs could fit existing 46-by-54-foot and 33-by-60-foot elevators.

After a conference at its plant, Grumman submitted a modified G-55E design, with a single tunnel gun instead of the belly turret. The basic design was approved in July 1943, leaving Grumman free to proceed with detailed design, using engineers released from the F7F (Tigercat fighter) program.[61] BuAer was impressed by Grumman's promise of delivery eighteen months from the date of contract for the first airplane. This TB2F would operate from *Midway*-class carriers without restriction, from the *Essex* class with restrictions. Production would be speeded by using the same engines and mountings as in the F7F.

The G-55E mock-up board met at the Grumman plant between 1 and 4 May 1944; it appeared that a prototype would fly late in 1945. Work on production aircraft would begin in the summer of 1945. By this time the airplane had grown considerably, to a gross weight of 37,027 pounds with 960 gallons of fuel (no drop tanks), far more than in the earlier versions. Maximum speed was 309 mph at sea level (331 at critical altitude), rate of climb 2,230 feet per minute at sea level, takeoff distance (zero wind over deck) 1,057 feet (no figure for the usual twenty-five-knot wind was given), and stall

In effect the alternative to the TB2D was a twin-engine Grumman TB2F, which never got past the mock-up stage. Note the retractable belly turret and the 75-mm gun in the nose. These photos are dated 11 June 1944. AUTHOR'S COLLECTION

speed without power was 94.3 mph. With internal fuel (no drop tanks) radius of action would be 393 nautical miles (452 miles). With an over-load of 1,560 gallons of fuel and four 2,000-pound bombs, range would be 3,050 miles. The radius calculation included time for rendezvous and combat time.

The mock-up showed a twin upper turret (modified Emerson Dumbo) and a Sperry A-13-A retractable lower turret abaft the bomb bay; each contained two 0.50-calibre machine guns with four hundred rounds each. The bombsight shown was optical, but it was assumed that a radar bombsight would be installed on production aircraft. There was provision for both a 75-mm cannon and a Mk 15 optical bombsight on the right side of the pilot's deck.

BuAer Military Requirements recommended cancellation.[62] The airplane would barely be able to operate even from *Midway*s. It was only slightly faster than the TBY (310 versus 293 mph), and much slower than the new VBTs. The takeoff run was far too long:

with reduced fuel (700 gallons) it was 449 feet, requiring that it be spotted 600 feet from the bow. With full internal fuel (960 gallons) takeoff run would be 546 feet, so it would have to be spotted 725 feet from the bow. It doubled the load of current torpedo bombers, but fewer than half as many airplanes could be carried.[63] Its increased range was pointless, because it did not have sufficient performance to fight unescorted; its effective range would be that of its fighter escorts. Nor could it rely on surprise, since the enemy now had radar. The design sacrificed too much to gain its long range.

The VTB Desk replied that the entire point of building the *Midway*s had been to operate larger airplanes like this one. It offered far better performance than existing B-25 (PBJ) and B-26 bombers, which could not carry the same bomb load and could not land on carriers (this was not quite true, see below). The takeoff roll was not very different from that of current aircraft: 480 feet for a TBM-1D, 485 for TB2D with two torpedoes, and 670 feet for an SB2C with a thousand-pound bomb. Cancellation would commit the Navy to short-range carrier aircraft. Grumman should be asked to reduce weight and to design within set dimensions. Most of the objections to the TB2F also applied to the TB2D; if the TB2F really was worthless, so was the TB2D.

E&D (Engineering and Design) pointed out that whether the TB2F should be canceled or not depended on whether the Navy wanted a twin-engine bomber of maximum capacity and range that could operate from *Midway*-class carriers. By this time the idea of adapting the F7F had apparently surfaced: "Any thought that the F7F can be modified to produce a torpedo plane carrying one torpedo internally and still having same overall dimensions should be discarded. Such an airplane would be approx. 49 feet long and thus not operable on CV [carrier] elevators. Its spot [deck-spot dimensions] would be practically the same as the TB2F."

The TB2F would not be desirable as part of the normal *Midway* air group, "but an air group containing some might be exceedingly useful for special purposes. In order to fully exploit all of the tactical capabilities of the CVB class it is believed that a moderate number of these planes should be built."[64] Later he wrote, "Admittedly it cannot be handled with ease from a CVB ["battle carrier," i.e., a *Midway*], but it can be handled. I believe that if the tactical conception of the CVB is sound it is considered that the TB2F is sound." In maximum overload condition (45,070 pounds) the airplane could not make rolling takeoffs from a *Midway*-class carrier, but it could be catapulted off in a thirty-five-knot wind.

On 1 June the acting bureau chief, Adm. Harold Richardson, ordered the airplane killed but also that Grumman be afforded the opportunity to discuss alternative projects.

Within a few days, Richardson informally approved the XTSF-1(Grumman G-66A) as such an alternative.[65] It was an F7F Tigercat fighter with a new fuselage incorporating a bay to carry a torpedo and bombs and with increased wing span and area (and increased tail span and area to match).[66] Its main advantage was that it could be developed quickly (apparently within six months) by making maximum use of parts of the maturing F7F. BuAer considered it a bomber counterpart to the fighter for shore-based (Marine) and *Midway*-class operations as well as for possible use on *Essex*-class carriers. The relevant BuAer memo declared, "It is considered desirable that a two-engine type be developed for operational test on board carriers and for use as a long range, high speed carrier or shore-based torpedo and search plane."[67] G-66A would use much the same power plant as the TB2F, two R-2800-22 "C" engines, but it had no power turrets to add drag. Gun armament was limited to four fixed 0.50s in the wings between engine and fuselage, with cannon as an alternative.[68]

It soon turned out that G-66A demanded considerably more engineering effort than had been expected; Grumman wanted out. By late 1944 it was working on what it considered the far more promising G-70 project. It had done so little on G-66A that there was no point in assigning the airplane to another company.

G-70 began with a mid-1944 request (now lost) to Grumman for a proposal for a new two-seat (pilot and radioman/gunner) torpedo bomber.[69] A sketch shows pilot and radioman/gunner back to back, with a fuel tank between them and torpedo below them. The flexible gun was in a remote-controlled turret (which BuAer disliked) abaft the gunner. With an R-2800-22 engine (as in the TBY-2), it would make 350 knots or more at sea level. Maximum load would be five 500-pound bombs. Takeoff weight would be 16,000 pounds (350-foot takeoff), and empty weight would be 10,000 pounds. Grumman submitted this design on 14 August, offering a sea-level speed of 316 mph (342 mph at critical altitude).[70] This was roughly what the Navy already expected to get in the TBY-2 Seawolf. The weight of a more powerful engine would roughly balance out added power, so that a 2,500- rather than 2,100-hp R-3350-24 would only buy 319 mph at sea level. Rate of climb at sea level would increase from 2,750 to 2,910 feet per minute. In both cases, carrying an APS-4 radar "bomb" would cost about 12 mph.

A Grumman team visiting BuAer on 29 August got the impression that the bureau considered their company too conservative, unwilling to use new airfoils or adopt jet power, more interested in "produceability" than in high performance. In this instance, BuAer wanted higher speed and all-internal tankage. It suggested that designing an airplane around the R-2800 and adapting it to the R-3350 would result in a better, lighter airplane than one designed around the R-3350 in the first place. Grumman moved the radioman/gunner to sit alongside the pilot and replaced the unwanted remote-controlled turret with a GE I-20 turbojet to supplement, not replace, the piston engine.[71] The maximum sea-level speed of this G-70D would have been 394 mph (417 mph at 13,000 feet), and takeoff run would have been 259 feet.[72] A 2 October 1944 Grumman drawing showed a G-70F, powered by an R-3350 and a Westinghouse 24C jet engine.[73]

During a 17 October 1944 visit, Grumman representatives were told that the mixed-power proposal was getting nearly unanimous praise, except that Armament (but no one else) wanted a turret. Its performance was prompting a suggestion that it be redesignated an ASBF (attack scout bomber by Grumman). The APS-30 radar, with a twenty-four-inch diameter radome, had been designated for use in the TB3F. The new proposal was liked because it had almost the same spotting factor as the TBM and was about 100 mph faster.

At the 30 October 1944 Military Requirements Branch conference, the VTB Desk officer, Cdr. Elwin L. Farrington, wanted to go ahead with the newly submitted TB3F if possible, although Grumman was currently a fighter plant.[74] The airplane might enter production in 1946. He wanted alternative proposals from Curtiss and Douglas, to keep them interested in such aircraft. He considered Douglas the better alternative; "If the BTD flops, Douglas will have nothing to keep them going." The BT2D, which became the Skyraider, did not figure in his thinking: "We should let (Leroy) Grumman say what he can do and then immediately get out a proposal to Douglas, regardless of what Grumman has to say. I think Grumman may be willing to drop it. I would suggest Vought as someone to back up Douglas." The highest VTB priority was to replace the TBM: "The TB3F is to be a high priority design. Douglas can get it out faster than any other company." It is not clear why the new single-seat attack bombers did not figure in the VTB discussion, inasmuch as all of them could deliver torpedoes.

BuAer was unable to interest either Douglas or Vought in a new torpedo bomber. By mid-December Grumman's TB3F was its priority, and a mock-up was due about January 1945. Grumman received a three-airplane contract on 19 February 1945 for airplanes to be powered by one R-2800-3W piston engine and one Westinghouse 19XB-2B (J30) jet engine. An August 1945 change order directed that the second airplane be powered by an R-3350 (and, later, a 24C4 jet engine) as the XTBF-2. The prototype of the first was expected to fly late in 1945, but the postwar slowdown delayed that to 19 December 1946. During its ground run, the titanium inlet ducts collapsed. Similar problems had been experienced

Late in the war, Grumman conceived a new torpedo bomber using a piston engine plus a jet in the tail, the TB3F. Note the intakes in the wing roots and the small tailpipe. The XTB3F never flew on jet power, and it was soon rethought as an ASW airplane. U.S. NAVAL INSTITUTE PHOTO ARCHIVE

with the Ryan FR-1 Fireball fighter. The Navy and Grumman decided, probably jointly, not to operate the jet engine in flight.

Without the jet, however, the TB3F could no longer deliver spectacular performance, and the project was rethought. In any case, priorities were now different. The TB3F never entered service as a torpedo bomber. The second and third airplanes were modified as ASW search and attack prototypes.[75]

BuAer also tried the B-25 as a carrier-based bomber.[76] It had taken off from the carrier *Hornet* in 1942 to bomb Tokyo, but only after drastic lightening and without any hope of returning to land on board. In 1944 the idea seems to have been to catapult such bombers off and recover them. The Marines' PBJ version was about the size of the abortive TB2F, with a gross weight of 34,000 pounds. Its R-2600 engines were somewhat less powerful (1,700 hp each), its span was somewhat less (sixty-seven feet), and its takeoff distance and stall speed were worse. Nevertheless, a carrier PBJ project seems to have begun in February 1944; in March, BuAer was negotiating with North American, the manu-

facturer. If carrier tests with one PBJ-1H were successful, a number of further conversions might be authorized. In April BuAer gave landing and takeoff characteristics to be expected of a carrier version of the airplane. Catapult and arresting-gear tests at Naval Air Materiel Center (NAMC) Philadelphia were intended specifically to determine carrier suitability.

In August, BuAer asked North American to quote a price to modify a hundred PBJ-1Hs. As the cover sheet cautioned, "This covers the quotation request . . . *and is not believed to be connected with the payment of the carrier model*" (emphasis added). In November 1944 a strengthened hooked PBJ-1H was flown from and recovered again by the new carrier *Shangri-La*, as were a tailhooked North American P-51D Mustang and an F7F-1 Tigercat. The bomber made two carrier landings (the ship having new high-capacity arresting gear) and two catapult takeoffs. The first landing and catapulting were both normal and caused no structural damage. The pilot's seat gave way on the second catapult takeoff. In the subsequent emergency landing the airplane struck nosewheel first, and the arresting hook hit the side

The enigma: in November 1944 a B-25 (PBJ-1H) medium bomber was successfully launched from and recovered onto the carrier *Shangri-La*. Here the PBJ is hooked to the catapult. It seems that adapting the PBJ-to-carrier operation was part of a project under which as many as a hundred aircraft might have been converted for a special mission, not otherwise specified. It seems likely that the mission was no longer important by the time the test was conducted. What was it? U.S. NAVAL INSTITUTE PHOTO ARCHIVE

of the ramp, the extreme after end of the flight deck. Even so, it hooked the third wire. There was minor damage. The plane was to be shipped to North American Kansas City for repairs, to allow further tests, but it is not clear whether either repairs or tests were carried out.

It appears that the project was intended to provide a special group of carrier-modified PBJs for a special operation that was never mounted. The entire project seems to have disappeared after mid-December 1944, either because the rationale was gone or because some officer who had supported the aircraft

in its own right (perhaps as an alternative to the TB2F) was gone.

The only other wartime VTB project seems to have been an unsolicited proposal by the Lockheed Vega division. In June 1942 it asked BuAer whether one would be accepted; the firm was already building the land-based Ventura patrol/torpedo bomber.[77] BuAer clearly preferred that it not dissipate energy on anything else, but at the company's request it provided outline carrier-torpedo-bomber requirements. Normal gross weight with a Mk 13 torpedo and a thousand miles of fuel should be within 21,000 pounds. Provision for two torpedoes was now desirable. The airplane should have maximum possible forward defensive firepower. If it had both upper and lower turrets, the upper one should hold two 0.50-calibre machine guns and the lower, one. Vega was told how important takeoff distance (350 feet with twenty-five-knot wind over deck) and stall speed (75 mph with power) were; the first could not be relaxed.

Vega offered three alternatives, each with the usual three-man crew.[78] V-141A was a conventional single-engine (R-4360) design. V-141B had a fuselage (with engine and upper turret) and two wing pods (forward-firing gun in the left one, pilot in the right). V-141C was a conventional twin-engine, twin-tail airplane powered by a pair of R-2000s. Nothing came of this proposal.

Douglas' Skyraider was by far the best of the new single-seat attack bombers conceived during World War II. These AD-4Bs, modified for nuclear attack, were assigned to VX-5, which developed the necessary tactics. This photograph was taken on 15 March 1953, as the Navy was developing an ability to attack using the new lightweight nuclear bombs.
U.S. NAVAL INSTITUTE PHOTO ARCHIVE

7

A NEW KIND OF ATTACK BOMBER

The idea of a single-seat dual-purpose attack bomber seems to have originated in December 1941 with Lt. Cdr. J. N. Murphy, who was then BuAer's VSB Desk officer. He wanted to modify the existing SB2D, which was eventually (but hardly immediately) done.[1] Murphy wrote on an internal document that although some officers thought the single-seater might be a fighter-bomber (VFB), it could not really be a good fighter: "The intent is to design the maximum striking power in an airplane." The Armament section pointed out that in consideration of higher load factors and braking flaps, the proposed VB would be a better torpedo bomber than a VTB. It should be designated *VBT*, dive and torpedo bomber, in that order, the *S* being eliminated because it would not be a scout. It would be far more practical than a two-seat fighter or a single-seat scout/observation plane, both of which had proponents within BuAer. The main objection was that no airplane would be left for which torpedo attack was the primary mission. Torpedo attack might be abandoned altogether.

The same 21–22 January 1942 conference that created the scout torpedo bomber approved the VBT. It set out a specification for an airplane powered by an R-3350 engine with a single-stage (two-speed) supercharger. Its bomb bay would hold one 500-pound or two 1,600-pound AP bombs or two 1,000-pound general-purpose (GP) bombs; or it could carry four 500-pound bombs (two external). It would carry its torpedo externally. Design strength would be 9 *g*s with one 500-pound bomb and a thousand miles of protected fuel and oil (less one-quarter). Limiting speed (in a dive) would be 525 mph. The airplane would have dive brakes sufficient to limit speed in a seventy-degree dive to 350 mph. Gun armament would be four 20-mm wing cannon, far more than current fighters. Protected fuel capacity would suffice for a thousand miles carrying a 1,000-pound bomb, with sufficient fuel in drop tanks for another five hundred miles with the bomb. Drop tanks might not be wanted when the torpedo was carried. The radio would be a standard fighter type. Maximum takeoff run in a twenty-five-knot wind with one 1,000-pound bomb and a thousand miles' fuel would be 250 feet (350 feet with maximum overload and torpedo).

An attempt to redesign the existing SBD as a single-seater failed.[2]

In January 1942, Curtiss and Douglas were the obvious choices for new naval attack aircraft. The BuAer conference split the work: Douglas got the scout torpedo bomber (TB2D) and Curtiss the single-seater. A 9 February letter to Curtiss emphasized the need for armor, since the airplane would have no rear gunner.[3] Since the airplane would not be used for high-altitude bombing, it would need only a single-stage supercharger; the R-3350 engine was suggested.[4]

Curtiss' BTC

BuAer found Curtiss' 30 April proposal extremely attractive. Curtiss promised a prototype within fourteen months, assuming the design was accepted by 1 June 1942. It would be powered by an R-3350-8 (2,300-hp takeoff, 1,900-hp military rating at 14,000 feet). An airplane powered by the alternative XR-4360-3 could be delivered within seventeen months. With one 500-pound bomb, the R-3350 version would make 388 mph at critical altitude (16,000 feet) and would take off in 233 feet (twenty-five knots of wind over deck). By the end of June 1942 BuAer had issued a letter of intent for two R-3350-8 airplanes and for two XR-4360-12A airplanes, the pairs to be designated, respectively, XBTC-1 and XBTC-2. The BuAer VSB Desk pointed out that below 15,000 feet the speeds and rates of climb of these airplanes would compare well with fighters. The BTC was so attractive that the Army decided to buy it, as the XA-40.[5]

The program began to go awry, partly because Curtiss was working hard to cure problems with its Helldiver (SB2C) and partly because it was also working urgently on the new SC-1 floatplane scout. The mock-up was delayed to December 1942; by that time delivery of the first XBTC-1 was scheduled for September 1943. Curtiss was so badly overloaded that it could not respond to repeated requests

for production data to enable BuAer to commit to funding necessary tooling. In March 1943 Curtiss finally submitted quotes for production tooling. By May BuAer wanted to initiate procurement of 250 BTC-1s. It was urged by the vendor to order BTC-2s as well, in order to gain a commitment for their R-4360 engines. It seemed, however, that the BTC-1 would have to be in production for at least eight months before the shift to the BTC-2 could be made. It seemed that Curtiss would be unable to deliver its first experimental airplane before the summer of 1944 or its first production airplane before February 1945—and even that proved rather optimistic. Curtiss proposed informally that the BTC-1 be dropped altogether in favor of the -2.

The VSB Desk urged that effort be concentrated on the BTC-2. Although Curtiss might be unable ever to place it in production, whoever built this extremely high-performance airplane would need the production design that Curtiss could create; transferring design work in progress would only cause further delay. The VSB Desk recommended that every effort be made to accelerate the XBTC-2 design at Columbus, while capacity was being found for a similar design at another company. Curtiss decided to use subcontractors to design (with its supervision) parts of the airplane, such as the outer wing panels, tail surfaces, and the tail wheel.

Curtiss' XBTC-2 was its approach to the new single-seat attack category. U.S. NAVAL INSTITUTE PHOTO ARCHIVE

The Helldiver lineage of the XBTC-2 shows in its retention of a bomb bay. Note its contraprop.
U.S. NAVAL INSTITUTE PHOTO ARCHIVE

BuAer wanted Curtiss to quote for twenty-three preproduction aircraft, but according to the 1 January 1944 VSB Desk report, prospective BTC-2 production was so far off that Curtiss found it difficult to do plan for it at all. At the company's request the delivery dates of the two XBTC-2s (XR-4360-8A) engines had been set back to 31 October and 31 December 1944.

At the beginning of July 1944, the VSB Desk reported that changes in requirements and progress in equipment and materials over the two and a half years of the project had already made the airplane obsolescent in many ways. Its cockpit would have to be rearranged, its armament modernized, its fuel system reworked, provision made for a two-thousand-pound bomb (internal or external), and radio and radar would have to be modernized. It would probably be necessary to increase wing area and fuselage length to meet current stability and controllability requirements. Curtis now proposed to limit diving speed further (to 275–285 knots) by using the airplane's landing gear in the extended position (wind-tunnel tests showed that fuselage dive brakes were not enough). The end of the war further slowed the program. The prototype XBTC-2 did not fly until July 1946, by which time it was completely obsolete.

In September 1944, Curtiss offered a single-seat version of the Helldiver powered by an R-3350 engine.[6] BuAer doubted that it could be produced any sooner than superior new single-seaters (the BTM and BK), because Curtiss was by then already so heavily committed to the SB2C and the new SC Seahawk surface-ship scout.[7] The new design would need a completely redesigned power-plant installation and modified landing gear to match its much larger propeller. The bomb bay and rear seat would be eliminated, and wing folding would change because the wingtips would be changed.

However, by early 1945 BuAer realized that none of the new single-seaters could enter combat before the spring of 1947. The only dive-bomber that could enter service during 1946, when it was expected that aircraft would urgently be needed for the invasion of Japan, would be a Helldiver derivative. Early in 1945 BuAer planned ten prototypes, to be delivered during the last half of 1945. Curtiss told BuAer that if production was authorized by 1 March 1945 (before a first flight), production could begin early in 1946. It pointed out that about half of the parts by number and 68 percent by weight would be common to the SB2C-5, so production could begin quickly using existing tooling and fleet personnel would need minimum training. It helped that this proposed BT2C-1 would retain the Helldiver's internal bomb bay. ComAirPac approved the idea—despite the Helldiver lesson that any production decision should be deferred until all problems had been solved.

The new airplane would be twenty inches longer in its tail than the Helldiver and would have more internal protected fuel (410 gallons). It would use an R-3350-24 (HD) engine. The pilot would sit under a bubble canopy, his seat raised two inches. Bullet-resistant glass would be integral with the windshield, used as a gunsight reflector. The bomb bay would be revised and the doors extended to fully enclose a torpedo. Control surfaces would be metal covered and the horizontal and vertical tails replaced by high-aspect-ratio elements. Given the increasing importance of airborne radar, the airplane had a "jump" seat buried in the fuselage for a radar operator. As in the Helldiver, fixed gun armament was two 20-mm cannon. BuAer gave Curtiss the expected contract for ten prototypes on 27 March. As expected, the program ran quickly, the prototype flying in January 1946, ahead of the BTC-2. Nine of the ten ordered were completed. As BuAer had imagined, the later designs were better, and with the war over it could afford to wait for them.

Alternatives

By the late fall of 1943, with the BTC and the alternative BTD (a modified SB2D) faltering, BuAer sought alternatives. The dive-bomber program was reviewed at an 11 November 1943 meeting.[8] Discussion centered on a single-seat, high-performance dive-bomber using the R-2800 engine. Cdr. J. B. Russell (Military Requirements, MR) urged an airplane specifically for the small escort and light carriers. The meeting decided to let three engineering-study contracts: to Brewster,

Curtiss' BT2C-1 gained support because in theory it could be developed rapidly using Helldiver components. U.S. NAVAL INSTITUTE PHOTO ARCHIVE

for the lightest possible R-2800 dive-bomber; to Fleetwings for a single-seat airplane combining an R-2800 with a jet engine, for maximum performance at moderate weight; and to Martin, for a single-seater powered by an R-4360 and carrying a two-thousand-pound bomb. The latter was intended specifically as an advance over the BTC.

By this time Brewster was nearly dead; it would soon close down. Fleetwings was a small aircraft company in Pennsylvania just bought by Henry Kaiser, the West Coast industrialist famous for mass-producing ships. Apparently it took over the planned Brewster study. BuAer's Fleetwings file shows three different proposals: one with an R-2800 engine, one with that engine plus a jet, and one with the smaller R-1820 plus a jet.[9] Fleetwings also offered a twin-nacelle "two-squirt" airplane, each nacelle carrying a Packard Merlin piston engine and a jet engine (it is the only one for which a drawing, dated February 1944, has survived). The jet proposals were dropped, and the airplane that emerged was considerably better than any of the R-2800 types proposed.

BuAer described Fleetwings' successful proposal as the smallest possible high-performance dive-bomber, intended specifically for small carriers. A 23 February 1944 letter to Fleetwings awarded the program an AA-1 priority rating. About March 1944 BuAer issued a letter of intent for two prototypes.

The airplane was designated XBK-1, a pure dive-bomber normally carrying a thousand-pound bomb with a four-hundred-mile combat radius. Its takeoff run under combat loading was expected to be 350 feet (twenty-five knots of wind over deck). It could lift a two-thousand-pound bomb and enough extra fuel to extend its combat radius to 550 nautical miles. It had a novel engine installation in which exhaust gas was used to pump cooling air through a tight cowling around its engine. As of March 1944 it was expected to fly that November. It was so attractive that BuAer was entertaining a Fleetwings proposal for twenty-three more aircraft and tooling for two hundred per month, using facilities expected to be available to Kaiser Cargo Inc. (Fleetwings received a contract for twenty more airplanes). The XBK and the big BTM (see below) seemed to cover the full range of performance the fleet wanted. Fleetwings' weight control was so effective that its airplane was consistently rated about two hundred pounds below design weight. As of September 1944, Kaiser Fleetwings thought it could build all twenty-two aircraft by September 1945. The VSB Desk cautioned that Kaiser had limited facilities and would need considerable production help. Later the Kaiser organization claimed that its specialists could recruit the required skilled labor.

BuAer was very impressed by Kaiser's compact and very clean BTK-1, which could operate from escort carriers. That ability lost its value postwar, when the small carriers were generally laid up. U.S. NAVAL INSTITUTE PHOTO ARCHIVE

On 1 December 1944 DCNO (Air) called for accelerated procurement and production of the BK. He saw it as a natural successor to the Helldiver. The designation was changed to BTK-1 in February 1945, to indicate the ability to carry a torpedo.

The XBTK-1 emerged from the shop not in November 1944 but on 15 March the next year and was turned over to a Navy inspector only on the 24th; it finally flew on 12 April. There were no problems with stability, control, or stick forces, and the airplane was four hundred pounds underweight. Wind-tunnel tests of a powered model showed very good results. The only problem, which seemed to be on the way to solution in the spring of 1945, was that the fully mobilized United States was badly short of skilled labor. The end of the war stopped the project. The unique capacity of the XBTK to operate from small carriers lost its value once most of them were laid up as demobilization began. Only four aircraft were delivered.

The Mauler

Martin's mock-up inspection was completed on 9 February 1944. BuAer's evaluator noted that Martin had deliberately kept the design conventional to avoid the delays associated with untried features. Prototypes could be built rapidly.[10] Unlike the BTC-2, it had conventional partial-span slotted flaps (split as dive brakes) rather than highly experimental full-span flaps. Bombs were all external: one 2,000-pound bomb (the first specified for a U.S. Navy dive-bomber) under the belly, with two wing racks for overload bombs up to a thousand pounds. Four wing guns could be 0.50- or 0.60-calibre machine guns or 20-mm

Initially the most attractive of the single-seat attack bombers was Martin's BM-1 Mauler, renamed the AM-1 postwar. It was powered by an R-4360. Proposing the airplane in November 1943, Martin offered some alternatives: a thrust augmentor (1,300 pounds of thrust at takeoff) with the R-4360, or a 3,200-hp Allison V-1710-F3OR liquid-cooled engine using radiators in the wing roots. BuAer chose the R-4360 without the augmentor. Here the big "Able Mabel" shows its load-lifting capability: three torpedoes (6,270 pounds) and either twelve rockets or twelve 250-pound bombs, plus four 20-mm cannon. This photograph was taken on 31 January 1949. U.S. NAVAL INSTITUTE PHOTO ARCHIVE

The Mauler shows its unique "finger-type" dive brakes. The upper and lower fingers meshed so that, with the upper part down, the combination formed a conventional flap. When separate, the two sections acted as dive brakes. An additional perforated brake, visible here, was located under the belly. The great span of the dive brake / flaps reduced the length of the ailerons. That was evident at low speed, and spoilers had to be added. The original Martin proposal called for wide-span perforated flaps. The airplane is carrying a 750-pound bomb on the centerline and an APS-4 radar "bomb" under the starboard wing. This photograph was taken on 26 April 1949. U.S. NAVAL INSTITUTE PHOTO ARCHIVE

cannon. "The bubble canopy [i.e., without vision-obstructing frames] and comparatively small fuselage resulting from the external bomb installation and other refinements contribute to the cleanness of the design in the get-away [after bombing] condition." The airplane was stressed to 7 *gs* in the "realistic" 19,450-pound condition, meaning the 2,000-pound bomb, full ammunition, three-quarters fuel, and 529 pounds of radio and radar equipment. This was equal to or better than existing service aircraft and the BTC-2. Gross weight was 20,200 pounds, and takeoff run in a twenty-five-knot wind was 392 feet. At sea level, in getaway condition (bomb dropped), speed was 350 mph. Rate of climb at sea level was three thousand feet per minute.

By this time the BTD-1 was in trouble, the BTC-2 program was moving far too slowly, and the Hell-diver was considered obsolescent. Martin was asked whether it could produce at least one prototype in about six months. Martin thought that was possible, the second airplane following two months later. The company's willingness to move very fast was extremely attractive. Acting bureau chief Admiral

Richardson approved the project on 8 January 1944, and on 11 January BuAer requested a letter of intent. The mock-up, completed about 7 February, was approved without major changes. Martin proposed to build twenty-three prototypes plus tooling sufficient for one, two, or three hundred aircraft a month. This airplane was designated XBTM-1. The production version became the Mauler.

This was the most promising VBT prospect, so it was expedited. As of March 1944 the goal was a first flight on 15 July 1944 (it slipped to 25 August). According to the 1 April 1944 monthly VSB status report, "If the XBTM-1 flies successfully, this airplane is expected to eliminate the XBTC-2 from the VSB picture." The BuAer VSB Desk was impressed with how quickly Martin had been able to build its prototype; test flights showed that it was a sound design. On 15 January 1945 the company received a letter of intent for 750 BTM-1s, all for delivery during 1946. BuAer emphasized that the schedule could not be met unless flight test and development were greatly accelerated. As of June 1945 it seemed that engineering for all production tooling would be approved by 1 July.

As of March 1944 the BTM was expected to take off in 360 feet (twenty-five knots of wind over deck), with a combat radius of 325 nautical miles on internal fuel. With a droppable three-hundred-gallon tank, its combat radius (with the two thousand-pound bomb) would be 700 nautical miles. Initial tests at Patuxent River late in 1944 indicated that despite an overweight of 480 pounds the takeoff run would be better than guaranteed. Speed would be slightly lower than guaranteed, stall speed slightly higher. The main changes after these tests were addition of a dorsal fin, which Martin had foreseen would be needed, and reduced dihedral. The prot type opened its balanced-flap-type dive brakes for the first time on 25 January 1945.

By late 1945 another single-seat bomber was flying, the Douglas Skyraider. The Navy did not need two parallel attack-airplane programs. By 1945, moreover, the Martin bomber was being characterized as suited primarily to *Midway*-class carriers, with some capacity to operate from *Essex*es.[11] At the end of the war Mauler production was cut back to two prototypes and 149 production aircraft. Production ended in October 1949. In 1950 all Maulers were assigned to reserve units.

Under the postwar system, the Mauler became the AM-1.

Eighteen aircraft were completed as AM-1Qs, the fuselage fuel tank being replaced by a station for an ECM operator. He had the same equipment as the operator in a TBM-3Q: an APR-1 search receiver, an APA-11 pulse analyzer, and an APA-38 panoramic adapter. There were two blade antennas under the body forward of the tailwheel and cavity-backed spirals on the sides roughly above the tail wheel. There was also a chaff chute. A door was cut into the side of the body for the ECM operator. (ECM Skyraiders had much the same arrangement.) It is not clear how many AM-1Qs deployed on board carriers.

The BT2D Skyraider

Douglas was not involved in the November 1943 BuAer VBT program, because it was already building the BTD-1. Through the spring of 1944 the VSB Desk was optimistic that the BTD would become operational, but its designer, Ed Heinemann, doubted that it ever would. He began to sketch alternatives late in 1943; copies survive in NHHC's Naval Aviation History Division file on the Skyraider. One concept resembled the Skyraider but had a jet engine exhausting under the airplane's tail. To accommodate it, the fuselage was deepened and an additional fuel tank (280 gallons) placed under the pilot; in another sketch, the fuselage tank was simply enlarged to 500 gallons. The air intakes were in the wing roots, as in the BTD-2. In another alternative design, the jet was buried in the fuselage, fed by an air intake from under the engine in the nose; in yet another, the jet could be placed below the pilot. A further possibility was a twin turboprop, the engines in nacelles, the nose cut back forward of the pilot. The single fuselage gas tank was moved farther back, presumably to maintain balance. In the most exotic version, the engine was moved abaft the fuel tank to drive contraprops in the tail, as in Heinemann's XB-42 bomber, and a jet engine was slung under each wing.

The first single-seat attack bomber was Douglas' BTD-1, modified from its SB2D-1 by, among other things, removing its two remote-controlled turrets. The failure of this project led Ed Heinneman to propose the very successful Skyraider.

At a 2 June 1944 BuAer conference, chaired by Admiral Richardson, to discuss the BTD program, Heinemann broached the idea of an entirely new airplane.[12] The meeting was told that the only real problem with the one in hand was takeoff distance: the BTD program was still attractive, because it offered an early service test of the single-seat dive-bomber concept. Heinemann nevertheless proposed canceling it and channeling remaining funds into a new dive-bomber, on which he had done preliminary work. He asked for thirty days to flesh it out; Richardson gave him overnight to sketch an airplane powered by the same R-3350 that powered the BTD. Heinemann and his team managed to present the necessary data and rough drawings the next morning. BuAer accepted his plan; the prototype had to be built in nine months.

New requirements set by Heinemann to himself to make the project worthwhile were shorter takeoff, increased combat radius and rate of climb, greater load-carrying capability, and greater stability and control, all of which required more lift and lighter weight.[13] Heinemann wanted to cut weight from 18,000 to 16,500 pounds, increase the maximum lift coefficient from 1.8 to 2.0, and halve fueling, arming, and maintenance time. He set a target weight 750 pounds below the guarantee and demanded of his team that any overweight in one part had to be balanced elsewhere in the design. All structural parts were strength-tested. Heinemann was willing to spend on average $10 to save each pound, based on the cost of tooling for five hundred aircraft; later he considered $15 or even $20 reasonable. He told his designers that every hundred pounds saved would reduce takeoff run by eight feet and increase combat radius by twenty-two nautical miles (or maximum range by sixty), sea-level rate of climb by eighteen feet per minute, and maximum speed by 0.3 mph. Ultimately he saved a thousand pounds.

To gain maximum lift, Heinemann abandoned the wing dive brakes of the BTD. Tests had shown that they caused buffeting and reduction of control in a dive. The alternatives were a reversing propeller (which did not exist, hence would have to be developed), a drag parachute, and fuselage dive brakes. Heinemann chose fuselage brakes, which were tested on BTD prototypes. Among their advantages was that they did not affect wing lift during a dive. The new higher-lift wing added drag, which Heinemann expected to balance by better detail design. Given the underweight, increased lift, and better stability and control at low speed, the airplane could carry a two-thousand-pound bomb or torpedo as normal load rather than overload, with a safe overload of six thousand pounds.

The winner was Ed Heinemann's masterly BT2D Skyraider, which became the AD series postwar. The XBT2D-1 was photographed on 29 August 1945. The carburetor air scoop is above the engine, the oil cooler air scoop below it. A big propeller spinner initially fitted was eliminated when it turned out not to be useful. U.S. NAVAL INSTITUTE PHOTO ARCHIVE

A 28 November 1944 BuAer summary of the experimental dive-bomber program showed that the BT2D was the fastest of the entries being considered, rated at 405 mph at sea level in combat condition, compared to 290 for the Martin's, 285 for Curtiss', and 343 for Fleetwings'.[14]

On 14 August 1944, after a successful mock-up inspection, BuAer approved production of fifteen prototypes (about March 1945 this was increased to twenty-five prototypes), to be funded from the balance of the canceled BTD program. Heinemann's drastic weight-control program was spectacularly successful: the first prototype was about 1,800 pounds—more than 10 percent—below the guaranteed weight, bettering an earlier prediction of about 1,500 pounds. In March 1945 a Navy conference was told that "because of Douglas' experience and capacity and the generally satisfactory appearance of the design, the BT2D appears to have the best chance of reaching the ready-for-combat stage in the minimum length of time." To accelerate production, Douglas was now issued a letter of intent to produce 558 BT2Ds (including the additional prototypes). The first airplane flew on 18 March 1945, two weeks ahead of schedule. Patuxent River test pilots judged it the best dive-bomber they had ever tested, with very good flight characteristics and performance and excellent wave-off characteristics (i.e., ability to abort, late and suddenly, a carrier-landing attempt and regain altitude). Heinemann had also succeeded with maintenance characteristics. Initially the airplane was called the Dauntless II; only later did it acquire the name Skyraider.

In mid-1945 BuAer settled on the BT2D as the standard dive-bomber for *Essex*-class carriers. The peak authorized production rate was then 200 per month, but the actual rate was thought unlikely to exceed 130 per month by the end of 1946. The contract was to be evaluated late in December 1946.

Under the postwar designation system, the BT2D became the AD.

Postwar Skyraiders

In September 1944, as he completed the BT2D design, Heinemann considered future versions that would keep the program alive.[15] BuAer had told him that it wanted to keep building naval aircraft at El Segundo.

An AD-1, the first production version, shows its three dive brakes: two on the sides and one in the belly. Weapons underwing are twelve 5-inch, high-velocity aircraft rockets (HVARs) and two 11.75-inch Tiny Tims. NHHC

Heinemann envisaged a gradual shift from specialized aircraft to a single type that could be used as either a fighter or an attack bomber. Although pure-jet aircraft might be practical if catapult takeoffs were accepted as general practice (as they would be), "the gas turbine [turboprop] should be pursued closely—perhaps even by means of a *contract* [emphasis in original] for an attack airplane with gas turbine propeller drive." An airplane designed to carry a 2,000-pound bomb could also carry a torpedo, or two Tiny Tim rockets, or two 1,000-pound bombs, or any gun arrangement yet have fighter performance in the clean condition (i.e., with no external stores to add weight and drag). Heinemann offered two possibilities: a twin-engine airplane (whose piston engines might be replaced by jets) that would leave the nose clear for a radar antenna (something he might have remembered from the early stages of the TB2D program), and a bomber-fighter using a turboprop. BuAer was interested in the bomber-fighter.

Heinemann offered sketches of BT2Ds powered by a variety of engines: Proposal "A" (R-2800-34W), "B" (R4360-14W), "C" (TG-100 turboprop), "D" (alternative with TG-100), and "E" (two TG-100s driving contraprops in a short nose). Of these, Heinemann saw B as an alternative in the event the R-3350 proved unsatisfactory. C had excessive wing loading; performance was not appreciably better: the wing would have to be moved forward, a larger tail designed, the fuel tank enlarged. D would be better, but performance would not be improved sufficiently to justify production. E offered enough improvement to be worth developing; it would have an excellent takeoff distance even with a two-thousand-pound bomb and could carry exceptional loads. It offered provision for a second crew member, who could be a radar operator. BuAer was interested enough to support development under the designation BT3D but had ordered no prototype when the war ended.

Because the Skyraider fuselage had been made deep enough to accommodate the big R-3350, it could accommodate crew members abaft the pilot. The first version to exploit this possibility seems to have been the BT2D-1N, two of which were converted from XBT2D-1s. They had a pair of radar operators side by side and a radar pod beneath the port wing. Matching it was a searchlight pod on the starboard wing. This version lacked dive brakes. One prototype was converted into an AEW (airborne early warning) airplane, the XAD-1W, comparable in theory to the TBM-3W Cadillacs of 1945. Like the night attackers, it had two fuselage positions, in this case for the operators of the big APS-20 radar and its data link. The pilot's canopy was faired into an opaque structure extending aft. Douglas converted one XBT2D-1 to an ECM version (BT2D-1Q) with a single fuselage crew position (below and abaft the pilot, for the jammer) and radar and chaff pods port and starboard. Another XBT2D-1 was converted into a photo reconnaissance prototype (BT2D-1P).[16]

Shortly after VJ-Day the order for 558 production Skyraiders was cut to 377 and then to 277. With the adoption of the *A* (for attack) category in 1946, the Skyraider was redesignated AD-1.

Production BT2D/AD-1s were strengthened, adding 415 pounds beyond the prototypes' weight, but otherwise unchanged. There were 242 production BT2D-1/AD-1. Another thirty-five aircraft were completed as AD-1Qs, similar to the XBT2D-1Q described above. These and later *Q* versions of the Skyraider had ALT-2 noise jammers in bomb-like pods carried underwing. AD-2 had a 2,700-hp R-3350-26W, more internal fuel (380 rather than 365 gallons), greater strength, and a revised cockpit. All 156 of this version were delivered during 1948. Twenty-one ECM versions (AD-2Q) were delivered between September 1948 and April 1949. One was converted into an airborne early-warning prototype.

Initially the AD-3 designation (1948) was applied to the turboprop version that became the A2D Skyshark described below. It was later used instead for an improved AD-2 with a redesigned canopy, further local strengthening, a longer-stroke undercarriage (which would allow larger-diameter stores to be carried), and an improved propeller. Production came to 125. In addition, there were fifteen night-attack versions (AD-3N), twenty-three ECM versions (AD-3Qs), thirty-one AEW versions (AD-3Ws), and two ASW variants, the AD-3E search and AD-3S attack version. The latter were converted from AD-3Ns and -3Ws. Douglas modified one AD-3S as a "single package" ASW airplane.[17] Some AD-3Qs were converted to high altitude high speed target towers.[18]

The AD-4 was built in greater numbers than any other version, owing to Korean War experience and demands.[19] The prototype was an AD-3 with a more powerful -26WA engine, a further-improved windshield, a P-1 autopilot, and a modified arresting hook. Production AD-4s had provision for APS-19A radar "bombs." Of 372 aircraft, 63 were modified (winterized) as AD-4Ls and 28 as AD-4Bs (nuclear bombers).[20] Another 165 were built as -4Bs.[21] There

were 307 three-seat AD-4Ns, of which 38 were winterized as -4NLs. Another 168 were AD-4W AEW aircraft, 50 of which were transferred to the Royal Navy as Skyraider AEW.1s. The -4N and -4W were fitted so that they could be used as hunter-killer teams for ASW, the -4W being the "hunter" (the APS-20 radar offered antisnorkel and antiperiscope capability).[22] Aircraft modified specifically for Korean operations (AD-4Ls and -4NAs) had four rather than two 20-mm wing cannon; the -4NAs were stripped of night-attack gear to increase their bomb loads. The -4B nuclear version also had the four wing cannon.[23]

A night-attack AD-3N on the deck-edge elevator of a carrier shows the door for the electronic operator (under the right-hand bars of the side insignia). This and other EW versions of the Skyraider could carry two observers or operators internally in addition to their pilots. The side dive brakes have been eliminated. Note the air scoop nearby for cooling air for electronics. Typically the AD-3N was used as a pathfinder and attacker (note the underwing stubs). ECM equipment was an APR-9B receiver and an APA-70B direction finder (there were no jammers). Aircraft typically carried underwing pods for the APS-31 radar and the AVQ-2A searchlight. Fifteen were built. This airplane is from VC-33, which provided night-attack detachments. By the end of June 1950 the unit included a single AD-3N plus eight AD-4Ns, three AD-4Qs, two AD-1Qs, and two TBM-3Ns. VC-33 used its AD-3Qs as radio relays. U.S. NAVAL INSTITUTE PHOTO ARCHIVE

An AD-3E (for ASW) of VX-1 is shown in 1952. Note the bulge of its underbody APS-20 radar, which turned out to be effective at detecting periscopes and snorkels. This airplane was converted from an AD-3W (for airborne early warning). The two were externally identical (as was the -4W). Only two were made, for tests. The second AD-3E operated in Korean waters in the winter of 1950 on board the carrier *Valley Forge.* U.S. NAVAL INSTITUTE PHOTO ARCHIVE

The AD-3S was the other half of the experimental hunter-killer team, converted from an AD-3N. Its main sensor was the radar in the big APS-31 radar "bomb" visible here. Under the far wing it had a searchlight, and it could carry a sonobuoy dispenser (it had the necessary receiver). It had a radar-relay receiver enabling it to work with an AD-3E. Note the door for the operator (between the *7* and the left bar of the national insignia), the cooling-air scoop, and the upper and lower radomes farther aft. This airplane and the -3E formed a hunter-killer team, the prototypes of AD-4 hunter-killer teams (AD-4W and -4N). As with the -3E, only two were made. One -3S was given an upgraded APS-31 to test the idea of a single hunter-killer airplane. U.S. NAVAL INSTITUTE PHOTO ARCHIVE

An AD-4 from VA-115, in a photo released just before the outbreak of the Korean War, 10 June 1950. This was the main production version that fought in Korea. AD-4s were wired and plumbed to take an APS-19A radar "bomb" under the starboard wing, in effect a successor to the World War II–era APS-4. AD-4Bs had four rather than two 20-mm cannon, and the last 28 AD-4s were upgraded. Four guns were also mounted in a hundred modified AD-4NAs and in thirty-seven modified AD-4NLs. VA-115 received its first AD-4s in December 1949, replacing AD-2s received in January. The squadron deployed on board the carrier *Philippine Sea* on 24 July 1950. At that time it had twelve AD-4s, two AD-4Qs, and one AD-3Q. The tail code (*V*) indicates not the squadron but the carrier air group (CVG-11). Marine attack squadrons had their own tail codes, as did special-purpose squadrons that provided detachments to carriers. U.S. NAVAL INSTITUTE PHOTO ARCHIVE

This AD-4L served as aerodynamic prototype for the AD-4N, which could be used as the "killer" element of a hunter-killer team, working with an AD-4W radar airplane. The winterization (*L*) feature is reflected in the rubber boots on the wing leading edge. Note that this airplane (BuNo 123986) does not show a door to the system operators' compartment or a cooling air scoop for operational electronics. The torpedo under the centerline was a stand-in for the centerline pod that AD-4Ns normally carried, with a heater and a homing torpedo inside. Not visible is the big AVQ-2 searchlight pod under the other wing, which also carried sonobuoys. Aircraft sometimes carried an Aero 2C sonobuoy/flare dispenser instead of the searchlight. The small radomes visible above and below the fuselage are for the APX-6 IFF system. Note the four 20-mm cannon of later AD-4s. DOUGLAS COURTESY U.S. NAVAL INSTITUTE PHOTO ARCHIVE

This AD-4NL, operated by VC-35, shows the searchlight. Note also the air scoop forward of the radio mast and the reflection off the glass of the system operator's door. The radar relay antenna on the tail fin is erect. VC-35 provided carriers with night-attack (VAN) teams, the first being on board the carrier *Boxer* between 24 August and 11 November 1950. This airplane is presumably part of Detachment (Det) 3, which served on board *Princeton* from 9 November 1950 through 11 June 1951; combat operations began on 12 December 1950. This photograph was released on 6 October 1951. U.S. NAVAL INSTITUTE PHOTO ARCHIVE

An AD-4W, not yet assigned to a squadron. The big radome contained the antenna of an APS-20 early-warning radar. The vertical antenna on the tail was for the data link to a ship or to a companion airplane in a hunter-killer combination. A ship could use the APS-20 as one of its radars, its radar picture being sent directly to its own CIC. Not until the advent of the WF-1 Tracer did an airplane share the fighter-direction function. AD-4Ws were transferred to the Royal Navy as the Skyraider AEW.1. Douglas dated this photograph 15 March 1951. DOUGLAS COURTESY U.S. NAVAL INSTITUTE PHOTO ARCHIVE

In 1948–49 Douglas proposed two new versions, collectively under an AD-5 designation. The first, powered by a turbo-compound version—that is, having an exhaust-driven turbine adding its power to that of the pistons, plus a supercharger—of the R-3350, was rejected because the heavier engine would have required considerable redesign. Douglas offered a second version in December 1949 as a single-package ASW airplane. It had the existing -26WA engine in a redesigned fuselage with side-by-side seating. BuAer was interested, but there was no money. That changed with the outbreak of the Korean War: the engine was moved forward eight inches, the vertical fin area increased by half, the two side dive brakes were eliminated, and the cockpit area greatly enlarged. The fuselage was lengthened by one foot one inch. This version had the four 20-mm cannon of the modified AD-4s. The remaining bottom dive brake alone was only 70 percent as efficient as the original three had been together, so dive angle

had to be limited to sixty rather than seventy degrees. Perhaps most importantly, the cockpit made for extraordinary versatility. Easily installed kits allowed conversion for target towing, photo reconnaissance, ambulance duty (four litters), VIP transport (six passengers, including the pilot), troop carrying (sixteen passengers), and attack. The added space made it possible to install the guidance system, homing aids, and other features desired by the Marines. A mock-up was inspected in October 1950.

Compared to the AD-4, at the design stage the AD-5 was nearly five hundred pounds heavier, seven knots slower in maximum speed, two or three in normal cruise, and a knot higher in stall speed. The AD-5 first flew on 17 August 1951, and 212 of the basic version were made.[24]

The AD-5N (of which 239 would be built) was a four-seat night-attack version with a radar pod and searchlight (the announcement of the designation mentioned production of 158 aircraft, delivery

The AD-5 was redesigned to provide side-by-side seating for two aircrew, with space for more behind them. Only the pilot had controls. The inner bomb pylons were noticeably enlarged and extended forward with a sharp rake. The engine was moved eight inches forward to maintain the position of the center of gravity, and the vertical tail was enlarged by 50 percent. A small air inlet (for cooling and ventilating the after cockpit) was added halfway up the vertical tail, just visible here. The fuselage air brakes were eliminated, the belly brake being retained. Douglas released this photograph on 17 September 1951. U.S. NAVAL INSTITUTE PHOTO ARCHIVE

The AD-5 airframe provided sufficient space for jammers and their operators. The ECM version (-5Q) and the -5W radar version both had an odd, arch-shaped window in the side of the opaque canopy for the two system operators. This AD-5Q operated by the Night Hawks of VAW-33 flew from USS *Independence*. It was the pathfinder (ECM) version, the big underbelly radomes housing antennas for the ALQ-33 auto jammer (forward) and APA-69 receiver (aft). The tail antenna (under the rudder) was for the ALQ-2 jammer. The big pod carried an APS-31C radar. On the centerline is a three-hundred-gallon tank. The bomb-shaped object under the starboard pylon contains an ALT-2 noise jammer. A second was carried under the port wing. This version also carried an MX-900A chaff dispenser under the port wing, not visible here. In an alternative radar-reconnaissance mode, the airplane carried a blade antenna for an APR-13 radar recorder between the two belly radomes (there was also an ALA-3 pulse analyzer), and all four pylons carried drop tanks. For both missions, the other pylons were omitted to reduce drag. The two radomes visible atop the airplane were common to the AD-5W version. The small one over the canopy was for the ARN-1 radio navigation (TACAN) antenna; the larger one near the radio mast was for the ARN-9 radio-direction-finder loop. The radio mast seems to have been for an ARR-2 radio-navigation receiver. The rubber boot at the top of the vertical tail covers a UHF radio antenna. The cavity-backed spiral antennas under the left bar of the national insignia are unidentified and were presumably added after the airplane was built; so too with the two small radomes abaft the big one for APA-69. U.S. NAVAL INSTITUTE PHOTO ARCHIVE

This AD-5W airborne early-warning version was operated by VAW-11. The underbody blade antenna replaced the radar-relay antenna atop the tail fins of earlier AEW aircraft. The blade visible under the right wing is a UHF antenna. Not visible atop the after (closed) canopy is a stub antenna for TACAN and IFF. Not present is the usual short rectangular structure just forward of the tail containing a radio direction finder. Of the two wire antennas, the upper one is for HF (note the wire linking it with the closed canopy). The other is the "sense wire" for the direction finder. Under the post-1962 system, this airplane became the EA-1E. U.S. NAVAL INSTITUTE PHOTO ARCHIVE

This AD-6 is from VA-104, operating in 1959 from USS *Forrestal* in the Mediterranean. In 1958 carrier air groups were assigned their own two-letter identifiers; CVG-10's was "AK." Later aircraft had the ship's name painted on. The AD-6 reverted to the AD-4 configuration but had the forward-swept pylons of the AD-5. Each could carry up to 3,000 pounds; the centerline pylon was rated at 3,600. Each of the six outer pylons under each wing could carry a 250-pound bomb, but the four outer stations were rated at 500 pounds each. The canopy was jettisonable, and the tailhook was hydraulic. There was no HF radio antenna; the wire antenna visible here was the sense antenna for the DF loop in the big, nearly rectangular radome. The main radio antenna was in the big blade forward of it. The smaller antenna is for IFF and TACAN (it matches an underbody antenna). The object forward of it is the fuel-nozzle grounding receptacle. A UHF direction-finding antenna was in the starboard wing, near the wingtip. The left wing carried a receiver for a marker beacon. The angled lines on the tail were to help an LSO judge height and attitude on landing. The final AD-7 was very similar. U.S. NAVAL INSTITUTE PHOTO ARCHIVE

to begin in April 1952). The AD-5W (218 built) was the AEW version. A conversion kit was developed to turn an AD-5 into the AD-5Q ECM variant (fifty-three kits made). There was also a single experimental AD-5S with MAD. Initially only N, S, and W versions were considered, but the AD-5 also showed considerable promise as a pure attack bomber. It carried all the armament of an AD-4, including the 3,500-pound nuclear bomb.

In October 1951 DCNO (Air) ordered that all future Skyraiders be AD-5s, -5Ns, or -5Ws. However, AD-6 (713 built) reverted to the earlier configuration; it was an improved AD-4B modified for low-level bombing and close air support.[25] The final AD-7 version (of which seventy-two were built, ending in February 1957) was an improved -6 with a -26WA engine, strengthened undercarriage and engine mounting, and wing panels. The Skyraider was still in service when the designation system was revised in 1962; it became the A-1.

The Skyraider continued in production so long because its planned successor, the A2D Skyshark, never entered service (as detailed in the next section). Through the mid-1960s, it was the Navy's low-level attack bomber, replaced only when the Grumman A-6 Intruder entered service in sufficient numbers. The need for such aircraft was so great that in 1961 AD-6 and -7 aircraft had to be extended from the originally planned three "service tours" (totaling about six years) to four tours (eight years) and then even more.[26] The situation was exacerbated by the need to transfer Skyraiders to the government of South Vietnam and then to the U.S. Air Force as well to fight the growing Indochina insurgency.[27]

The last Navy single-seat Skyraider combat sortie was flown on 20 February 1968, the last countermeasures sortie on 27 December. The very last Skyraider in U.S. Navy service, a test airplane at China Lake, was retired on 16 July 1971.

AD-6s of VA-115 on board *Kitty Hawk* line up at the after end of the flight deck for a rolling takeoff; "NH" indicates Carrier Air Group 11. Aircraft sometimes showed the carrier's name as early as 1959, but the practice seems to have become common (though not universal) about 1962. U.S. NAVAL INSTITUTE PHOTO ARCHIVE

The Skyshark (A2D)

On 4 June 1945 CNO directed BuAer to develop a turboprop attack plane.[28] A turboprop promised shorter takeoff than a pure jet and a heavier load, presumably because the propeller generated airflow over the airplane's wings even when it was not yet moving. It also promised much greater responsiveness to power changes at low speed, such as during a carrier landing. Preliminary studies of airplanes using turboprops promised great advances in bomb load, takeoff run, cruising speed, rate of climb, maximum speed, and combat radius. Immediate development of such an airplane was required to permit "steady growth in carrier striking power." As in 1938–39, the object was both to accommodate possible much-heavier weapons and to contribute to a homogenous air group in which the fighters and bombers would have comparable performance. The project had high priority.[29] BuAer had been interested in turboprops at least since 1939, and in 1945 it stated that they were more likely to be useful for naval aircraft than pure jets. In 1943 BuAer let engine-development contracts to Chrysler, Westinghouse, Pratt & Whitney, and Allison (General Motors). By 1945 it favored Allison's T40.[30]

On 25 June 1945 BuAer issued Douglas a letter of intent for preliminary studies of dive-bombers with three alternative turboprop arrangements: Lots I (two TG-100s in separate nacelles), II (two TG-100s geared together), and III (one Westinghouse 24D [T30]). They became Douglas' D-557A, -557B, and -557C.[31] All three were mocked up and model-tested in a wind tunnel; the contract was later extended to include design of the reduction gear required for D-557B. Mock-ups were inspected in September 1946, detailed specifications were completed, and wind-tunnel data were delivered in October. A Douglas proposal for a fourth design based on the Allison XT40 turboprop was rejected. Douglas favored a dual turboprop such as the T40, because half the engine could be shut down for long-endurance cruising while the other drove both elements of the contraprop. By this time BuAer was making further studies of turboprop bombers, as were companies other than Douglas.

Formal requirements for a turboprop attack bomber were restated by CNO on 21 April 1947: sufficient performance for self-defense against jet fighters, capability to operate from a *Casablanca*-class escort carrier (the smallest of that type), and combat radius of six hundred nautical miles with reduced bomb load. Douglas offered an airplane similar to both the Skyraider and to the D-557B study, powered by an Allison XT40-A-2 driving contraprops. Although at least one other company submitted a proposal, there was no formal competition.[32] A letter of intent was signed out to Douglas on 11 June 1947; the new model designation was made official by a 24 October 1947 memo. It fulfilled Edward Heinemann's vision of a bomber with sufficient performance to function at need as a fighter. Heinemann wanted to incorporate as many Skyraider parts as possible, which is why the airplane was initially to

have been designated AD-3. That proved difficult. For example, the wing and tail had to be made thinner for higher speed and the landing-gear strength and stroke increased. The cockpit was moved forward to improve the pilot's view. Canopy lines were set by the use of single-curvature laminated plate glass, which was wanted to withstand expected flight stresses.

The memo announcing the designation described the A2D as a bomber with a combat radius of three hundred nautical miles and also as a fighter with a maximum speed of 439 knots at 30,000 feet. The mock-up was inspected on 15–19 September. A procurement contract was issued on 25 September 1947. First flights of the two planned prototypes were expected in March and June 1949. At this stage service ceiling would be 50,250 feet, and takeoff run would be 231 feet. Minor changes, such as moving the wing aft and adding dihedral to the horizontal tail, were made after wind-tunnel tests.

In addition to the two prototypes, the Navy ordered ten service test aircraft (letter of intent issued 30 June 1950), five of which were completed to flight status; four were neither test flown nor delivered to the Navy.[33] Douglas planned to use the two prototypes and four of the service test aircraft for its own flight tests, demonstrations, and for prototyping planned changes in other aircraft. Allison, which made the T40, used some of the aircraft for developmental engine tests. As of 1951, plans called for ordering another two batches as part of a threefold increase in naval aircraft production ordered after the Korean War broke out: eighty-one (letter of intent 18 August 1950) to begin deliveries in June 1952, and another 250 (letter of intent 2 February 1951) to be delivered beginning in April 1953.

All of this fell apart, because of engine problems. The XT40 was to have been completed in July 1947, to have passed its fifty-hour test in mid-September, and to be delivered for the A2D prototypes in October 1947. In the event, no engines were delivered to Douglas until August 1949, and those were cleared

The unfortunate Skyshark. U.S. NAVAL INSTITUTE PHOTO ARCHIVE

only for ground running; flight engines were not delivered until March 1950. The first flight was promising, but the XT40 encountered continuing issues, to the point that in October 1950 serious consideration was given to installing a Pratt & Whitney T34 instead—but the T34 had no contraprop. An engine problem caused the 19 December 1950 crash of the first prototype.[34]

The second airplane was almost complete when the first crashed. A series of engine failures postponed its first full flight. There were "liftoffs" in June 1951 and January 1952, but the first full flight was made only on 4 April 1952. A production T40-A-6 replaced the prototype engine, and the first service-test airplane flew on 10 June 1953. A near-crash in

October revealed that stresses on the propeller and gearbox exceeded those allowed for: the airplane was not rigid enough between engine and gearbox. The fuselage was stiffened, but by that time production aircraft had been canceled.[35]

The project had one important by-product. Heinemann realized that his airplane would be faster than the terminal velocity of the existing two-thousand-pound bomb. He therefore developed a new streamlined bomb shape, which would not contribute undue drag on the aircraft. He claimed that an A2D with the three new type bombs would be 50 mph faster than one carrying three of the old type. The new shape was adopted for the Mk 80 series bombs that U.S. aircraft still carry.[36]

The Big Fighter

At least one company, Boeing, tried to sell the Navy a fighter with bomber features, the company's F8B-1. The proposal seems to have been unsolicited, and there is internal evidence (for example, that it did not refer to glide bombing) that Boeing had not been in close contact with BuAer. After submitting several fighter designs, Boeing offered Model 398, a redesigned version of earlier heavy fighters with a more modern bubble canopy. It had a new feature: it could accommodate a bay for a two-thousand-pound bomb. Design 400 (1 April 1943) was a further refinement, presumably what the Navy ordered as the XF8B-1. It was not a dive-bomber; Boeing characterized it as a level or skip bomber (low-level release, literally skipping the bomb across the water's surface).

Boeing claimed a sea-level speed of 351 mph with normal rated power (421 mph at 23,800 feet). With military rated power, sea-level speed would be 375 mph (435 mph at 23,000 feet). Stall speed at sea level with power off would be an acceptable 74 mph, and takeoff run into a twenty-five-knot wind would be 183 feet. Rate of climb at sea level would be 4,460 feet per minute using military power. Range at 15,000 feet was given as one thousand miles, which in terms of combat radius was hardly extraordinary. No weight was given.

Boeing received a letter of intent on 1 May 1943; the mock-up inspection was held 20–24 September 1943 (the power-plant mock-up followed, from 13 to 15 December). Its aircraft was described as a conventional single-engine monoplane designed with emphasis on range; slightly later it was credited with a combat radius substantially longer than that of any Navy fighter available or projected. Somewhat later again, it was claimed to offer extreme tactical flexibility, as it could deliver bombs, guns, and rockets. The big fighter first flew on 27 November 1944, showing good engine and airplane performance and excellent handling characteristics.

BuAer did not need a long-range fighter, but it was interested in attack bombers, so it asked Boeing to convert the big fighter into a low-altitude attack bomber and scout. Boeing offered its proposal in January 1945.[37] It seems still to have been under consideration in May 1945 (according to a 1 June 1945 BuAer progress report), but the idea petered out. At that time the airplane (empty) was about 1,000 pounds overweight (total 17,213 pounds).

Vought too might be mentioned in this category. It sketched a big fighter-bomber powered by an R-4360: its V-334, Vought's drawing showed two torpedoes side by side.[38]

The General-Purpose Attack Bomber

At least one aircraft company thought there was a market for a pure-jet (meaning higher-speed) light attack bomber. The BuAer proposal file includes the Martin 246, offered unsolicited in May 1949; it had the T-tail the company was then offering the Air Force in its B-51 tactical bomber and a semidelta wing. the inboard part of which was more sharply swept (and thicker) than the outer part. Its selling point was claimed supersonic combat speed (Mach 1.07), which doubtless it would not have attained. The payload was one 2,000-pound bomb and four 20-mm cannon with their ammunition. Claimed combat radius was 524 nautical miles, most of it at 510 knots at about 36,000 feet going out and about 40,000 returning. At the end of the cruise out it would descend to 20,000 feet at high speed (50 nautical miles at 570 knots) and then dive to 12,000 feet to toss its one bomb. Before the final descent it would spend five minutes at Mach 1.07 and another five at Mach 0.96. There was also a close air support mission with similar fly-out-and-back but combat at sea level. The airplane could run in at sea level, in which case radius was 323 nautical miles (run-in speed was 510 knots). Martin's selling point was that the airplane would enjoy fighter-like climb with sufficient range and load for naval attack missions.[39] The copy of the brochure in the BuAer file includes a handwritten evaluation (17 November 1949). Martin had somewhat overstated performance, but the main problem was a failure to meet catapult requirements; the only catapult that might possibly launch the 246 was the XC-7, then being developed for the *Midway* class.

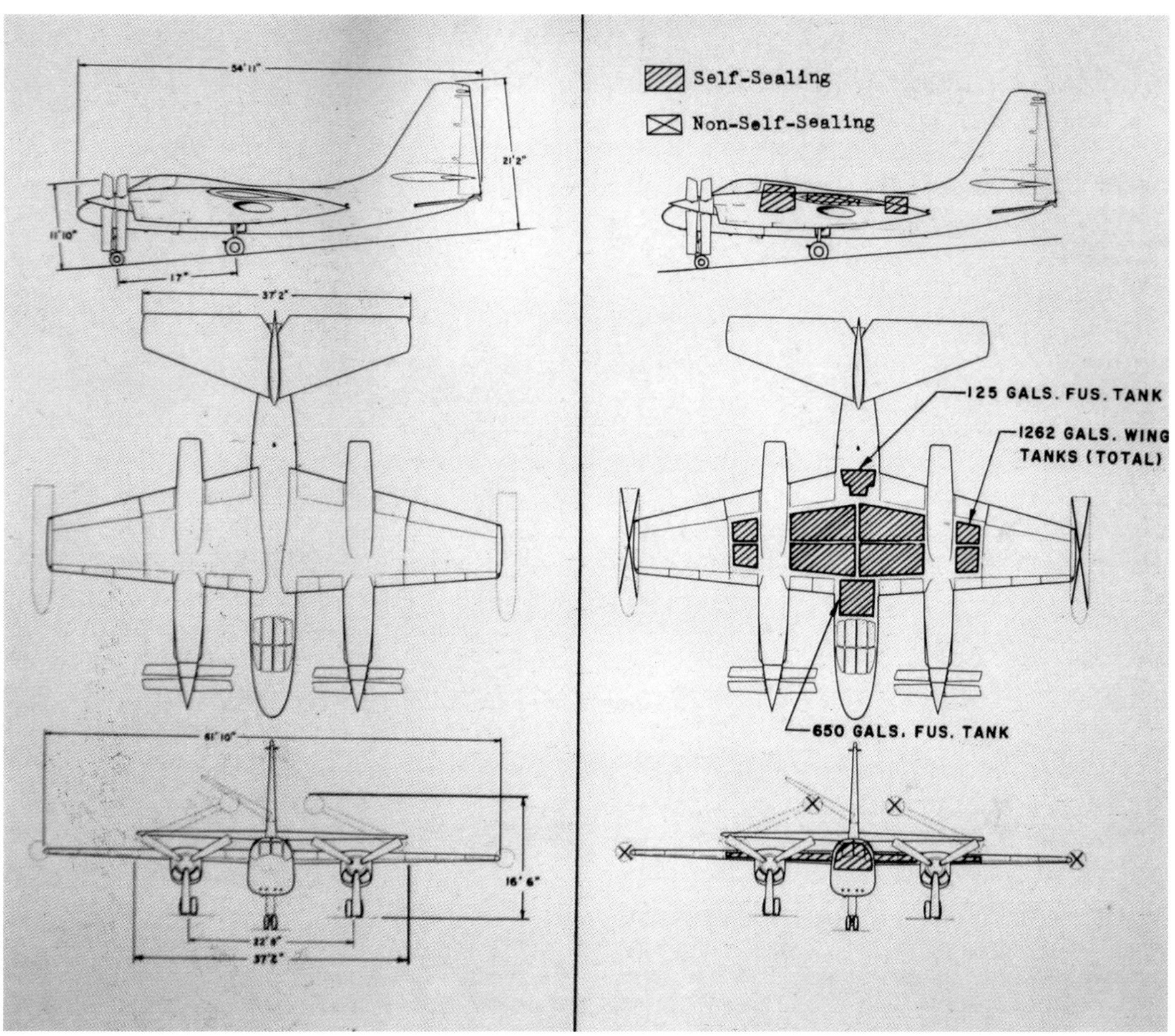

North American tried to develop the Savage into this general-purpose attack bomber. This is the three-view drawing from the usual "standard aircraft characteristics" proposal form. According to the form, the engineering study was begun on 18 May 1951. Like the A2J, this airplane was powered by a pair of XT40-A-8 turboprops rated at 6,825 shp for takeoff. In a close-support role, the airplane would take off at 58,280 pounds. Maximum speed would be 466 knots at 25,000 feet, and combat radius would be 775 nautical miles. Payload would be 6,000 pounds. Alternatively, with a 3,500-pound (i.e., nuclear) bomb the airplane would have a combat radius of 1,106 nautical miles at an average cruising speed of 412 knots. In a 25-knot wind the takeoff run would have been 570 feet in the close-support condition and 670 with the nuclear bomb—meaning that the airplane had to be catapulted.
AUTHOR'S COLLECTION

With work on the A2D proceeding, in October 1950 a further Operational Requirement (CA-01501) directed BuAer to develop a successor general-purpose attack aircraft, to be in the fleet by 1955.[40] The requirement was issued to guarantee continuing development of attack aircraft suitable for carriers or for small airfields; the airplane would also be used by the Marines for close support. Although aircraft already in existence met current requirements, "new developments in power plants will increase the ability to carry in offensive and defensive armament and will make them obsolescent." Presumably that meant new turboprops.

Minimum requirements were a payload of four 20-mm guns, three bombs, and twenty rockets; at least four hours' endurance on close-support missions (of which at least two hours would be on station at 5,000 to 25,000 feet); and cruising speed, altitude, and service ceiling in combat-loading condition compatible with those of contemporary escorting general-purpose fighters. The droppable armament load would be at least 6,000 pounds, with one 4,000-pound bomb on one rack as an alternative. Provision had to be made for complete winterization. Size would be such that it could operate from a modernized *Essex*-class carrier. Speed and maneuverability should be such as to protect it against enemy fighters. The airplane should be versatile enough for special missions: ECM, target towing, photography, and airborne early warning.

To this end the director of BuAer's Piloted Aircraft Division called a predesign conference for 23 February 1951. The airplane was tentatively designated XVA(GP). The four-hour close-support endurance changed to four hours at high speed. That exceeded what the A2D could do. Possible engines were the T40-A-8 and the T34 (an upgraded T34 would not be available in time). A preliminary design study showed that it would take a 23,000-pound airplane to provide the desired four-hundred-mile radius, 490-knot maximum speed, and an eight-hundred-foot takeoff run carrying one 2,000-pound store. The VA-class desk considered these design figures proof that Military Requirements wanted too much. A second conference was scheduled for 2 March 1951; unfortunately, the BuAer file does not describe its deliberations. It does not appear that any formal competition resulted.

The BuAer proposal file includes a North American general-purpose attack bomber offered in October 1951 as a derivative of the A2J nuclear bomber (Model ESO 47651).[41] A drawing shows two Allison XT40-A-8 turboprop engines in two nacelles driving contraprops, the nose almost in line with their spinners. It seems likely that North American had the new GP requirement in mind, since among the roles indicated in its brochure were land-based Marine close support and close support from a carrier. In the shore-based Marine role the airplane would be able to operate at a much greater weight, 67,899 pounds. The brochure included a graph of on-station time (up to three hours) plotted against combat radius: with tip tanks, the airplane could orbit on station for three hours at a range of about three hundred nautical miles. Most of these figures were so much worse than those BuAer had envisaged that it is unlikely that the proposal went very far. Nothing else in the BuAer proposal file seems to correspond to the General Purpose Attack project.

In April 1952 Commander in Chief Atlantic Fleet (CinCLantFlt) pressed for a decision, and its earliest possible implementation, that the standard air group comprise three VF and two VA squadrons; in November 1951 he had urged the development of a higher-performance attack bomber.[42] The timing coincides roughly with North American's presentation of an all-weather version of its proposed general-purpose attack bomber, but it is not clear whether CinCLantFlt ever heard about it.

Bullpup

The Korean War featured extensive bombing against an enemy using massed antiaircraft guns with radar fire control. U.S. Navy dive-bombers offered the only major precision-attack capability in the United Nations force. Antiaircraft fire forced Skyraiders to drop their bombs from 16,000 to 17,000 feet, which precluded great accuracy. Somehow precision had to be regained from above the reach of massed light antiaircraft fire. Studies of means of reducing aircraft losses to ground fire began in 1951, one of them a formal BuAer study of what became the Bullpup missile being in April–October 1952. The weapon was conceived as an extension of dive-bombing. An operational requirement was issued in June 1953. BuAer was responsible for this missile: it held the usual design competition.[43] Bids were requested on 10 August 1953. On 15 March 1954 the missile concept was formally approved by the assistant secretary of defense for research and engineering.

BuAer wanted a complete weapon system. It emphasized simplicity and low cost. The missile should be compatible with any naval carrier aircraft having sufficient space and strength to carry external stores of its size and gross weight. It would be usable at all dive angles up to seventy degrees. The attacking airplane should have sufficient freedom of maneuver to avoid antiaircraft fire. Launch range (slant range to the target) would be from 15,000 to 8,000 feet, but proposals were to indicate expected performance at greater and lesser ranges. The missile would be launched at speeds between 250 and 500 knots, although higher launch speeds (up to Mach 0.95) were very desirable. Use of the standard Mk 51 bomb rack was desirable, as was provision for ripple fire (half-second intervals). Although good performance was looked for, it was secondary if it imposed undue limits or penalties. There should be a minimum need for new, unique, or complex components or support facilities, and vulnerability to enemy countermeasures should be the least possible. Reliability should be such that no fewer than 75 percent of the demonstration missiles functioned completely. A solid-fuel engine was desirable but not mandatory. The warhead would be either a standard type of the 250-pound category or something in the same weight class. An appended table of aircraft bomb characteristics gave the dimensions of the 250-pound GP bomb envisaged and also of a 260-pound fragmentation bomb. BuAer envisaged two alternative guidance techniques, inertial and command (direct view).

Competitors were Bell, Chance Vought, Consolidated, Douglas, Fairchild, Hughes, Martin, McDonnell, Republic, and Sperry.[44] The demand for simplicity and low cost eliminated all but Douglas and Martin.[45] They had very similar guidance systems and offered essentially the same accuracy (Douglas would be slightly better). Both used a battery to power the missile's electronics. Douglas used the aircraft's radio to command the missile; Martin used a new secure command-link transmitter, something like that already developed for the Regulus strategic missile. Jamming would not be a very serious problem, as the missile's command antennas would be directional, pointing back to the airplane. The main difference was that Martin used rollerons (small ailerons on the missile's fins that passively stabilize roll), a roll-rate gyro, and bang-bang (on/off, toggle-type) controls; Douglas did not require roll or roll-rate control and used proportional (using a "feedback loop" for smooth response) controls. Either could use the inertial system BuAer had developed. Martin won, and it seems to have done so because its missile was far less expensive than Douglas'. The engineering recommendation for the winner was dated 27 January 1954; Martin's missile was designated ASM-N-7 and later AGM-12. It entered fleet service in April 1959 (the first missile was delivered on 4 February 1955). The first successful air launch was made on 15 June 1955.

The ASM-N-7A (1960) had a storable liquid-fuel engine for greater range. It was later redesignated AGM-12B. Bullpup "B" (AGM-12C) combined Bullpup control surfaces and guidance with a much heavier warhead (1,000 versus 250 pounds) and a new rocket motor giving a range of ten rather than eight nautical miles (although the guidance system limited effective range). It began operational evaluation in January 1962; it, the AGM-12D, and the -12E were Air Force versions, the -12D with interchangeable nuclear and conventional warheads. About 1959 BuAer planned a twenty-nautical-mile version. The missile's peacetime test CEP (circular error probable, the radius of a circle within which half the missiles would strike) was 30 feet, and in squadron service CEP was 56 feet; in combat, however, CEP was 100 to 150 feet. Bullpup turned out to be dangerous to use because of its all-the-way command guidance. According to a 1968 OSD report, Bullpup proved disappointing in combat owing to the highly vulnerable straight flight path the launch airplane had to follow to track a missile to impact.

Nevertheless, Bullpup was built in huge numbers: 45,350 Bullpup A/Bs and 15,400 Bullpup Bs, plus over 8,000 Bullpup As built in Europe from 1963 on. It armed not only attack aircraft but also P-3 maritime patrol aircraft; until the advent of Harpoon it was the only antiship missile they had. Bulldog (AGM-83A) was an early Bullpup modified by China Lake with a laser seeker for the Marines to use as a close-support weapon (Maverick was adopted instead).

The Marines and Close Air Support

The Marines, in part, saw their aircraft as an extension of their artillery: a precision arm to be used under ground control.[46] That vision led them to develop special means by which ground commanders could control bombers as they would control their artillery. Initially tactical aircraft in Korea bombed blind using the ground-based MPQ-2 radar, whose screen showed the airplane. A controller based verbal commands, including that to drop bombs, on what he saw. Accuracy was limited in part by the delay between when the pilot heard the command and when he pressed the bomb release. Moving at about two hundred yards per second, an attack airplane might miss by up to eight hundred yards. To solve that problem, Marine Air Support Radar Team 1 (MASRT-1) at the Naval Air Material Test Center at Point Mugu, California, developed an alternative, MPQ-14. It used a standard antiaircraft-gun tracker to track the incoming bomber in such a way that data, including course and speed, could be inserted directly into an analog computer, which would calculate the appropriate drop point. The computer could also correct the airplane's flight path if it was seen to deviate from its planned track. Commands were transmitted directly to an Automatic Bombing Control Unit working with the P-1 autopilot standard in Marine night fighters (F4U-5Ns and F7F-3Ns). Development was completed in June 1951, and units were sent to Korea for evaluation. Units were installed on board eleven Corsairs and two F7F Tigercats. After initial problems arising from a flaw in the ground computer, in January 1952 the system managed to place 62 percent of bombs within one hundred yards of the target. Even so, as used during 1952–53 MPQ-14 was not considered accurate enough to be relied on to hit enemy frontline units. Its main defect was the short tracking range, 16,000 yards, of the radar component. It had to be emplaced very close to the front line (in one case, its operators had to beat off North Korean attackers), so it could not strike deep targets, nor could it cope with a fluid tactical situation.

MPQ-14 seems to have been the first ground system to control a tactical airplane via its autopilot, predating the carrier automatic landing system that also worked that way. Its air component was the one installed on Skyraiders adapted for close air support. They did not have P-1 autopilots, so the airborne element was attached to their Instrument Landing Systems (ILS). Pilots "flew the needles" displaying system commands: the system was not fully automatic.

The successor to MPQ-14 was the TPQ-10 radar bomb director. The TPQ-10 operator maintained a current picture of both Marine positions and of the incoming airplane, which the operator would pick up and direct. The Marines understood that when Link 4 was introduced for carrier-controlled landings and also for fighter control, it could be applied to attack control as well. By about 1965, they were experimentally using Link 4 in conjunction with the TPQ-10. Properly equipped aircraft could be controlled, and could drop bombs, without intervention by the pilot. TPQ-10 and Link 4 were used in Vietnam to control not only Navy and Marine Corps aircraft but also B-52s making Arc Light strikes. TPQ-10 was also used to control helicopters—for example, in Army operations. It was regarded as the most precise available bombing aid. Between 1966 and 1971 "air support radar teams" of Marine Air Support Squadron 3 controlled more than 38,010

TPQ-10 missions, directing more than 121,000 tons of ordnance onto 56,753 targets. TPQ-10 was considered to have a fifty-foot CEP and to be capable of supporting 105 missions per day. The system was apparently particularly valuable during the 1968 siege of Khe Sanh, when weather conditions often precluded visual bomb delivery. On the other hand, computer control was sometimes ineffective, and (as in Korea) voice often had to be used. Also, pilots could opt out of control.

The Marines' follow-on TPB-1D (1984) allowed aircraft to be controlled by a unit requesting close air support under visual conditions. During the First Gulf War (1991) the Marines withheld their AV-8Bs from the unified air-attack plan so that they could continue to function as artillery. More recently, systems based on GPS and on large-scale tracking have replaced dedicated bombing control.

The big punch: A3D strategic nuclear bombers lined up on board the carrier *Saratoga*. The airplane taking off is an F3H Demon fighter. The "GM" tail code was part of a system introduced in 1956, in which the first letter indicated Atlantic (through *M*) or Pacific. "GM" indicated the squadron (VAH-9) rather than the air group, because heavy-bomber detachments of squadrons were assigned to ships. The colors indicate the Navy's nuclear emphasis, the glossy insignia white undersides having been adopted specifically to reflect away the heat and flash of a nuclear explosion. Upper parts were light gull gray, which was considered inconspicuous at altitude. The Air Force also adopted glossy white undersides, for the same reason, though the upper parts of its airplanes were left metallic. This scheme was adopted in February 1955, after various alternatives were considered in 1953–55; all aircraft were to be repainted by 1 July 1957. BuNo 138955 was an A3D-2 accepted in April 1957. It and the others all still had their tail guns when this photo was taken in September 1957 during a NATO Strikeback exercise. In 1958 the heavy attack aircraft were integrated with the carrier air group, taking the same tail codes as other aircraft on board (here, "AC"). Note steam rising from the port bow catapult. NHHC

8

WHAT NOW?
Strategic Attack

By March 1945 the naval air community was looking toward a future generation of carriers and aircraft. On 27 April 1945 DCNO (Air) set up an informal advisory board on future carriers and seaplane tenders.[1] Officers newly returned from Europe told board members that the smallest effective bomb was a 4,000-pounder; some targets could not be damaged by anything short of the British 12,000-pound "Tallboy." In fact, almost the greatest relevant World War II lesson had been the need for the heaviest practicable bombs, to deal with reinforced structures such as submarine pens—second only to the lesson that carriers needed maximum reach. This understanding that future carrier aircraft needed greater range and payload seems to have been reached well before anyone involved was aware of the existence of nuclear weapons.

After the nuclear attacks, President Harry S. Truman rejected the Army Air Forces' bid for a monopoly over U.S. atomic weapons; the Navy would be a full participant. The Navy immediately created a DCNO for Special Weapons, whose office included an atomic weapons branch.[2] For naval aviation, a key personality was Cdr. (later admiral) John T. "Chick" Hayward, a wartime member of the Manhattan Engineering District and in 1947 director of plans and operations for the Armed Forces Special (i.e., nuclear) Weapons Project under the Atomic Energy Commission. He later commanded the first Navy atomic-bomber unit, VC-5 (the designation indicated a composite, i.e., multitype, squadron).

A Heavy Carrier Bomber

Sometime in the summer of 1945, the BuAer Design Research Branch began studies of a heavy carrier bomber.[3] The surviving one, dated 12 October 1945, described an airplane capable of landing at the present carrier weight limit of 30,000 pounds and arresting-gear limit of 90 mph; it would carry a 7,000-pound bomb. Limiting weight (no bomb, two hours' fuel) would be 27,794 pounds, gross weight 40,085. The power plant would be two turbocharged R-2800s and three 24C (J 34) jets (one in each nacelle and one in the tail). Armament would be two fixed 20-mm guns. This airplane could take off in 570 feet in a twenty-five-knot wind. It would make 525 mph at sea level, 545 at ten thousand feet. Landing speed would be 89 mph. Rate of climb would be 5,970 feet per minute at sea level. Combat radius would be three hundred nautical miles, the airplane cruising out at 335 mph. Span would be 72.5 feet and length 52 feet (wing area 700 square feet), which would be large for existing elevators.

The three-hundred-nautical-mile range was the minimum acceptable. One or more jets in a carrier air wing could be traded for more fuel. Since much of the drag would be caused by compressibility (the compression of air ahead of the aircraft), that extra fuel might not cost much speed; the airplane might gain 25 or 30 percent in range at the cost of no more than 20 mph. The BuAer preliminary designers knew about German work showing that sweeping wings back would delay the onset of compressibility, but they were unable to develop a successful swept-wing design owing to very serious balance problems.

In December 1945 the informal advisory board pointed out the need for both much heavier bombs and much greater striking distance; carrier aircraft had been handicapped by their limited range through the war.[4] By this time BuAer's preliminary studies were already showing that the best combination of bomb load, speed, and range could be attained using turboprops rather than either pure jet engines or a combination of piston and jet engines. All studies envisaged an 8,000-pound bomb as normal load with sufficient space for a 12,000-pound bomb at a cost in speed and range. No 8,000-pound bomb then existed; it was presumably a surrogate for a nuclear bomb (existing ones weighed about 10,000 pounds).

This project seems to have been personally sponsored by Adm. Marc C. Mitscher, the wartime fast carrier task force commander. There were three categories, all powered by turboprops:

(a) Aircraft that could operate from unmodified *Midway*-class carriers
(b) Aircraft that could operate from *Midway*s in a restricted way: incapable of being struck below but capable of landing in light condition and taking off fully loaded
(c) Aircraft that would require a new class of carriers.

An airplane in Category (a) would take off at 30,000 pounds and land at 20,000. Speed at sea level would be 362 mph, and combat radius would be three hundred nautical miles. Category (b) would make it possible to deploy 45,000-pound aircraft (30,000-pound landing weight) with a speed of 500 mph at 35,000 feet and a combat radius of a thousand nautical miles. The unlimited airplane in Category (c) would weigh 100,000 pounds for takeoff and 65,000 pounds for landing, with the same speed but with a combat radius of two thousand nautical miles.

No satisfactory turboprop yet existed, so it was proposed that as a first step a Category (b) bomber would be developed, powered by piston and jet engines. It would take off at about 41,000 pounds (landing weight 28,000 pounds), and with jet engines it should achieve 500 mph at 35,000 feet. Combat radius should be three hundred nautical miles. When turboprop development was mature enough, a turboprop, Category (b) bomber should be produced while work proceeded on Category (c). Meanwhile work should begin on appropriate escort fighters.

CNO quickly approved the proposed aircraft program.[5] An Outline Specification (OS-106) was drawn up for a heavy attack airplane capable of operating from *Midway*-class carriers, as a stopgap pending development of the favored turboprop. A prerequisite was a new catapult; a very heavy bomber could not make a rolling takeoff from a carrier.[6]

On 25 January 1946 BuAer circulated OS-106 to twelve manufacturers.[7] The bomber would be able to deliver an eight-thousand-pound bomb from 35,000 feet. Its bomb bay should be 5 feet in diameter by 16 feet long, and radius of action was to be not less than three hundred nautical miles. This specialized airplane would be taken on board a carrier as a special load. It would be highly desirable that it go into the hangar, but it would not have to.

During the fall of 1945, Cdr. Frederick J. Ashworth of the CNO staff (Op-36) had visited BuAer to find out which of its current aircraft projects could be associated with nuclear weapons.[8] There he met Capt. Joseph N. Murphy, who like him was convinced that it was vital to develop a carrier nuclear bomber. Murphy had previously been involved in the conception of the single-seat attack bomber. Ashworth was aware of the sixty-inch diameter of the "Fat Man" bomb. He drafted a letter for Secretary of the Navy James Forrestal to sign asking President Truman to approve modification of the bomber design to accommodate a nuclear weapon. The letter was short-stopped in the chain of command: Forrestal personally approved the project, considering the modification within his authority.[9] The BuAer Armament Division advised bidders for OS-106 that the dimensions of the "ultimate primary bomb" (presumably a nuclear bomb) had changed; the bomb bay had to be six inches wider and a foot deeper than originally specified.

The OS-106 bomb-bay dimensions corresponded with those of nuclear bombs: the larger of the World War II atomic bombs, "Fat Man," was 60 inches (5 feet) in diameter and 128 inches (10 feet 8 inches) long. Access to arm the bomb manually in flight required that the bomb bay be somewhat longer than the bomb.

To obtain operationally balanced designs, BuAer deliberately imposed difficult flight conditions: ten minutes of combat over the target after a fifty-mile approach and before retirement at maximum speed. Eliminating "time over target" would increase operational radius with protected fuel to four hundred nautical miles. With two 300-gallon tip drop tanks as overload, radius might increase to seven hundred nautical miles. Even with this overload, at 46,200 pounds, the takeoff run into a twenty-five-knot wind should be about five hundred feet, well within the originally specified six hundred feet and entirely within the capabilities of the *Midway* class.

Bids were due on 1 May. Three manufacturers offered designs—North American, Consolidated, and Douglas; their proposals were evaluated at a

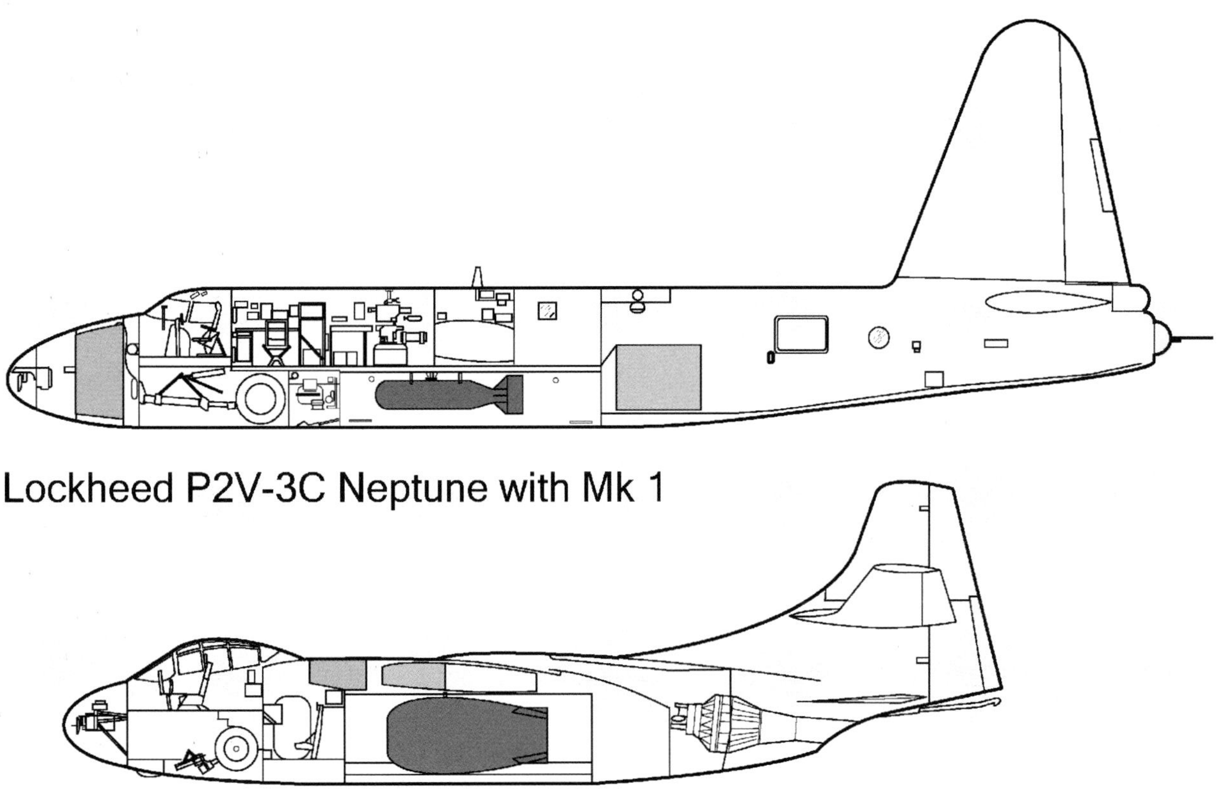

Carrying a nuclear bomb was a matter of dimensions as well as weight. As this comparison shows, the Neptune lacked bomb-bay height to accommodate the fat Mk 4 bomb. The Mk 1 that it could carry used scarce U-235; only a few were made. The fat Mk 4 and its fat successors used plutonium, which was easier to obtain. The weights of the two bombs were similar. TOMMY H. THOMASON

29 May 1946 meeting.[10] The Douglas (D-566) and North American proposals were both very attractive, offering about the same performance with the same power plants (two R-2800 and an I-40/J33 turbojet in the tail). Consolidated's was judged least promising in either weight or performance.

The Savage

North American won: its airplane was lighter and was expected to be substantially less expensive. The firm received an initial contract for its XAJ-1 on 24 June 1946. The project enjoyed BuAer's highest priority.[11] A production contract for the first twelve was let on 25 September 1947, almost a year before the airplane first flew (on 3 July 1948); the first production aircraft were accepted in August 1949. The new bomber was named the Savage. The first squadron, VC-5, commissioned in September 1948, received its first airplane on 1 September 1949.[12] The Savage made its first carrier takeoff, from USS *Coral Sea*, on 21 April 1950 and the first landing, on board

the same carrier, on 31 August. The delay between squadron delivery and carrier takeoff was largely owing to the squadron's movement from California, where it was training, to Norfolk, to become fully operational   the carriers able to operate its nuclear capable Neptunes (below) being in the Atlantic Fleet. The first overseas deployment was by the squadron's six Savages and three Neptunes, to Port Lyautey in French Morocco, beginning on 5 February 1951. Because the presence on board ship of these aircraft precluded normal flight-deck operations (they could not be stowed below), the initial deployment included only nineteen days at sea.

The requirements levied on the AJ demanded a very design that had to be beefed up after static tests. Structure accounted for about 30 percent of "all-up" (gross) weight, an unusually low proportion (for the Skyraider it was about 50 percent).[13] Presumably to save weight, North American did not incorporate power wing folding. Instead, external hinges and hydraulic actuators had to be placed

The first airplane conceived from the outset as a carrier nuclear bomber was North American's AJ Savage, shown here. U.S. NAVAL INSTITUTE PHOTO ARCHIVE

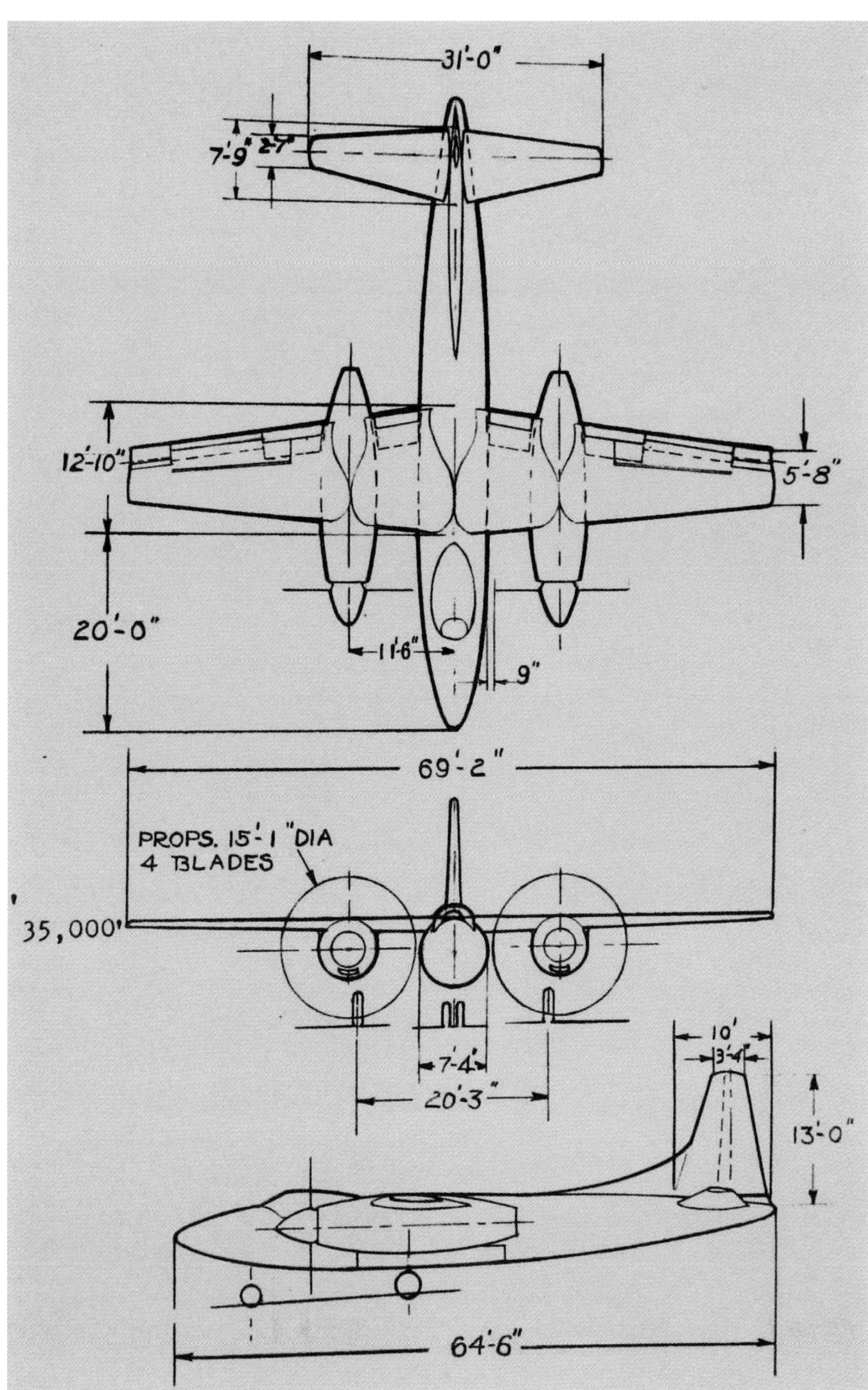

Convair's alternative to the Savage was offered in April 1946. A cutaway showed alternative bomb loads: an 8,000- or 12,000-pound bomb or alternatives including two torpedoes. The bomb bay had the desired envelope, five-foot diameter by sixteen feet, and a canopy covered a two-man crew. This is the data sheet. AUTHOR'S COLLECTION

on the wings and tail; during Board of Inspection and Survey (BIS) trials it took a four-man crew sixteen minutes to fold these surfaces, even without the inconvenience of normal shipboard conditions. Sacrifices in structural weight may explain why two of the three prototypes were lost to in-flight failures during company tests. The AJ continued to suffer structural failures in service: during its first seven years, nearly four crashes each year. Many were attributed to a very, perhaps overly, complex hydraulic system; its jet engine was suspected as a cause of in-flight fires.

With a 7,300-pound Mk 15 bomb but without drop tanks, takeoff weight was 47,000 pounds rather than the specified 43,000. The airplane was unusually large, with a seventy-five-foot wingspan

Two Savages from VC-5 prepare to launch from the new supercarrier *Forrestal*. The Savage to the right has the air intake for its auxiliary jet engine, above its wing, open in preparation for takeoff. U.S. NAVAL INSTITUTE PHOTO ARCHIVE

including tip tanks (the wings and the tail could fold for hangar stowage). With 60 percent fuel and no bomb (40,000 pounds), it landed (power on) at about 108 knots. There was no defensive armament. Maximum speed at 34,000 feet was 390 knots (449 mph) rather than the desired 500 mph at 35,000 feet, and radius on internal fuel was 460 nautical miles, not the desired 600. With tip tanks radius increased to 720 nautical miles, and with the 3,600-pound Mk 5 and a five-hundred-gallon bomb-bay tank, radius with tip tanks was 1,010 nautical miles (takeoff gross weight was 51,000 pounds).

Hopes that speed would be a sufficient defense soon evaporated; BuAer found itself running a design contest for a dedicated escort fighter—which was not built.[14] BuAer also considered replacing the jet engine in the tail of the AJ with a gun turret.

The AJ-2 had a rearranged cockpit, with a single compartment for the three-man crew rather than the previous two compartments.[15] Its third crewman sat under the same canopy as the other two, facing aft. Also, the hydraulic system was simplified, the fuselage was two feet longer, the vertical tail was somewhat taller, and the horizontal stabilizer had no dihedral. The AJ-2 could be catapulted at up to 54,000 pounds. Maximum indicated air speed, without tip tanks, was 420 knots at 11,000 feet or below, 230 knots at 40,000 feet. That compared to 380 knots with tip tanks at 17,000 feet or below for the AJ-1, which also had a maximum catapulting weight of 54,000 pounds. There was a photo version (AJ-2P).

Total production, including prototypes, was 143 aircraft, including 55 AJ-2s, the last in June 1954.

The Savage was often an aerial refueler. Here an AJ-2 of VAH-7 fuels an FJ-3M Fury fighter from the carrier *Essex*, on 15 June 1958. NHHC

Reportedly the Savage was extremely unpopular in the fleet because of its many problems. Because of its size, most of its pilots were recruited from the land-based-patrol community; they were not used to carrier operation, and they performed poorly. The A3D had much the same problem when it was introduced. The Savage suffered particularly badly in comparison with light attack aircraft (Skyraiders and nuclear-capable fighters) that could deliver the new much smaller weapons now available more effectively.

In November 1952 North American proposed a more compact further development of the concept it called AJ-3.[16] Nothing came of this idea, presumably because by 1952 the much more attractive A3D program was well along. North American also proposed to modify an AJ with T34 turboprops.[17]

That impressed Commander Air Forces Atlantic Fleet, who pointed out on 15 January 1954 that, like other jets, the new A3D Skywarrior would lose range at low altitude, where the turboprop would do better. He also suspected that the A3D would do poorly in bad weather as a result of its relatively low design-load factor. The AJ had a 50 percent higher load factor. Too, there was now interest in operating from advanced bases with short runways, and such interest was likely to grow—probably a reference to Korean War experience. The AJ but not the A3D was suited to such runways. If the North American proposal was accepted, ComAirLant wanted the firm also to investigate the feasibility of interchangeable tails having either turrets or jet engines. The commander of Heavy Attack Wing 1 was similarly enthusiastic. The A3D, he argued, depended completely on the new steam catapult, but the turboprop AJ, like the original AJ, could operate from *Essex*-class carriers. The AJ airframe had been largely debugged; it would be some years before the new A3D was as reliable.

BuAer was interested too. Converting AJs to turboprops would cost about half as much as building new airplanes. The idea failed to come to fruition, however, apparently because AJs were needed for adaptation as tankers instead. The jet air groups envisaged for 1956–60 would find bad weather difficult without emergency refueling; jet fighters had particularly short cycle times unless refueled in the air. The Savage was by far the most attractive tanker candidate. For initial tests, the jet engine was removed and the fueling drogue let out through the former jet exhaust. A special adapter was placed in the bomb bay to carry an extra fuel tank. The bomb-bay doors were notched for the drogue, which protruded from a short pylon. The tanker conversion was operational by late 1954, when Savages were still operational as heavy attack bombers. The aircraft involved were AJ-2s; the tanker variant was a temporary conversion, hence not specially designated. Also, there were no special tanker units; Savages were withdrawn from tanking as they were replaced in heavy attack squadrons by Skywarriors.

The Interim Bomber: The Neptune

While the Savage program proceeded, the Navy sought an interim capability. In 1946 it had one long-range high-performance airplane that could carry a heavy bomb: the land-based P2V Neptune. In 1947 a BuAer civilian pointed out that it could rival the performance of the Air Force's much larger B-36.[18] It is not clear when or how this insight turned into the idea that the Neptune could offer the Navy a useful nuclear capability; in any case, work on a carrier-launched Neptune was well under way by April 1948.[19] Nose guns, the dorsal turret, and the usual ASW equipment would all be removed. Only the tail turret would remain. Stripping would leave a crew of four (pilot, copilot, radio operator, and navigator). Electronics would include an APS-31 radar in the nose and APA-5A bombing equipment.

The first carrier nuclear bomber was an adapted P2V Neptune conceived as a land-based patrol bomber. Neptunes of VC-5, the first Navy strategic squadron, are spotted for takeoff on board the carrier *Coral Sea* at Norfolk, 7 March 1949. The "NB" tail code indicates squadron VC-5. These aircraft would have taken off using rocket assistance (JATO). NHHC

Aircraft, P2V-3Cs, would operate at a gross weight not exceeding 73,700 pounds, which was far beyond the normal maximum takeoff weight of 60,000 pounds and thus would tax airframe strength. Far beyond the limits of a ship's catapults, they would be launched using jet-assisted takeoff (JATO) rockets—eight 1,000-pound (2,000-pound-thrust) units. BuAer calculated that at 73,000 pounds a Neptune would have to start at least five hundred feet from the forward end of the flight deck. Taking off, one wingtip would pass only ten feet from the island of the carrier. Estimated combat radius operating from a carrier was 2,235 nautical miles.[20] Tailhooks were fitted in the hope that the aircraft could be recovered when they returned, but that proved impractical.[21] On completing nuclear missions, Neptunes were expected either to ditch near a carrier or other U.S. ship or to land on friendly territory. Aircraft were fitted with "hydroflaps," intended to keep their noses up in the event of ditching.

As proof of concept, two stripped Neptunes were taken on board the carrier *Coral Sea* on 27 April 1948; the next day they successfully took off, retuning to Norfolk. Neptunes had already practiced ashore, making numerous short takeoffs. The first full demonstration was the 7 March 1949 launch from *Coral Sea*, off the East Coast, of a P2V-3C piloted by the VC-5 commander, Captain Hayward. At 74,000 pounds this was by far the heaviest airplane yet launched from a carrier. The airplane took off, flew across the country to the El Centro bombing range in California, dropped its dummy 10,000-pound nuclear bomb, and then flew back to and over *Coral Sea* before landing at Patuxent River, twenty-three hours after taking off.

A prototype P2V-3C carrier nuclear bomber and eleven production P2V-3Cs (taken randomly from the P2V-3 production line) were delivered to Captain Hayward's VC-5 between September 1948 and August 1949.[22] VC-5 went to the Atlantic Fleet as soon as it was operationally ready, to operate from the *Midway*-class carriers assigned there. Neptunes were replaced by Savages as the latter entered service.[23]

Flight decks on the three *Midway*s and on active *Essex*-class carriers were strengthened to accommodate Neptunes, and they were modified internally so that bombs could be assembled on board.[24] Only the three *Midway*s ever took Neptunes on board.

The A2J

In December 1947 BuAer asked North American for an engineering study of a turboprop bomber derived from the AJ-1, to be powered by Allison T40 turboprops.[25] It appeared that such an airplane could be adapted from the existing AJ-1 in two-thirds the time and for half the cost of a new design. A contract was issued on 1 April 1948, and a letter of intent for two prototypes was issued on 28 September.

North American soon found that it could not simply adapt the AJ-1 airframe; in May 1948 the contract was changed to a new airplane that would make maximum use AJ-1 components. It was to be a "(b)" airplane: 62,146 pounds gross takeoff weight, a speed of 442 knots (508 mph) at 35,000 feet, and a combat radius of 1,200 nautical miles— nearly twice that of the AJ-1. Like the AJ-1, it had a turbojet engine in its tail to boost speed over the target and would be able to operate from *Midway*-class carriers. A mock-up was inspected at North American's Los Angeles plant on 11–15 September 1948.[26]

BuAer judged the design deficient in respect both of carrier landing and takeoff characteristics and of excessive vulnerability to enemy fighters. Fixing the first required addition of very complex high-lift devices in the wing. To handle the second, the airplane was to be redesigned in such a way that an armed escort version could be spun off. Takeoff gross weight increased in the process to 72,000 pounds, but maximum speed improved to 452 knots. The new mock-up was inspected on 4–8 April 1949. The resulting report stressed carrier-suitability problems (a later BuAer history suggested that they were particularly important once the big USS *United States* was canceled later that month and the airplane had to operate from smaller *Midway* flight decks).

In 1946–48 it was assumed that long range could be achieved only by using turboprops. North American developed the follow-on A2J, which was initially conceived as a turboprop Savage. It was dropped because the much higher-performing jet A3D was available at about the same time. U.S. NAVAL INSTITUTE PHOTO ARCHIVE

The airplane was again redesigned. The design bomb load (not actual capacity) was reduced from 10,500 to 8,000 pounds. The boost jet engine was eliminated; maximum speed was cut to 410 knots. Gross weight was reduced to 58,000 pounds; the airplane could now operate from *Midway*- and *Oriskany*-class carriers. Because of reduced weight, operating altitude increased to 38,600 feet. The escort version was abandoned, the airplane being given a twin 20-mm tail turret. The revised mock-up was inspected on 12–22 September 1949.

At the same time, the mock-up of the revised version of the all-jet Douglas A3D (see below) was being inspected at Douglas El Segundo. The A2J was further modified to match the A3D cockpit arrangement and escape chute. Initially the A3D had been seen as a long-term successor to the Savage, but now it was a contemporary of the near-term successor, the A2J. As a result, when money for heavy carrier bombers became available after the outbreak of war in Korea, the Navy could choose between the two heavy carrier bombers. Neither had yet flown. The A3D was expected to be considerably faster (564 versus 425 knots), with a slightly longer combat radius (1,075 versus 1,025 nautical miles). The only important A2J advantage was that it could operate more easily from converted *Essex*-class carriers (*Oriskany*s), but that became less important once it became clear that the Navy would receive a new class of large carriers (the *Forrestal*s). It did not help that the T40 engine was suffering serious problems (in February 1951 North American offered the T34 as an alternative). The A2J development schedule initially envisaged delivery of the first engine in November 1949, but that did not happen until March 1951, so the airplane first flew on 4 January 1952. Thereafter, engine problems badly slowed the flight test program. Meanwhile, the U.S. Navy adopted steam catapults, which could launch even the A3D from modified *Essex*-class carriers. Not surprisingly, the A2J program was canceled on 8 June 1953.

The Heavy Bomber

The A3D resulted from the next-generation heavy-attack requirement set out by DCNO (Air) in January 1948.[27] As in 1946, the weapon would be a 10,000-pound bomb. Escort would be impracticable: the airplane's main protection would be its maximum speed, which should be at least 525 knots

(603 mph) at 40,000 feet (with the highest possible cruising speed). Operating altitude should be 40,000 feet or more. Radius of action should be 2,000 nautical miles or more, but range could be sacrificed for speed, as long as it was at least 1,700 nautical miles. The airplane would fly the last hundred miles into the target and the first hundred back out at maximum speed. Defensive armament would be tail guns. Size and weight should be minimized. A BuAer preliminary design (ADR 62) seems to have shown that the requirements could be met at a weight of about 100,000 pounds, corresponding to the catapult BuAer was designing—in short, Admiral Mitscher's Category (c).

The requirements were revised in March.[28] Speed for the run in and run out was now Mach 0.9 at 40,000 feet or above. All other flight was to be at Mach 0.85 and at 35,000 feet or above. Electronics would be limited to that needed for accurate navigation and bombing; there would be no defensive armament and no armor other than for crew protection. Contractors were to submit proposals for four different gross takeoff weights: 62,000 pounds, 100,000 pounds, 200,000 pounds, and any weight between 62,000 and 200,000 pounds. Aircraft should be able to land on board having expended their bombs and all but reserve fuel. Combat radius should be between 1,700 and 2,000 nautical miles. The weights involved were well above the nominal limits for existing carriers, but those limits had been calculated for rapid and repeated operations by large numbers of small aircraft. *Midway*-class carriers would soon park small numbers of Neptune bombers weighing considerably more than the nominal figures.

Design studies were to show airplanes with and without landing gear; landing gear might be ditched to allow for high landing speed. To reduce takeoff weight, the airplane might be fueled in the air near the carrier after takeoff. If it turned out that the desired speed, weight, and range were completely incompatible, additional studies would be made using "drone" and "control" aircraft.

As a yardstick, the BuAer preliminary designers produced a new ADR-64A design employing a highly loaded wing swept back at forty-five degrees. The power plant was two J40 engines. This design could not meet the combat-radius requirement. No other details seem to have survived. The combat-radius problem led the preliminary designers back to turboprops. They returned in July with two turboprop designs (ADR-65A and -65B).[29] The ADR director, Ivan H. Driggs, wrote that turboprops offered much the same cruising and maximum speed as jets, with a much greater combat radius. Because they were relatively conservative designs, turboprop bombers might be available sooner than ADR-64A. Their only untested feature was propeller efficiency beyond 500 mph.

All of these aircraft were linked with a huge new carrier, the USS *United States* (CVA 58), which was included in the Navy's 1948 program and laid down in 1949. Douglas' Ed Heinemann later wrote that when he visited BuAer to discuss such bombers, the bureau's view was that the airplanes would weigh about 100,000 pounds, presumably, he thought, on the basis of ADR studies. Heinemann later wrote that he had doubted the big carrier would ever be built. If it were canceled, as he expected, bombers would have to be scaled back to fit on board existing carriers. Heinemann was already interested in cutting aircraft weight.[30]

When Heinemann brought his jet sketch design to BuAer in March 1948, he was told that it was impossibly small. He later wrote, however, that his sketch convinced BuAer to base its competition on jet rather than turboprop engines. He also wrote that there was already considerable sentiment favoring jets but that range was a real problem. He seems to have shown a swept-wing design with two podded jet engines and a semi-externally carried bomb.[31] An alternative with two turboprops weighed ten thousand pounds more. According to Heinemann, Driggs angrily confronted him, charging that the only way he had managed to design a smaller but longer-range jet was to have changed basic assumptions, presumably the mission model BuAer was using.

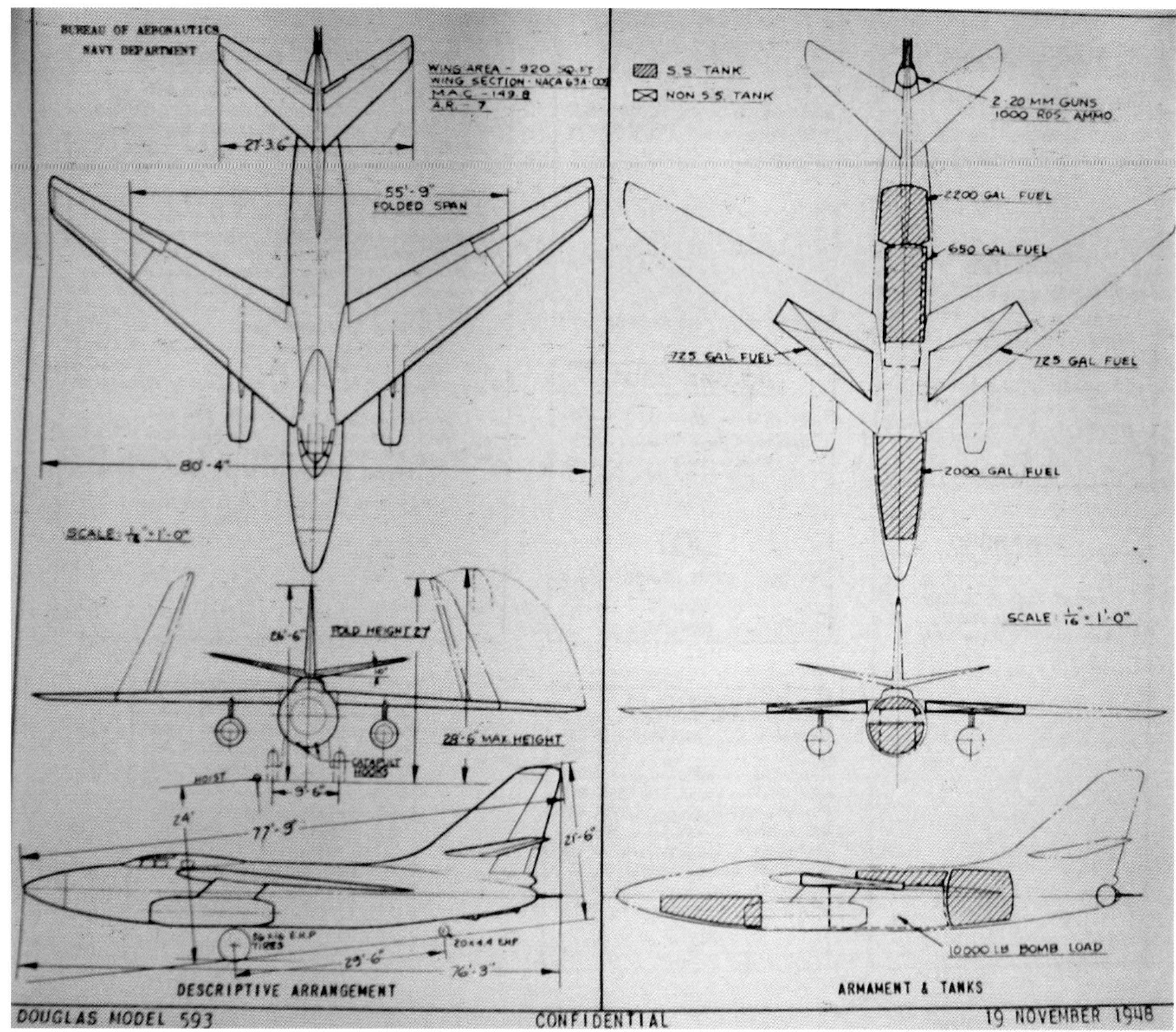

Edward Heinemann's proposal for the twin-jet Model 593 bomber that became the A3D Skywarrior, 19 November 1948. At this stage the engines were XJ40-WE-12s rated at 8,100 pounds military thrust (7,300 pounds normal). This version incorporated the desired internal bomb stowage; an initial sketch in the package of designs (alongside this one) shows Heinemann's original idea of carrying the single bomb semisubmerged. Note the tail-down undercarriage. Takeoff weight was 87,500 pounds. With a projected double H4-1 catapult (i.e., H8) on a *Midway*-class carrier, the airplane would need JATO assistance for takeoff, and it might be unable to take off at all in hot conditions. With the projected double H9 catapult on board the abortive carrier *United States*, it would be able to take off without JATO on a standard day in a 27.7-knot wind. Combat radius was given as 1,615 nautical miles at 489 knots, altitude at the target being 40,700 feet. These figures applied to estimated future performance of the XJ40 engine. Radius was 1,562 nautical miles at 489 knots. With current performance, operation from anything but the huge projected carrier was impossible. AUTHOR'S COLLECTION

BuAer requested proposals to Outline Specification OS-111 on 16 August 1948. They were submitted that November through early December. Heinemann's 78,000-pound design became the basis of Douglas El Segundo's proposal. Other contenders were Curtiss-Wright, Douglas Santa Monica, Republic, Martin, Fairchild, and Consolidated-Vultee (Convair); Lockheed submitted a design but not a cost estimate; it did not want to bid, because it could not meet the requirement. All of the companies used J40 and J46 turbojets. For evaluation purposes, the J40 was rated at 8,100 pounds thrust and the J46 at 4,200 pounds at sea level, both expected future ratings. The formal evaluation was issued in February 1949.[32]

Martin submitted its Model 245 for both the jet-bomber and supersonic-bomber competitions, although its entry did not really meet the requirements of either. This Martin drawing showed the airplane taking on a bomb at a harbor in the United States, cruising out 1,550 nautical miles at 401 knots to land in the sea to refuel from a submarine, then refueling again after another 1,550 nautical miles, dropping its hull, and then flying 1,200 nautical miles to the target. This drawing showed how it could drop its seaplane hull, leaving an upper body that could land on a carrier. Another drawing showed it taking off from a carrier at a gross weight of 105,040 pounds and flying 1,617 nautical miles to a target. It would drop its bomb while cruising at 522 knots at 49,000 feet. Takeoff weight after fueling would be 121,715 pounds. Another possibility was to use this type of airplane as a tanker, extending range from a carrier to 2,350 nautical miles by fueling when 100 nautical miles out from the ship. Length would have been 87 feet even, span 105 feet 9.96 inches, and height 31 feet 0.7 inches; there would have been four XJ40-WE-6 engines (7,500-pound thrust at takeoff). It could take off normally using a catapult, as takeoff speed was 110 knots. It was, however, too large for any carrier short of the hoped-for new one. Its bomb would be dropped normally from a rotatable bomb bay. AUTHOR'S COLLECTION

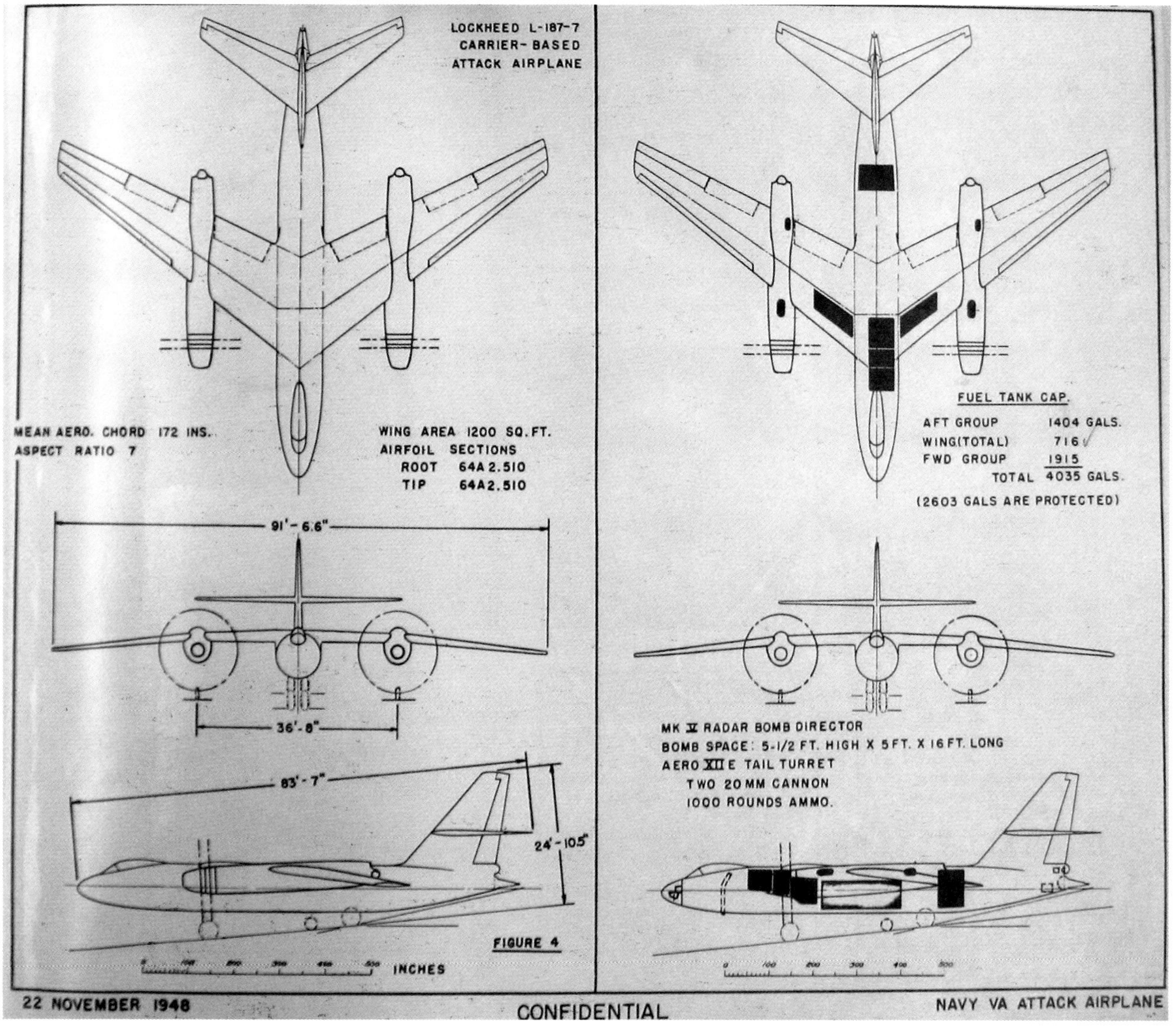

Lockheed offered both jet and turboprop attack bombers. This turboprop L-187-7 was dated 22 November 1948. Like the other 1948 submissions, it had a high takeoff weight, in this case 100,000 pounds. The engines were two Allison XT44-As rated at 11,250 shp for takeoff. Lockheed estimated that with a 10,000-pound bomb combat radius would be 1,350 nautical miles at an average speed of 488 knots. JATO would not be needed for takeoff using the H-9D catapult planned for the carrier *United States*. An alternative had two pods, each with two J40 jet engines. AUTHOR'S COLLECTION

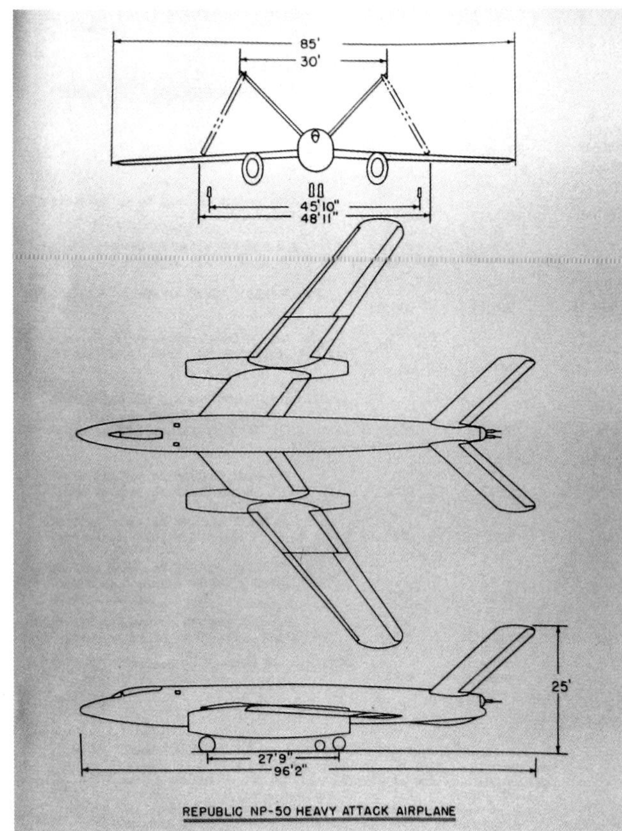

REPUBLIC NP-50 HEAVY ATTACK AIRPLANE

Republic's NP-50 submission was unusual for its V-tail. The company's report was dated 20 November 1948. Span was eighty-five feet (forty-five feet ten inches folded), length ninety-six feet two inches, and height twenty-five feet. Design flight gross weight was 86,855 pounds (maximum was 107,100). The bomb load was 10,000 pounds. Republic claimed a cruising speed of Mach 0.85 (491 knots) and a maximum speed of 550 knots (Mach 0.95) at combat altitude (40,000 feet); combat radius would be 1.670 nautical miles. AUTHOR'S COLLECTION

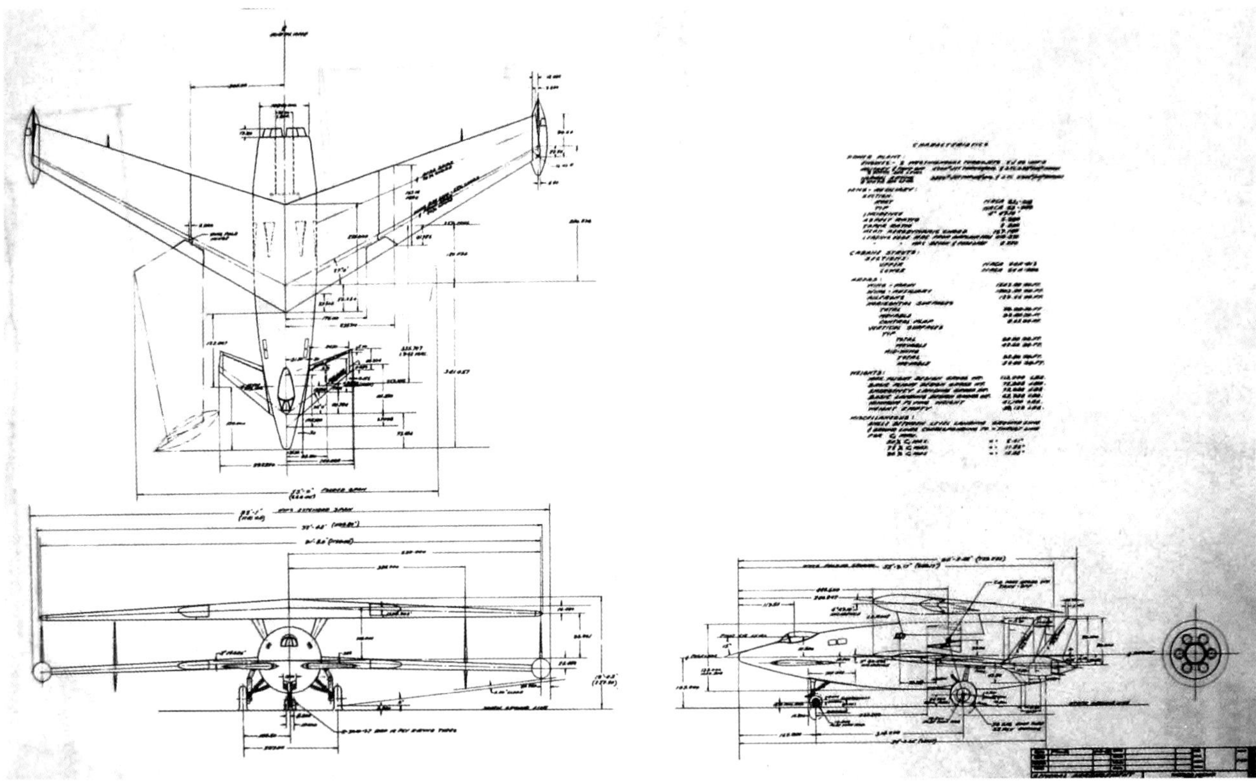

Fairchild's Model 128 must have been the most remarkable of the subsonic proposals, with an upper (biplane) wing to provide initial lift (it would have been discarded after takeoff). BuAer found it the heaviest, least conventional, and most expensive of all. Span over the upper wing would have been ninety-two feet (fifty-three folded) and ninety-five feet one inch over the tip tanks; length would have been sixty feet two inches. Each wing had an area of 1,203 square feet. Gross weight without JATO would have been 113,000 pounds, including 57,600 pounds of fuel. The power plant would have been six J46-WE-2s, with a normal rating of 3,220 pounds thrust (4,100 pounds military rating). BuAer estimated combat radius as 1,080 nautical miles at a combat altitude of 40,000 feet. At design gross weight (72,928 pounds), speed would have been 591 knots (680 mph) at sea level and 519 knots (597 mph) at 40,000 feet. The airplane would have taken off at 113 knots (130 mph). AUTHOR'S COLLECTION

Martin, Lockheed, Republic, and Fairchild were immediately eliminated.[33] Convair and Douglas Santa Monica both offered three-engine designs that were both already at the 100,000-pound limit hence had no margin to deal with defects or with service growth.

That left twin-engine designs offered by Douglas El Segundo and by Curtiss-Wright. Both were substantially smaller than the other designs. Douglas used podded engines. Curtiss-Wright used a flexible wing (as in the B-47) and avoided engine-pod problems by burying its engines in the wing roots. Both were marginal in speed, climb, and altitude performance, particularly when guaranteed engine ratings rather than better, expected ones were used.

Douglas' design was considered clearly superior. Curtiss offered slightly higher speed (by eight knots at sea level and twelve knots at 40,000 feet), but Douglas offered greater combat radius (1,650 versus 1,535 nautical miles). It enjoyed marked advantages in stall speed, takeoff speed, and takeoff distance (Curtiss slightly exceeded the desired 115-knot stall speed). Curtiss' takeoff run was so long that it could not take off without JATO; it needed 70,000 pounds of JATO thrust on an atmospherically standard day, compared to 33,000 for Douglas. Douglas offered superior stability and control. BuAer considered that Curtiss' design would require considerably more development to achieve satisfactory flying characteristics. It foresaw lateral and combined lateral-directional problems owing to the flexibility of Curtiss' wing. Curtiss used a leading-edge flap instead of an automatic slat, incurring flap-up stall problems.

Curtiss' bomber used a bicycle landing gear with outriggers (i.e., extending from the wing itself via the fuselage), which BuAer doubted would be suitable for a carrier. If the aircraft were placed on a *Midway* catapult, one of the outriggers would hang beyond the edge of the deck. The catapult shuttle would attach forward of the nosewheel, which it was practically guaranteed to foul. Given the position of the bomb bay, it would be difficult to move this attachment to a better place. Douglas' narrow-tread but conventional landing gear presented no problems. Its bomber could fit a *Midway*-class carrier if its

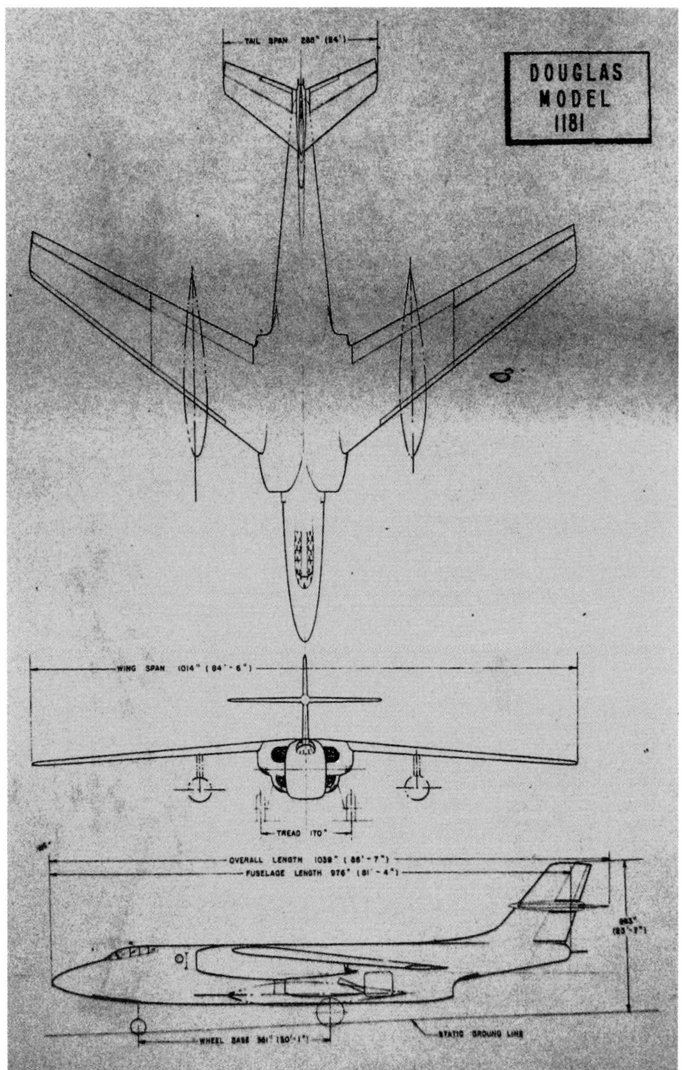

Douglas Santa Monica prepared designs in parallel with Heinemann's El Segundo division. Model 1181 was its entry for the 1948 carrier bomber. AUTHOR'S COLLECTION

wing fold was modified and its vertical tail made to fold. If the carriers were modified for greater deck strength and fitted with the projected slotted-cylinder catapult and higher-capacity arresting gear, they could operate the Douglas bomber.

Douglas' bomber was likely to be more survivable. Both airplanes had unprotected fuel cells, but Curtiss' were in the fuselage and had no means of being purged of fuel fumes. Douglas' were in the wings and could be purged. Curtiss placed the radar for its tail turret atop its tail fin, where it might be affected by vibration and would encounter parallax problems. Also, Curtiss used an expendable bomb-bay cover, dropping which might be dangerous.

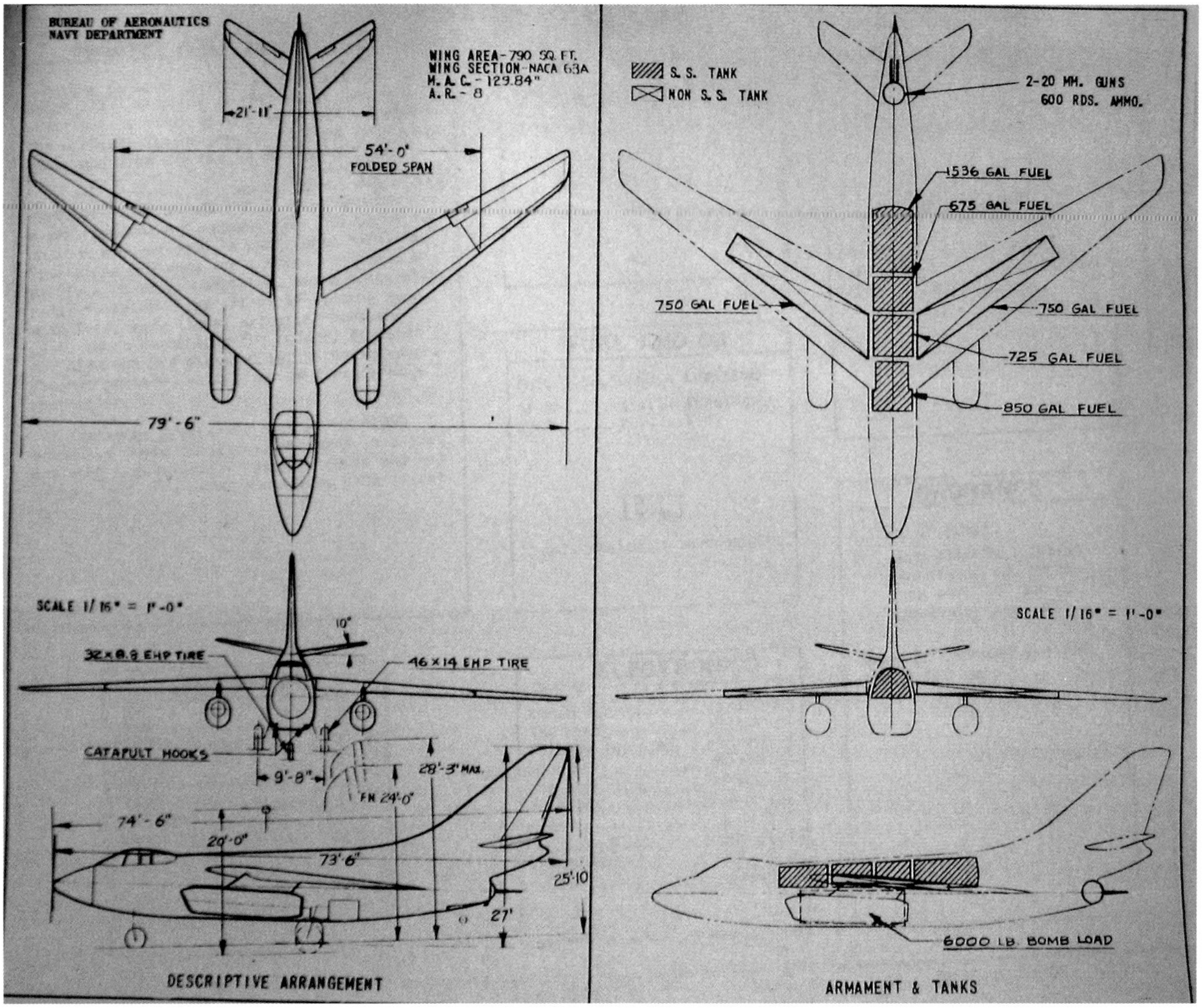

This considerably refined sketch design (Model 593-7) was submitted on 25 March 1949, just before cancellation of the supercarrier. The only obvious difference from the A3D as built was the rooster tail. At this stage the payload was 6,000 pounds of bombs. Takeoff weight had been reduced to 72,500 pounds. On that basis the airplane could take off from a *Midway*-class carrier using a catapult plus JATO in a twenty-five-knot wind; without JATO but with the catapult it could take off in a 43.9-knot wind. Maximum sea-level speed was 563 knots (Mach 0.852); combat radius was 1,700 nautical miles at Mach 0.832. Model 593-8 (13 May 1949) was substantially the A3D as built but still with J40s. Heinemann managed to cut takeoff weight to 68,000 pounds. Combat radius was 1,500 nautical miles at Mach 0.85; maximum speed at sea level was 568 knots (Mach 0.860). At 40,000 feet that fell to 511 knots (Mach 0.889). AUTHOR'S COLLECTION

Curtiss was runner-up to Douglas with this shoulder-wing bomber, its engines in its wing roots. Span was eighty-one feet, length eight-three feet one inch and height twenty-three feet five inches. Normal gross weight was 87,000 pounds. North American submitted a somewhat similar design. NHHC

Curtiss placed the pilot on the centerline, and it seemed that the crew members' ejector seats might interfere with each other when the crew had to escape. BuAer preferred Douglas' staggered crew arrangement and its escape chute; ejector seats were still experimental. (In retrospect, ejector seats were better. It turned out to be difficult for crew members to get to the escape hatch, so much so that the A3D was sometimes called "All Three Dead.")

Douglas' weights were considered more optimistic than Curtiss', but BuAer's experience with the Skyraider indicated that Heinemann could probably do even better than he promised. Curtiss quoted a considerably lower price, but BuAer's assistant chief for E&D wrote that this airplane was so important that "compromise in an engineering choice because of cost of the experimental airplane is certainly not in the best interests of the government." Douglas' record in designing and producing naval carrier aircraft "is far superior to all other competitors in this competition."

Both finalists were asked for refined designs. Douglas submitted its D 593-7 on 25 March 1949. It closely resembled the A3D as built, except that its tail turret was in a notch under its vertical tail rather than (as later) directly under the trailing edge of the tail fin.

Like the ADR projects, this airplane was tied to the supercarrier *United States*, with its high-energy catapults. The carrier had been conceived in tandem with the Navy's nuclear bomber project. Funds for it under the 1948 program were part of a wider American rearmament program to match a Soviet threat rendered newly urgent by the overthrow of the elected Czech government that February. In 1949, however, military spending had to be cut sharply. That did not reflect a softened American position but rather an understanding that the Soviet threat was at least as much political as ideological. President Truman saw investment in European economic recovery (the Marshall Plan) and in NATO as aspects of U.S. security spending, funds taken from the same pot as defense. He gambled that the U.S. nuclear monopoly would deter the Soviets from any armed aggression until Western Europe recovered. The one element of the 1948 rearmament program that survived was accelerated nuclear-weapons production, creating an "era of nuclear plenty" in the 1950s. For the moment, the U.S. nuclear arsenal was tiny, and U.S. security was unable to conceal that reality from Stalin.

It became necessary to cut the budget, but in 1949 discussions of the FY51 budget the three services (i.e., including the newly independent U.S. Air Force) were unable to agree on how to do so. Both the Army and the Air Force pressed for cancellation of the new carrier. Secretary of Defense Louis Johnson agreed. He seemed contemptuous of both the Navy and the Marine Corps (President Truman had personal animus against the Marines, based at least partly on his World War I Army experience). The April 1949 carrier cancellation prompted the public "Revolt of the Admirals" against the Air Force's nuclear-only strategy. It much embarrassed President Truman and his secretary of defense but did not revive the carrier project.

Johnson hedged the bet by ordering the Navy to downsize the bomber planned for the new carrier enough to operate from the existing *Midway* class.[34] Speed and altitude were not to be compromised, but combat radius could be reduced if necessary. In retrospect it seems that the supercarrier was probably too great a technological leap. The catapults conceived for it, which were key to operating its heavy bombers, proved extremely difficult to develop. They would not have been ready in time for the ship's completion (their type was never successfully developed). It also seems unlikely that the ship's flush-deck configuration, with its need to exhaust an enormous volume of smoke at deck level, would have succeeded. New technology was needed; it would become available a few years later.

Truman lost his strategic bet. Thanks to well-placed spies, Soviet dictator Joseph Stalin knew just how thin the U.S. nuclear stockpile was. His spies also provided what he needed to build his own bomb well before anyone in the West expected it. He now saw a window of opportunity and felt safe starting a local war in Korea. Thinking that Korea might be the opening phase of World War III, the United States and other Western countries rearmed. Suddenly Congress was willing to run large deficits. Korean War appropriations paid for much of the crucial naval and military technology of the 1950s and early 1960s. Among other things, the new funds paid for a generation of supercarriers nearly as large as the *United States*—the *Forrestals*.

The Supersonic Bomber

Alongside the conventional OS-111 bomber, DCNO (Air) asked BuAer to look at a higher-performance piloted missile or one-shot bomber. He assumed that it would involve intensive aerodynamic research and studies of operational techniques. BuAer began with preliminary design studies by Martin and Grumman and then drew up an outline specification, OS-115. It framed specific requirements in February 1948: the bomber would run in the last thousand miles to the target at Mach 1.2 at 40,000 feet. It would retire at Mach 0.92. Maximum weight would be 100,000 pounds. The airplane should be able to take off from a *Midway*- and land on an *Essex*-class carrier. It could be a composite airplane (escape vehicle plus main expendable body). Proposals were solicited from Grumman, Martin, Douglas, Chance Vought, Consolidated- Vultee, and Fairchild. The winner would be chosen on the basis not only of engineering merit but also on demonstrated ability to grasp the broad concepts of the problem.[35] Only Grumman and Chance Vought failed to offer anything. Presumably Grumman's preliminary work convinced it how difficult the problem was.

Both Douglas Santa Monica (Model 1186) and Douglas El Segundo (Model 594) offered composite airplanes, the escape vehicle being mounted on, respectively, the vertical fin and a pylon of the expendable part. Both would be launched from dollies using JATO. The Santa Monica version employed four J40 engines with afterburners (10,000-pound thrust) for flyout, after taking off at 194 knots. It would retire on one J46 (thrust 6,200 pounds). If given

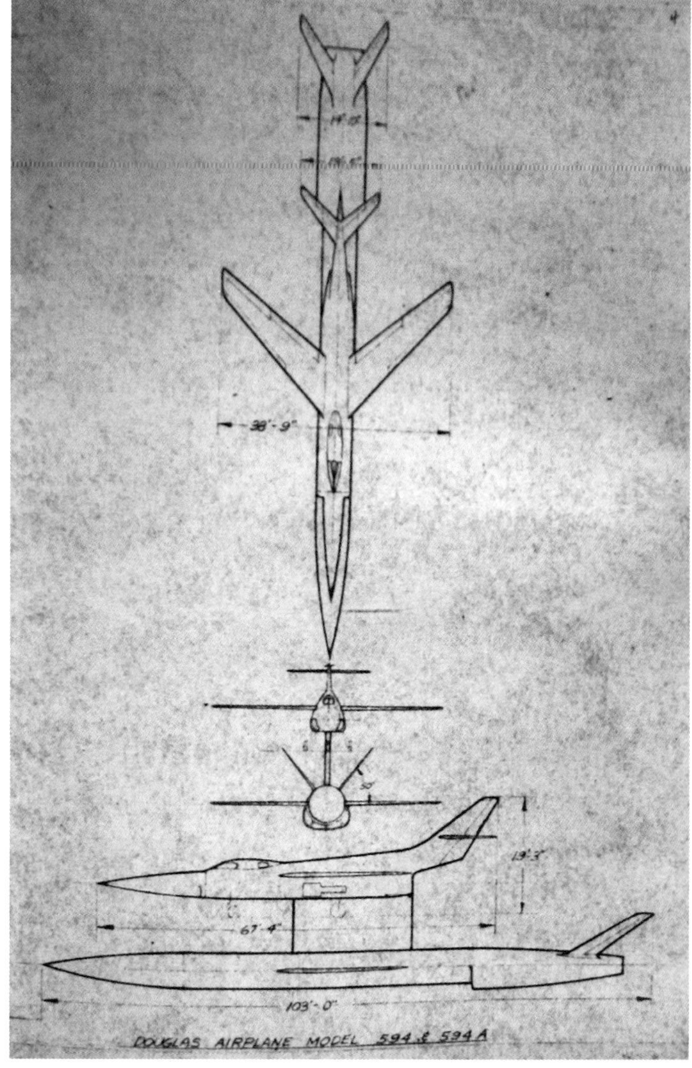

Douglas El Segundo's Model 594 supersonic-bomber proposal was a composite airplane. Span of both components would have been 38 feet 9 inches; overall length would have been 103 feet (upper component 67 feet 4 inches). Takeoff weight would have been 100,000 pounds, but the escape vehicle (the upper component) would have weighed 19,595 pounds. The escape vehicle (the upper component) would have had one engine, the expendable lower component (store) two J40s, each modified to provide 11,000 pounds static thrust (15,950 pounds with afterburning). En route to the target the engines in the store would use afterburners. Because it was not expendable, the engine in the escape vehicle would operate mainly at normal thrust. Douglas assumed that the usual 5-foot-diameter bomb could have its length reduced to 8 feet by eliminating unneeded empennage. The combination would take off from a launching dolly using either a catapult or JATO. Model 594-1 (about 60,000 pounds) was similar, but the lower component (carrying the bomb) had only one turbojet. Model 594-2 had an unpowered lower component. Its upper component had a planform similar to that of the Skyray fighter. AUTHOR'S COLLECTION

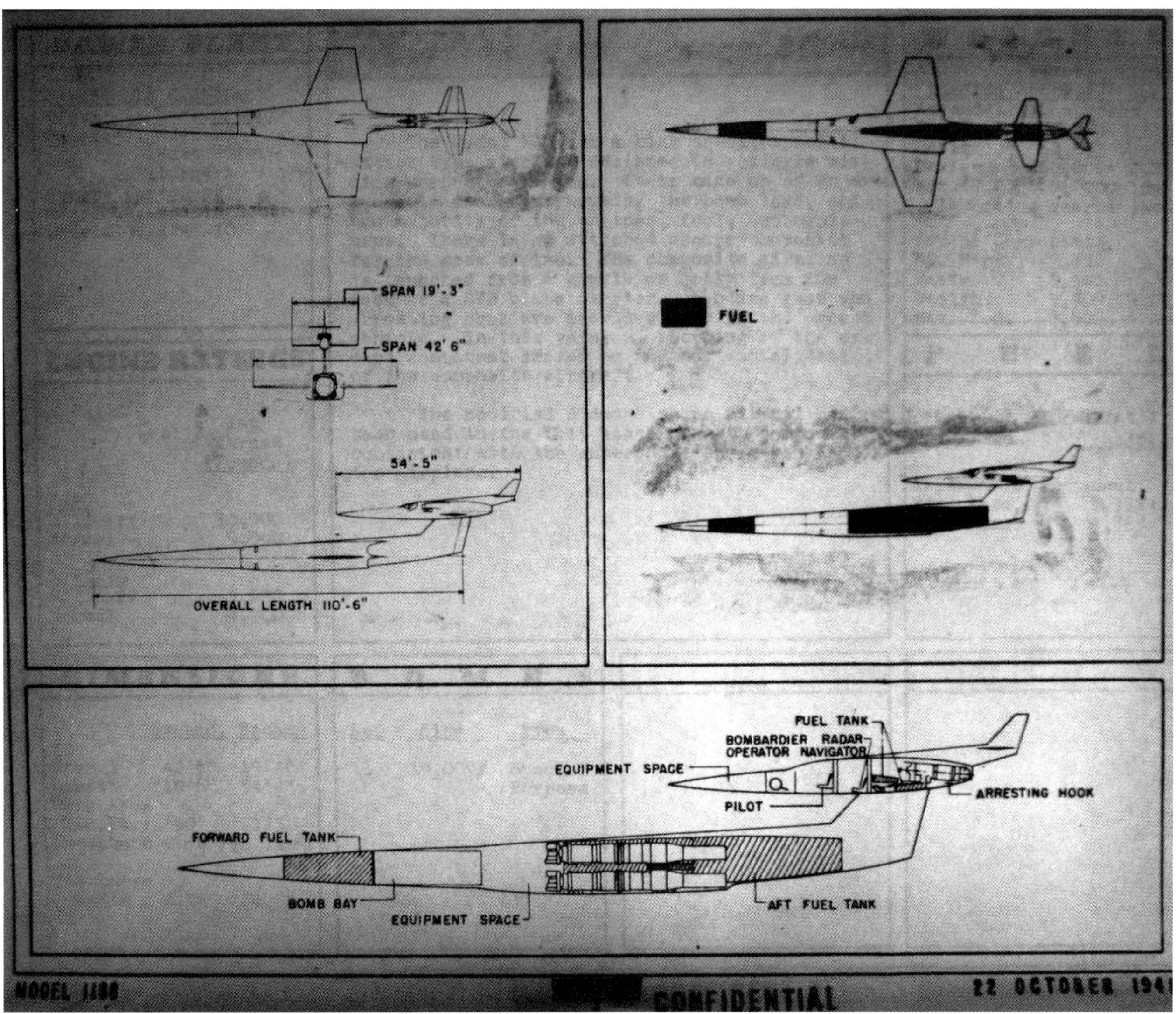

Douglas El-Segundo's Model 1186, 22 October 1948. Note the similarity between the escape vehicle and the company's X-3 Stiletto research airplane. Power would have been provided by four XJ40-WE-10 engines in the expendable store and one X24C-10 in the escape capsule. The escape vehicle had a span of 19 feet 3 inches and a length of 54 feet 5 inches; the main vehicle had a span of 42 feet 6 inches and was 110 feet 6 inches long. Takeoff weight would have been 100,000 pounds (74,203 pounds over the target). Estimated speed over the target was Mach 2 (1,149 knots); the escape capsule would cruise at 487 knots. Combat radius would depend on the percentage of the mission flown at Mach 2.0. It would be 621 nautical miles for a wholly subsonic mission, 326 nautical miles for a mission flown 69 percent at Mach 2.0. These figures could not have been approached in reality. AUTHOR'S COLLECTION

slightly more wing area, the manned part could land on an *Essex*. A sketch suggests that the manned element of the composite was based on Douglas' X-3 Stiletto research airplane. Combat radius was only 459 nautical miles (run in 133 nautical miles), but the run in would be at 43,000 feet and Mach 2.0. Over the target, total weight would be 74,203 pounds, and speed would be 1,149 knots at 54,600 feet. With the expendable part of the airplane gone, it would retire at Mach 0.85 at 9,500 pounds' thrust

for 133 nautical miles. Douglas El Segundo used two afterburning J40s (11,000 pounds thrust each) for the run out and one for return. It would meet the range requirement (1,700 nautical miles), but the high-speed run in would be 570 nautical miles; speed over the target would be Mach 1.2. The airplane would retire at Mach 0.92. At the target the airplane would weigh 47,606 pounds and would be flying at 944 knots at 50,000 feet. It would weigh 18,595 pounds after dropping its bomb-carrying pod, flying at 765 knots at 50,000 feet. It could land on an *Essex* but would be too long for an elevator and too high for the hangar deck.

Convair used a delta-wing escape vehicle buried in the top of the expendable pod. The pod would be powered by three J40s (9,000 pounds of thrust) and would weigh 96,200 pounds on takeoff. It would have neither wings nor tail. The escape pod would be powered by a single J40. Convair met the requirements of 1,700-nautical-mile range and a 1,000-nautical-mile run in (at Mach 1.37) at 40,000 feet. Unlike the two Douglas airplanes, this one would jettison bomb-bay doors in the expendable pod to drop its bomb; the rest of the pod would go on and crash. The combination of expendable pod and escape vehicle would take off at 224 knots, using jettisonable landing gear and JATO. At the target the airplane would weigh 57,300 pounds and fly at 865 knots at 40,000 feet. After dropping the pod, the single-engine escape airplane would weigh 28,550 pounds and would fly at 585 knots at 40,000 feet. The escape airplane could land on an *Essex*, but it was three feet too high for the class' hangar deck.

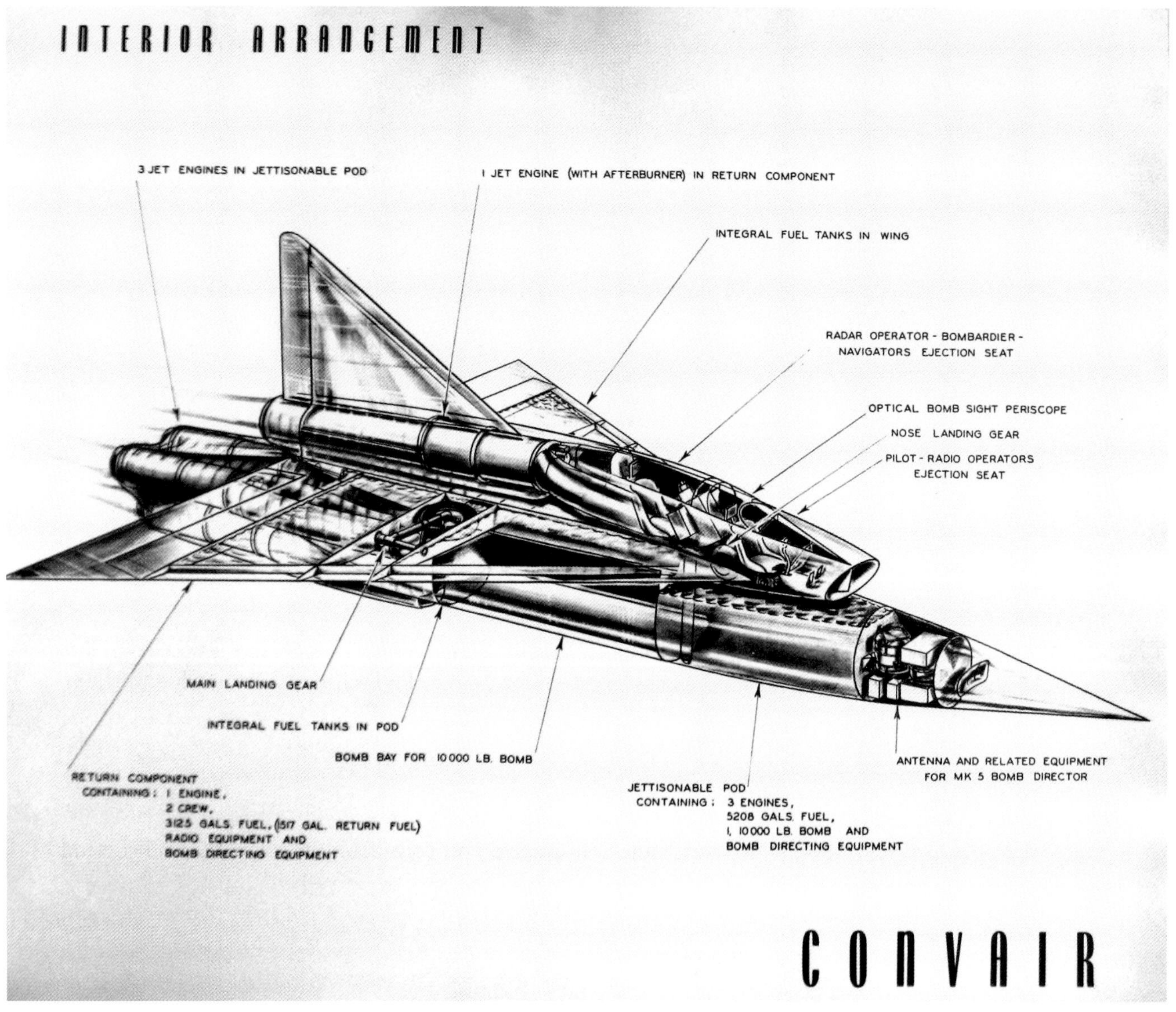

The Convair two-component supersonic bomber. Its upper component resembled the company's XF-92A delta-wing fighter. The expendable pod contained the bombing radar (ASB-1), integral fuel tanks, a bomb bay, and three jet XJ40-WE-10 engines, one with an afterburner. The upper component had a fourth engine. Span was forty-five feet seven inches (area three hundred square feet), and overall length was seventy-eight feet three inches (forty-three feet four inches for the upper component alone). Takeoff weight would have been 96,260 pounds. Estimated combat radius would have been 1,700 nautical miles at 785 knots (cruising at 530 knots). AUTHOR'S COLLECTION

Fairchild's M-130 was a tailless airplane with a forty-five-degree swept wing and vertical surfaces at its wingtips, launched using a dolly and JATO. For takeoff it would have an auxiliary wing, which would be jettisoned soon afterward, so initially it would be a jet biplane. Powered by ten J46 engines (5,160 pounds thrust each), it would weigh 100,000 pounds at takeoff. It would not meet the range requirement (1,350-nautical-mile combat radius, with a 650-nautical-mile run in at Mach 1.1). It would retire at Mach 0.92–0.85. It could land only on the supercarrier.

Martin entered the same low-aspect tailless design it had entered in the heavy attack competition, a 100,000-pound seaplane with a detachable hull. It failed the speed requirement.

Convair won. It offered the only design proposal that met or exceeded the requirements and reflected an adequate understanding of what was involved. Its escape vehicle had a configuration similar that of the company's XF-92 delta-wing fighter, which was already flying. Convair claimed that it already had enough test data to justify a normal development program, but no funding was planned until FY51.

BuAer's evaluator clearly liked Douglas El Segundo's proposal but could not recommend giving the company further work, because its engineering staff was already heavily committed. It too showed a good grasp of the problem, and its design was superior to the others. It achieved its 1,700-nautical-mile combat radius by reducing the run in from 1,000 to 570 nautical miles. However, its engine ratings were considerably more optimistic than those of the other companies or of BuAer itself. Douglas Santa Monica tried for a Mach-2 run in, because it had considerable test data from another project, presumably the X-3. BuAer saw the short radius of action as proof that a very fast run in was not worthwhile.

Fairchild was rejected for underperformance. Its unusual feature, a retro-bomb installation (in which the bomb's forward momentum is counteracted so that it falls straight down to the point over which it was dropped), would require a long development program.

Military Requirements Division recommended not awarding any contract. It was difficult enough to fund even more urgent projects. The last R&D review had made ADR-64 (OS-111) a Priority I project, but the supersonic attack bomber had been only Priority III. The bureau chief wanted to keep the supersonic project alive on a reduced basis (about $250,000), but the R&D Review Board killed it.

The Skywarrior: A3D (A-3)

Heinemann's heavy attacker survived because it could be modified to fit a *Midway*. Its shoulder-wing (at the top of the fuselage) design was easily modified to meet a requirement for a wider (sixty-six inches rather than fifty-eight) bomb bay. Heinemann mounted his main wheels in the fuselage to avoid using inordinately long struts, which might have been too weak for carrier landings; Douglas had to prove, however, that the resulting narrow track (hence concentrated weight) was not a problem on board a carrier.

The wing was a compromise. High speed demanded sweepback. Long range demanded a narrow wing. Strength without too much weight required a broader wing. Carrier stowage favored a shorter wing. Heinemann chose a thirty-six-degree sweep and an aspect ratio of 6.75, with zero dihedral for good lateral stability. To save weight, Heinemann designed a much less rigid airplane than usual—including the wing. Podded engines, chosen for easy access, affected the way the wing vibrated—its "flutter." Unfortunately experiments and calculations showed that flutter would become unstable below the limiting speed of the airframe. At Mach 0.8 flow around the pylon would become supersonic, setting up shock waves that could wreck the airplane.

Notwithstanding, BuAer found the design so attractive and considered its intended role so vital that in July 1949 it issued Douglas a detailed design-and-prototype contract. On 10 February 1951 Douglas received a letter of intent for twelve initial production aircraft.

Plans called for delivery of the first airplane in July 1951, with a first flight the following month, but Douglas did not receive the two J40-WE-3 engines until August, and then further delays ensued while the wing structure was redesigned to cure the flutter problem. The static-test wing suffered a major failure in January 1952, but the cause was resolved without major redesign. The airplane first flew on 28 October 1952, fourteen months late. By then, because of problems with the J40, the Navy had switched (in April) to the Pratt & Whitney J57 for production aircraft. Its different weight helped considerably to reduce flutter. Later the trailing edges of the pylons were extended. In April 1954 the flutter problem, the only major design issue, was considered solved. Carrier suitability trials began in April 1956. The A3D Skywarrior (A-3 under the post-1962 designation system) became the standard heavy U.S. Navy attack bomber of the 1960s.

As the first flight of the A3D approached, CNO pressed BuAer for the required new-generation catapult. BuAer kept reassuring OpNav that its internal-combustion slotted-tube project was progressing on time. In fact, unfortunately, the program was failing badly. As early as 1950 BuAer had been by no means sure how to build its slotted-cylinder catapult. By 1951 DCNO (Air) realized that the BuAer catapult program was not moving nearly fast enough.

The American naval attaché in London, Rear Adm. Apollo Soucek, a naval aviator, had become aware of an alternative: the British steam catapult. BuAer had only reluctantly listed it as an alternative to its favored internal-combustion device but now had to admit that it could not deliver its own catapult in time for early A3D carrier tests. Apparently it did not want to adopt something "not developed here." Crucially, however, it was DCNO (Air), not the chief of BuAer, who was pressing for a catapult that could launch the A3D. He could override the "not invented here" attitude. Soucek arranged for a demonstration of the British catapult in American waters. The U.S. Navy adopted it in time for the A3D to enter service, and its derivatives are only now being superseded by electromagnetic catapults. It seems that the adoption of the steam catapult convinced key American officers that the British had important naval-aviation innovations to offer.

Another, and perhaps the most important was the angled deck, which made jet operation far safer. That was soon coupled to a third British innovation, the mirror landing sight.

Without the steam catapult, the A3D might have operated on board carriers using JATO, just as the Neptunes did, but that would have been difficult and dangerous and would have made for very slow operation. There was a vast difference between what would have been a marginal heavy-bomber capability and what actually happened: A3Ds could operate from all U.S. fleet carriers, including modernized *Essex*es. Catapults that could launch these heavy bombers made it possible to design further heavy carrier aircraft, such as the Intruder, the Vigilante, and the Tomcat. Several decades ago, British naval constructor David K. Brown saw steam catapults as the floating equivalents of thousands of feet of concrete runway ashore.

Of fifty A3D-1s (A-3As post-1962), one became a prototype photo airplane (YA3D-1P) and five became prototype ECM aircraft (YA3D-1Q or A3D-1Q). Another was tested with J75 engines. The main version (164 built, accepted between April 1956 and January 1961) was the A3D-2, with a strengthened airframe (load factor 3.4 versus the -1's 2.67 gs), an enlarged bomb bay, and a more powerful version of the J57. As the -2 entered production, the Navy and Douglas discussed alternatives—either a liquid-fuel rocket or ECM—to the unsatisfactory tail turret. The rocket was abandoned, because buffeting would limit aircraft speed in any case. An ECM installation (initially ALQ-19, -32, and -35 but later ALQ-41 and -51) replaced the turret in the last twenty-one aircraft and was retrofitted to many others. These aircraft also had ASB-7 (rather than -1) bombing systems with their radars in a flat nose radome sometimes called a "duck butt." This radar was retrofitted to other aircraft.

The Navy asked Douglas to study both a tanker version of the new bomber and a kit for temporary conversion (to allow "buddy tanking"). This work dramatized the need to modify the A3D so that it could be launched at a greater weight (originally the limit was 70,000 pounds). To that end Douglas developed the "cambered leading edge" (CLE), with inboard slats and wing area increased from 779 to 812 square feet. It cost three or four knots in maximum speed but increased combat ceiling, cruising altitude, and combat radius; reduced wind over deck needed for catapulting and landing; and lowered stall speed and reduced the intensity of buffeting at the stall. A CLE airplane could be launched at 84,000 pounds, which is still the heaviest among U.S. carrier airplanes. CLE was incorporated in the last forty-one A3D-2s, twenty of which were fitted for buddy tanking.

An A3D-2 Skywarrior lands on a carrier, probably *Forrestal*, in 1957–58. This airplane was accepted in August 1957. The "GK" tail code indicated the squadron (VAH-5, the redesignated VC-5), not yet integrated into a carrier air wing. The tail turret is installed, but the two 20-mm guns seem to be missing. The initial radome was later reshaped. U.S. NAVAL INSTITUTE PHOTO ARCHIVE

An A3D-2 of VAH-13 on board the carrier *Kitty Hawk* shows the two main service modifications. This photograph was probably taken in 1962–63, as the squadron did not switch to an air-group (wing) tail code (NH) until 1963. The nose radome was reshaped and the tail gun replaced by jammers (ALQ-41 and -51 track breakers). Typically ALQ-41 receiving antennas were placed atop the tail and under the nose. The tail receiving antenna was probably the small circular one visible against the red of the fin tip. The tail ALQ-41 transmitting antenna used a small radome flush with the after side of the boat tail, near its top. The nose receiver was in a big rectangular antenna (which was shared with ALQ-51) abaft the nosewheel, not visible here; the transmitting antenna was in the nose itself. The object visible under the nose is an antenna used for the Automatic Carrier Landing System (ACLS). The side near the national insignia shows JATO attachment points. This airplane was accepted in December 1957. U.S. NAVAL INSTITUTE PHOTO ARCHIVE

In addition to 214 bombers, Douglas delivered 30 A3D-2Ps (RA-3Bs), 25 A3D-2Qs (EA-3Bs), and 12 A3D-2T (TA-3B) trainers. The special versions all shared a common airframe, unofficially designated A3D-2B. They had the 3.4g wings and CLE of late-production A3D-2s but also a pressurized compartment instead of the forward fuel cell and the forward half of the bomb bay. Two air-turbine motors to power generators were fed by side-by-side air scoops under the cockpit (the bombers had two turbines on one side). Radars in the photo and SIGINT (signals-intelligence) versions had ASB-1B radars optimized for search and navigation.

The EA-3B (A3D-2Q) was a SIGINT conversion intended to provide a fleet commander with organic carrier-based collection capability.[36] It was feasible

A KA-3B shows the characteristic refueling probe installed later in these airplanes' lives. This airplane has the snub nose of later A-3s. The prominent blade antenna under the nose was for the ACLS introduced in the early 1960s. Note also the nosewheel-door radome. This is one of three KA-3Bs of VAH-2 Det Mike assigned to the carrier *Ranger* after the Gulf of Tonkin incident: combat operations over Vietnam demanded tanking. The ship arrived in the Gulf on 18 September, and the first combat tanking mission Dowas flown on 17 December 1964. The cruise ended on 3 May 1965. This airplane is being hooked to the catapult; the bridle wires are visible. U.S. NAVAL INSTITUTE PHOTO ARCHIVE

because the fuselage had so much volume. This version was so important that in 1987 a TA-3B was converted into an EA-3B to replace a crashed airplane. By December 1956 there was also a proposal for a military-transport version of the A3D-2Q, seating up to ten. After the RA-3B was retired as a photo airplane, some were converted into ERA-3B ECM trainers (as "aggressors," emitting "threat" signals).

Prior to the Vietnam War, the only Skywarrior tankers were those temporarily fitted (installation took about four hours); Skywarriors were primarily heavy attack aircraft. However, when heavy attack squadrons deployed to Vietnam, they found their aircraft used mainly as tankers. They were valued, for example, for their ability to fuel damaged aircraft so that they could make it to their carriers. This experience inspired dedicated KA-3B conversions, from which bombing equipment was removed. Ninety aircraft were converted by the Naval Air Rework Facility (NARF) Alameda. Of these, five were converted as countermeasures-strike-support tankers (EKA-3Bs).[37] Later, thirty-four KA-3Bs were converted into EKA-3Bs; three EKA-3Bs became KA-3Bs. Surviving EKA-3Bs were converted into KA-3Bs after their tactical jamming role had been taken over by EA-6Bs.

"Heinemann's hot-rod" or "scooter." The Skyhawk was conceived as a light nuclear bomber, an alternative to modified fighters. This A-4C (A4D-2N) is shown on the catapult on board the carrier *John F. Kennedy* in the Atlantic, 23 December 1969. Like this one, later aircraft of this type had paintable radomes, with only the tip left black, as here. The object under the second *5* of the nose number is the angle-of-attack vane. The object on the base of the windshield wiper is a "total temperature" sensor. Note the undernose ALQ-51 (track breaker) antenna, the small blade antenna (for TACAN) farther back, and the Bullpup command-guidance antenna protruding from the nosewheel door. In shadow, to the left of and below the 20-mm barrel, is another ALQ-51 antenna. NHHC

9

THE ERA OF "NUCLEAR PLENTY"

President Eisenhower entered office in 1953 convinced that sustaining the big Army created during the Korean War would be unaffordable. He considered nuclear weapons an essential equalizer against massed Soviet and Chinese forces; he had understood as much as Supreme Allied Commander Europe in 1950–52. As president, he declared that in the event of Soviet or Chinese aggression, the United States would retaliate at a place and time of its choice. Carrier-based nuclear weapons were deployed around the Eurasian coast.[1]

Eisenhower's "New Look" strategy was possible because nuclear material was available in great quantity, thanks to the program begun in 1948. As importantly, it had turned out that bombs could be much smaller. Early bombs had to be hand-armed, hence could not be carried externally. For carriers that meant heavy attack bombers, which could not be carried in any numbers. Too, the first bombs were mainly derived from the big (sixty-inch-diameter) "Fat Man" that had destroyed Nagasaki.[2] The first step down was Mk 5. As conceived, it would have weighed 6,000 pounds, a figure embodied in the changing OS-106 requirement. As completed, the Mk 5 weighed only 3,175 pounds, light enough to be carried by Air Force fighter-bombers.[3] From a Navy point of view, the most important consequence of widespread Air Force deployment of tactical nuclear weapons on fighter-bombers was probably that service's development of a low-altitude attack technique known, with its supporting avionics, as the Low-Altitude Bombing System (LABS).

The next bomb (Mk 7) was streamlined so that it could be carried by fast jets; it armed Navy light bombers such as the FJ-4B.[4] It could be carried at high altitude and also dropped from low altitude by dive- or glide bombing. Its large diameter (to accommodate its implosion mechanism) and its considerable length were awkward for a carrier weapon. Fighter-bombers initially armed with the Mk 7 lacked sufficient centerline clearance; they had to carry it underwing as an asymmetrical load. The shape of the Mk 7 and the expectation that future bombs would be similarly shaped accounted for the unusually high undercarriage of the Skyhawk, the first specially designed light nuclear carrier bomber.

Implosion bombs were considered somewhat delicate. The Navy was vitally interested in a bomb strong enough to penetrate concrete or earth or armor, as in a submarine pen.[5] The Navy requested a gun-type weapon after the big underwater test at Bikini in July 1946, despite its inefficiency in terms of uranium required. The resulting 3,230-pound Mk 8 bomb could be carried externally. Its weight explains the strength of the centerline bomb rack on AD-4Bs. Because it was torpedo-shaped, however, the Mk 8 was poorly suited to fast jets.

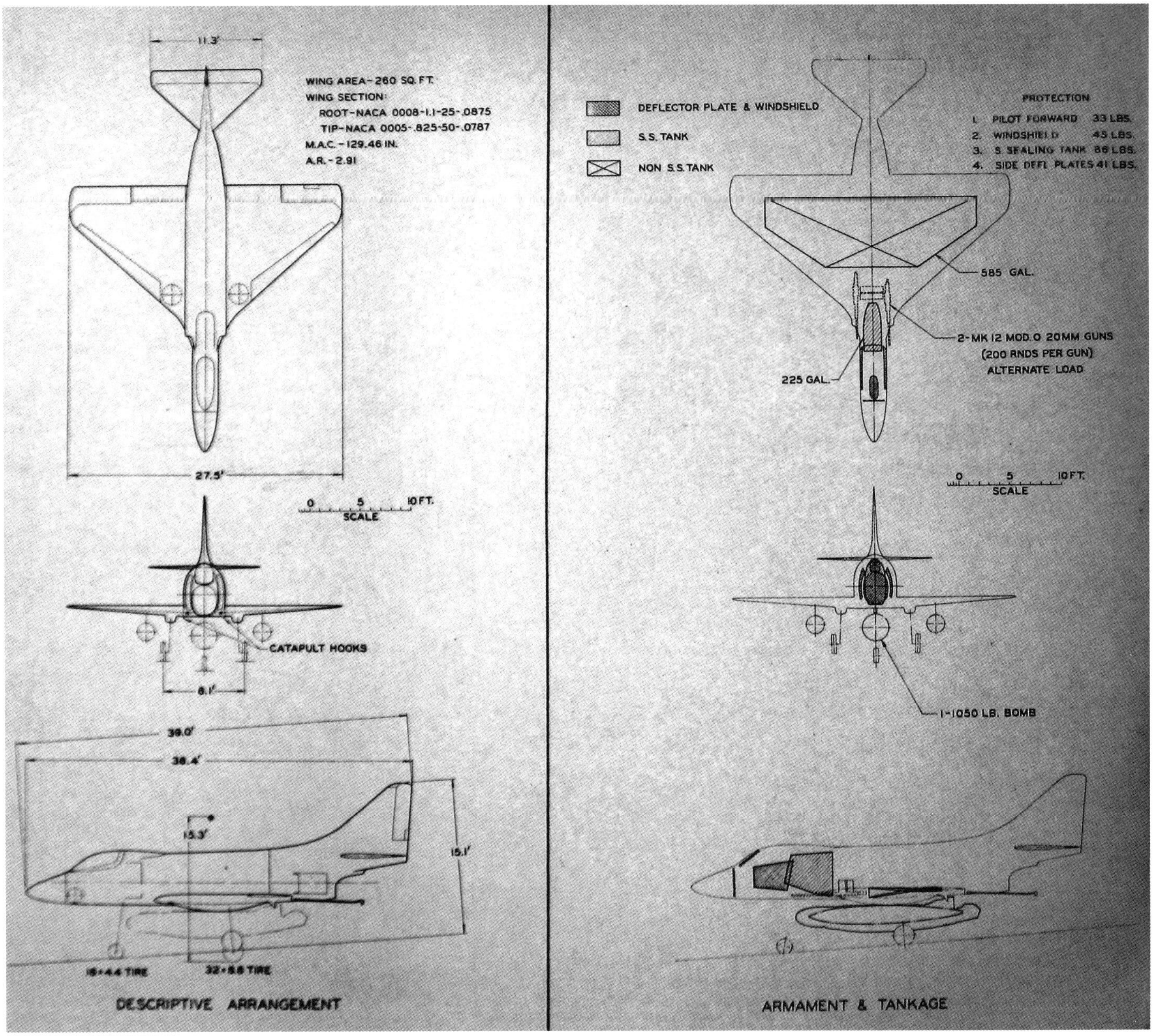

WING AREA – 260 SQ. FT.
WING SECTION:
ROOT–NACA 0008-1.1-25-.0875
TIP–NACA 0005-.825-50-.0787
M.A.C. – 129.46 IN.
A.R. – 2.91

11.3'
27.5'
SCALE 0 5 10 FT.

CATAPULT HOOKS
8.1'

39.0'
38.4'
15.3'
15.1'

15×4.4 TIRE   32×8.8 TIRE

DESCRIPTIVE ARRANGEMENT

DEFLECTOR PLATE & WINDSHIELD
S.S. TANK
NON S.S. TANK

PROTECTION
1. PILOT FORWARD   33 LBS.
2. WINDSHIELD   45 LBS.
3. S SEALING TANK   86 LBS.
4. SIDE DEFL PLATES 41 LBS.

585 GAL.
2-MK 12 MOD.O 20MM GUNS
(200 RNDS PER GUN)
ALTERNATE LOAD
225 GAL.

SCALE 0 5 10 FT.

1-1050 LB. BOMB

ARMAMENT & TANKAGE

A sketch of what became the Skyhawk from the Standard Aircraft Characteristics form included in Heinemann's proposal dated 12 September 1952. Only the rooster tail distinguishes it from the airplane as built. Although the sketch described the object between the wheels as a 1,050-pound bomb, its contours approximate those of the much heavier Mk 7 bomb, with its pair of upper fins. AUTHOR'S COLLECTION

The first light nuclear bombers were modified fighters. This Banshee (F2H-4) has a Mk 7 shape underwing; note the limited clearance. The sheer size of the Mk 7 explains why the Skyhawk needed its tall undercarriage. THE HOOK

The step beyond such fission bombs was the H-bomb (thermonuclear, or fusion), hundreds of times more powerful.[6] The first such weapon small enough to be delivered by a naval bomber was the Mk 27. Where earlier naval bombs had yields of up to about eighty kilotons, Mk 27 had a yield of a few megatons—millions, rather than thousands, of tons of TNT equivalent. It was possible to imagine a protracted war fought with fission bombs, but an H-bomb war would be one horrific pulse of attacks. The shared understanding that no such war was acceptable shifted the Cold War to the periphery of Eurasia and toward a new era of limited conventional warfare. Through the 1950s the Navy came to think increasingly about the limited, nonnuclear (at least at first) wars that might be fought in the shadow of deterrence. Like fission bombs, H-bombs began very massive and then shrank, in this case remarkably quickly.[7]

Targeting and Tactics

As the Soviet nuclear arsenal grew, it became more difficult to imagine using NATO nuclear weapons as an equalizer. By the late 1960s NATO had a strategy of "Flexible Response," in which an initial Soviet attack would be met by nonnuclear forces but, should that defense fail, NATO would escalate, initially to demonstrate its resolve and then to destroy the attackers, using tactical weapons. NATO fighter-bombers armed with nuclear weapons were seen, accordingly, as an essential means of deterring a Soviet ground attack. In Western Europe, that meant "quick-reaction" aircraft sitting on runways. On the NATO "Southern Flank," their equivalent was naval aircraft on board Sixth Fleet carriers. In the mid-1970s, for example, a former commander of Allied Forces South pointed out that the Sixth Fleet carriers' nuclear weapons were the only counter he had to a massed Soviet ground attack.

It was one thing to fight a war with a few nuclear-armed airplanes that would probably be far apart at all times. It was quite another when airplanes might find themselves flying through or near the nuclear bursts created by other aircraft. H-bomb bursts were, moreover, far larger than those of the earlier atomic bombs. The Air Force successfully argued for creation of an integrated plan for all U.S. nuclear weapons, the SIOP (Single Integrated Operational Plan). In it short-range naval aircraft were generally assigned to destroy air defenses that might otherwise oppose heavy bombers—for example, as a naval aviator later wrote, a fighter airfield in Romania. The SIOP required not only that carriers maintain nuclear-armed aircraft on constant alert (as did Strategic Air Command) but also that carriers assigned to SIOP missions always remain within defined areas, "boxes," from which they could fly their attack missions. That considerably reduced their flexibility in other missions and made them more vulnerable. Carriers were removed from the SIOP mission when it was taken over by the Polaris/Poseidon submarine force about 1963. They continued, however, to deploy with substantial numbers of nuclear weapons on board.

An early heavy nuclear bomber simply dropped its weapon from high altitude then turned away sharply to get clear of the blast. For example, the A3D-2 was intended to run in at about 41,000 feet, flying when at the target for fifteen minutes at normal thrust, including two minutes for postattack evasion and eight for escape, then leaving the target at the original altitude. This was good enough for a fission bomb. The ASB-1 radar bomb director was developed specifically for this tactic; it (and the improved ASB-7) equipped late AJs and all A3Ds. The important virtue of the ASB-1 was its clear radar picture of the target, which made release accurate.

High-altitude delivery tactics became obsolete as Soviet air defenses improved. Any low-altitude tactic had to enable an airplane to escape blast and radiation. By 1953 Navy light attack (VAL) aircraft such as Skyraiders and Banshees were expected to dive-bomb for accuracy. A sufficiently high pullout kept the airplane clear of blast (which would take some time to move outward from the explosion) and radiation.[8] A Maj. John A. Ryan Jr., USAF, invented (or reinvented, "loft bombing" having appeared in

1943—see chapter 4) "loft," or "toss," bombing. The bomber approached low to evade radar detection, flying over a chosen "initial point" (IP). There, or at a precomputed point beyond, the airplane pulled up at a specified acceleration. The combination of speed and angle to the horizon when the bomb was released determined its trajectory. Typically the bomb would be released automatically at about a forty-five-degree angle of climb and would be thrown three miles.[9]

VX-5, which worked on the new tactics, realized that it might not always be possible to find the IP during the run in. In its "over-the-shoulder" attack, the airplane pulled up over the target itself, releasing its bomb after passing the vertical (at about 110 degrees), then dove, and escaped as in toss bombing. The bomb would continue upward to about ten thousand feet before falling back onto the target.

Airplanes were fitted with LABS bomb directors (typically AJB-3s). As a pilot passed either the IP or the target, he pressed his bomb release and began a 4g pull-up. The bomb director measured how much he was deviating from the required 4 gs and determined whether his wings were level; crosshairs indicated any error and showed how it should be corrected. A whistling tone in the earphones stopped when the bomb was released, telling the pilot to begin his escape maneuver.

Low-altitude delivery placed a considerable load on the wings of the attacking airplane. The low load factor in the A3D seemed to preclude the full LABS maneuver, so instead an A3D would pull up after a low-level approach and release the bomb at about fifty degrees pitch angle. Instead of continuing into a half–Cuban eight, the airplane would bank sharply to reverse course and escape. The bomb flew forward. This was equivalent to the loft tactic developed during World War II.

LABS became less important once "laydown" bombs (initially versions of Mk 28) appeared, with retarding drogues. They could be dropped in level flight, their parachutes slowing their descent so that the attacker could escape without looping.

A low-altitude approach required much better reconnaissance than did high-altitude tactics, ideally a strip map the pilot could use on his way in and out. In a presatellite world, that was likely to be difficult to create for targets hundreds of miles inland, and for a target approached over the sea there would be no recognizable landmarks on the way.

Light Attack

Much as single-seat fighters had been adapted as the first dive-bombers, jet fighters were the obvious way to deliver the new lightweight nuclear weapons. The first such airplane was the F2H-2B version of the Banshee fighter. At least twenty-seven aircraft were converted, their wings strengthened and fitted with a single large pylon (different for the two weapons) so that they could carry a Mk 7 or Mk 8 bomb under the left wing.[10] The inboard flaps were notched to clear the bomb fins.[11] The capacity of the left-hand tip tank was reduced from 200 to 139 gallons to compensate for the large inboard load. To make up for that, in turn, aircraft were given an air-to-air fueling probe in the upper-right-hand gun tunnel; the external line from the probe to the fuel tanks was covered by a belly fairing. LABS had not yet been invented. The very small number of these conversions suggests that they began as a way of testing the Mk 7 or Mk 8 bomb (without the nuclear part) before the A2D Skyshark became available. As the A2D program continued to be delayed, the F2H-2B became a stopgap operational nuclear bomber. Hence its initial assignment was to the VX-5 service test squadron at Kirtland, charged with developing nuclear-attack tactics. Operational airplanes were assigned to composite squadrons (VC-3 and -4). VC-4 was the first to take the airplane to sea.[12]

The F2H-3 was ordered with nuclear capability already installed and so never had a *B* suffix (the same applied to the AD-6 and -7).[13]

Although BuAer was interested in developing a successor to the A2D, it failed to develop formal requirements through early 1952. Yet it still badly wanted something that could be operational by 1955, presumably by the time the new lightweight nuclear weapons became available in quantity (this was not stated). Without such an airplane, the only light nuclear bomber would be the A2D, whose future was problematic. The assistant bureau chief for research and development asked which current fighters could be modified as bombers to fill the gap. The head of the Attack Aircraft Branch responded in a 16 May 1952 memo. The key requirement, to carry a 3,500-pound store, eliminated the F9F-5, F9F-6, F4D-1, and the FJ-2.[14] No remaining fighter was completely suitable. For a time Chance Vought's proposed A2U-1 version of its F7U Cutlass fighter seemed to be the best alternative. It was proposed for the FY54 program but not bought, because the Cutlass itself was in serious trouble.[15]

BuAer was much more impressed with Edward Heinemann's proposal for what became the A4D Skyhawk (see below), but it needed a stopgap: conversions of the F9F-8 and FJ-4.[16]

F9F-8B Cougars were given two strong pylons, one for the bomb and the other for an Aero-1C 150-gallon drop tank. The model designation was formally established on 19 December 1955.[17]

The last 221 of 371 FJ-4 Fury fighters were reconfigured as light nuclear bombers (there was also a prototype). They were given additional speed brakes, spoilers for better control during the pop-up maneuver, an improved longitudinal control system, and a boosted rudder. Performance above 20,000 feet was similar to that of the FJ-4, but at lower altitude it would be superior, thanks to improved lateral and longitudinal control. The modifications added about 250 pounds. The FJ-4B first flew in December 1956. Fleet deliveries began on 18 June 1957. These aircraft were deployed only to Pacific Fleet attack squadrons, replacing Cougars, beginning with VA-214 on board USS *Hornet* in September 1957. The last went to VA-216. These aircraft in turn were replaced by Skyhawks.[18]

Both interim light bombers had the Aero18C LABS low-altitude bombing system.

This ended purchases of modified fighters; a 1956 Grumman proposal of a bomber version of the F11F Tiger it called the A2F-1 was unsuccessful.[19]

The situation presented Douglas' Ed Heinemann with a great opportunity. For years he had preached extreme simplicity and drastic weight reduction. In January 1952 he presented BuAer with a fighter study, conducted at Douglas' expense, that was interesting but not of immediate concern.[20] BuAer was much more interested in an attack-bomber study Heinemann presented in March.[21] BuAer now gave Douglas a small (about $36,000) contract to develop an attack bomber, an outline specification of what it now called the XA4D-1, dated 12 July 1952.[22] Douglas offered a 14,000-pound airplane capable of Mach 0.92 at 35,000 feet. Estimated unit cost was far below that of any other airplane of comparable performance.

Heinemann was already thinking about extremely small attack aircraft. He had an ONR contract to sketch a submarine-launched light nuclear bomber (D-640 in the Douglas El Segundo series). It was a manned alternative to the Regulus missile, which had guidance problems. Designed to carry a Mk 7 bomb, it was powered by a single Westinghouse J34-W-36, probably the engine Heinemann was offering in his fighter. It had small, swept, foldable wings to fit the planned submarine pressure container. There was a "rooster" tail, the jet exhaust being as far forward as possible (as in the Grumman Panther). D-640 was interesting enough for the Assistant Secretary of the Navy for Air to ask for designs to modify existing fleet submarines to launch it.

Heinemann's new entry for BuAer was the D-641. The first sketch of the new bomber resembled D-640 but with a delta wing (a swept wing was an option) and a wide-track, inward-retracting landing gear. The October 1952 mock-up of the A4D also had the rooster tail, the engine exhausting forward of most of the vertical tail (the after end of the fuselage was extended before the prototype was built).[23]

When he returned to El Segundo to begin work on the new design, Heinemann set his team two requirements: sufficient ground clearance for a Mk 7 on the centerline and a maximum wingspan of twenty-seven feet six inches, to avoid the weight associated with wing folding.[24] Eliminating the folding mechanism made possible a lighter-weight (yet strong) structure extending from wingtip to wingtip. Heinemann chose a delta, which offered maximum wing area (lift) for limited span, albeit at a cost in range due to reduced aspect ratio.[25] The wing had to be clipped because of the span limit, and the airplane needed a conventional, separate tail.[26] The delta was simpler structurally, hence inherently stronger, than a swept wing. The range penalty required Heinemann to provide more than the usual amount of fuel per pound of airplane, not only in the fuselage but also in tanks filling most of the wing. Transonic drag (i.e., at or near the speed of sound) depended on, the ratio of wing thickness to average chord (wing width): "thickness ratio." With its great average chord, a delta could be thicker for the same thickness ratio, 8 percent in this case. Unfortunately, fuel in partly empty wing tanks could slosh as an airplane maneuvered. Baffles in the tanks did not solve the problem.

The airplane generated transonic shock waves, because air flowing over it could become supersonic. The shock from the wing distorted airflow over the horizontal tail. It could make the elevator ineffective: when the shock hit the elevator hinge line, it could cause controls to buzz or flutter. Standard solutions, which Heinemann rejected on weight grounds, were to place the horizontal tail either well below or well above the wing; to use an all-moving horizontal tail; and to isolate the elevator from the stick using an irreversible actuator (which would eliminate the pilot's "feel," so that artificial "feel" would be needed).[27]

Heinemann emphasized that every pound saved directly cut overall weight by several pounds. For example, he saved about fifty-five pounds by consolidating some avionics into an integrated "biscuit." However, he made no allowance for tactical avionics such as LABS, which had to be added before armament tests.

Heinemann simplified his airplane by having it carry its weapon on the centerline. That in turn demanded an unusually high undercarriage (BuAer had to abandon the requirement that airplanes not require an external ladder). To minimize loss of wing fuel-tank space to undercarriage, Heinemann pulled the legs inboard, burying only the wheels in the wings.[28]

The J40 program was falling apart, so Heinemann sought an alternative engine with growth potential in the seven-thousand-pound-thrust class. He chose Wright's J65, a license-built version of the British Sapphire. It had already been adopted by the Air Force and was used in the FJ-3 and -4 Fury fighters.

No specific requirement for Heinemann's airplane had been drafted, but it met a well-understood need for a light nuclear-attack bomber. In 1952 the target date for such an airplane was still 1955, which could be met only by adopting an existing design and negotiating for its production. The assistant chief of engineering at BuAer wrote that the interests of both government and industry would best be solved by approaching Douglas. There was no point in a design competition, since to be fair Heinemann's ideas would have to be given to the other companies. Even then those companies would be at a gross disadvantage. Because the outline specification would have to be written around the Douglas design, it would appear to the other companies that it had been "wired" to favor Douglas. The aerodynamics of the Douglas proposal were largely based on those of the F4D Skyray fighter, for which BuAer had already paid.

Surviving Navy proposal files include Douglas' 12 September 1952 report on the proposed A4D-1, the drawing of which is recognizably the Skyhawk.[29] Heinemann offered much the same weapon payload as the A2D: one centerline pylon and two underwing pylons (for loads up to 1,000 pounds) but no cannon and no rocket stubs. Each underwing pylon could carry a hundred-gallon drop tank. The airplane could be fitted with two 20-mm cannon in the interdiction or day-fighter role, or it could carry rocket packages (seven Zuni rockets) underwing. Carrying a 3,250-pound bomb, the airplane would need 1,048 feet to take off with 25 knots of wind over the deck: it needed a catapult. Sea-level speed was 546 knots, much faster than the A2D. Climb rate at sea level was 7,340 feet per minute. Combat radius with the 3,250-pound bomb was 340 nautical miles.[30]

By this time BuAer was acutely aware of how quickly aircraft technology was changing. If it waited

Contemporary with the Mk 7 was the torpedo-shaped Mk 8 "light case" penetrating bomb, here also under an F2H-4.
THE HOOK

BuOrd developed the powered BOAR (BuOrd Aircraft Rocket) so that Skyraiders could escape nuclear effects despite their limited speed. This one is under the wing of a Banshee fighter at China Lake (NOTS Inyokern).
THE HOOK

to complete prototype tests before ordering production, airplanes might be obsolescent before they were available in quantity. Under a new FIRM (Fleet Introduction of Replacement Models) process, low-rate production aircraft were ordered for fleet tests before flight tests, saving several years. The A4D was ordered into production off the drawing board, its prototype built with production tooling—much as the Helldiver had been but this time with much happier results. The mock-up was inspected in October 1952. The Navy ordered nineteen aircraft late in 1952, before the detailed type specification had been settled, and another fifty-two in October 1953, well before the prototype flew. The prototype was rolled out in February 1954, but as a result of late delivery of its J65 engine, it did not fly until 22 June 1954. Only minor modifications were required at that point.[31]

The Skyhawk

The A4D-1 (A-4A) was permanently fitted with the two 20-mm cannon in its wing roots, and it had the three weapon pylons of the Douglas proposal. It ran carrier qualification trials on board the carrier *Ticonderoga* in September 1955. Production totaled 165.

An A4D-1 of VA-93 based at NAS Alameda. It was the first West Coast squadron to receive Skyhawks, in November 1956; it shifted to the A4D-2 in May 1958. This version was distinguishable by its unreinforced tail. Note the "sugar scoop" fairing at its base. This Skyhawk shows a 20-mm cannon in its wing root; many A4D-1s and -2s lacked cannon. This version lacked the refueling probe of later Skyhawks. As in versions through the A-4E, there were only three pylons, enough for one bomb and two of the big drop tanks shown here. The painted nose indicates the absence of any radar. The "NG" tail code indicated CVG-9 (an earlier code was *N*). This group was on board the carrier *Ticonderoga* for a September 1957–April 1958 cruise. U.S. NAVAL INSTITUTE PHOTO ARCHIVE

The A4D-2 (A-4B) had a new single-skin rudder with dual hydraulic boost, a strengthened rear fuselage, single-point fueling, and a revised cockpit. The engine was a somewhat more powerful J65-16A (rather than -4 or -4B).[32] Aircraft built after the first few had in-flight refueling probes on the starboard side of the nose. The centerline bomb rack was adapted to take a three-hundred-gallon buddy-refueling tank. Provision for JATO takeoff was added (widely used by the Marines). Of 542 built, some were reconditioned for foreign sale as A-4P (Argentine air force), -4Q (Argentine navy), or -4S (Singapore).

A4D-3 (1957) was a proposed strengthened all-weather-attack version with an APG-53 ranging and terrain-avoidance radar.[33] It had an Aero 26A or Mk 9 weapon sight, a three-axis autopilot with automatic LABS (AJB-3) capability, a fixed cambered-wing leading edge (instead of leading-edge slats, to deal with high-speed buffet and increase maximum altitude), and increased elevator effectiveness at high speed. It would have been wired to work with the Marines' TPQ-10 bomb-control system. The engine would

have been a J52-P-2. BuAer preferred the less expensive -2N, essentially an A4D-2 with A4D-3 avionics but without the new engine. The four planned prototypes were canceled.[34]

The A4D-2N (A-4C) was recognizable by its black dielectric radome housing the antenna of the APG-53 radar. About 15 percent of the earlier structure was modified, including a nine-inch nose extension. Slat improvements eliminated the need for the cambered wing. Many -2Ns had more-powerful J65-W-20 engines (8,400- rather than 7,800-pound takeoff thrust). This version first flew on 21 August 1958. Some of the 638 built were reconditioned as A-4Ls for reserve service.

A4D-4 was another abortive project, proposed by Douglas in 1958.[35] It had moderately swept folding wings with Whitcomb bodies (for "area ruling," a way of reducing trans- and supersonic drag) near their tips. There was a new tail. The engine would have been the J52 proposed for the A4D-3. The new wing would have carried four more pylons. The airplane would have been about eight feet longer than the A4Ds. A sketch shows a full-bubble canopy longer than that in earlier A4Ds and air intakes moved aft. The enlarged fuselage carried much more fuel, 48 percent more internal fuel than in an A4D-2, which would have given a third more combat radius.

A4D-5 (A-4E, "A-4D" having been omitted as confusing) was powered by the Pratt & Whitney J52-P-6A engine, which had 27 percent lower fuel consumption than the J65. Two additional pylons were mounted underwing, for a total of up to 8,200 pounds of weapons. This addition was reportedly associated with a shift away from the nuclear-strike role. The navigation system was given Doppler input, which presumably solved the earlier dead-reckoning problem. This version had the Mk 9 toss-bombing system, an updated LABS, and a radio altimeter. The nose was lengthened by twenty-two inches to accommodate avionics. Air intakes were moved away from

A formation of VA-83 A4D-2s (A-4Bs) from the carrier *Forrestal*, April 1960. The -2 version could be recognized by its reinforced ("tadpole") rudder, which eliminated an occasional severe flutter problem. This version introduced the refueling probe (a few early aircraft lacked it). Note the absence of wing-root 20-mm cannon. Vortex generators are visible on the wing, both near the leading-edge slat and forward of the aileron (as a line of bumps). Many A4D-2s lacked the wing-root 20-mm cannon, as here. VA-83 operated A4D-2s between December 1957 and September 1960, joining *Forrestal* on 28 January 1960 for a Mediterranean cruise that continued until early September 1960. *Forrestal* had two Skyhawk squadrons on board, the other being VA-81. U.S. NAVAL INSTITUTE PHOTO ARCHIVE

Two A-4Cs (A4D-2Ns) from VA-144 and VA-146 share an elevator of USS *Constellation* on 1 September 1964, showing how compact they were even though their wings did not fold. The only obvious indication that they have radomes is the way the antiglare panel forward of the cockpit ends abruptly. The blade abaft the canopy is the UHF radio antenna. Note the rows of vortex generators on the wing, including a row forward of the aileron. While operating over Laos during this cruise, VA-144 became the first unit to drop Snakeye bombs in combat. U.S. NAVAL INSTITUTE PHOTO ARCHIVE

the fuselage so as not to ingest boundary-layer air. Some were later fitted with fuselage humps containing additional avionics, and many were given more powerful J52-P-8s (9,200 rather than 8,500 pounds of thrust). The prototype flew on 12 July 1961; 499 were made.

A4D-6 was a failed contestant in the VAL design competition.

TA-4F (238 made) was a trainer version. However, A-4F (147) was a single-seat attack bomber produced to make up for losses in Vietnam after A-4 production was to have stopped in favor of the follow-on A-7. It was powered by the J52-P-8A engine (9,300 pounds of thrust) and had wing spoilers, a zero-zero ejector seat (i.e., enabling survivable ejection even from a stationary aircraft on the ground), and nose-wheel steering. After the first few, the Shoehorn ECM system, which had been mounted in the ammunition bay, was relocated to a hump abaft the cockpit. The modified A-4E prototype flew on 31 August 1966. One hundred refitted with J52-P-408s (11,200-pound thrust) were called "super Foxes." They had squared-off vertical tails carrying ALR-45 "hotdog" antennas.

An A-4C of VA-15 approaches *Forrestal* during a Mediterranean deployment, 22 July 1968–29 April 1969. Note how the 20-mm guns were staggered so as to interlace their ammunition belts. Also evident are the Bullpup antenna on the nosewheel door and the blister for the radar altimeter near the tip of the port wing. The four-hook rack on the centerline could accommodate two triple ejector racks in tandem. The flat plate forward of the nosewheel door covers a drum-shaped ARA-25 UHF/DF antenna. U.S. NAVAL INSTITUTE PHOTO ARCHIVE

The two extra pylons of the A-4E (A4D-5) offered a lot more carrying capacity. The first released photo of the new version showed Bullpups on the outboard pylons, 500-pound bombs in tandem triple ejector racks on the other wing pylons, and 750-pound bombs on the centerline pylon. This was the second production airplane, with the new long nose. U.S. NAVAL INSTITUTE PHOTO ARCHIVE

A-4M (158 built) was a new-production Marine version powered by the P-408 engine. The APG-53 radar was deleted, and IFF and ECM antennas were moved to the tip of the vertical tail. Aircraft were later modified with head-up displays, bomb directors with laser and television trackers (for the Angle-Rate Bombing System, or ARBS), and new ECM. Twenty-three TA-4Fs brought to A-4M standards (but without bomb directors) were delivered as OA-4Ms for forward air controllers. The Marines bought this airplane in preference to the A-7.

There were several export versions.[36] As of 1990 the A4D had been in production longer than any other Free World military airplane. Nicknames

ASW carriers often operated detachments of A-4s as defensive day fighters. This A-4E from Det 1 of VA-45 (USS *Intrepid*) is shown escorting a Soviet naval Badger reconnaissance airplane, about 1970. The centerline pylon carries small practice bombs. The nose (but not the tail) ALQ-51 antenna seems to be missing. Some A-4Es were retrofitted with A-4F humps like the one visible here during rework (some were fitted before delivery), similar to those featured in the next version, the -4F. The hump contained an engine oil tank and the boxes of the ALQ-51 system, moved from the gun magazines. NHHC

included "Heinemann's Hot Rod," "Tinker Toy," "Bantam Bomber," "Mighty Mite," "Ford" (which was also applied to the F4D Skyray), and "Scooter."

By 1954 BuAer was increasingly interested in missile-armed all-weather fighters to defend carriers against Soviet jet bombers. The weight of relatively light nuclear bombs being comparable to the total weight of the usual four Sparrow air-to-air missiles, BuAer's Aircraft Design Research group pointed out, the all-weather fighter and light nuclear bomber missions could be combined in a single airplane. That is probably why the airplane that became the McDonnell Phantom (F4H, later F-4) began its design life under the designation AH-1. By the time the Phantom flew, nuclear attack was no longer a design role. Even so, it had an AJB-3 that provided LABS capability.

The Fast Bomber: Vigilante

BuAer seems not to have sought a replacement for the A3D. The future Navy envisaged by the first Long Range Objectives Study (1956) was built around ships firing supersonic Regulus II cruise missiles. Without any need to launch big high-performance bombers, there would no longer be a need for very large carriers.

However, the Navy found itself buying the spectacular A-5 (A3J) Vigilante. It arose out of a North American Aviation proposal. The company's Autonetics Division had developed the inertial navigation system for the huge SM-64 Navaho ground-based cruise missile. North American's "General Purpose Attack Weapon" (NAGPAW) bomber was an alternative application. The project seems to have begun in 1953, when a Navy AJ-2 squadron commander told North American's Columbus, Ohio, design team that he had entirely evaded radar detection when flying from Barcelona to Paris and back at low altitude.[37] What he needed was a way to locate targets. A bomber would need at least transonic speed to escape the nuclear explosion. The team reasoned that accuracy would become more important as weapons shrank and their yield decreased. Doppler-corrected navigation would give away the presence of the low-flying bomber.

High altitude was an alternative way to evade interception: the North American team was aware that it was supposedly proving difficult for interceptors to reach Soviet Tu-95 Bears at high altitude.

The team was also aware that the company's B-45s were finding it difficult to drop bombs at high speed: air rushing into the cavity created when bomb-bay doors opened caused excessive turbulence inside. The team proposed a linear bomb bay, the bomb being ejected from its after end. The linear bay could at the same time also accommodate drop tanks, without paying any price in drag. They and the bomb would be ejected together, the empty tanks stabilizing the bomb.

The North American Columbus design team began work in November 1953. It envisaged a small, high-transonic, single-seat bomber with moderately swept wings, a low tail, and a linear bomb bay between two jet engines. Its inertial navigation system would be aligned before launch, and it could be updated using a small radar observing a shoreline checkpoint. This correction would be good enough to enable the inertial system to guide the airplane into its LABS release point. Afterburning engines would accelerate the bomber away from the explosion.

North American patented this concept.[38] The fore end of the airplane contained a "flight intelligence system" communicating with the pilot via, for the first time in a production airplane, a head-up display. It was connected to the forward-looking radar and to a checkpoint computer. The latter was also connected to a course computer, a bombing computer, and an autopilot. A command stick controlled the airplane through a device described as an "Autonav," in a very early form of fly-by-wire; the pilot would fly mainly on instruments. The drawing showed a "ship's positioner," the prelaunch input to the inertial system. The patent drawing showed the linear bomb bay.

The single-seat NA-233 NAGPAW was presented to BuAer in January 1954. It included a rocket engine to accelerate it away from the target. The bureau was interested enough to consider inclusion in the FY55 program. North American submitted a formal proposal on 5 April 1954.[39]

NAGPAW had a small wing to limit its response to low-altitude gusts. They had to be enlarged to meet a DCNO (Air) requirement that carriers at anchor be able to launch their aircraft, to meet sudden demands for strategic attack.[40] The change that made protracted low-altitude flight difficult. In January 1955 DCNO (Air) decided to emphasize high-altitude Mach-2 penetration, although it still wanted better than Mach 1 at sea level.[41] The two modes were incompatible. Given its relatively large wing, the airplane was limited to Mach 0.95 at sea level.[42] It could have reached Mach 1 with a smaller wing, which would also have improved the ride at low altitude, but that would have reduced lift and hence maximum altitude and payload. The mock-up had two vertical tails, but as built the airplane had one. It was powered by J79 engines, the most powerful then available, used in the Phantom and in the Air Force's F-104 fighter.

In response to a BuAer request, North American submitted a detailed design summary (its Report NA55H-65) on 1 April 1955. It received a design contract on 26 October. On 29 June 1956 North American received a letter of intent to proceed. The first two aircraft were ordered on 29 August. The final specification and configuration had been established by February 1956, and the mock-up review was held in March. Prototype fabrication began in August 1957, and the first airplane was rolled out on 18 May 1958. It first flew on 31 August. It was designated A3J-1 (A-5 after 1962) and named "Vigilante." The Navy promoted it by using the prototypes to set several world records, among them a new altitude record (91,451 feet with a thousand-kilogram payload, 13 December 1960).

North American's ASB-12 attack system was built around its Versatile Digital Analyzer (VERDAN) computer and its Radar Equipped Inertial Navigational System (REINS).[43] At this time BuAer was actively promoting digital computers (the slightly later A-6 Intruder was designed around one). The digital system could store up to five positions, which could be input during flight. The inertial system constantly computed position, presenting distance to go to the target and the appropriate heading. No previous airplane had had anything like it. The computer could fly the airplane via an Automatic Flight Control System (AFCS)—that is, via one of the first "fly-by-wire" systems (in which digital interfaces rather than mechanical linkages actuate control surfaces and systems) in service, in this case with manual backup. It automatically dropped the bomb.

Navigation would be largely inertial. Even if it was not precise enough, it would minimize vulnerable radar search time around a predicted target position. At its best it would enable an attack on a target that might not easily be seen.

The Vigilante largely reflected the NAGPAW concept, except that it had a second crewman, the bombardier/navigator ("bomb/nav"). The bomb/nav operated the computer that controlled television and radar cursors to indicate waypoints or target location. The bomb/nav had two monitors, one of which could switch between radar and television pictures. North American proposed an information projector alongside the radar/television scope to show an intelligence reel. For example, for an attack on Leningrad it might show (as final corrections to navigation) radar photos of the Norwegian coast the bomber would cross, radar photos of prominent checkpoints the bomb/nav might see on the radarscope, a radar map of the Leningrad area, and a radar photo of the target itself. The end of the reel would show the usual checklists and emergency instructions. As built, the A3J-1 had a terrain map display to the right of the radarscope. The reconnaissance version had an optical viewfinder (for its cameras) on the right.

The inertial system could guide an airplane over the featureless sea, something no ground-mapping radar system could do, and it did not give away the presence of the airplane by radiating. For the alternative radar-navigation mode, North American offered a radar picture reflected up onto a combining glass atop the instrument panel. This was the first head-up display in an American production airplane.[44]

The ship operating the bomber had to initialize the airplane system, to indicate to the system where the airplane was on takeoff. That requirement had figured in the original NAGPAW patent, and it seems likely that North American pressed BuAer to sponsor the necessary system. Its prototype was installed on board the carrier *Enterprise*—which was the first and only carrier to operate the Vigilante in the attack role. By the time the Vigilante was operating more widely as a reconnaissance plane, other large-deck carriers had inertial systems, for other purposes.

North American proposed a version with a liquid-fuel rocket booster at the after end of the linear bomb bay to accelerate the airplane for its Mach-2 run over the target. Two tanks in the bomb bay would carry the hydrogen peroxide oxidant, and the rocket would take its JP-4 fuel from the bomber's main tanks. The rocket would accelerate the airplane while it climbed from 50,000 to 70,000 feet for a hundred-nautical-mile run to the initial point. Once there, tanks, rocket motor, and bomb would all be ejected together, the empty fuel tanks stabilizing the weapon. North American tested the rocket on board a modified FJ-4 fighter. However, in August 1958 BuAer ordered the rocket engine deleted from production aircraft; Op-55 concurred in a 24 January 1959 memo.[45]

In its 1957 brochure, North American claimed that the combination of countermeasures and high maneuverability (thanks to the large wing and all-moving tail) would ensure survival. The airplane could outrun many interceptors, particularly if it detected them early enough. It offered a very sophisticated ECM system. Soviet interceptors were voice controlled from the ground, so a layered defense began with a voice jammer and progressed to dealing with the interceptors by breaking the lock-ons of their fire-control radars. Voice jamming would cover the "forward hemisphere" (the arc from wingtip to wingtip across the nose, both higher and lower), from which interceptors would come. North American planned IR detectors to cover the rear hemisphere (they failed in tests). A tone in the Vigilante pilot's headset would indicate the pulse repetition rate of the detected radar; in theory, that would provide

Two A-4Fs of VA-55 from the carrier *Hancock* fly over Mt. Fuji. The airplane in the background shows a reshaped refueling probe, its head bent away from the airplane's body. All production A-4Fs had the hump, uprated J52s, spoilers on their wings, and steerable nosewheels. U.S. NAVAL INSTITUTE PHOTO ARCHIVE

some sense of what sort of radar it was. Initially Vigilantes had only the APR-18 radar warning system. By 1963, however, they also had ALQ-41 and -51 track breakers (X- and S-band, respectively) and the ALQ-55 fighter-control-net jammer.

While the prototypes were being built, the design was revised to provide pylons for drop tanks that would add 428 nautical miles more radius.[46] These pylons were configured to carry other stores and weapons as well.

Nuclear certification tests in May–December 1962 using simulated Mk 27 and Mk 28 bombs showed that the linear bomb bay was not dependable.[47] This problem was never solved, because in 1961 the Joint Chiefs of Staff had withdrawn carriers from the SIOP. The Vigilante operated only during 1962.[48]

To overcome limitations on range due to insufficient fuel, North American proposed an A3J-2 with a fuselage hump carrying more fuel. Weapons would be carried externally. The Navy agreed to relax the zero-wind launch requirement. "Blown" leading-edge flaps (with increased lift achieved by channeling of airflow) replaced the earlier blown trailing-edge flaps. Using much more powerful J79-GE-8 engines, the airplane could take off at 60 percent greater weight (over 80,000 pounds) and could enjoy greater range. The last eighteen A3J-1s were built as -2s. The first flew on 29 April 1962. By that time the reconnaissance mission was becoming much more important. The A3J-2 (A-5B) became the RA-5B (NA-269) and as such served extensively during the Vietnam War.

McDonnell proposed that a modified Phantom fighter replace the Vigilante, prompting the Office of the Secretary of Defense to ask why the Navy needed two apparently comparable airplanes, each supersonic, each powered by two J79s.[49] McDonnell argued that the Phantom could operate in either fighter or nuclear-bomber roles; a carrier with Phantoms embarked could swing much of its air group

The other new nuclear bomber of the 1950s was the big Vigilante shown here during the type's brief career as a bomber rather than as a reconnaissance airplane. This one was assigned to VAH-3, identified by the tail code. The object above the tail number is a tail warning antenna; another ESM antenna is under the nose. U.S. NAVAL INSTITUTE PHOTO ARCHIVE

to either mission. For the bomber mission, the Phantom would use a NASARR attack radar instead of its air-to-air radar, a new Doppler navigational radar, and a transistorized (hence lightweight) analog bombing computer. These could be packaged to be interchanged with the air-to-air missile systems in the fighter's nose. The Phantom could carry a single centerline Mk 28 or Mk 43 bomb, with drop tanks outboard. With the weapon on board the fighter would make Mach 1.10 at sea level and Mach 2.29 at 50,000 feet. However, its range would be shorter (696 nautical miles with two 370-gallon drop tanks), and the Phantom would not have an inertial navigator. The normal fighter load (four Sparrow IIIs, 1,520 pounds) was less than the 1,885 pounds of a Mk 28 bomb. The Phantom was slightly faster (1,220 knots at 36,000 feet versus 1,147 at 40,000) but had a shorter combat radius (410 versus 685 nautical miles, albeit to a different set of requirements). In September 1959 a BuAer evaluator pointed out that the Phantom would be limited to visual attacks except for level bombing; the Vigilante was a true all-weather airplane. The hybrid analog-digital computer of the Vigilante would be more flexible than the pure analog computer in the Phantom. The Vigilante had a larger radar antenna generating a narrower and hence more precise beam, and it had ECM equipment the fighter lacked. BuAer also pointed out that much of the equipment McDonnell proposed

was not yet available, so that the modified Phantom could not be available in time. That killed the idea.

About 1959 BuAer produced a book of advanced projects it wanted to pursue, including one it called AWS (Air Weapon System)-202A, which it hoped to introduce into fleet service in 1968, given a 1964 go-ahead. By that time, it was presumed, Soviet targets would be so heavily defended that any bomber would need a long-range "standoff" missile (i.e., fired from beyond the range of a target's defenses: see below). AWS-202A would launch 200- or 400-nautical-mile missiles using a Vigilante-style linear bomb bay to minimize drag. The most dramatic new feature would be very high speed: Mach 4. Typically the airplane would accelerate to its engine limit at 36,000 feet then climb to 90,000 feet. After firing it would climb to 100,000 feet to glide home. Strike range would be up to 1,250 nautical miles.

BuAer had already contracted with North American to study a Mach-4 carrier bomber it called AWS-404A employing a new J58 engine with about twice the thrust of the J79.[50] North American was to study both a bomber for 1961 powered by a J58 (or a scaled version) and a 1963 bomber powered by either an improved J58 or by a hybrid turbojet-ramjet ("turboramjet"). Takeoff weight would be 100,000 pounds, cruising altitude 80,000 feet, and fuel would be conventional JP-5 (special high-energy fuel might be used in an afterburner). Requirements

A Vigilante shows its big wing, adopted to meet a requirement for zero-wind takeoff. In service it gave the bomber (and its reconnaissance version) a notoriously bad low-altitude ride.
U.S. NAVAL INSTITUTE PHOTO ARCHIVE

included zero-wind catapulting using existing catapults. If more advanced catapults and arresting gear were assumed and the plane could use afterburners at takeoff, weight could be considerably reduced—to 59,000 pounds for the 1961 airplane. The airplane would have an all-weather bomb/nav system (with inertial, Doppler, and stellar inputs), and ECM, including communications countermeasures (CCM), and IR countermeasures. BuAer envisaged a thousand-nautical-mile mission, slightly shorter than what a Vigilante could do. As in the Vigilante, North American envisaged a linear bomb bay. For 1961 North American offered an 80,000-pound airplane that would cruise at Mach 3.5 on two J58s. For 1963 it offered a Mach-4 airplane powered by two turbo-ramjets. Given the weight limit, radius was limited to seven hundred nautical miles. The 1963 airplane could achieve the desired one thousand nautical miles if allowed to grow to 112,000 pounds. Improvements were possible if wing area and power plant could be optimized for cruise rather than for zero-wind takeoff. A more advanced turboramjet could drive a 49,500-pound airplane at Mach 4.5 and yet offer a combat radius of a thousand nautical miles. Nothing came of these studies, which were surely the high-water mark of BuAer long-range-attack-bomber thinking. The J58, which powered the SR-71, seems to have been the most important fruit of the project.

Antiradar Missiles

Given the perception that the Soviets were developing effective air defenses, on 1 April 1955 CNO ordered BuAer to develop a standoff missile, initially named Raven (ASM-N-8). The missile would be delivered by a heavy attack bomber or by an attack seaplane. Target operational date was 1962. By that time Raven should have a range of a hundred miles and a CEP of six hundred yards, carrying a nuclear or thermonuclear warhead. By 1965 range might be two to three hundred miles, with the same CEP. In 1956 ADR produced sketch designs with speeds of Mach 0.9 or Mach 2.0 and ranges of fifty, two hundred, or four hundred nautical miles. For it BuAer considered simple command guidance (as in Bullpup), radar relay (command using the missile's radar picture), and inertial guidance. The alternative (now

Corvus, instead of Raven) was to home on radar signals, on the theory that any important target would be surrounded by defensive radars. The attacking missile could also home on reflected radar illumination provided by an attacking airplane, or on a cooperating beacon.

About 1956, antiradar homing was chosen. Given its expected precision, the missile could make do with a new low-yield XW40 warhead. Bids to meet Type Specification TS-3002 were requested on 27 March 1956. Five companies responded: Crosley, Douglas, Fairchild, Sperry, and Temco. The BuAer engineering recommendation, in favor of Temco, was issued on 4 October 1956 (the contract was signed in January 1957).[51]

Maximum missile range was 170 nautical miles. Because its seeker (interchangeably X- or S-band) could detect emissions at greater ranges, the operator would tune Corvus to the appropriate frequency and pulse rate before launching it. The missile could be launched at any altitude between 400 and 50,000 feet (so long as it was within line of sight of the target) and at any speed between 150 knots and Mach 1. Maximum range corresponded to 40,000-foot altitude. To attack a nonradiating target, the missile would be commanded (by 100-nautical-mile link) into a "homing basket," near enough to the target to detect radar signals from the launch airplane reflected by the target. The missile would generally follow an "up and over" flight path. Temco claimed that at least 80 percent of shots would burst within a cylinder with a radius of six hundred feet and a height of four hundred.

A projected inertially guided Corvus B (Corvus II) would have had a higher-yield warhead.[52]

BuAer also evaluated an unsolicited proposal submitted by Vought and its guidance partner Hughes for a standoff missile based on their unsuccessful submission for the Eagle long-range air-to-air competition.[53] Nothing came of it.

In February 1959, with work on Corvus under way, the CNO, Adm. Arleigh Burke, released a Long Range Objectives study (LRO-59) that envisaged two possible futures for the Navy's attack bombers. One emphasized penetration of enemy air defenses,

using high-performance and electronic countermeasures. The attack would be supported by antiradar missiles. A quarter to a third of attacking bombers would get through. Alternatively, the bombers could attack from outside enemy defenses using a standoff weapon, either AWS-302B (Bullpup equivalent, effective to twenty nautical miles) or AWS-303B (inertially guided, effective at two hundred nautical miles, equivalent to Corvus B).

Corvus first flew on 18 July 1959; tests continued through March 1960. The program was canceled in July 1960, after the Navy agreed to leave long-range standoff-missile development to the Air Force. This decision may have been connected to the decision to withdraw Navy aircraft from the SIOP.

Thereafter, Navy strike aircraft would have to rely on antiradar missiles to smash through enemy defenses. For that mission China Lake conceived Shrike in 1957. As with other China Lake initiatives, the formal operational requirement (February 1960) was issued well after development began (1958). Shrike combined a modified Corvus seeker with the existing Sparrow air-to-air missile airframe and an enlarged (145-pound) blast-fragmentation warhead. The motor was somewhat smaller but sufficient for a maximum speed of Mach 2 and a range of twenty-five nautical miles. Like Corvus, Shrike had a pre-tuned seeker; reportedly there were ultimately thirteen different seekers. Like China Lake's Sidewinder, the missile communicated with the pilot through his earphones. When it detected a radar to which it had been tuned, a tone indicated the pulse rate of that radar. The pilot's attitude direction indicator (ADI) indicated radar bearing.

Initially named Cobra (a name taken over by an antitank missile), the missile was designated ASM-N-10 (AGM-45 under the post-1962 system). Preproduction test missiles were deployed during the Cuban Missile Crisis in 1962; they would have been used by China Lake A-4C and A-4E test aircraft for attacks on Cuban SA-2 sites (Soviet-built surface-to-air missile [SAM] system, range perhaps twenty-eight miles, introduced in 1957). At the time, the developmental Shrike was the only U.S. missile capable of attacking a radar missile site.[54] The aircraft would have found the sites using APR-24 radar receivers. As the crisis wound down the attack plan was shelved, but while it lasted the Skyhawks flew fourteen missions to measure the frequencies and other parameters of radars associated with SA-2.

Shrike was considered an improvised weapon; something with longer range was wanted. In March 1962 an operational requirement was stated for second-generation antiradar missiles (ARMs): ARM I would be a Shrike replacement effective at up to fifty nautical miles. ARM II would outrange early-warning and ground-controlled-intercept (GCI) radars, which suggests about twice that range. Neither was built.

Shrike was produced (by a consortium headed by Texas Instruments and Sperry) beginning in 1963. It entered fleet service in 1965. The first combat mission was flown by VA-23 Skyhawks from USS *Midway*, on 25 April 1965.

The seeker was rigidly mounted in the missile: the missile had to be pointed directly at the radar in order to home. That made for a sudden diving strike that in itself warned the radar operator of imminent attack. To solve that problem, Shrike was often lofted so that its ballistic path would bring it close enough to the radar for its seeker to take over. If the radar shut down suddenly, the missile went ballistic; it had no memory as a backup to carry the missile all the way to the shut-down radar. Guidance lasted as long as the motor burned (ten seconds). If released precisely a Shrike could hit within twenty feet of the target radar. Typically, Shrike destroyed the radar antenna but not the associated antiaircraft site, so Shrike aircraft were accompanied by other aircraft using Zuni rockets and cluster bombs. Shrikes often had white phosphorous smoke markers to indicate the radars they hit, which might otherwise not be very conspicuous. By the spring of 1967, at least some North Vietnamese SA-2 operators had learned to recognize the characteristic loft-launch maneuver. They would shut down their radars.

Despite its limits (including range much shorter than that of a SA-2), Shrike was widely and successfully used both in Vietnam and in the Middle East. Production continued through FY80; 18,500 were made.

By mid-1966 both services wanted something better. The secretary of defense chose the Navy's proposal to modify the new antiaircraft Standard missile over an Air Force proposal to modify its air-to-air AIM-47 Falcon. Missiles were first ordered in September 1966. Maximum range was sixty miles, well beyond that of the SA-2. Standard ARM (AGM-78) had a much larger warhead than Shrike (215 versus 149 pounds), and it could operate against a much wider frequency range, including those of some of the long-range radars.[55] Unlike Shrike's, its seeker was gimballed to allow aircraft a wider range of maneuvers without losing contact with the target. The missile could turn 180 degrees after launch (at a cost in range): the airplane did not have to fly toward the target radar. An impact marker enabled follow-on attacks to be mounted against a concealed target.

The missile had strapdown inertial guidance, so that (at least in theory) it could continue to fly toward a target even if the target shut down. Some launching airplanes were fitted with a computerized TIAS (Target Identification and Acquisition System) that could localize an emitter by tracking it as the airplane flew, hence could direct an attack against a radar even if the radar shut down. TIAS proved less than successful in combat.

The missile uplink, which did not require TIAS, offered a degree of bomb-damage assessment. It stopped when the missile exploded. If the target radar continued to emit, it had survived. If both target and missile stopped emitting at the same time, the missile was judged to have hit. On this basis the Navy claimed a 64 percent kill ratio, but the feature was not considered altogether reliable and was dropped in 1973.

Standard ARM was three times as heavy as Shrike and five times as expensive. It could be carried only by the A-6 Intruder. It therefore went out of production before Shrike did. The follow-on High-Speed Anti-Radiation Missile (HARM, the AGM-88) in effect offered Standard ARM features in something close to Shrike weight and dimensions (798 pounds, 10-inch diameter, length 163 inches).[56]

If the Skyhawk was shaped by the single-bomb nuclear mission, its successor the A-7 Corsair II was shaped by the need to haul large numbers of conventional bombs. This A-7A was from VA-147 (on board USS *Ranger*) in the Seventh Fleet. NHHC

10

THE RETURN OF LIMITED WAR

Even while the Navy was preparing to fight a strategic, or "central," war using A3Ds, its idea of the most likely future warfare was changing dramatically. In December 1955 its future study group (the Ad Hoc Committee to Study Long-Range Shipbuilding Plans and Programs) argued that mutual deterrence made central war unlikely.[1] The contest between the Soviets and the West would likely be fought on the Eurasian periphery, in the collapsing European colonial empires. The French had just lost such a war in Indo-China, and the United States was helping prop up what was left of it. The United States had fought its own peripheral war in Korea. The British were fighting a counterinsurgency war in Malaya.

Limited war would include nonnuclear attacks, not least for close air support, so at least half of a carrier's attack aircraft should be able to deliver them, on a substantial scale. Aircraft might use missiles (Corvus was being developed) to deliver high-yield weapons in the face of heavy air defenses. However weapons were delivered, strike reconnaissance would be a major problem.

The focus was now on sustained operations, rather than the short, sharp pulse of nuclear attack. With relatively few airplanes being produced each year, it was essential to limit losses, at least when they attacked targets having conventional defenses. Attackers would be unescorted, protected only by low altitude and ECM.[2] With few aircraft attacking, precision would be important. Conversely, against heavily defended strategic targets high losses and considerably reduced precision would be acceptable.

The Ad Hoc Committee (soon replaced by the Long Range Objectives Group, or LROG, Op-93) envisaged an all-weather attack airplane that would penetrate enemy defenses at low level at a speed of at least five hundred knots (clean). It would have two, alternative missions: nuclear attack at eight-hundred-nautical-mile range, the airplane flying at low altitude for the last three hundred, in and out; and close air support, carrying a four-thousand-pound load three hundred nautical miles and remaining on station for an hour.[3] "All-weather" meant locating and attacking targets at night and under instrument conditions. Since 1953 BuAer had been sponsoring work on an airborne moving-target indicator (AMTI) radar that would enable a bomber to find mobile targets, such as ships or even trucks. The future bomber would combine that with a semiautomatic navigation system and an ability to deliver both missiles and nuclear weapons from low altitude.

The new airplane should be available about 1963.

The first Long Range Objectives Group report (1956) explicitly included Marine aircraft operating ashore, as in Korea, from improvised airstrips. Since the Marines would use the same attack aircraft as the Navy, this report added a requirement that the new bomber be able to operate from short strips. The Marine mission entailed nonnuclear capacity.

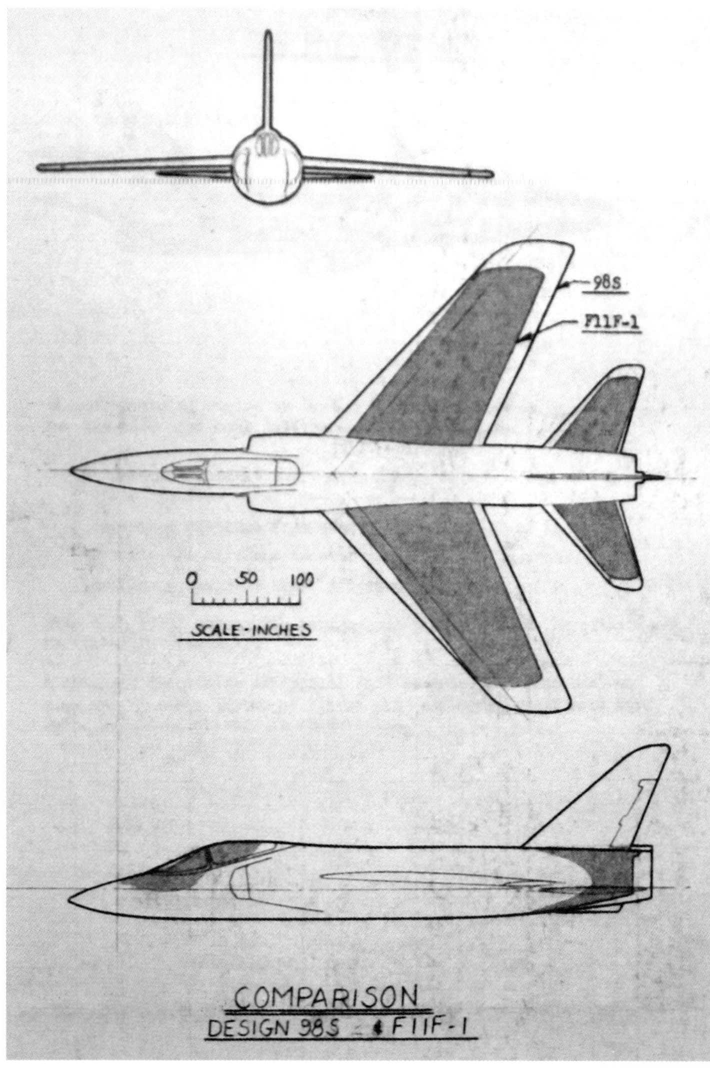

Prior to the VAX competition, Grumman offered a version of its Tiger (F11F) fighter as a light bomber under the designation A2F-1, before that was used for the Intruder. Its diagram compares the bomber (G-98S) with the fighter (shaded areas). AUTHOR'S COLLECTION

The Operational Requirement and Type Specification

OpNav issued a new operational requirement (CA-1504) based on Ad Hoc and Long Range Objectives reports.[4] Nuclear-strike range was stretched to a thousand nautical miles, but the three-hundred-nautical-mile sea-level dash was retained. This mission would be flown at maximum-range cruising speed (apart from a three-minute run in and a two-minute run out at maximum power). The load would be one Mk 28 hydrogen bomb and external fuel. For close air support the airplane would take off from a

1,500-foot runway (over the usual 50-foot obstacle), fly three hundred miles, loiter on station at five thousand feet or below for an hour, and return. The specified load was two Mk 83 1,000-pound) bombs. The weight goal was a maximum gross weight of 25,000 pounds in the close-air-support mission. The first eight airplanes would be delivered during FY59 (ending 1 April 1960); a briefing chart shows thirty on hand in 1961. Whoever wrote CA-1504 seems not to have been aware of how difficult it would be to develop the necessary weapon system.

CA-1504 translated into Type Specification TS-149, issued in February 1957 as the basis for a design competition for what was called "VAX," the experimental attack airplane.[5] The airplane would have five weapon stations, the centerline one stressed for the heaviest externally carried nuclear bomb, the 3,575-pound Mk 91. Two stations would be stressed for at least 2,000 pounds and two for at least 1,000 pounds. Both the centerline station and the two 2,000-pound stations were to be compatible with a 20-mm gun pod under development. The airplane would be compatible with the two Navy standoff missiles (Bullpup and Corvus) under development. It should be able to carry a defensive missile (which might also be used against targets of opportunity): Sidewinder or its semiactive radar version (Raywinder).[6]

For the first time, the aircraft manufacturer would be responsible not only for the airframe but also for the avionics. TS-149 described what the aircraft system should do and roughly how it should function but left details to the bidders.

The airplane would normally attack at an altitude of fifty to five thousand feet. The AMTI search radar would find fixed and moving targets and would enable the bombardier to select an aim point, "cueing" (steering onto the target) the stabilized monopulse (i.e., jam-resistant) attack radar.[7] Altitude would be measured based on ground returns. The attack display would be similar to that of a fighter's interception radar. The radar boresight could be slewed, the pilot using moving cross-hairs to correct the aircraft's position for an attack. Maximum CEP was something slightly over 2 percent of the distance from an aim point to the target. Target speed could be up to a hundred knots, and the system should handle dive angles up to seventy degrees.

The search radar would be used for terrain avoidance prior to an attack. Modes were analogous to those of the APG-53 in the A4D-2N (terrain profile and terrain above a fixed height). The airplane should be able to fly at five hundred feet over general terrain when free to maneuver vertically and in level flight at two hundred feet (fifty feet over water).

The core of the system would be a hybrid analog/digital solid-state computer. It would process radar data, provide bombing solutions, and navigate the airplane. Using digital technology would minimize the need for precision-machined parts. Computer memory was sufficient for track-while-scan (in this case, for two aim points and a target) and AMTI. The aim points made it possible to use offsets up to twenty nautical miles from the target. Required mean time between failure (MTBF) was fifty-five hours for the search radar system, seventy-five hours for the tracking radar, and twenty hours for the computer.

Navigation modes would be Doppler-inertial (principal mode), Doppler only, and inertial only. For the first time for an attack bomber, ECM was specified. Previously some aircraft had been specially modified for ECM duty, but this bomber would be operating alone.

The airplane had to be operable from angled-deck *Essex*-class carriers, in line with the LROG expectation that supercarriers would not be needed in future: at slightly below its gross weight it had to be launchable by the H-8 hydraulic catapult on board some *Essex*es, with ten knots of wind over the deck. Carrier approach speed would be 15 percent above stall speed.

Bids were requested on 15 May 1957. They were offered by Chance Vought, Douglas, Grumman, Martin, North American, Lockheed, Boeing, and Bell.

Chance Vought's V-416 Vigilante (a name not yet chosen for the A3J) was a twin-engine design (engines in nacelles) with a straight shoulder wing whose outer panels folded. To operate from the short temporary runways the Marines envisaged, each engine would have a three-position jet deflector. As in the company's Crusader fighter, the main landing gear would retract into the fuselage. V-416A (25,490 pounds) was a minimal airplane, and -416B (29,500 pounds) met all requirements.[8]

Douglas El Segundo offered D-715 and D-725.[9] D-715 had low swept wings with two underslung Pratt & Whitney J52 (JT8B) engines. Douglas claimed

Lockheed's CL-364 proposal. Span was 41 feet 2.7 inches (27 feet folded), length 45 feet 9 inches, and height 16 feet 5 inches. The engine was a single J52. According to Lockheed, maximum speed at full gross weight was (33,275 pounds) was 637 mph. Several bidders offered air intakes on top of the fuselage, apparently to avoid ingesting debris when operated by the Marines from improvised air strips.
AUTHOR'S COLLECTION

that it met the two range requirements and that its sea-level speed was 570 knots. Variations were 715A (water injection for thrust augmentation), 715B (close-support loading not compromised by the long-range requirement), 715C (blown flaps), and 715D (thicker wing and tail surfaces). In the basic design, close-support gross takeoff weight was 33,500 pounds without JATO boost. D-725 was an A4D-3 with a new nose section and a modified cockpit. To meet the long-range requirement, it would have needed buddy refueling. There was apparently no effort to meet the BuAer avionics requirement.

Lockheed's CL-364-2 was straight-winged, with a dorsal air intake feeding a single engine, probably a J52, and a crew of two in tandem.[10]

Martin emphasized that its Models 345 and 346 were conventional designs for simplicity, reliability, and operational flexibility.[11] Both had low, nearly unswept wings. Martin 345 had J52 two jet engines in its wing roots, the crew riding forward of the engines; Model 346 had a single engine with intakes above the wing roots, like a Skyhawk. Model 345 had a T-tail, Model 346 a conventional tail. The 35,500-pound Model 345 would exceed all performance specifications; it could fly at 450 knots on one engine. Model 346 met the weight requirement (25,000 pounds) and could operate from an *Essex*-class carrier. Mission radius on internal fuel was 435 rather than 895 nautical miles; loiter time was 1.7 rather than 3.4 hours. Against the required 1,500-foot takeoff run, it needed 1,930 feet (Model 345, 1,467 feet), but with water injection it would come close (1,550 feet).

North American's Vigilante was powered by a single J52, its designers having found that two were not needed to meet the high-speed performance requirement. It would need a second only to meet the 1,500-foot takeoff requirement and to provide sufficient acceleration after catapult takeoff without an afterburner. It seemed that the emphasis on simplicity favored a single afterburning engine. It was given a fuselage-top intake abaft and above its cockpits to avoid ingesting missile exhaust or debris. The straight wing, tapered on both edges, was chosen for its lift characteristics; it allowed a short, simple

landing gear. It could accommodate extra fuel in tip tanks. BuAer evaluation notes suggest that a modified A3J was also considered.

Boeing Wichita offered its Model 806, a high-wing airplane with an air intake on top, well abaft its cockpits.[12] It claimed superior low-speed handling and takeoff but also Mach 2.0 at 60,000 feet and a radius of action of 1,775 nautical miles. The engine

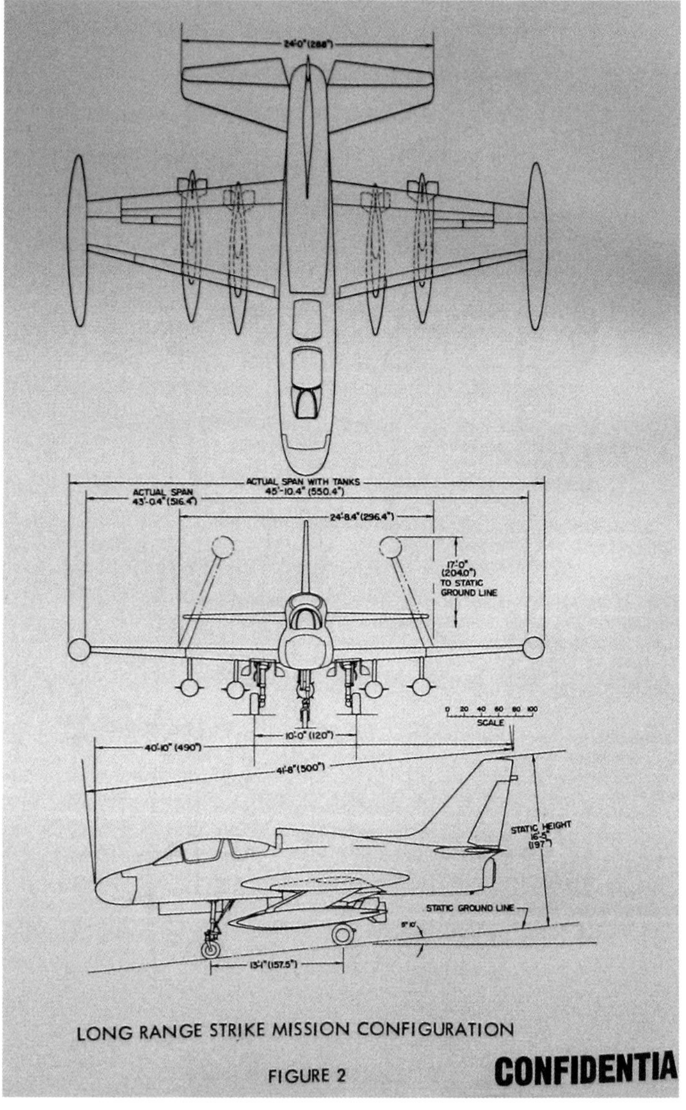

North American offered this design under the name Vigilante, which had not yet been assigned to its supersonic Vigilante bomber. As with other contestants, the engine was Pratt & Whitney's J-52 (B series), in this case credited with a military rating of 8,000 pounds of thrust (12,500 at takeoff). Span was 43 feet 0.4 inches (not including tip tanks; 24 feet 8.4 inches folded), length 40 feet 10 inches, and height 16 feet 5 inches. Gross weight at takeoff would be 21,800 pounds, well below the 25,000-pound limit given in TS-149. Maximum sea-level speed would be 512 knots, and combat radius would be the required three hundred nautical miles. AUTHOR'S COLLECTION

Boeing's Model 826 light-heavy attack bomber was the culmination of its work on a supersonic attack bomber. In the low-level attack role, it would take off at 40,230 pounds (29,440 in the ground-support role), carrying one Mk 28 nuclear bomb. Total radius of action would be 1,040 nautical miles, including a 300-nautical-mile dash. Length would have been fifty-four feet ten inches, span thirty-eight feet (twenty feet ten inches folded), and height fourteen feet six inches. The power plant would have been a J79-10 engine (19,300 pounds thrust with afterburner). Maximum sea-level speed was given as 647 knots (744 mph). In supersonic-attack mode, maximum speed would have been Mach 2.0 (1,150 knots / 1,323 mph), and Boeing claimed a total radius of 975 nautical miles including a 235-nautical-mile dash. AUTHOR'S COLLECTION

was a single J79-GE-X207A. A drawing showed one centerline weapon station and two more under each wing. Boeing claimed that it met the radius-of-action requirements (the 1,775 nautical miles was at high altitude and Mach 0.93). Boeing's basic brochure did not dwell on the avionics that TS-149 emphasized.

Boeing later modified Model 806 as the Model 826 "Light/Heavy Attack" airplane, which it claimed was a new concept in naval aviation. Advanced blowing boundary-layer control could offer higher lift with less blowing air. Avionics and nuclear weapons were both being miniaturized. Given these advances, Boeing claimed its bomber offered superior low-altitude speed and ride and supersonic performance comparable to that of the A3J yet was small enough to operate from modernized *Essex*-class carriers without afterburner. Speed would be 647 knots at sea level and Mach 2 at 65,000 feet. There were five external store stations, three of which could accommodate four-hundred-gallon drop tanks. Model

826 was heavier than Model 806. It had a larger vertical tail, which BuAer analysts thought would offer only marginally acceptable stability at subsonic speeds (Model 806 had been judged unstable at all speeds). Some features were unacceptable: the exposed tailpipe offered too much IR radiation, and the airplane took its blowing air through its wheel wells, making it vulnerable to runway debris. In any case, by this time (probably unknown to Boeing) high-altitude supersonic attack was far less important.

Boeing also offered a twin-turboprop Model 807.[13]

Bell offered a D2001 VSTOL attack bomber. It was then working on a VSTOL fighter, the most prominent feature of which was two swiveling wingtip pods, each carrying two J85 engines. Another four such engines, paired and pointed vertically, were in the fuselage. The airplane would have a centerline pylon and two pylons under each wing.[14] The Navy showed considerable interest in the fighter version of this concept but not in the bomber.

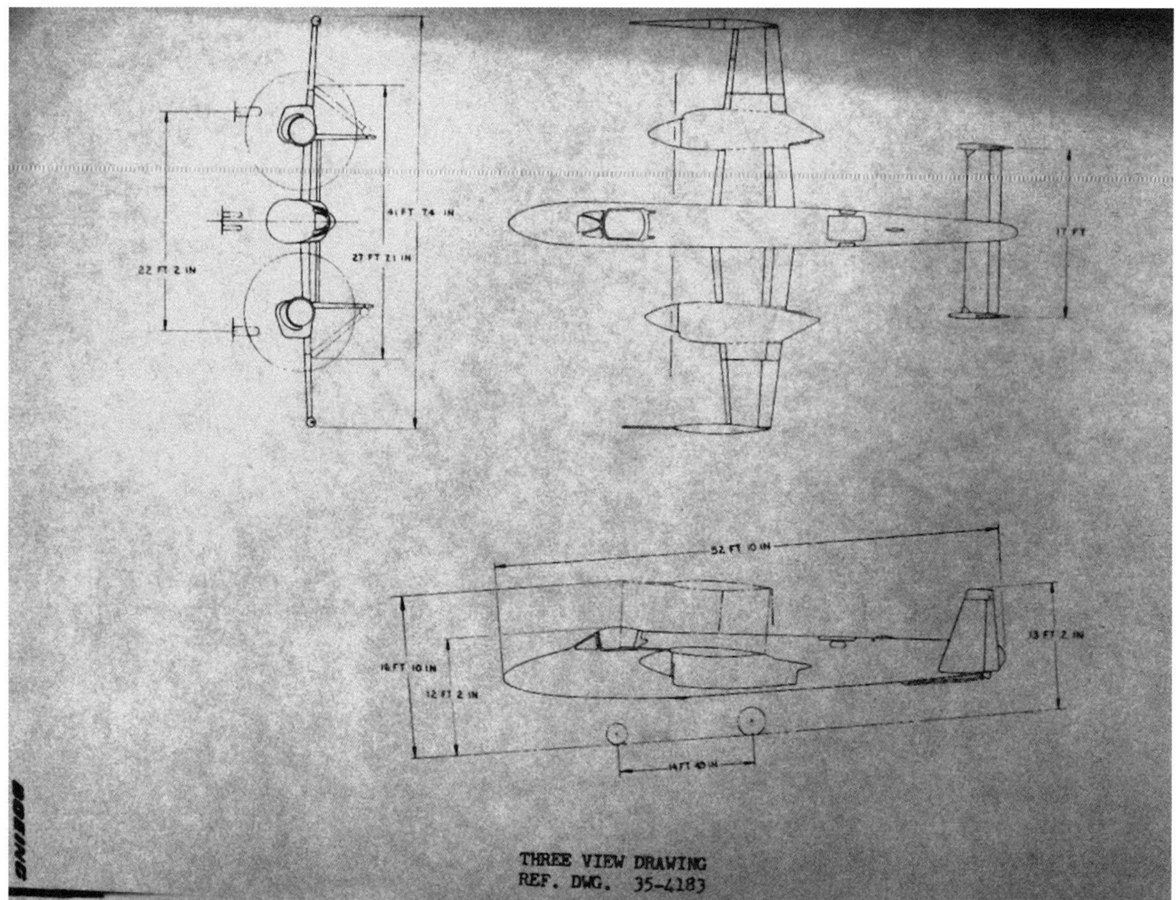

Boeing Wichita's Model 807 turboprop proposal for TS-149. The engines were two T56-A-7s. Boeing gave maximum sea-level speed as 518 knots (596 mph). Span was 41 feet 7.4 inches (27 feet 7.1 inches folded), length was 52 feet 10 inches, and height was 16 feet 10 inches. Gross flight weight was 26,317 pounds, but design catapulting weight was 38,053 pounds. Boeing did not give estimated combat radius but claimed that it met the 300-/1,000-nautical-mile requirements of TS-149. AUTHOR'S COLLECTION

BuAer received separate bids for the attack system itself, from Honeywell Minneapolis and from Texas Instruments. They illustrate how far attack-bomber design had gone from the days of airplanes with limited electronic elements.

Grumman's Intruder

Grumman's G-128 won, in December 1957; it became the Intruder (A2F-1, later A-6).[15] Its avionics rated highest technically and least risky. Its pylon arrangement was also judged superior, since none of its pylons would be beyond the wing-fold line. By October, the only viable alternatives had been Douglas and Vought.[16] Grumman was fastest of the three, if only by a few knots. It came closest to the desired takeoff distance: only 181 feet too long and would

have done better with water injection. It alone met the requirement for landing distance. None of the designs met either the short-mission one-hour loiter or the 1,500-foot takeoff (Vought came closest but would lose control if one of its engines failed on takeoff). Vought met the thousand-nautical-mile radius requirement, but Grumman exceeded it (albeit with four rather than the desired two) fuel tanks. Douglas failed altogether.

Grumman decided to act as avionics system integrator, having learned about complex systems in developing airborne early-warning and antisubmarine aircraft (WF-2 and S2F). Its WF-2 team would manage the project. As integrator, Grumman could take elements from several manufacturers, some of whom were already supplying other Grumman programs.

BuAer liked G-128's crew arrangement: uniquely among the contestants, Grumman placed the two crewmen side by side. Most pilots seemed to prefer having their teammates alongside and not behind them. Each could see what the other was doing, not guess at it. Each could point and draw the other's attention. Many master controls could be shared: it seemed that this was a better way to use cockpit volume. This arrangement also made for maximum forward visibility. Pilot visibility to the side was improved by staggering the two seats vertically and horizontally. With the big radar and two engines, the fuselage was already wide enough.[17]

For carrier operations the airplane needed excellent stability and ample control power, so it was given a long "tail arm" (distance between tail and center of lift) for, in effect, leverage. Engines and avionics all had to be forward both for efficiency and to balance the long tail arm. Length (fifty-six feet) was dictated by carrier elevators. The pod carrying the tail was as slender and clean as possible, producing the characteristic A-6 tadpole shape.

Grumman considered long cruising range rather than high maximum speed the driving factor. To that end it had to reduce wetted surface (which creates drag). It therefore rejected podded engines. Keeping the engines well forward divorced them from the basic airframe almost as well as if they had been podded and simplified engine replacement. Placing the engines in the fuselage made for good single-engine performance and control. Tailpipes were angled slightly out and down to minimize single-engine trim requirements. Placing engines under the wing structure made for maximum separation between the weapon pylons. Grumman chose to house the landing-gear wheels in the wing leading-edge "glove" (nearest the fuselage) to clear the pylons. That gave a wide wheelbase for ground stability.

The cruising-range requirement also dictated the choice of moderate sweep (twenty-five degrees at quarter-chord) and high aspect ratio (at least 5) with a modest thickness ratio (9 to 6 percent at the tip).[18] The sweep was enough to eliminate transonic buffet and postpone transonic trim change to beyond maximum speed (Mach 0.9). Moderate sweep also reduced gust loads. The thickness ratio (9 percent) was chosen to limit structural weight and prevent the low-lift buffet that a thicker wing would experience transonically. The wing was thick enough, however, to accommodate the necessary fuel.

Folded width (twenty-five feet four inches) was set by a requirement that two airplanes clear each other within the hangar-deck fire doors on *Essex*-class carriers.

To meet the 1,500-foot takeoff requirement, Grumman's tailpipes' after ends could be tipped down twenty-three degrees to add lift and thus reduce the speed needed for liftoff (from 86 to 78 knots) and thus the takeoff run. Grumman considered this idea, which offered a short-field benefit with little cost to the carrier attack mission, a significant contributor to its win. In fact, the low approach speed, speed brakes, and spoilers involved actually helped in the carrier mission. It turned out that vectoring the tailpipes had only a minimal effect on takeoff at anything but very light weights. The approach speed with pipes up was good enough, so the feature was eliminated after the seventh airplane.

Grumman called its computer-based system DIANE (Digital Integrated Attack Navigation Equipment). It was the Navy's first computer-integrated weapon system. Grumman chose a Litton computer for light weight and also because it was compatible with the Litton central inertial platform, already approved by BuAer for the company's radar-early-warning aircraft (WF-2 and W2F).[19]

Digital technology made it possible to share data between functions—for example, to provide the product of track-while-scan processing of the search radar to the tracking radar and also to fire control. The same computer data drove the pilot's vertical and horizontal displays and the radar operator's display. It also fed into an automatic flight control system. The single computer operated on a time-shared basis, shifting between functions but maintaining all data in a single memory available to all functions. The computer received all sensor and communications (including data link) data. Software control offered unprecedented flexibility both in tactics and in adopting different weapons in the future. Over the entire operating range, ballistic equations were solved with an RMS (root-mean-square) accuracy of 1.3 mils (the specification requirement was 10 mils). Weapons were released automatically, the computer taking into account factors such as dive angle, pull-up gs, wind, airspeed, range, and target velocity. Options were level drop, low-angle loft, and high-angle release.

For flight safety, some functions, such as the stability section of the flight-control system, were isolated from time-sharing. Navigation was entirely automatic, as in the A3J/A-5.

Grumman had envisaged a head-up display like that in the Vigilante, but had to settle for a television-like screen displaying video created by the digital computer—a first for a U.S. combat airplane. The digital computer could provide pilots with multiple different synthetic displays. (This was the distant ancestor of modern "glass cockpits," in which multi-function electronic displays replace dials and gauges.) The pilot's screen was called the "visual display indicator" (VDI). In terrain-clearance mode (search radar terrain clearance, or SRTC), the search radar created a terrain display covering a fifty-three-by-twenty-six-degree window centered on the airplane. For example, in hilly terrain it showed valleys as notches through which the airplane could fly. A horizontal display under the VDI could show terrain profile like the radar in an A4D-2N. The bomb/nav had a map-like view, like that of the A4D-2N, on his "direct view indicator" (DVI).

In an alternative Contact Analog mode, the pilot saw a path heading toward the horizon, which corresponded to that on the usual artificial-horizon instrument. Three lines below it pointed to the horizon. Symbols on the "ground" indicated obstacles or provided a sense of motion (crews called dark spots "cow flops"). The "sky" could show heading, in the form of a compass bearing. A dot indicated a target, an open square the aim point; the pilot and bomb/nav steered to place the square around the dot. White T-shaped objects in the "sky" above the horizon were "bugs" indicating altitude (on a scale of

radar altitude) and speed. Horizontal lines above the "ground" could indicate required pull-ups.[20]

No inertial navigation system was small enough, so Grumman chose a downward-looking Doppler radar that could measure the airplane's speed over the ground, both in the forward and the cross (drift) directions. Both the main radar antennas and the Doppler antennas had to be stabilized.

The pair of radars initially envisaged (APQ-71 and APG-46) were soon superseded by APQ-92 search radar and APG-88 attack radar (itself later replaced by APQ-112). The dish of the search radar, which scanned from side to side, was mounted above that of the attack radar. It had a Q (multipurpose) designation because it also provided terrain-avoidance data.[21] The radar change required an enlarged (more bulbous) nose.

The prototype flew on 19 April 1960. During flight tests it was found that extended dive brakes (e.g., for landing) changed the downwash on the tail; the stabilizer had to be moved aft. The dive brakes were not effective enough for either dive-bombing or for carrier approach, where instantaneous drag control was needed. The brakes had to be moved to the wingtips, where they would neither cause buffeting nor interfere with movable surfaces or weapons nor cause huge trim changes (side brakes were later eliminated). Rudder chord was enlarged at its base to provide more area for spin recovery.

Fleet introduction was delayed both by teething problems and by the complexity of the bomb/nav system, which produced serious maintenance and reliability issues.[22] The Ku-band radars suffered from rain interference, and it was difficult to fly over mountainous terrain at altitudes below a thousand feet. The computer system was not as accurate as required, particularly for level, dive/glide, and rocket delivery. The airplane did not meet its guaranteed range, but with four 300-gallon drop tanks it could deliver a Mk 28 at 1,216 nautical miles, flying 100 nautical miles in and out "on the deck."

During the first deployment to Vietnam, only 73 percent of aircraft launched completed their missions. That may have been due in part to computer

An A-6A of VA-42, the Atlantic Fleet Intruder Replenishment Squadron, carries a practice bomb on its inboard pylon. The black rectangle is a dive brake, which was inactivated (but retained) after flight test and Navy evaluation. The plating of the brakes, but not their internal mechanism, was retained until it had to be replaced because of cracking or corrosion. The brakes were eliminated altogether in later aircraft (from BuNo 154170 onward). They were replaced by brakes in the wingtips. A few aircraft flying in 1991 still had the visible brake panels. The triangular wingtip patch was part of the ALR-15 radar warner. Avionics represented 43 percent of the cost of an A-6A, compared to 20 percent of the contemporary F4B Phantom, which was considered highly automated. U.S. NAVAL INSTITUTE PHOTO ARCHIVE

Intruders like this A-6A from VA-75 on board USS *Independence* were sometimes used as buddy tankers. Forward of the cockpit is the massive fueling probe. U.S. NAVAL INSTITUTE PHOTO ARCHIVE

failure caused by the shock of catapulting, a problem some other early computer-controlled airplanes also experienced. Humid conditions near and over Vietnam did not help.[23] According to a 1968 OSD report, the attack reliability of the A-6A radars was only 40 percent; the reliability of the whole system only 26 percent. Targets were often difficult to see in strong clutter. It was estimated that in combat an A-6A could achieve a 900-foot CEP from an altitude of 1,000 feet against a good target (1,500 feet from 3,000 feet). However, against a difficult radar target a typical CEP would be closer to 4,800 feet. A-6s found it difficult to hit Viet Cong trucks at night. Between September and November 1967 the best A-6A squadron (VA-196) managed only 0.6 trucks per sortie, based on observation of secondary explosions. Even so, this was an improvement; in 1965 the score had been ten times worse, 0.07 trucks per sortie—hence the TRAM program (see below) to fit the A-6 with higher-resolution, nonradar sensors.

Many of the 488 A-6As delivered through 28 December 1970 were modified: seven as EA-6As for the Marines, nineteen as A-6Bs, three as EA-6Bs, and twelve as A-6Cs.

The A-6B was an antiradar version, delivered between August 1967 and August 1970. The first ten were modified to launch Standard ARM antiradar missiles. The next three had passive angle tracking (PAT), designed to localize a radar site by tracking the way its bearing changed as the airplane moved and so make it possible to fire even if the enemy radar was turned off. The last six aircraft had the Target Identification Acquisition System. Five A-6Bs were shot down, and the remainder were converted into A-6Es.[24]

The A-6C had a three-thousand-pound belly pod carrying IR and television cameras for the TRIM (Trails Roads Interdiction Multi-Sensor), to attack vehicles on the Ho Chi Minh Trail. An internal Black Crow device detected truck ignitions. A developmental prototype (NA-6A) that flew in December 1968 had FLIR and low-light-level television in wing pods. Other sensors were in fuselage pods. Reportedly the belly pod was disliked, for its weight, drag, and limited reliability. Presumably, however, experience with its IR sensor helped inspire the IR turret in later A-6Es. Twelve A-6As were modified as -6Cs. One was lost, the others being converted into A-6Es between October 1975 and February 1980.

Approaching the carrier *John F. Kennedy* during Fleet Ex 1-90, a KA-6D tanker displays its characteristic wingtip air brakes. It has the glove fairings of the ALQ-126 jammer near its wing root and the pods of APR-25 under its wingtips. Like an A-6E, it had ALE-29A chaff dispensers under its belly. Unlike an A-6 doing buddy tanking, the KA-6 did not have a special store with an air turbine on its centerline. The nozzle for its refueling hose is visible just above its undercarriage wheel, at left. U.S. NAVAL INSTITUTE PHOTO ARCHIVE

Two VA-145 Intruders from USS *Ranger* show the difference between the standard A-6A, in the background, and the A-6C, with its massive turret carrying a low-light-level television and a FLIR. The A-6C was modified under the TRIM (Trails Roads Interdiction Multi-Sensor) program. Prominent underwing on both aircraft is the tandem triple ejector rack on the outboard pylon. Both aircraft show the characteristic boom (on the outboard pylon) of the ALQ-100 countermeasure. It was an airframe change dated 29 February 1968, added on production airplane number 310. Both airplanes show pods under their wingtips associated with the APR-25 radar homing and warning system, an airframe change ordered on 30 April 1968 and incorporated in production aircraft number 359. A few airplanes had both the pods and the earlier ALR-15 wingtip antennas. The blade antennas under the nose of the A-6C are for TACAN and, abaft it, for UHF communications (as was a blade aft). The short back-swept antenna is for Bullpup missile control. The A-6A does not seem to show the TACAN antenna, because it is banked slightly toward the camera. The big underbody fairings were for the downward-looking Doppler navigational radar, which measured aircraft motion over the ground (or the water). The A-6A in the background has a truncated fairing to make space for two ALE-29A chaff dispensers, a change ordered in 1969–70. A-6As initially had the small slit of an ALE-18 chaff dispenser forward of the fairing. U.S. NAVAL INSTITUTE PHOTO ARCHIVE

KA-6D was a tanker version first flown on 16 April 1970. The computer attack system was removed, but an optical bombing capability was retained. Hose and reel were internal. Loitering 150 nautical miles from a carrier, a KA-6D could carry 15,000 pounds (2,300 gallons) of transfer fuel (5,000 pounds at 450 nautical miles). By the end of 1988, ninety had been converted, including twelve A-6As that had been reworked to A-6E standard.

A-6E was the major upgrade. In June 1966 Grumman began design work on a version with a revised electronic system and an internal cannon (which did not make it into the A-6E). It had a new, solid-state ASQ-133 computer and a new APQ-148 multimode radar. A late-production A-6A became system prototype, flying on 27 February 1970. Grumman switched to the new type in 1970, delivering the first A-6E in 1971. Many A-6As were rebuilt as -6Es. Major changes during the production run were replacement of the inertial platform by the ASN-192 Carrier Aircraft Inertial Navigation System (CAINS), addition of a FLIR/laser (TRAM, see note 25) turret under the nose, and installation of universal wiring that enabled the airplane to carry a wide variety of missiles, including Sidewinder and Harpoon.[25] A further Systems/Weapons Improvement Program (SWIP) included a new wing. Maximum gross weight of late A-6Es was 12.5 percent greater than that of early A-6As. The last A-6E was delivered in April 1992.[26]

The subsequent history of the A-6 program (rewinging and the A-6F and -6G) is covered in chapter 12.

This TRAM A-6E was built as an A-6A (from the last batch of those aircraft) and modified as an A-6E. The AJ tail code indicated Carrier Air Wing 8; A-6E TRAMs of this wing first deployed on board USS *Nimitz* in September 1979. The airplane is carrying both an underbody drop tank and an underwing drop tank. The air scoop near the vertical tail was associated with the CAINS carrier aircraft inertial navigation system. The glove fairing on the leading edge was associated with a radical change in ECM (the first few A-6Es retained the earlier boom). To accommodate the glove, A-6Es had their inboard wing fences relocated slightly, from just outboard to just inboard of the inboard pylon. The inboard part of the stall indicator strip was retained. The blade antenna visible under the air intake was for UHF communication. Abaft it, visible on the airplane in the background, was a TACAN blade antenna. The new radar and computer were credited with a 21 percent improvement in operational readiness and a 76 percent improvement in full systems capability. The new computer and radar reduced radar CEP by 49 percent and visual CEP by 53 percent. A-6Es modified for CAINS, the carrier aircraft inertial navigation system, could be distinguished by an additional air scoop forward of the tail fin. It reduced navigational CEP by 28 percent. U.S. Naval Institute Photo Archive

A non-TRAM A-6E on board the museum carrier *Midway* gives a sense of its load-carrying capability, each pylon carrying six 500-pound Snakeye bombs—a total of 15,000 pounds. This airplane is painted to represent the one in which the last A-6 flight crew to die in Vietnam were killed, on 13 January 1973. Dr. Raymond Cheung

Counterinsurgency and Ultralight Attack

While the A-6 was being developed, the limited-war scenario began to shift toward guerilla warfare, which would require relatively small airplanes carrying light weapons, especially guns. In 1958 BuAer prepared to hold a design competition for such an airplane. It was already buying light attack airplanes for the Army and for the Marines, such as the OV-1 Mohawk and the OA-10 Bronco.[27]

The BuAer proposal file includes a March 1959 Vought proposal (V-433) for an unsophisticated turboprop carrier attack bomber. Vought argued that existing jet attack aircraft were too expensive but that current propeller aircraft (Skyraiders) were too slow and vulnerable. Its proposal offered a flyaway cost under $500,000 and all-weather capability. V-433 was considerably smaller than a Skyraider.[28] Powered by a T64-GE-4, it could lift 3,100 pounds

of bombs, with four underwing pylons and another two under its wing roots. It could take off after a short run, and it could be catapulted at 20 percent overload. It would be instrumented for all-weather flight, and it would work with the Marines' TPQ-10 command bombing system. Carrying 3,100 pounds of bombs for close air support, it could fly 150 nautical miles at optimum altitude. It could also deliver a special weapon at a range of 1,010 nautical miles, flying the last 50 in and out at sea level and the rest at 1,500 feet. Maximum sea-level speed would be 410 knots (clean).[29]

BuAer was interested.[30] In April and May 1959 it was looking at a number of limited-war visual-attack aircraft, not only V-433 but a stripped-down Intruder (A2F-VW), versions of the Skyhawk (A4D-2N and -3), and the drastically modified A4D-4, which Douglas was proposing as an alternative to the Intruder.

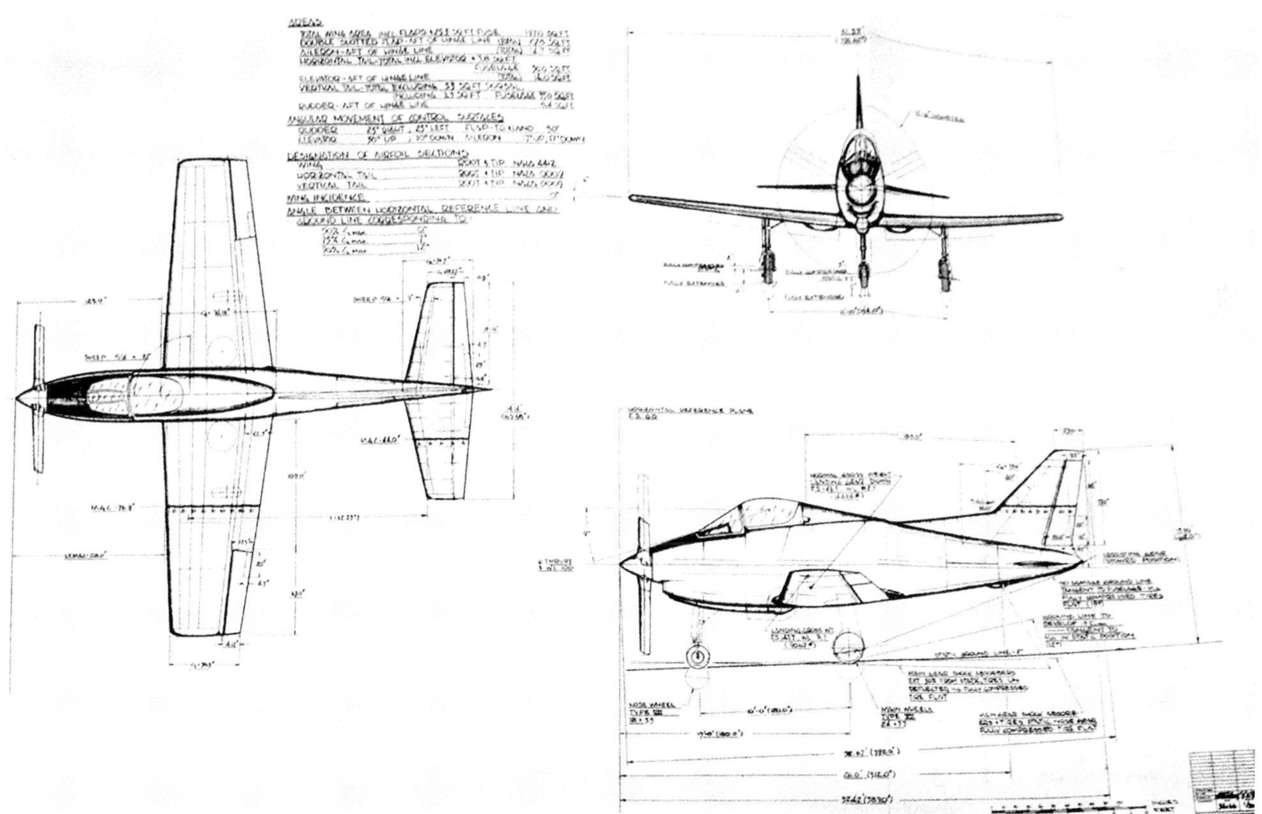

Vought's V-433 limited-war attack airplane was first proposed in March 1959 but was revived during the Vietnam War. Design gross weight was 14,500 pounds, with a 3,100-pound bomb load (six pylons). The engine was a T64-GE-4 turboprop (2,570 shp, with projected growth to 3,800 shp). Span was thirty-one feet five inches (twenty-two feet six inches folded), length thirty-two feet, and height twelve feet eleven inches. The airplane would have a tricycle undercarriage. Vought's brochure showed a basic close-support mission at a range of 150 nautical miles, including 120 minutes of combat time (forty at maximum speed). There was also a special-weapons (nuclear) mission at a range of 1,010 nautical miles, flown at sea level, with an alternative 1,590-nautical-mile mission flown at altitude except for a 50-nautical-mile run in and out. Maximum sea-level speed would be 410 knots (Mach 0.72 at 25,000 feet). In effect this was a Skyraider replacement, although it was much smaller, about the size of a Skyhawk (but shorter). Vought's proposal included a comparison with the Crusader; V-433 was about half as long. AUTHOR'S COLLECTION

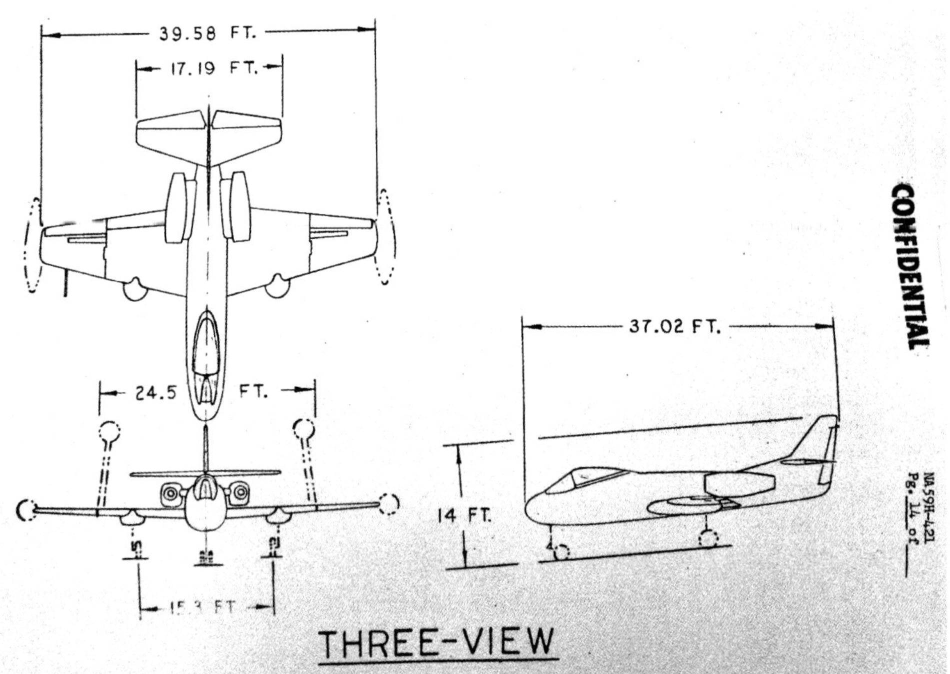

North American Columbus Div. proposed this Lightweight Versatile Attack Aircraft (LVA) roughly in parallel with Vought's V-433, probably in the expectation that BuAer would hold a design competition. Take-off weight would have been 25,983 pounds for a close air support mission, with six Mk 83 bombs on individual pylons (there was an alternative long-range mission carrying an Mk 43 thermonuclear bomb). The two engines would have been J12A-7s (J-60s), as in the company's Sabreliner (T39), with a static thrust of 3,300 pounds each. North American offered a variety of radars, including the APG-53A already used on the A4D-2N (A-4C) and its own NASARR. A diagram of the maximum-range mission (1,153 nautical miles) showed the airplane flying out at 377 knots (initially at 46,600 feet) and returning at 379 knots at somewhat lower altitude (the brochure did not show corresponding data for the close-support mission). AUTHOR'S COLLECTION

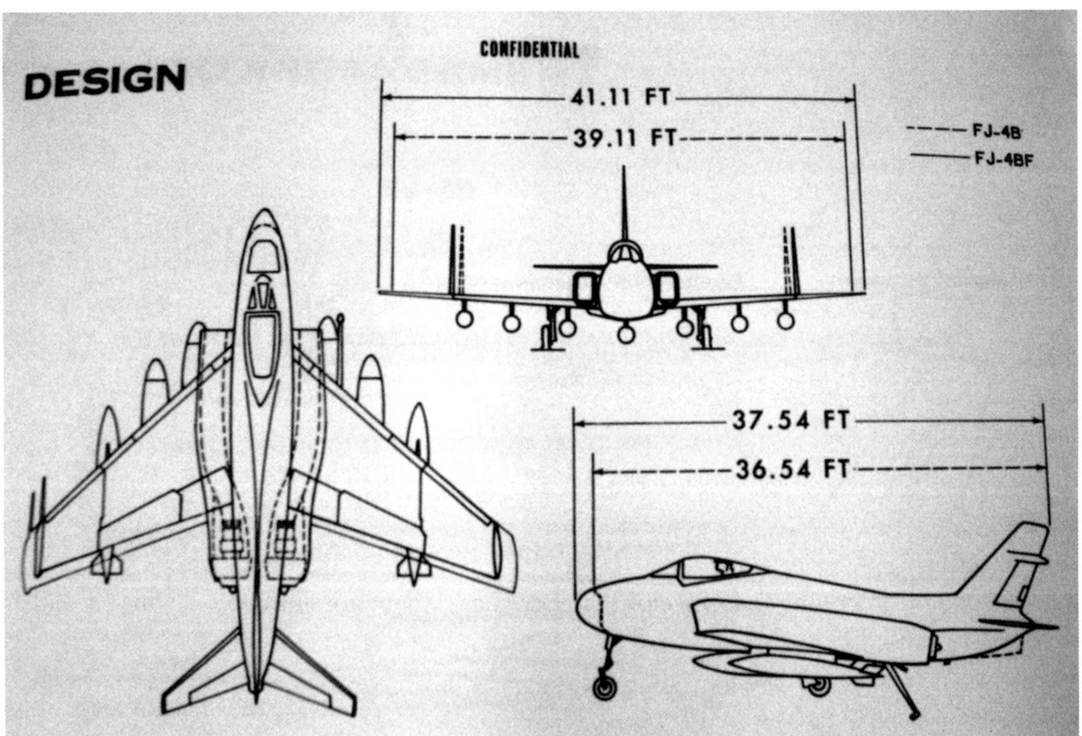

North American Columbus also proposed this FJ-4BF, in a brochure dated 6 December 1960. It was described as an LVA-type airplane for limited-war visual attack, with sufficient fuel and payload for long-range strikes. The FJ-4B was modified to incorporate two General Electric CF-700-28 turbofans and additional store stations (seven in all) plus a terrain-clearance-and-search radar (APG-53 or its equivalent). It used the FJ-4 wing but added a nose radome and side-by-side engines with side intakes. A centerline store station was added. Empty weight would be 12,963 pounds compared to 13,787 for the FJ-4B; takeoff weight clean would be 15,127 pounds (compared to 16,172). The turbofans were rated at 4,300 pounds of thrust for takeoff, a figure expected to grow to 4,850 pounds. At this time three prototypes were running. Production engines were expected in a year. The new airplane would be somewhat slower than the FJ-4B with its J-65 engine. Close-support radius of action would be 410 nautical miles with 5,000 pounds of bombs (fifty-minute loiter, ten at maximum thrust), and a special store (a thermonuclear bomb) could be delivered at a radius of 1,140 nautical miles. Maximum sea-level speed would be five hundred knots. Airplanes could be in the fleet early in 1963. AUTHOR'S COLLECTION

In December 1959 North American Columbus proposed a Lightweight Versatile Attack Aircraft (LVA).[31] The company touted maximum use of off-the-shelf equipment and twin-engine reliability. LVA was a small airplane with two jet engines mounted on the after fuselage, abaft and above the wing, much as in the later A-10. The straight wing could carry tip tanks for extended range. For a close-support mission, takeoff weight was 25,983 pounds, much as in the TS-149 competition. As in the A-6, a computer would drive a pilot's displays: an altitude indicator above a horizon indicator, with a rectangular (hence presumably computer-driven) radar and terrain-clearance display to the right. The main sensor was the APG-53A radar the A4D-2N used. The LVA designator had previously been used generically in connection with comparisons between versions of the A4D and the stripped-down Intruder.

Any plan of the Bureau of Weapons (BuWeps, established in October 1959 to replace the Bureau of Aeronautics) for a design competition was killed when CNO ruled that any Navy light bomber had to be able to fight in the whole range of warfare.

Given CNO's decision, Douglas offered a Sky-hawk adapted to conventional attack, with five rather than three store stations and greater range: A4D-5 (290 versus 210 nautical miles). The two inboard stations were rated at a thousand pounds, the two outboard at five hundred. With a four-hundred-gallon tank on the centerline, the airplane could carry out a 605-nautical-mile attack (flyout at about 40,000 feet, with five minutes of combat at sea level). With the same load, it could perform close support at 415 nautical miles (including an hour at 5,000 feet). However, without the centerline drop tank range would shrink. With a 1,000-pound bomb on the centerline and six 250-pound bombs on each of the inner ones, close-support range would be only 125 nautical miles.

Douglas tried several alternatives in a D-774 series. D-774-B and -H were counterinsurgency turbo-props (using, respectively, one T64 of 2,570 equivalent shaft horsepower [eshp] or two 1,464-eshp T53s), with a maximum speed of 380 knots at sea level. D-774-D was better adapted to the new policy,

powered by the desired TF30 turbofan for better range. It had an air intake on either side under its swept shoulder wing. With 10,000 pounds of thrust, it was as fast as the A4D-2N (580 knots at sea level), and it could carry a lot more ordnance, up to 10,000 pounds compared to the 7,100 of the A4D-5.[32] Avionics would have included a new APQ-89 radar developed from the APG-53, an inertial platform, flight data transducers, a central digital computer, a servo sight, and a display. BuWeps considered D-774-D a reasonable base case for a study leading to a new light attack airplane.

Since there was interest in twin engines, Douglas followed up with D-774-E, with two 4,400-pound-thrust JT12TPs. It could carry the same weapon load as -774-D on its seven weapon stations. With external fuel, it could deliver a Mk 28 thermonuclear bomb at a radius of 1,465 nautical miles, well beyond the usual 1,000-nautical-mile requirement. Maximum speed at sea level would be 550 knots (520 knots at 35,000 feet) while carrying two 1,000-pound bombs. There was enough weight for avionics required for all-weather navigation and limited all-weather attack.

V-433 resurfaced in 1965, in connection with the Vietnam War.[33]

VAX

BuWeps interest in D-774-D reflected, as explained, the shift toward a nonnuclear day-attack mission, which would require both more range and more payload than the Skyhawk offered. The John F. Kennedy administration, which took office in January 1961, took this view. The new secretary of defense, Robert F. McNamara, was aware that the Air Force was also developing a new bomber (TFX), and he ignored deep differences between that service's requirements and the Navy's. He tried to cut costs by forcing the Navy to adopt the Air Force's airplane. As a way out, the Navy offered to help finance TFX by incorporating its key technologies in its own new attack bomber: a variable-sweep wing ("swing wing") and the TF30 fanjet. Both would expand its flight envelope, and the turbofan promised much better range as a result of higher efficiency.[34]

The Navy airplane was soon called "VAX." A junior member of OpNav's TFX-VAX team realized that a modified Skyhawk powered by the TF30 engine might meet the new requirement; in a week Douglas assembled a proposal.[35] It was set aside, because in fact an all-new airplane was wanted. (The idea of mating the engine with an existing airframe returned later.) For that new program, and notwithstanding the TFX-technologies connection, CNO emphasized simplicity and low unit cost. It would be desirable to use either a single TF30 or two smaller turbofans (not yet fully developed). A swing wing should be considered only if no fixed-wing airplane could meet the specifications.

On 1 April 1961, DCNO (Air) issued Development Characteristics CA-01502-3, the capabilities desired for the new attack airplane.[36] As with the 1957 VAX project, there were two main missions. For close support (300-nautical-mile radius, thirty-minute loiter at five thousand feet) the airplane would lift ten thousand pounds of bombs; it would land with a thousand pounds of fuel left. The long-range mission (with drop tanks and a single two-thousand-pound nuclear weapon) would be conducted at 1,000-nautical-mile radius including a 200-nautical-mile run in (at not less than Mach 0.85) and then out at sea level. Maximum sea-level speed should be at least Mach 0.9 with a two-thousand-pound store on the centerline and full internal fuel. The airplane should have two engines optimized for low specific fuel consumption (i.e., per unit of thrust) at sea level.

With low-level attack the norm, the airplane needed the sort of terrain clearance offered by the APG-53, an objective being to allow flight at two hundred rather than five hundred feet. In contrast to past aircraft, it would have a nonradiating (presumably inertial) automatic navigation system (error not more than 1.5 percent of distance flown), with provision for in-flight position correction using visual checkpoints. Since this was a day (visual) bomber, it could rely on a visual sight; it should be adapted to dive, glide, roll-ahead (a rather complex maneuver involving visual acquisition upside down), and laydown techniques. The Marines' TPQ-10 remote

bombing control system would work through the autopilot. Of at least seven external store stations, the centerline one would be rated at 3,500 pounds, four other stations at 2,500 pounds each, and two at least at 1,000 pounds. Five would be capable of carrying multiple bomb racks or three-hundred-gallon drop tanks. There would be at least two 20-mm guns (at least two hundred rounds each). The airplane should also be able to carry a standard external Navy–Marine Corps reconnaissance pod for tactical reconnaissance.

With zero wind over the deck, the airplane would be launchable from a C11-1 catapult with full fuel and a ten-thousand-pound external load; it would land using Mk 7 Mod 1-3 arresting gear with a six-thousand-pound load on board, also with no wind over the deck. For land operation, it would take off in 1,500 feet over a 50-foot obstacle with four thousand pounds of stores on board. With thrust augmentation (JATO) it would take off with a ten-thousand-pound load. The landing roll with 2,500 pounds of fuel on board should not exceed 1,500 feet. The weight at which the airplane could be waved off on landing (i.e., could abort the landing to keep flying) with one engine running would include a six-thousand-pound load, including remaining fuel.

Development should be completed by 1 January 1968. Procurement would begin in FY64 with 12 aircraft, followed by 24 in FY65, 60 in FY66, 120 in FY67, 206 in each of FY68 and FY69, and then 192 annually in FY70–75—a substantial program amounting to 1,780 airplanes.

The Navy managed to convince OSD that its program was sufficiently distinct from (i.e., nonduplicative of) the TFX for the OSD Committee on Tactical Air to recommend the two different programs (both to be joint) on 19 May 1961. The Navy in effect bought control of the light attack airplane by accepting Air Force control of the TFX (F-111).[37] That made sense, since the Air Force would be buying far more TFXes than the Navy, which would use the airplane (in a modified F-111B version) only for fleet air defense. VAX was sometimes advertised as a half-scale TFX.[38]

BuWeps bought parametric studies from Boeing, Douglas, and North American (and perhaps others). There may have been some feeling that to compete with TFX a new Navy attack bomber needed the same supersonic dash speed (but the characteristics called for a subsonic airplane).

Boeing Wichita offered alternatives under the designation "Model 837": single engine or twin engine with either fixed or variable-sweep wings.[39] As with the 1957 proposals, Model 837 had an over-body air intake. The engines were generally versions of the TF30 (in the twin-engine design they were scaled down). The twin-engine Model 837-638 (8:1 ratio turbofans with afterburning, thrust reversal [for short-field landing], and IR suppression, with 12,500-pound sea-level thrust) had a gross weight of 31,610 pounds. It could take off in 1,900 feet (over the usual 50-foot obstacle). Boeing's studies translated into VAX proposals.[40]

Douglas, like Boeing, tried variable-sweep wings but also a delta with canards, using projected 1968 technology.[41] Douglas envisaged a dual-purpose, variable-sweep Mach 3 airplane that might perform combat-air-patrol missions. It might fly a high-altitude attack mission at Mach 3.0 and 72,000 feet; a CAP mission (cruise at Mach 0.81) at a radius of 150 nautical miles, loitering for two to five hours; and a low-altitude attack (800 nautical miles total, cruise out at Mach 0.81, 300-nautical-mile dash at Mach 0.92 to Mach 1.2). Its twin-engine designs were powered by GE MJ-270As (expected to be in service in 1968) rather than the TF30, sized to accelerate the airplane from CAP loiter to combat speed in five minutes. The airplane would take off at 60,000 pounds with a 7,540-pound military load. The delta required slightly more airframe. It offered a shorter Mach 0.92 dash (190 rather than 233 nautical miles) and a much shorter CAP loiter time (0.74 versus two hours) but shorter takeoff and landing runs. Douglas also considered the impact of reducing maximum speed to Mach 2.3 (at 60,000 rather than 72,000 feet) at the same takeoff weight and also of different low-altitude attacks (low-altitude all the way, a high-altitude approach and low-altitude dash). A variable-sweep wing would suffer considerably less from gusts at Mach 1.4 than the Intruder did at Mach 0.82, but the delta would do considerably worse, presumably due to its larger area.

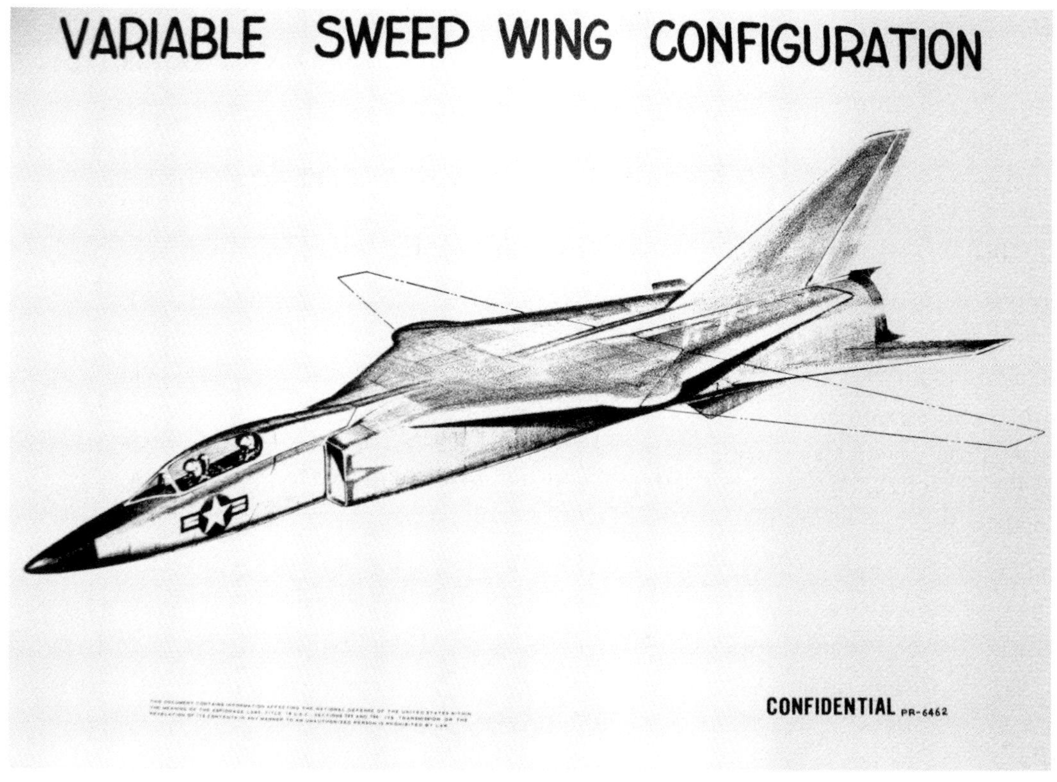

Douglas and Boeing conducted parametric studies for NAVWEPS of supersonic swing-wing attack fighters. This is the Douglas version. There was also a canard configuration. Assumed takeoff weight was 60,000 pounds with a military load of 7,540 pounds. Douglas compared a variable-sweep design with a delta. In both cases, design speed was Mach 3.0. High-altitude attack radius was 680 nautical miles. Takeoff weight of the variable-sweep design was 50,500 pounds (wing area 468 square feet); for the delta it was 60,000 (wing area 875 square feet). Design altitude in both cases was 72,000 pounds. An alternative low-altitude mission would be conducted at Mach 0.92. The variable-sweep design offered longer dash range (233 vs. 190 nautical miles) and more CAP loiter time (2 vs. 0.74 hours). Another table compared Mach 3.0 and Mach 2.3 designs. AUTHOR'S COLLECTION

North American offered two variable-sweep designs. A subsonic design was optimized for sea-level penetration at Mach 0.9, with a single nuclear weapon (600 nautical miles total, including a 300-nautical-mile Mach-0.9 dash in and out). The other was a Mach-3.0 design for supersonic high-altitude attack and also for supersonic combat air patrol, built of steel rather than aluminum. Weapons would be carried in a rotating bomb bay; external store stations could be added to the lower fuselage corners ahead of the main landing gear, to the fixed glove surrounding the wing leading edge, and to movable outer wings. The subsonic airplane was powered by two scaled-down TF30s, the supersonic one by two scaled-down advanced versions of the JT8D-30.[42]

The studies showed that high-altitude supersonic speed would not be too difficult. BuWeps recast requirements.[43] VAX would be a relatively small (less than 27,000 pounds) twin-engine (for reliability and survivability) single-seater with a spotting factor of no more than 1.3 (compared to a Skyhawk). It would be less than fifty feet long, and its cost would be minimized. Combat speed would be at least Mach 2.0 and ceiling over 60,000 feet. It would accelerate from Mach 0.9 to 1.8 in less than two minutes. A new interdiction mission (4,000 pounds of bombs, 600-nautical-mile range including a 150-nautical-mile sea-level dash in and out) was added to the earlier close air support (300 nautical miles, thirty-minute loiter, 6,000 pounds of bombs). No long-range nuclear strike mission was specified. For interdiction the airplane would carry two 300-gallon drop tanks. A new reconnaissance mission included a high-altitude supersonic segment. A two-seat trainer version might be modified for all-weather attack.

There would be five store stations (total of 13,500 pounds) plus, if possible, two Sidewinder stations and two 20-mm cannon. In February 1962 store capacity was specified as 3,500 pounds (i.e., a nuclear bomb) on the centerline and 2,500 pounds on each of the other store stations. At that time an overload close-support load of five 2,000-pound bombs (loiter time twenty minutes) was specified. The normal long-range strike load was two 2,000-pound bombs.[44]

As in the earlier development characteristic, this would be a day-attack airplane with terrain-avoidance radar, as in an A4D-2N. The nonradiating (inertial) navigation system would be accurate to within 1 percent of distance flown. An ECM receiver would be installed, with provision for a jammer. Radar and IR signatures would be minimized. The ashore takeoff run was extended to two thousand feet. Airplanes would have all-weather automatic-carrier-landing capability.

Airplanes would be designed for thirty-five flight hours per month.

In mid-1962 McNamara demanded that VAX become a triservice program. Convinced that duplication among the services was unaffordable, he had only the vaguest idea of the differences involved. The Army had been without high-performance aircraft since the creation of the Air Force in 1947, and it is not clear to what extent it believed that the ban would now be reversed. An undated BuWeps sheet summarized the services' requirements, including those of the Marines. Compared to the Navy's preferred maximum gross weight of 30,000 pounds, the Army wanted no more than 25,000 and the Marines 26,000; the Air Force preferred its 60,000-pound TFX. The Navy wanted a 10,000-pound payload, compared to 7,500 for the Marines and 3,000 for the Army. The Army wanted an hour of loiter at a point 200 nautical miles from a base, but the two Navy missions were 300-nautical-mile close support and 1,000-nautical-mile nuclear strike. The Air Force TFX requirement was a radius of 1,900 nautical miles and an ability to carry 10,000 pounds for four hours. By this time the Navy wanted a speed of Mach 0.9; the Army would accept three or four hundred knots. The Air Force wanted Mach 2.2 at altitude and Mach 1.2 at low level. The Air Force was also willing to accept a much longer takeoff run, 2,700 feet, whereas the Army, the Navy, and the Marines all wanted 1,500 feet.

The Air Force wanted nothing to do with a subsonic VAX, and it did want a VTOL capacity. Although the development characteristic had called for a supersonic airplane, VAX could easily go subsonic. In mid-1962 OSD ordered the basic mission load reduced from 6,000 pounds to 4,000, on the ground that more-efficient guided missiles could replace bombs.[45] BuWeps cut basic mission weight from 30,000 to 27,000 pounds. Multiple high-drag stores would be counted as an overload. Maximum external load was 10,000 pounds (with a thousand gallons of internal fuel). By this time BuWeps would accept both supersonic and subsonic STOLs but desired maximum speed was Mach 1.0 at sea level and Mach 1.8 at 35,000 feet, both in clean condition. The short burst of supersonic speed at high altitude may have been seen as a way of overcoming a sudden threat, such as an enemy fighter or anti-aircraft missile.[46] The stress requirement was set at 7.33 gs.

Boeing proposed a range of variable-sweep Model 837 airplanes with intakes above their bodies. Model 837–665 met the requirements, and -666 and -667 showed the effect of changing gross weight and wing area. They had Pratt & Whitney STF 182 engines (8,500 pounds of thrust), scaled down as necessary. Model 837-665 would operate at 17,230 pounds (takeoff weight 26,700 pounds). Spotting factor was 1.06. Maximum span was 48 feet (24 with wings drawn in), and overall length was 45.9 feet, well within the set limit. Height was 15.8 feet. Sea-level design speed was Mach 1.2 (cutting it to 1.0 would cut weight to 25,900 pounds). Close-air-support takeoff run would be 840 feet, well under the 2,000-foot requirement.

Douglas offered two variable-sweep-wing designs.[47] A single-engine high-wing airplane with intakes under the wing roots had a takeoff weight of 27,427 pounds. It met the other requirements. The alternative twin-engine airplane would weigh 29,500 pounds on takeoff. Both were rated at Mach 0.92 at sea level, with military thrust and all ordnance on board, and both met the takeoff-run requirement. Both could fly continuously at Mach 2.0 (limited by airframe in the one-engine case, by engine in the two- engine case), with maximum speeds of at least Mach 2.2. Acceleration time was 2 minutes for the twin-engine version but 2.6 minutes for the single engine. Neither met the 60,000-feet requirement (52,800 feet for single-engine,

56,000 for two). Spotting factors were 1.2 and 1.25; length was fifty feet in each case. With two stores stations under the fuselage, they had six stations in all: two on the fuselage rated at 3,000 pounds, two inner wing at 2,000, and two outer at 1,500. Both had about 50 percent more internal fuel than the required thousand gallons. Both met the 7.33g requirement. The overall penalty for using twin engines was 2,037 pounds.

Chance Vought offered a modified Crusader fighter (V-456) in June 1962 as an interim VAX.[48] Because this was an attack airplane with fighter capability, carriers using it could merge their day-fighter and light-attack squadrons. Vought began testing a Crusader with underwing pylons (i.e., in attack configuration) on 1 May 1962. V-456 was essentially the existing Crusader with a new larger (400 square feet), somewhat more sharply swept (thirty-seven degrees) wing carrying four stores stations. Two more, which normally carried Sidewinders, were on the sides of the fuselage (each could carry a two-round Zuni rocket pod instead). Each underwing pylon could carry eight Mk 81 (250-pound) bombs, or six 500-pound Mk 82s, or three 1,000-pound Mk 83s (total 12,000 pounds), or a 2,000-pound Mk 84, or a nuclear weapon, or a Bullpup or Shrike missile. This was not too far from what was wanted. Each station could also carry a three-hundred-gallon drop tank. For close support, V-456 would fly in and out at about 450 knots; it would loiter for thirty-two minutes. For interdiction, it would fly out 450 nautical miles at altitude, then 196 nautical miles in and out at sea level (Mach 0.8). It would fly out at 475 knots and back at 495, far from the desired Mach 2. V-456 was heavier than desired, with a design gross weight of 37,000 pounds, and also somewhat larger (length 54 feet 2.75 inches, span 40 feet [27 feet 6 inches], and height 15 feet 9.1 inches). A slightly later turbofan-powered V-456B was considerably lighter (25,750 rather than 36,200 pounds on takeoff, with 910 rather than 1,600 gallons of internal fuel: loiter time was 0.5 rather than 1.3 hours). Ordnance load was reduced from 12,000 to 8,000 pounds. Speed at sea level was Mach 0.95 (Mach 0.97 at 35,000 feet, clean). Spotting factor was 1.1.

In the spring of 1963 Vought offered an improved V-461 powered by a non-afterburning JT 8D-2 engine, by the turbofan version of the J52, or by the JTF10A-8 (non-afterburning TF30). Like V-456, it retained the original Crusader fuselage. Its smaller wing (300 square feet, aspect ratio 4, thirty-seven-degree sweep, span 34.6 feet) carried six stores stations, each with 3,000 pounds capacity (for example, four with 500-pound bombs (six per station) and two 300-gallon drop tanks). An alternative wing with the same area and taper ratio would have an aspect ratio of 6.3; to maintain stability, sweep would be reduced to twenty-five degrees. That would improve "specific range" (distance flown per unit of fuel consumed) by about 30 percent. Design gross weight was 26,000 pounds, about what was wanted. Maximum speed (clean) at sea level would be Mach 0.946; at altitude it would be Mach 0.968. With twelve 500-pound bombs, it would be Mach 0.872 at sea level and Mach 0.907 at altitude. With an afterburning JT8D, V-461 could exceed Mach 2.0. An inboard profile showed an APG-53A radar in the nose and two 20-mm cannon in the sides of the fuselage. The airframe was strengthened for longer service life.

In 1961 Grumman offered an A2F Intruder powered by an improved (turbofan) J52 engine.[49] It argued that with a turbofan version of the J52, an otherwise largely unmodified Intruder would offer much better range performance, better IR suppression, and shorter takeoff. Intruder design requirements already provided the desired extreme accuracy: a navigational error of four nautical miles after an hour and a ten-mil fire-control accuracy. In the three-hundred-nautical-mile close-support mission, the Intruder/VAX could deliver ten thousand pounds of bombs after flying out on internal fuel. For comparison, the Phantom could match its three-hundred-nautical-mile combat radius, but the Skyhawk could not get to two hundred nautical miles. The Intruder could meet the long-range mission requirement, including a two-hundred-nautical-mile dash in and out at Mach 0.85. Combat radius would exceed the thousand nautical miles required. Neither the Skyhawk nor the Phantom could come close.

Grumman also developed a series of G-303 designs bearing no resemblance to the Intruder. Work began on the basis of a May 1961 development characteristic. By January 1962 Grumman's engineers were reporting that using a scaled-down version of the General Dynamic swing-wing (which would appear in the F-111), they could obtain Mach 1.1 at sea level and Mach 2.2 at altitude. The Navy was not asking for a VTOL version, but, like Boeing, Grumman was aware of its occasional interest in one; Grumman therefore developed a VTOL design using four Continental lift engines and two cruise engines. BuWeps certainly was interested in IR reduction, and Grumman estimated that a turbofan engine (its MF 295) would reduce the usual turbojet radiation by a third to a half. For STOL (G-303-16) Grumman tried a vectored-thrust design similar to the Intruder.

Grumman considered variable sweep most valuable for gust alleviation at low altitude at high speed; it would also save weight (the variable-sweep 303-28D was 1,120 pounds lighter than the fixed-wing -30 for the same basic mission). However, variable sweep had not been tried in service, so Grumman also looked at conventional fixed wings. Another question was whether a single engine, in this case a Pratt & Whitney JTF-10A (18,500 pounds thrust with afterburner), would be better than twin engines.

The smallest acceptable airplane (6,000 pounds of stores) would have a 290-square-foot variable-sweep wing and a takeoff gross weight of 32,100 pounds. It would be powered by a pair of half-size (52 percent) MF 295 engines. G-303-19 and -22 were designed to meet the initial development characteristic; drawings were dated 10 and 19 January 1962. When it received the revised development characteristic, Grumman estimated that it would mean a slightly larger (291 square feet) wing and more powerful (63 percent) engines. This 303-28D met the revised development characteristic except for takeoff weight (32,829 pounds) and spotting factor (1.29). It was a midwing, variable-sweep airplane with air intakes for twin engines above the wings and well aft. A drawing of 5 July 1962 showed it with a span of forty-three feet six inches (twenty-seven feet with wings swept as far back as possible), a length of fifty feet nine inches, and a height of fifteen feet two inches. The wings carried four stores stations, and the airplane could carry a three-hundred-gallon buddy tank under its centerline.

North American Columbus offered a minimally modified FJ-4B powered by a TF-30 engine.[50] It stressed savings gained by reactivating the FJ-4B program, having retained most of its tooling. The proposed airplane had the same wing (extended by lengthening the carry-through structure, which would be strengthened), a fuselage deepened eleven inches to take the new engine, with a new radome on the upper lip of the air intake (for an APG-53A radar) and an added fuselage stores station. There would be two 20-mm guns. An ASN-19 navigational computer would be added, and the airplane would have another 150 gallons of internal fuel. Takeoff gross weight would be 30,000 pounds. A drawing showed three pylons under each wing. The brochure showed a load of twelve 250-pound bombs or six 500-pound bombs, at a takeoff weight of 28,300 pounds. The airplane could run out at altitude and then loiter at five thousand feet: radius would be 318 nautical miles for a sixty-minute loiter, 457 nautical miles for thirty minutes. Close-support radius (8,000 pounds of bombs and internal fuel) would be 300 nautical miles (thirty-minute loiter). Interdiction radius (4,000 pounds of bombs, two 200-gallon drop tanks) would be 600 nautical miles (150-nautical-mile run in at Mach 0.8, run out at Mach 0.85). Radius for an alternative nuclear-strike mission (2,000-pound bomb, two 150- and two 200-gallon drop tanks) would be 1,330 nautical miles, including five minutes' run in and out at sea level (635 if the airplane flew at sea level all the way). The airplane would cruise at 330 knots at sea level but would attack at 500 knots.

Lockheed proposed CL-706, for which it considered conventional, STOL, and VTOL versions between January 1962 and May 1963. The versions considered for submission all had variable-sweep wings and takeoff weights of about 32,500 pounds.[51]

Northrop's 8 August 1961 "VA(X)" was based on its N-156 lightweight fighter (which became the F-5 and T-38). It was attractive because of its cost and availability. The BuWeps chief, Adm. Robert Pirie, showed considerable interest, although the airplane did not meet the desired characteristics. N-156D would have five stores stations, three of them with two-thousand-pound capacity and two with one-thousand-pound capacity. Any guns would be podded. External fuel would be carried in tip tanks. Radius of action would be 250 nautical miles.

On 28 November 1962, McNamara said that no money would be budgeted for VAX or any other A-4 replacement pending a Navy study of the light-attack-aircraft requirement itself. If the study did show such a need, he would be willing to earmark FY64 R&D money. The VAX requirement had been framed as the result of an internal Navy study, which was now to be duplicated. OSD system analysts had recommended cancellation, partly because they saw no point in the high-altitude supersonic speed embodied in the VAX requirement. Ironically, the Mach-2 requirement had probably never been badly wanted in the first place. OSD also wanted to terminate A-4 production, leaving the Navy without a new light-attack bomber.[52]

Between February and May 1963 DCNO (Air) supervised a Navy Sea-Based Air Strike Forces Study.[53] OSD System Analysis insisted that total system cost, not just procurement cost, be taken into account. Its measure of effectiveness was the cost per ton of bombs on target. "Cost" meant not just the airplanes but the entire weapon system, all the way from the carrier to the target. A new weapon system would be approved only if was at least 10 percent better than existing ones; the ideal was a quantum jump in cost-effectiveness.[54] Counting total cost highlighted the potential of new technology.

Taking into account the cost of the carrier air group emphasized the trade-off between fighters required to defend a carrier and strike aircraft it could accommodate (numbers depending in part on spotting factors and fuel requirements). In a limited war, the air threat to carriers would be light bombers that the Soviets were transferring to their Third World allies. Standoff from the target would reduce the threat against which the carrier had to be defended. How far inland strikes would be needed could be predicted given likely areas of conflict around the

Eurasian periphery. The study concluded that a 600-to-700-nautical-mile strike range would allow a carrier to attack likely targets while 90 percent of potential enemy attackers would be unable to reach it.

Given recent studies, BuWeps had sufficient data to compare twenty-seven existing and future aircraft, including fighters and five versions of VAX.[55] Only four attracted real attention: a modified A-4 powered by a TF30 turbofan (which Douglas called the A4D-6), a drastically modified FJ-4 (which North American called the FJ-5), Vought's V-461, and a stripped-down Intruder. The two best, indistinguishable in cost-effectiveness terms, were the modified A-4 and FJ-4. The stripped-down Intruder was too expensive. V-461 was not fully evaluated, because it was offered late in the process. The fleet was already used to the Skyhawk, and much fleet equipment was already adapted to it. The BuWeps comparison recommended buying the A4D-6.

That alternative combinations of avionics and ordnance were considered suggests that expected bombing accuracy was taken into account. References to the relative value of supersonic dash suggest that there was an explicit estimate of how probable it was that airplanes would survive to get to the target, probably in the face of the new Soviet-supplied SA-2 missile. The study concluded that supersonic performance was too expensive and bought too little: three light, subsonic attack aircraft could be bought for the price of one supersonic VAX, which would not be anything like three times as effective.[56] Such an airplane would take too long to develop, and it would not be much less vulnerable.[57] Although the A-4E was currently adequate, it would not meet likely future requirements. The study favored a non-afterburning turbofan in a modified version of an existing airframe.

The new attack airplane was designated "VAL," to distinguish it from VAX. McNamara approved the decision the day he received the study report, 17 May 1963. The Navy immediately issued Specific Operational Requirement (Follow-On Light Attack Aircraft) W11–26. It specified the TF30 turbofan

engine and called for the simplest possible (preferably existing) avionics pending completion of the current project for an Integrated Light Attack Avionics System (ILAAS). There would be a two-seat trainer version.

Type Specification 157 was issued 14 June 1963. The competitors were the Douglas A-4F (A4D-6), the North American NA-295, Grumman G-128G-12 (-10 had been offered for VAX), and LTV V-463, of which only the Grumman entry did not use the TF30 engine (like the Intruder, it had two J52s).

The A4D-6 had been proposed during the VAX competition. According to a BuWeps comparison sheet, it was lighter than its competitors (21,553 pounds on takeoff), carrying 8,850 pounds of internal fuel. It met the half-hour loiter requirement, but unlike the others it needed two 300-gallon drop tanks to reach 960 nautical miles. With five pylons, maximum external load was 7,000 pounds. Avionics was essentially those of earlier Skyhawks, with Doppler navigation. Sea-level speed would be Mach 0.92.[58]

Grumman's G-128G-12 was a single-seat derivative of the Intruder, with a smaller radar. The required pair of 20-mm cannon was in the nose under and abaft the main radar. The airplane would have had six wing stations plus a centerline station (the model showed a large drop tank there).

North American offered FJ-4 "Attack" (it sometimes referred to the modified FJ-4 as the FJ-5). No brochures describing the other VAL alternatives seem to have survived. North American offered a flying prototype by May 1964. The airplane resembled the FJ-4, except that it had a radome protruding from the lower lip of its nose air intake, carrying an APN-149 terrain-clearance radar (which could also be used for air-to-ground ranging and ground mapping). Avionics included Link 4A (for automatic carrier landing) and also the Marines' TPQ-10 precision bombing system. There would have been five stores stations, with an option for two more, each of which could each carry the standard Navy ALQ-31 ECM pod. Span would have been 39.33 feet (27.5 feet folded), length 40 feet, and height 16.4. The engine would have been a TF30 (JTF 10A-8). North American claimed that it would meet the load-factor requirement.

DCNO (Air) apparently considered it politically impossible to award the contract to Douglas without competition, although it was widely believed that Douglas was the likely winner.[59] That was partly due to congressional unhappiness about the award of the TFX contract to General Dynamics (in Texas) rather than to the more experienced Boeing; Texas was the home state of Vice President Lyndon B. Johnson. Remarkably, there was to be no similar outcry when LTV (Ling-Temco-Vought), which was also in Texas, won the VAL competition.

LTV certainly came from behind. Initially it seemed that its V-463 was a distant third behind the A-4 and the upgraded FJ-4. The company was highly motivated: its F8U (F-8) program was tailing off, with nothing else in sight. Its marketer argued that the Navy would have to open the competition, because an A-4 powered by a TF30 would be virtually a new airplane. Vought saw the upgraded A-4 as its main competitor. Since VAL had to be a modified version of an existing design, LTV's airplane had to have a family resemblance to the F-8 fighter, although otherwise it could be a new airplane, powered by the TF30 engine. That meant a single chin air intake, a high swept wing, and a low horizontal tail. The new airplane could not be as light as a Skyhawk, but it could have minimum cost and spotting factor. It had to be more maintainable than the F-8. Since nothing resembling an F-8 could be inexpensive, LTV decided to emphasize performance beyond what was specified. The company's cost-effectiveness study led it to use eight rather than six external weapon pylons and 10,000 rather than 6,500 pounds of internal fuel. Vought claimed that similarity to the F-8 would make the airplane simpler to design and support, and it cited the F-8's flying qualities and carrier suitability.[60]

In keeping with its new regime, the Defense Department undertook its evaluation in two stages: first to choose the best design and second to decide whether it was cost-effective enough to be worth developing and building. The first stage was elaborate. In addition to the usual two attack missions (close air support and interdiction) it considered air-to-air combat. Eighty-five range/payload combinations were tried, presumably because computers now made that possible (in the past, evaluation had been based on ten range/payload combinations).

The capabilities of the North American and LTV designs were very close, the Douglas design's less than either. LTV's design was substantially less expensive than North American's.[61] In November 1963 BuWeps issued its engineering report: the LTV airplane was "clearly superior" to the existing A-4E.[62] The report was approved by the Secretary of the Navy on 13 November, LTV's selection was formally announced on 11 February 1964, the contract was let on 19 March 1964, and the mock-up was approved on 22 June. Congress became involved, because the Defense Department asked that $34.4 million in FY64 money be reprogrammed to begin work (about $15 million in emergency funds had already been spent).[63] The airplane was designated the A-7 and named "Corsair II."

V-463 was little modified. It retained the two cheek Sidewinder rails of the F-8 fighter. The four pylons Vought offered on each wing were reduced to six total, because the Navy questioned whether stores would be sufficiently separated.

VAL was intended to have the same simple avionics as an A-4, the most important improvements being a much better radar (APQ-99, soon replaced by APQ-116) and Doppler navigation. The radar display could show a Walleye picture and ECM data. An analog bombing computer was fed airspeed (estimated from air pressure), attitude (from the stable inertial platform), and acceleration. It was designed to handle dive and LABS (toss) attacks, with adjustments for dive angle, slant range to the target, and airspeed. However, the system could not measure wind, because it had no automatic means of comparing movement through the air and over the ground. Weapon release was automatic. The only command indicated on the optical sight was "Pull up." No single display integrated the information

An A-7A from training squadron VA-122 at Naval Air Station Lemoore, California (the "NJ" tail code indicated RCVW-12 and, later, Pacific Fleet land bases). This version could be distinguished by its 20-mm cannon in the nose and by the ECM radome at the base of the tail fin (without the ECM fairing that in other versions was atop the rudder). Behind the pilot are, left to right, the UHF radio mast and the stubby radome of the radio direction finder. The short blade antenna below the fuselage is for TACAN, with an IFF antenna abaft it in a radome. Note the rail against the fuselage for a Sidewinder defensive missile. The bombs are Snakeyes. According to a 1967 data sheet, early A-7As like this one had the ALR-15 radar warner, which triggered a chaff dispenser. The dark area in the middle of the wingtip was presumably also for an ALR-15 receiver, as in early A-6As. In theory the A-7A could also carry a variety of countermeasures pods, such as ALQ-31 and ALQ-76 on its pylons. ALQ-31 could accommodate the two standard self-contained track-breaking jammers, ALQ-41 and -51. In contrast to an A-4, carrying one such pod would not for an A-7 entail a great sacrifice in bomb-carrying capacity. Photos of A-7s carrying such pods have not, however, emerged. U.S. NAVAL INSTITUTE PHOTO ARCHIVE

the pilot needed, and the pilot could not hope to do that in his head.[64] Even so, the Board of Inspection and Survey later credited the A-7A with a bombing accuracy of 10.3 mils, compared to between thirty and fifty for the A-4E.[65] That degree of accuracy seems to have been unexpected; in the development stage the unintegrated analog system was credited with a system accuracy of 20 to 40 mils, compared to a potential accuracy of 10 for an integrated digital system.

The pilot had an attitude display indicator and a horizontal status indicator. The digital position indicator produced by the navigation system could point a light at the current aircraft position on a cockpit roller map. There was a Link 4A receiver (ASW-25) for carrier-controlled landing and also for ground-controlled bombing.

Meanwhile, BuWeps was working on the Integrated Light Attack Avionics System. Sperry was prime contractor. ILAAS would use four digital computers to gain the necessary flexibility in attack tactics and choice of weapon. Different radars could be used as inputs. An inertial system would provide continuous position and speed data. The air data computer would indicate the track ordered by the controls, and the difference between that and the actual track would give current wind. The system would sense the motion of the airplane, which would become an input into its bombing solutions. Once a pilot selected an aim point, the system would provide a continuous bombing solution for any of a variety of tactics. The computers would present pilot commands (and key data) on a head-up display, freeing the pilot from any need to consult his instrument panel dur-ing an attack. At least in principle, ILAAS offered both much better accuracy in attack and much more flexibility in tactics. The OSD Director of Defense Research and Engineering approved the ILAAS program in April 1963, during the Sea-Based Air Strike Forces Study.

BuWeps did not want to repeat the TS-149 experience, in which the successful airframe contractor

An A-7B (BuNo 154370) shows China Lake's Walleye missile under a pylon, with Snakeyes outboard. Abaft the pilot are the short TACAN antenna and the tall UHF communications antenna, with the short dome of the airborne radio direction finder farther aft. The small bulge above the "Danger" sign near the nose is part of the ALQ-100 track-breaking ECM system introduced in this version. Another small radome is barely visible in the lower part of the air intake. Other antennas were in the tail. This preserved airplane is on board the museum ship *Midway*. DR. RAYMOND CHEUNG

had also been the weapon-system contractor. This time it hoped that the airframe builder would integrate a complete ILAAS into later versions of its bomber. The 201st airplane, the first A-7B, was to use some version of ILAAS. However, ILAAS turned out to be far too massive for the A-7.

Pilots found the A-7A underpowered and less maneuverable than the Skyhawk. Heavily loaded airplanes slowed entire strike groups. The rolling-map display proved unreliable. The flight-control system, similar to that of the F-8 fighter, proved vulnerable, because its duplicate elements were routed together; in later versions the hydraulic actuators were armored and the hydraulic lines rerouted. The automatic weapon-delivery system had reliability problems. Catapult steam entering the low air intake could cause brief compressor stalls and loss of power during launch. The engine was modified, and catapult tracks sealed more completely.

The Navy was also interested in more power. By October 1965 it was discussing adding a low-power afterburner. It decided to adopt the P-8 rather than the original P-6 engine and to add an afterburner for takeoff only. Aircraft with this engine were designated A-7Bs (they also had better ECM, in the form of an ALQ-100 track breaker [false-target transmitter] and an ALE-29 chaff dispenser). Unfortunately, the engine developed severe turbine cracks, so all P-8 engines were grounded in January 1968 and all deliveries stopped.

A projected two-seat A-7C was abandoned, the Navy buying the TA-4F instead.

Having become involved in the program in a roundabout way, the Air Force introduced an alternative engine, the British Spey, which was license-produced as the TF 41.[66] The Navy considered an upgraded TF30, but ultimately it too adopted the Spey. The Air Force and Navy versions of the higher-powered A-7 were, respectively, the A-7D and -7E.

The Air Force also initiated the other great change in the airplane, a computer attack system. It became a truly joint effort. The Air Force leadership explicitly rejected an integrated system, presumably

An A-7C assigned to China Lake carries a HARM missile on one of the underwing pylons, 22 December 1977. The A-7C was effectively an A-7E without the new TF41 engine. Note the bulge of the Gatling gun below the angled danger lines on the air intake and the antenna housing above the upper of the two lines (with the letters "JE," of the word "Jet," just below it). Compared to the arrangement on the A-7A or -7B, the topside antennas have been considerably rearranged. Those closest to the cockpit are for missile-launch warning (APR-27); the mast much farther aft is for TACAN and the data link associated with the digital weapon system. Note the bulge of the downward-looking Doppler navigational radar just abaft the nosewheel door. The usual strike camera farther aft is missing. U.S. NAVAL INSTITUTE PHOTO ARCHIVE

for its cost; rather, it wanted a better analog system, using a servo (computer-pointed) sight to achieve an accuracy of fifteen to twenty mils.[67] A pilot looking through the sight would fly the path ordered by the bombing computer. By this time the Navy had accepted that ILAAS would never be fitted in an A-7. Some in the program office feared that ILAAS could sink the A-7 program itself. In July 1966. Cdr. Robert F. Doss was assigned as Navy deputy project manager to the joint Navy–Air Force A-7 office. He had previously worked on conventional weapon delivery at China Lake.[68] Doss convinced his Air Force counterpart, Col. Robert E. Hails. that the analog system would not work; Hails agreed to press for an integrated digital system.

The project benefited from rapid computer development. DDR&E was impressed by the shrinking size and growing reliability of microelectronics, which promised its desired quantum enhancement in A-7 system performance. It ordered a joint study of avionics improvements. It seemed that integrating a digital computer with an attack radar (to give accurate slant range) and a head-up display could provide ten-mil accuracy (CEP 140 feet from 10,000). The A-7E attack system was built around a multimode APQ-126 radar and an ASN-91 tactical computer, which produced the video for the head-up display as well as the horizontal and attitude displays.[69] A much-improved map display was projected onto a screen below the radar screen, where it could easily be compared to the radar image. The A-7E was the most accurate Navy bomber to date, managing four to six mils during Armament Service Acceptance Trials in 1970.[70] Perhaps the most important consequence of adopting an all-digital system was flexibility. It proved relatively easy, for example, to add a self-contained FLIR pod on one of the weapon pylons, its display piped to the cockpit radarscope. This image could also be shown on the head-up display.

Navy aircraft also had the M61 Gatling gun the Air Force preferred.

Production Speys were not immediately available, so the first digital Navy aircraft had TF30s. They were designated A-7Cs.

The most important improvement in the A-7E was its digital computer, which enabled it to exploit entirely new sensors, like the FLIR reconnaissance pod hung from its inboard pylon. This VA-72 A-7E was taxiing onto no. 2 catapult position for launch from the carrier *John F. Kennedy* during Fleet Ex 1-90. Note the "low-vis" paint scheme, with national markings in a slightly contrasting shade. The bulge visible just abaft the nosewheel door was for a downward-looking Doppler navigational radar. The object just visible abaft the landing gear wheel aft is the housing of a strike camera. The two antennas above the body are, respectively, a small one for the APR-27 missile-launch warner and a larger one for TACAN and a data link. The small antenna just forward of the nosewheel door is also for APR-27. Just visible above the right end of the upper air-intake warning stripe is an antenna of the ALQ-126 jammer. There is no UHF mast, because UHF antennas have been relocated to the top of the vertical tail, with a Loran array on the leading edge of the vertical tail (not visible here). U.S. NAVAL INSTITUTE PHOTO ARCHIVE

Although the A-7 was optimized for conventional bombing, with much greater capacity than an A-4, its nuclear mission was still important. This A-7E shows a simulated B61 bomb outboard of a FLIR pod. Its tail code and squadron identification have been censored, presumably in accord with a policy of never identifying which units were nuclear capable. EMIL BUEHLER LIBRARY, NATIONAL MUSEUM OF NAVAL AVIATION

An A-7E on board the carrier *Coral Sea* in the Indian Ocean, April 1980, showing its attack radar. The dark patch under the side number (303) includes the muzzle of the single Gatling gun, the bulge for which is visible slightly farther aft. Note the bulges on either side of the air intake. One was part of the ALQ-126 system, the other part of the ALR-45 radar-homing-and-warning system. The waveguide visible on the side of the airplane, part of which is under the Sidewinder missile, connects nose and tail parts of ALQ-126 (at the base of the rudder, not visible here, was the radiating part). The object atop the rudder was part of ALR-45, with a spiral antenna on either side. NHHC

In company with an A-6E, an A-7E (*City of Olongapo*) shows the bulge of its Doppler navigational radar (under the gap between the rows of bombs) and the angular housing of its strike camera (just at the after end of the shadow cast by the wing). Both aircraft were from CVW-9 on the carrier *Constellation*.
U.S. NAVAL INSTITUTE PHOTO ARCHIVE

A-7s on board *Constellation* line up to launch in the Gulf of Tonkin, 9 May 1972. Carrier air wings typically included two light attack squadrons and one heavy attack squadron; these aircraft are from VA-146 and VA-147. NHHC

VFAX

The 1963 Sea-Based Strike Forces Study distinguished between VAL (A-7A) and a future VAX. In June 1963 DCNO (Air) issued a tentative SOR (TSOR) for a "High Performance" General Purpose Attack Aircraft (W11-06T). Both single- and multirole aircraft should be studied, as well as conventional and VTOL types. BuWeps bought studies from three aircraft manufacturers and one engine manufacturer. Between March and November 1963 it studied trade-offs, as well as airborne target acquisition, attack (advanced bombing techniques and terrain avoidance), and ILAAS.

In 1965 BuWeps had the required quantum improvement in mind: to combine the fighter and attack missions in one "VFAX" airplane—not merely a common airframe but a single airplane. It would replace the A-6 from 1972 on. The key was a truly multimode radar.[71] Different missions required very different radar waveforms. Ranging and ground-mapping radars needed moderate pulse rates (which set maximum unambiguous range). For air-to-air combat, and particularly for the ability to look down, very-high-pulse-rate Doppler radars were best.

In past radars, the dimensions and shape of the power tube (magnetron or klystron) had dictated the signal sent out. In a new kind of radar, a wide-band device (typically a "traveling wave tube," or TWT) amplified a weak computer-generated signal shaped by software. This new concept transformed the ongoing study. A single airframe with the same avionics could be both a fighter and an attack bomber. As of February 1966, Hughes Aircraft was negotiating a contract to modify its existing APQ-72 fighter radar into a high-resolution, low-pulse-repetition-frequency (PRF) radar. By June 1967 a flight-tested feasibility model was expected of the multimode radar using an advanced antenna, a coherent transmitter (i.e., computer-generated waveform and TWT), a beam-sharpening mode, automatic terrain following, and terrain-avoidance and target-designation circuits. Ideally the radar would translate what it received into digital terms that could be processed by another computer and fed into the airplane's central computer. BuWeps hoped to limit risk by marrying the ongoing multimode radar to applicable parts of the 1967/1970 ILAAS attack-aircraft-avionics package.

True multimission aerodynamics seemed to require the flexibility offered by a swing-wing and lighter-weight engines, ideally turbofans for long range. Fortunately, the Air Force was already investing heavily in a new-generation turbofan to be the lift/cruise engine of a projected VTOL aircraft, for which light weight was essential. Through a significant increase in compressor-blade loading, the number of compressor stages would be approximately half that of the TF30 engine in the F-111 and the A-7. New forced-air-film/convection-cooling techniques would allow operating temperatures about four hundred degrees Fahrenheit hotter than in existing engines, which would greatly increase efficiency.

The BuWeps Advanced Systems Office conducted computerized simulations as a basis for cost and effectiveness studies, especially a two-carrier five-day attack on shore targets four hundred nautical miles away (a model of a two-carrier support of an amphibious operation seems to have been less important).[72] The main conclusion of the strike analysis was that VFAX would buy much more capability at a small increase in price.

There was considerable interest in a VSTOL version in other contexts, and the Marines wanted a STOL airplane to operate from relatively short airstrips. In January 1965, BuWeps issued its report R-5-65-1, "An Evaluation of Attack and Attack/Fighter Aircraft for the 1970s." DCNO (Air) issued Specific Operational Requirement 11-06 on 8 October 1965 (the SOR seems to have originated in September). A Preliminary VFAX Technical Development Plan (TDP 11-06) was dated 18 February 1966.[73] By this time the multimode radar dish was expected to have a thirty-two-inch diameter, which would set the width of the airplane; the crew would accordingly be seated in tandem. Unlike earlier aircraft, in this one the electronic warfare systems would be fully integrated into the combat system. Because the TDP was written in 1966, as Standard ARM was becoming a vital anti-SAM weapon, it included explicitly an ARM targeting capability. VFAX would be capable of operating from modernized *Essex*-class carriers.

Earlier studies had shown that to reduce drag, stores should be carried either semisubmerged or tangentially (not on pylons); the envisaged maximum load was twenty-four Mk 81 bombs and two Sparrows. Initially airplanes would have a one-way data link so that targets could be designated to them; later a two-way link might allow them to download reconnaissance data. Of existing aircraft, only the F-111B and the F4J approached VFAX in multimission capability, and in both of those the air-to-air aspect was dominant.

Maximum speed would be Mach 2.2 (sea-level speed at least Mach 0.85), and combat ceiling (air to air) would be 58,500 feet (cruising altitude 36,000 feet). Acceleration time from Mach 0.8 to Mach 1.8 was 3.3 minutes (later reduced to 2 minutes). Later a requirement for buffet-free maneuverability at 35,000 feet at Mach 0.85 and at least 2 *gs* was added. Spotting factor (compared to the A-4) would be about 1.25. Mission-success probability was to be 80 percent (90 desired).

Designed service life would be 6,000 hours, and in peacetime the airplane would fly 50 hours per month (this figure became standard for U.S. tactical aircraft). The uninstalled weight of the entire avionics package would be 2,366 pounds (in a 66.2-cubic-foot volume), a significant improvement over existing systems. Special efforts would be made to improve system reliability, the goal being an MTBF of 33.2 hours.

In November 1966 the DCNO (Air) planning chief called for air-to-air capability superior to that of the current F4J and air-to ground capability superior to the "design goals" of the A6A Intruder. DCNO (Air) warned against trying to replace the A6A; VFAX should complement it.[74] The F-111B was already in trouble, but DCNO (Air) cautioned against adapting VFAX to fire that program's heavy Phoenix missile as an alternative. Others argued that since the Navy might be forced to choose between the F-111B and VFAX, it would be unwise to ignore the heavy missile in the VFAX project. In October 1966 the two F-111B/Phoenix contractors, Grumman and Hughes, completed a contract study of arming VFAX with it.

Alternatives considered late in 1966 were VFAX (two-place, two-role), VFX (one-place all-weather fighter), VAX (one-place attack), and two-place variants of VFAX with day-fighter or attack capability. There was also a version sized for long-range interdiction.[75] Minimum time should be needed to switch between air-to-air and air-to-ground missions, ideally less than an hour and a half.

An OSD attempt to merge VFAX and the Air Force next-generation FX fighter study (which became the F-15) showed mainly that they were incompatible. The Navy wanted considerably greater radar capability. Because it was multipurpose, VFAX required a general-purpose digital central computer instead of the Air Force's lighter analog one. The Navy demanded radar-guided air-to-air missiles rather than shorter-range IR ones. OSD directed the Air Force to proceed with an engine suitable to both services and the Navy with avionics. Progress in avionics was slow, partly because ILAAS had no specific line in the budget (usually avionics was paid for by a particular airplane program). The Navy feared that once the Air Force had a definite design, it would overwhelm the Navy making VFAX appear to be a poor choice. The Navy created a special task group that sponsored the sort of studies OSD would accept. This effort explains the level of detail in the May 1967 SOR.

The Sea-Based Air Strike Forces Study had shown an interrelationship between the ability to conduct air-to-air offensive and defensive operations, on one hand, and air-to-surface attack on the other. Various studies had already shown that the F-111B was useless for fighter roles other than fleet air defense.

The May 1967 SOR restored the long-range nuclear strike mission. Close-support combat radius would be the usual three hundred nautical miles with thirty minutes of loiter at ten thousand feet and (a new requirement) five minutes of combat at sea level. The load would be either nine Mk 82 bombs or two air-to-air missiles plus six bombs. Interdiction combat radius would be six hundred nautical miles, including a two-hundred-nautical-mile run out and run in at, respectively, Mach 0.8 to Mach 0.85; there would be five minutes of combat at the target. The load would be twelve Mk 82s or eight Mk 82s and two air-to-air missiles. For the nuclear mission (one Mk 43 bomb), combat radius would be the usual thousand-nautical-mile including a four-hundred-nautical-mile sea-level run in and out and five minutes of combat. Ideally drop tanks would not be needed.

The airplane would be able to attack in all weather. Its multifunction radar would provide both "A-6 like" and fighter capability. As in the A-6, the bomb/nav system would provide for visual and all-weather bombing against prominent fixed and moving targets, presenting the target as an aim point. Overall bomb-delivery accuracy should be no more than ten mils, with six mils desired. Capacity would be at least twenty-four free-fall weapons and two air-to-air weapons. The pilot would have a head-up display to facilitate transitions between all-weather and visual aircraft and weapon control. Specified error in the integrated navigation and attack systems should be no more than 1 percent of distance flown (one-half of 1 percent was desired), with provision for inflight correction using visual and radar aim points.

The airplane and its pilot would be protected against nuclear light and heat flash. The airplane would minimize radar and IR signatures and deploy IR countermeasures. It would be capable of buddy tanking. Every effort should be made to make the airplane quickly repairable after light combat damage. Since it would spend much of its life below five thousand feet, it needed a high "fatigue life" (i.e., before its structural strength degraded), as well as the ability to fly at high speed and deliver weapons despite high gust loads. Spotting factor should be no more than 1.2, and a maximum gross takeoff weight of 45,000 pounds was a design goal.

NavAir hoped to provide sufficient documentation to insert VFAX into the Defense program by 1 June 1967. It would have to be distinguished not only from FX but also from a combined U.S.-German

VSTOL of similar size (but with avionics traded off against its greater propulsion weight). The existence of this latter project explains why VTOL considerations appear so often in VFAX papers.[76] The VSTOL was not expected to complete its tests until mid-1972, so production quantities would not be available until 1974–75, about two years after the planned fleet introduction of VFAX. In the event, it never became operational.

The Navy hoped that the first three airplanes would be procured under the FY68 budget, the next three under FY69, then sixteen under FY70; total procurement through FY77 would be 1,164 airplanes, to form thirty VFA squadrons, eight VFS squadrons, fifteen Marine (VFMA) squadrons, and the usual training units. Procurement would be completed at the end of 1979.[77] By this time the rising cost of the F-111B had cut expected purchases to between six and eight aircraft per carrier (down four to six from the figures in the Five Year Defense Plan).

The NavAir Aerodynamic Design Section produced a reference design against which industry proposals could be evaluated. It had two engines derived from GE studies, and its airframe was based on the sketch designs various companies had already submitted.[78] Maximum weight was set at 45,000 pounds with four Sparrows on board. Up to twelve Mk 82 Snakeyes could be carried on individual (hence low-drag) racks under the airplane's body. Two pylons might be mounted on the gloves of the swing-wings, and there might be a centerline position for a 450-gallon drop tank. The radar antenna would be thirty-two inches in diameter, and the canopy would be shaped for low drag. Maximum speed was Mach 2.2, but there was interest in increasing that to Mach 2.4. A graph showed range desired under various weight conditions for an attack mission with twelve Mk 82 bombs and 13,500 pounds of internal fuel. With five minutes at sea level, radius would be 800 nautical miles; including a sea-level dash at maximum speed would drop radius to 650 nautical miles; requiring an hour's loiter would cut it to about 600. Another graph showed an attack mission with six bombs and a 450-gallon belly tank;

maximum radius would then be 1,100 nautical miles. In the long-range strike mission, VFAX would carry a single Mk 43 thermonuclear bomb plus 6,120 pounds of external fuel. In this case radius would be 910 nautical miles.

Thus the supersonic VFAX offered a better radius than the subsonic A-6 and A-7; only the A-7 was smaller and weighed less. Its multimode radar ensured that VFAX attack capability would be vastly superior to that of the F4J and better than the F-111B's for the basic interdiction mission. Further microminiaturization (integration of circuits) would increase reliability and reduce weight and allow the system to warm up much more quickly.

Grumman and NavAir were still trying to modify the big F-111B enough to make it an acceptable carrier fighter, though it was widely understood that it could not achieve any sort of dogfighting performance. NavAir produced vu-graphs that asked the obvious question: If a version of VFAX could carry and employ the Phoenix missiles planned for the F-111B, was that large airplane really necessary? A comparison of alternative carrier complements showed that if a version of the VFAX could be modified to carry Phoenix missiles, the carrier would not even need any A-7s, let alone F-111Bs.

By late 1967 it was clear that the naval aviation community saw little point in the F-111B program. Secretary McNamara, exhausted by the demands of the Vietnam War, would resign on 29 February 1968. Late in 1967 VFAX was dropped in favor of an Advanced Fighter (VFX) that would replace both F-111B and VFAX, meaning that it had to accommodate the former's long-range Phoenix air-to-air missile. Several companies offered designs. Grumman won with an evolved G-303 that became the F-14 Tomcat. The Tomcat's VFAX heritage showed in its multifunction radar, the first in any airplane.

For Navy attack airmen, Vietnam was very much an electronic war, pitting them against surface-to-air missiles and ground-controlled interceptors. This A-4E "super Echo" (because it was reworked, including installation of the hump) of VA-106 on board USS *Intrepid* is about to launch in the Gulf of Tonkin, September 1968. Because the "scooter" had limited internal fuel stowage, aircraft typically carried a large drop tank on the centerline, which left the outboard pylons for weapons. The airplane had provision for in-flight refueling via the prominent right-side probe. The pointed antenna under the nose was part of the ALQ-51 defensive jammer; reworked aircraft also had ALQ-100 jammers. NHHC

11

VIETNAM

Unlike World War II, the Vietnam War profoundly affected later U.S. Navy attack aircraft development. The reason was that this time there was no great technological revolution immediately after the war. Lessons learned (or mislearned) during the war therefore continued to be directly relevant afterward.

For American airmen, probably the key tactical reality of the Vietnam War was rooted in the strategic belief in Washington that Vietnam was only one of a series of insurgencies the United States would face around the Eurasian periphery. It had to be fought with limited resources and without trying to mobilize American public opinion—"in cold blood." No one in Washington understood that the war was unique: that the North Vietnamese had effectively blackmailed the reluctant Soviets into supporting it. They played the Soviets off against the Chinese, threatening that if Moscow failed to support the war, Beijing would win control of the world revolutionary movement. The same deep Sino-Soviet rivalry led the Soviets not to supply the North Vietnamese with their best antiaircraft systems, for fear that they would fall into Chinese hands.[1]

The Lyndon Johnson administration was much affected by nuclear theorists who saw military moves as signals in a larger confrontation. Once the United States began its bombing campaign, Secretary of Defense McNamara demanded control over the details of target lists. President Johnson once boasted

that U.S. forces could not "bomb an outhouse without my permission," but that really meant McNamara. McNamara and his deputies sometimes even set days and times for attacks and chose the numbers and types of aircraft involved. This micromanagement was possible because long-haul communication was far better than in the past. Fear of escalation drove extraordinary efforts to avoid attacking Soviet and Chinese personnel in North Vietnam working on the air-defense systems or repairing and maintaining railroads and ports.

For attack pilots and aircrew, this was a war against a Soviet-style integrated air-defense system. For the Soviets, its success or failure had implications for the effectiveness of their own strategic air defense, even though what they provided the North Vietnamese did not represent their best equipment.

Naval aircraft were engaged mainly against North Vietnam rather than communist forces in the South.[2] There were two main campaigns. Rolling Thunder began in 1965, largely ending on 31 March 1968 when President Johnson announced negotiations with the North Vietnamese. For months thereafter strikes continued against areas in southern North Vietnam from which an invasion of the South could have been mounted (a zone known to strike planners as "Route Package I"), but all bombing of North Vietnam stopped on 31 October 1968. Johnson's successor, President Richard Nixon, reopened the bombing campaign in 1972 as Operation Linebacker, with much less restriction. This campaign was suspended during peace talks, then resumed (as Linebacker II) when it became clear that the North Vietnamese were stalling. Linebacker II included attacks on Hanoi itself, a denied target through the earlier part of the war (the Air Force called these attacks "going downtown"). The previously denied port of Haiphong was mined.

The pause between the Rolling Thunder and Linebacker campaigns allowed the Americans to field technology developed under considerable pressure during Rolling Thunder and also to improve their tactics. The North Vietnamese may have improved their tactics as well, but due to the technology transfer restrictions imposed by the Soviets they had no real hope of obtaining anything new.

Strike Tactics

The Navy contribution to Rolling Thunder comprised attacks mounted from Task Force 77 carriers operating on Yankee Station in the northern Tonkin Gulf. Normally each of three carriers on station conducted air operations for twelve hours and spent the next twelve repairing and replenishing for their *next* twelve, so that targets were covered continuously. Carriers either launched deck-load (Alfa) strikes for shock effect or to penetrate heavy defenses or operated "cyclically," launching and recovering twenty-five to forty aircraft every hour and a half (eight cycles, or "events" in twelve hours). Aircraft from one attack would refuel and rearm while the second was in the air, then launch before the second recovered. In this mode fifteen to twenty aircraft could be committed to one target. Sometimes as many as five carriers operated together on Yankee Station (at the peak of the 1972 offensive there were seven). The carriers' offshore position was selected to minimize the time strike aircraft spent over North Vietnam ("feet dry"). Some jamming aircraft orbited offshore, out of reach of North Vietnamese interceptors.

The air wing on each carrier was self-contained; the pilots knew each other, briefed together, and talked tactics together. The Navy also valued simplicity and flight discipline. Strike packages were small and tactics as simple as possible, to lessen the usual fog and friction of war. For the Air Force, tactics were much less flexible and loss rates much higher.[3] Navy pilots flew about two-thirds of the 1972 Linebacker sorties but suffered only 41 percent of the losses. As an example of the integration of various roles, on 10 May 1972 USS *Constellation* launched a thirty-three-airplane Alfa strike: sixteen bombers (ten A-7s and six A-6s), a nine-Phantom CAP to protect them from MiGs, four Phantoms for flak suppression (bombs and Rockeye cluster bombs), and two A-7s for SAM suppression, armed with Shrike missiles.

From 1967 on, the Navy maintained a "strike support ship" (SSS) on station in the Tonkin Gulf near a radar buoy code-named Red Crown (the ship was sometimes called "Red Crown" as well, by extension).[4] The ship could warn strike groups against the approach of North Vietnamese fighters. Its area of coverage was called the PIRAZ (Positive Identification Radar Advisory Zone); it worked with airborne-early-warning radar aircraft that could vector strike aircraft toward their coast-crossing points, account for aircraft (to identify enemy fighters among or near them), maintain traffic separation, and arrange rendezvous.

Each strike was assigned a strike controller, who might be considered analogous to a fighter controller.[5] Unlike the "strike leader" (in one of the aircraft), the controller had a broad tactical picture. He could see and track all the strike aircraft, tankers, and other support aircraft, and he could bring them together as necessary. He might also be able to see and track hostile MiGs, as well as "strangers" coming within ten miles of the strike group. He could keep the strike aircraft out of unauthorized airspace, an important consideration given political limitations on operations over North Vietnam and the prohibition against entering Chinese airspace. The SSS normally provided advisory control to the target CAP (TARCAP, protecting strike aircraft while over the target), and could take over close control as requested by the strike leader or directed by the fleet antiair warfare (AAW) controller when MiGs threatened.

All aircrew were briefed before takeoff as to target, route, and tactics. An Alfa strike on a single target mixed aircraft from several squadrons, so it was generally led by the carrier air wing commander. All radio communication to the group went through him, strike aircraft sharing a single radio frequency. RESCAP (Rescue CAP) and MiG CAP (which arrived over the target before the ECM or strike aircraft) might have their own frequencies, typically communicating with the strike controller.

Cyclic strikes were often for armed reconnaissance, in which case aircraft would spend longer over land. Because the strike controller had access to communication from reconnaissance flights and previous strikes and armed-reconnaissance missions, he could identify lucrative targets, enter their locations as reference points into the shared digital tactical data system, and quickly divert cyclic-strike aircraft to them. He normally provided assignments and vectors to armed-reconnaissance and strike aircraft. He could also help bring tankers together with returning strike aircraft.

Given rapidly changing weather and the possibility of aircraft casualties, there were generally alternative targets. The strike controller had prior knowledge of these plans and routes. When the strike leader was informed of a change via a code word, the controller assisted him in responding. He diverted the MiG CAP to cover the new target and pointed out the relevance of general warnings to his strike group. For example, a general warning of a MiG concentration might cite a geographic position. The strike controller would translate that into a range and bearing from the strike group. The same applied to other kinds of warnings whose import might not be immediately recognized by the strike pilots.

Marine aircraft, mainly at Da Nang, contributed to air attacks inside North Vietnam, in the Demilitarized Zone (DMZ), and in the southern part of the county.

Integrated North Vietnamese Air Defense

At the outset, in July 1964, the North Vietnamese had a rudimentary air-defense system not too different from what the United Nations faced in Korea in 1950. Lacking the large fighter force the North Koreans built up, they relied heavily on antiaircraft guns effective up to about five thousand feet (but with few fire-control radars).

The slow process of American escalation gave the North Vietnamese time to build up a sophisticated national air-defense system.[6] Early in 1965 the Soviets agreed to provide them with SA-2 missiles (see chapter 9). The first SA-2 site, fifteen miles southeast of Hanoi, was identified in a photo taken on 5 April 1965 by an RF-8 Crusader from USS *Coral Sea*. The administration vetoed a planned joint Navy–Air Force strike to destroy it before it became operational, on the grounds that an attack might kill Soviet technicians. Its first success (against an Air Force RF-4C) was credited to a Soviet lieutenant colonel, F. Illinykh, on 24 July 1965. The ban against attacking sites before they became operational persisted nevertheless, making anti-SAM measures vital.

Between May and December 1965 two full regiments of SA-2s were formed; fifty-six SA-2 sites (not all occupied at the same time) were built around Hanoi. Antiaircraft artillery (AAA) grew from twelve regiments and fourteen battalions to twenty-one regiments and forty-one battalions (eight of the latter mobile). The air force grew from one to three regiments of MiG-15/17 subsonic fighters and also received supersonic MiG-21s.[7] Air defenses grew further during a cease-fire between Christmas 1965 and January 1966, fighter strength increasing to 125 and the SA-2 network growing by a third. An increasingly effective command-and-control system was evidenced by the first night intercept, of Marine Corps F4Bs, by MiG-17s (which had no inherent night capability) on 3 February 1966. Later that month the North Vietnamese netted multiple radars to enable an SA-2 to shoot down an RB-66C, which had been fitted specifically to detect and jam the missile system. In November 1966, the North Vietnamese introduced what the United States called "SAMbush" tactics: a MiG-17 or a transmitting SA-2 site lured U.S. aircraft into overlapping SA-2 engagement zones, the missile systems silent, using optical trackers only, until they fired. In one case, when the U.S. aircraft arrived in the engagement zone, six missiles were fired and a flight of four MiG fighters vectored in. This ploy failed, but it demonstrated a high level of sophistication. It was possible because the North Vietnamese fighters had been fitted with IFF, so that their air-defense system could distinguish them from U.S. aircraft. In December 1966, the North Vietnamese began using barrages of SA-2s, a technique that might be read as a response to jamming. At the least, proliferation of SA-2s forced back barrage (noise) standoff jammers (such as the EKA-3B), reducing their effectiveness.

In 1970 the Defense Intelligence Agency reported that the North Vietnamese had 471 radars (including missile and gunfire control and ground control/interception), 35 to 40 SA-2 battalions, 1,847 antiaircraft guns (37-mm and above), and 257 fighters.

The North Vietnamese system was controlled by a national air-defense headquarters in Hanoi fed by local filter centers connected to radars and observers.[8] The system was credited with the ability to determine within five minutes the direction and probable targets of approaching American formations. The North Vietnamese air-defense headquarters allocated targets and ensured that fighters were used efficiently and also that they were not directed into missile engagement zones. It was assisted by Soviet intelligence-gathering ships that dogged U.S. carriers (they also operated off Guam, where Air Force B-52s were based). The Chinese linked their air-defense system to that of the North Vietnamese, so that no U.S. strike could completely blind air defenses. China also provided a refuge for North Vietnamese pilots, as well as reserve airfields when North Vietnamese fields were put out of action.

At the time the war broke out the U.S. Navy was emphasizing low-altitude tactics, mainly to avoid enemy radar detection. The only low-altitude threat was antiaircraft guns, and the expectation seems to have been that gunners would find it difficult to hit airplanes flashing past them at high speed. Thus strike groups flew at four thousand feet or below, climbing to seven thousand feet at a designated point to identify their target before diving to attack. But the North Vietnamese used their guns—most of which were aimed visually—quite effectively, accounting for about 80 percent of aircraft losses. No effective countermeasure was developed, although there was interest in flashing devices to blind or confuse gunners. Defense suppression was usually a matter of dropping antipersonnel cluster bombs near missile sites in hopes of killing gun crews or at least forcing them to take cover. Since missiles, for their part, forced aircraft into the guns, it might be argued that they were effective whether or not they actually shot down U.S. aircraft themselves.

Guns and missiles defended point targets; fighters operating between these point defenses tied the system together. They prevented attackers from using standoff weapons to break the system. Although they accounted for a relatively small proportion of U.S. aircraft shot down, the fighters helped break up attacking formations and thus frustrate complex forms of jamming involving multiple aircraft. When MiGs vectored by long-range radars made high-speed runs at approaching bombers, the attacking pilots jettisoned their bombs so that they could execute violent evasive "jinks."

As in Korea, enemy interceptors were tightly controlled using voice links. Ground controllers had to ensure that these short-range aircraft did not run out of fuel. In at least one case, a North Vietnamese pilot was so intent on completing his interception that he ignored low-fuel warnings and crashed. Despite claims that the interceptors were far more effective dogfighters than their American opponents, the Vietnamese lost about half their fighter pilots in March–June 1967 dogfights.[9]

The Missile

The SA-2—as known in the West, in the Soviet Union the S-75 Dvina—was first seen in a Red Square parade in November 1957. It was then widely deployed in Central Europe, so basic information was collected fairly quickly. Because it was conceived as a strategic defense against high-flying heavy bombers, SA-2 was ineffective below about four thousand feet. Its command-guidance system turned out to be a key weakness, unsuspected prewar. Initially Navy aircraft were equipped with track breakers, a generic countermeasure. The Soviet tracking technique was very different from Western ones, but the track breakers could handle it.[10]

Pilots learned to exploit system weaknesses, some of which were corrected during the war.[11] Radar warning receivers on attack airplanes could often distinguish the Fan Song's tracking mode from its engagement mode. Fan Song tracking was clumsy, so an airplane that detected it could often evade it. It might take Fan Song as much as forty-five seconds to reacquire its target, by which time the target might well be out of range. War experience showed that the main shortcoming of SA-2 command guidance was the noticeable time lag (about five seconds) between a command to alter trajectory and a maneuver by the missile—so long that if a pilot saw the missile in time, he could often evade it. Too, the booster created a very conspicuous blast, and the missile did not begin maneuvering until it dropped away after about six seconds. Overall, the combination of high speed, considerable size, and small control surfaces limited the missile's maneuverability; an Air Force pilot pointed out that a fighter could take it on just like it would another airplane, outturning it once it had committed itself.[12] The missile might save the attackers' targets, however, even if it missed, by requiring them to jettison their bombs. Preferably, from the U.S. perspective, the system would simply be jammed.

The specified SA-2 kill probability of 80 percent for a three-missile salvo was never approached in Vietnam. In 1966 Russian advisors estimated that American electronic countermeasures had more than tripled the number of missiles required to shoot down an airplane, from one to between three and four. This estimate was about seven times too low. Of 7,658 missiles supplied to the North Vietnamese, 852 were available at the end of the war. In all, out of 5,804 missiles fired in combat, about 600 were defective and over 200 were destroyed in air attacks; others were expended in training. In 1965 it took 9.9 missiles to shoot down an airplane; that figure went to 17.3 in 1966, to 31.6 in 1967, and

34.2 in 1968, the worst performance. Improvements reduced the number to 27.2 in 1971 and to 27.8 in 1972; in 1973 it was 24.0.[13]

During the war fifty-six of the ninety-five batteries supplied by the Soviets were destroyed, despite effective camouflage and the North Vietnamese practice of moving batteries from one site to another.[14]

The U.S. Navy, its carrier capacity limited, did not form specialized defense-suppression units. It saw suppression as a way of making particular operations possible. Any attack airplane could carry and launch Shrike, for example; it was soon clear that Shrike could not destroy enemy radars but that it could shut them down for a time (such missiles were rated in terms of how long repairs would take). It was an aid to penetration. Direct attacks on SAM sites were called "Iron Hand" operations. The first such Navy mission was carried out by four A-4Es led by one A-6 "pathfinder" on 17 October 1965 against a site near Kep Airfield northeast of Hanoi. By early 1967 standard Navy tactics employed two Iron Hand (site-attack) plus two Shrike aircraft to precede or accompany strike aircraft. Known SAM sites were studied and plotted ahead of time. Typically, a Shrike airplane had an APR-23 radar warner on board. It used the APR-23/Shrike direction-finding capability to detect radiating Fan Song missile-guidance radars; the pilot maneuvered into estimated Shrike range and launched.[15]

To protect their "clients'" sites from antiradar weapons, Soviet advisors told the North Vietnamese to limit transmission time. That also made it much more difficult for U.S. electronic intelligence aircraft flying offshore to localize the radars.[16] However, once the radar was turned off, it took too long (about a minute) to warm back up to full power. Systems were accordingly modified with "dummy loads" onto which power could temporarily be switched. There was also a manual tracking technique.

Over Hanoi in August 1967, 66 percent of SA-2s lost control altogether, and 6 percent flew into the ground, causing significant damage.[17] The North Vietnamese shifted to untested track-on-jam techniques, which proved ineffective. During the intense air battles of 24–27 October 1967, they managed to shoot down only 5 U.S. aircraft, although they claimed 22. They shot down 8 more between 17 and 19 November, but over Hanoi on 14 December all SA-2s crashed shortly after launch. The next day the most experienced North Vietnamese missile regiment fired eight missiles, all but one of which crashed. Of twenty-nine fired by another regiment, eleven immediately went out of control. It was apparent to the North Vietnamese that U.S. aircraft could jam the missile's control link. Unfortunately for the Americans, an F-105 carrying the QRC-160-8 pod involved was shot down on 14 February 1968. The Soviets analyzed the pod and modified the missile's transponder beacon by which the QRC-160-8 tracked the missile. Apparently, this change was unknown to the U.S. Air Force, which suffered as a result during the Linebacker II campaign. On the other hand, about 40 percent of the 1,500 SA-2s in North Vietnam had by then been made effectively inoperable, and the system's optical adjunct was useless because missions against Hanoi were flown at night. American losses on the first two nights of Linebacker II can be attributed to both a failure to modernize countermeasures and extremely poor tactics on the part of the B-52s; the bombers did far better on their third strike.

Airborne Countermeasures

As can be seen in the development of the A-6, by 1956–57 the Navy was well aware of the need for protective jammers. Proposed equipment for the A-6 included an ALR-15 radar warner and direction finder. The pilot's indicator lit one of six lights to show that a threat signal was coming from, respectively, up, down, left, right, nose, or tail. As the *L* indicated, ALR-15 also provided information on the character of radars it detected. Its receivers could distinguish a locked-on airborne interception radar from a slowly sweeping search radar. The pilot had two countermeasures, chaff and a new kind of repeater jammer. The latter exploited a broadband amplifier (a traveling wave tube) that enabled it to react instantly to a radar it detected. Grumman envisaged four receiving and four transmitting antennas around the airplane. It argued that size and weight constraints ruled out the alternative tunable noise generator (which the Air Force preferred): such a jammer needed an operator to tune it to the appropriate frequency and could handle only one radar at a time.

It was assumed that a bomber needed two complementary jammers, one handling the S-band signals of ground-based search and fire-control radars,

This A-4E at Patuxent River, 29 August 1967, shows the initial -4E configuration, without the later hump. Under the nose and under the wing root are foundations for ALQ-51 antennas, apparently without the antennas themselves. The small object under the *3* of "130" forward is a forward APR-25 (radar homing and warning) antenna; another is mounted below the "sugar scoop" aft. The objects visible in front of the canopy are a total-temperature sensor and a pitot tube (the bent tube). The major change from earlier versions was an additional outer pylon under each wing. The left-hand one carries a Snakeye bomb, and the centerline pylon carries two triple ejector racks (with similar bombs) in tandem. The weight limit for the centerline pylon, 3,575 pounds, was an artifact of the original nuclear mission. For the inner pylons the limit was 1,200 pounds; for the outer ones, 500. NHHC

An A-4F of VA-144 from *Bon Homme Richard* (so spelled) over the Pacific, 13 April 1970. Pointed ECM antenna covers (for ALQ-51) are visible under the nose, in the wing root (under the 20-mm gun), and under the jet pipe. Because ALQ-51 was a deceptive ECM system, it had to be able to listen to the enemy radar while it jammed. The jamming signal could not be allowed to corrupt the received signal, so active and passive antennas had to be shielded from each other by the bulk of the airplane. In many cases the cones were absent. The mast atop the hump is for the airplane's UHF radio (air-to-air and air-to-ship). The outer pylon carries a Zuni rocket pod with a frangible nose, which would be discarded when firing rockets. U.S. Naval Institute Photo Archive

The Marines' A-4M was the final version of the Skyhawk, with much more elaborate electronic defense. ALR-45 and -50 replaced the earlier APR-25 warner; this system could recognize what it heard. The radome atop the vertical tail carried ALR-45 antennas, with a radar beacon atop it. The small nose radome covered a pair of ALR-45 antennas. The visible spot on the nosewheel door was an ALR-50 L-band warning antenna. Abaft the usual UHF radio mast was another ALR-50 L-band antenna. The ALQ-51 track breaker was replaced by the reprogrammable ALQ-126. Below the nose was a radome containing mid- and low-band antennas for the ALQ-126 self-protection jammer. More ALQ-126 antennas were mounted on the after end, from the base of the rudder down. Uppermost, and not visible, was the after ALQ-126 low-band antenna. The upper of the two visible antennas was for IFF. The lower is actually a pair of ALQ-126 antennas: high-band on the right side and low-band on the left. The spot on the wing was a high-band ALQ-126 antenna. Earlier A-4Ms had APR-25 and ALQ-100. This version lacked the APG-53 nose radar of Navy versions. Earlier A-4Ms had a missile-guidance antenna forward of the cockpit; this one shows only the pitot tube. What looks like an antenna protruding from the nose is the head of the refueling probe, bent back away from the fuselage. The object below the tailpipe is the braking parachute, provided because this version would operate mainly from shore bases. U.S. Naval Institute Photo Archive

and one the X-band radars of enemy interceptors. In 1956 BuAer contracted with Sanders Associates for an S-band TWT deception repeater (track breaker). That year the device was successfully demonstrated against an SCR-584 fire-control tracker. An airborne version was successfully tested during the summer of 1957: 150 of these ALQ-19s were immediately ordered. The Naval Air Warfare Center in Indianapolis developed a corresponding X-band deception jammer (ALQ-32).[18] In 1960 the new Intruder was intended to have the APR-15, ALQ-19, and ALQ-32, as well as an ALE-18 chaff dispenser.

ALQ-19 and -32 entered service in July 1960 in eight A3Ds of VAH-9 on board USS *Saratoga*.[19] These aircraft needed jammers because they typically flew alone and unlike the Intruder had no radar warners. Initially, their pilots apparently preferred to fly without their jammers, using the weight for fuel; the jammers were stored on the carrier but not tested. Shorter-range aircraft were not assigned jammers.

By 1960 the Navy was contracting with Sanders for a new range of deception jammer using more reliable TWTs and additional jamming techniques. The most important of the new jammers were the ALQ-41 (X- band) and ALQ-51 (S-band).[20] Enemy interceptors were expected to have X-band radars, and enemy fire-control (including SAM) radars were expected to operate in S-band. The Intruder had deception jammers for both. Both received enemy signals, amplified them, shifted their timing, and sent back the distorted signals. Because they had to process signals they received while emitting deception signals, they needed separate receiving and transmitting antennas. Paired ALQ-51 (and successor) antennas are visible in photographs of many Vietnam-era attack aircraft.

The next-generation ALQ-100 jammer covered both S- and C-bands; its development was a direct response to the detection of C-band Fan Song signals in Cuba. It was also considerably more powerful than the ALQ-51 (it was derived from the interim ALQ-51A). ALQ-51 and -100 were so closely related that they used much the same operator-warning

panels. ALQ-100 still needed the additional ALQ-41 to cover the X-band, but ALQ-100X (later designated ALQ-126) covered all three bands. It was installed on all but the earliest A-6Es and also on late-production A-4Ms.

In theory a small bomber could carry jammers in an external ALQ-31 pod hung on a standard bomb pylon, but it was preferable to have specialized jamming aircraft that could either stand off at a distance or escort strike aircraft. Carrier air groups typically included a four-plane detachment of EA-1Fs/AD-5Qs. In exercises they located enemy airborne radars and also functioned as standoff jammers, but they had limited jamming power.[21] Also, once Skyhawks replaced Skyraiders, the Skyraider escorts were too slow to accompany strike aircraft.

The Marines adapted thirty-five Skyknight night fighters as escort jammers (F3D-2Qs [EF-10Bs]).[22] They may have been considered vital because Marine aircraft supporting troops would probably have to fly within range of enemy antiaircraft guns, many of which might be radar controlled. The Navy apparently assumed that its aircraft could fly above most such fire. It is not clear when the Skyknights were ordered modified. During the spring and summer of 1965, before the arrival of Air Force barrage jammers, Marine EF-10Bs supported Air Force strikes against North Vietnam. Range was limited, because they could not refuel in the air and their external stores created considerable drag. Typically, three or four would orbit over the Tonkin Gulf, disrupting both long-range radars and Fire Can gun-control radars. They could provide SAM warning and jam Fan Song radars. They operated inland, circling at 20,000 to 25,000 feet, until about October 1965, when the SAM threat pushed them too far from their targets.

Apparently as part of the emergency fielding of Shrikes during the Cuban Missile Crisis, the Navy sponsored development of a radar homing receiver, Melpar's APR-23 "Redhead." The project lapsed afterward but was revived in 1965 when aircraft operating over Vietnam faced (and had to avoid or evade)

SA-2s. The APR-23 used broadband (crystal video) detectors covering the C-, S-, and X-bands, providing audio and visual indication of threat signals. The pulse repetition rates of the radars modulated the audio signals, so that, for example, pilots could hear fire-control radars switch to lock-on (higher-PRF) mode.[23] By 1967, APR-23 was being supplemented by the APR-27 missile-launch warner. These devices must have been specific to aircraft operating over Vietnam, since Standard Aircraft Characteristics charts for attack aircraft do not list them.

The Navy bought a next-generation warner developed under an Air Force crash program: the APR-25 (Vector IV).[24] Its revolutionary feature was a small CRT that showed threat direction, based on the relative strengths of signals in its four antennas, and an indication of distance to the SAM. In a 1968 air battle, an A-4F pilot who received a SAM warning looked down a strobe displayed on the CRT and saw an approaching SA-2 missile. Two other SA-2s had been launched slightly earlier. Without seeing the strobe representing the third missile, the pilots involved might have ignored it altogether as they evaded the first two. Typically APR-25 was supplemented by the APR-27 missile-launch warner.

In A-6 Intruders, APR-25 replaced the ALR-15 under a 1968 order.

The crystal video technique used in all the detectors limited sensitivity and hence range. Because the devices could not distinguish emitters operating simultaneously in the same bands, display strobes averaged their directions. The next step was to provide the receiver with a clock that tagged pulses with time of arrival. Pulses arriving close together from the same direction could be associated with each other and a radar's pulse rate measured; the radar could be identified using a threat library. Different radars could be distinguished despite operating at similar frequencies. This technique was implemented in ALR-45 (1970), which took account of the "Mediterranean threat," presumably meaning that it could recognize signals generated by Soviet warships there. Presumably its having an *L* rather than *P* designation

indicated its ability to recognize particular radars.[25] Typically ALR-45, which could trigger later jammers, was supplemented by the L-band ALR-50.

The jammers were paired with chaff dispensers, the idea being that a radar led astray by the jammer had to be given something other than the airplane onto which to lock. The earliest in the Intruder, ALE-18, cut up a roll of chaff to the appropriate length before dropping it through a slit. That limited the rate and volume of chaff dispersed. ALE-18 was superseded by cartridge dispensers like ALE-29 or -39 (in the A-6E), which dropped chaff or flare packages in a matrix of small chutes.

The Navy's deceptive ECM contrasts with the noise jamming the Air Force favored. Navy pilots found that though deceptive ECM could not prevent the enemy from firing, it slowed their reaction time, making it easier to evade attack.[26] Overall, the Navy had lower loss rates than the Air Force.

In 1965 Capt. (later rear admiral) Julian Lake was in charge of air ECM in OpNav. He considered it urgent to provide Skyhawks with jammers. With only three pylons, the Skyhawk could not sacrifice one for an ALQ-31 pod, especially since at least one might be needed for a drop tank. A Skyhawk jammer thus had to be internal, but Heinemann's attempt to minimize size and weight had left little free space. Lake crawled around an A-4 and found room in the 20-mm gun bays. Fitting the ALQ-51 there left space for the guns and half the ammunition. This was Project Shoehorn (because everything was "shoehorned" in).[27] There was also a radar warner. It took fifty missiles to shoot down a Shoehorned airplane, compared to the usual ten. Also, since aircraft using ALQ-51 could fly higher, they were much less subject to antiaircraft fire. The necessary equipment was installed at Naval Air Station Cubi Point, in the Philippines, late in September 1965, and the first modified Skyhawks flew from USS *Constellation* the following month. Shoehorn was later extended to F-4 and F-8 fighters. Later Skyhawks had humps over their upper fuselages into which the ALQ-51 black boxes were moved.

A pair of VA-128 A-6As show the initial Intruder configuration. The big fairing atop the rudder housed aft-pointing transmitting antennas for both the ALQ-41 and -51. The receivers were in the tail cone below the rudder. Other ALQ-51 antennas, which are not visible, were a small receiving antenna near the lip of the left-hand air intake and a combination of horizontally and vertically polarized horns in a small radome in the upper lip of the right-hand air intake. The tail code indicated RCVW-12, a replacement air wing (hence no carrier name on these aircraft). The squadron received its A-6As in September 1967; they were replaced by A-6Es in December 1973. U.S. NAVAL INSTITUTE PHOTO ARCHIVE

A-6As had their ALQ-51s replaced by more-powerful ALQ-100s, which covered both low (S-) and middle (C-) frequency bands. These A-6As show booms carrying ALQ-100 antennas, mounted on their outer pylons, the transmitter on the right wing and the receiver on the left. Installation of ALQ-100 was an airframe change dated 29 February 1968; it replaced ALQ-51 at production number 310. The ALQ-41 in the tail remained; as before it shared tail radomes with ALQ-100. These aircraft show the wingtip pods associated with the APR-25 radar warmer. They were from VA-34, operating on board USS *John F. Kennedy* as part of Carrier Air Wing 1. U.S. NAVAL INSTITUTE PHOTO ARCHIVE

This TRAM A-6E was probably photographed on board the carrier *John F. Kennedy* about the time of the Gulf War (tail code "AC" for CVW-3, in a style used at the time). Typically the bombardier/navigator found a target by radar, then designated it to the FLIR in the turret, which had a zoom capability, and marked it with the laser in the turret. He also had a laser-spot detector, which enabled him to see a target designated by a forward observer or by another airplane. TRAM reduced A-6E radar CEP by a third, to about a third of the original A-6A's CEP, or about three hundred feet, and 71 percent of the visual CEP of the A-6E. The thick glove at the wing root covers three antennas for the three bands of the ALQ-126 jammer. As with ALQ-100, the left wing carried the receiving antennas, the right wing the jammers. Because ALQ-126 covered the X-band as well, it replaced the ALQ-41 in the tail. Inboard of the glove is a stall strip. Only the earliest A-6Es retained the booms of ALQ-100. The pods visible under the wingtips carried antennas and amplifiers for the ALR-45 radar warner. NHHC

The alternative to jamming is simply to destroy hostile radars. The A-6B version of the Intruder was conceived for that purpose. The diamond-shaped antennas on the nose radome are homing antennas for the APS-107A system. The wingtips and the tips of the horizontal tail carried APS-107A warning antennas. The airplane also had the usual self-defense systems, such as ALQ-100 (note the boom on the pylon) and APR-25 (note the wingtip pod). BuNo 151820 was photographed in November 1971. *THE HOOK*

A VA-147 A-7 is readied for launch on board the carrier *Ranger,* January 1968. Its tail shows both the original ECM radome at its base and also the fairing associated with the APR-25 radar homer and warner. Note the defensive Sidewinder on the fuselage launch rail. U.S. NAVAL INSTITUTE PHOTO ARCHIVE

Combat experience soon showed that Navy attack aircraft also needed standoff jamming support, which meant something with higher performance than a Skyraider. The senior carrier commander in the Tonkin Gulf pointed out that his Skyhawks also badly needed tanker support. Lake chose the big A-3, which was available and fast enough to accompany Skyhawks. It was not difficult to fit a tanker package in the big bomb bay. EKA-3Bs could be distinguished by blisters on their sides housing ALQ-92 VHF voice jammers (to disrupt fighter control) and by canoes housing ALT-27 S-band barrage jammers.[28] The ALT-27 signals were fed into a steerable high-gain antenna. It was pointed manually as indicated by the ALR-28 receiver, whose scope showed the direction of whatever it picked up. Its antenna was in a small blister atop the tail fin. In addition to the jammer, the canoe carried an APA-69 direction-finding antenna. There was a single ECM crewman, who had an ALA-3 or ULA-2 pulse analyzer. For self-defense the airplane had warning receivers (an ALR-29 L- and C-band receiver, an ALR-30 S-band receiver, and an APR-32 SAM warner) and self-defense jammers (dual ALQ-41 and -51), as well as two ALE-2 chaff dispensers.

Despite their deception jammers, EKA-3Bs were considered too vulnerable to be risked over North Vietnam. Instead, they typically took up racetrack paths offshore at 20,000 feet, barrage-jamming the SA-2s while the strikes proceeded inland. The EKA-3 was in any case never more than an interim solution to the larger Navy problem of jamming Soviet anti-carrier forces. The ideal solution was a dedicated EW airplane that could escort attackers.

Before war broke out in Vietnam, Grumman was working on a dedicated jammer version of the A-6A. It seems to have been developed to meet a Navy requirement to deal with enemy electronic capabilities in the years 1965–70, which suggests that it paralleled the A-6, the attack airplane of that era.[29] BuWeps sought an integrated electronic-warfare system capable of rapid and selectable detection and jamming. It intended the winner to become the basis of its next-generation EW airplane, which would be designed around it. This was much the way in which it had recently selected the weapon system for a long-range fleet air-defense fighter, the abortive F6D Missileer.

Loral won, with its semiautomatic ALQ-53 "Giraffe." As proposed, it could operate in eight bands (ten by 1962, between 40 MHz and 26.5 GHz), five of which it could handle and display simultaneously, as separate traces. On each band the system automatically measured and either semiautomatically or manually displayed parameters (including direction) for the operator's analysis and action. Modes were passive (reconnaissance) and active (jamming). In passive mode, ALQ-53 continuously scanned each band, automatically stopping when it detected a signal, then measuring and recording digitally its frequency, pulse width, pulse repetition interval, and direction. Signals were tagged with the aircraft position at which the signal was detected. After analysis (about half a second), scanning resumed. Data were all logged digitally. In semiautomatic operation, the signal was presented for operator analysis and the screen photographed; in "manual," the operator selected a single channel and visually analyzed the signal. Above this analyzer display was an intratarget display (showing details of a target signal), below it a digital display, all with controls alongside. The ECM operator in the right-hand seat selected a "victim" radar and monitored jamming; against a tunable radar, the operator shut down his jammer periodically and "looked through" to see if the target radar was still transmitting on the same frequency.

Grumman won the 1961 competition for the airframe to carry Giraffe. It argued that the Intruder airframe was uniquely suited in terms of its equipment volume, power, and performance. The Intruder mission-profile requirement was substantially that of the escort jammer BuWeps envisaged. Modification of an Intruder airframe into what would be the "EA-6A" configuration was authorized in March 1962. BuWeps wanted the store capacity of the original Intruder retained, plus Shrike and laydown (nuclear) capacity.

Grumman sought to limit redesign.[30] The most prominent change was a big tail-fin pod, which carried the midband ALQ-53 receivers. The radar, used primarily to identify and locate targets and navigation checkpoints and secondarily for terrain clearance, was derived from the Intruder search radar. Its antenna was moved forward to accommodate a new compartment forward of the cockpit, the fuselage lengthened accordingly by eight inches. ECM controls and displays replaced the bomb/nav display for the right-hand crewman. The Intruder's self-defense suite was retained: two ALQ-51s and one ALQ-41, plus a chaff dispenser (ALE-18) and four warning receivers (ALR-15s). A prominent port-side underbody blade antenna abaft the air intake served an ALQ-55 communications jammer. Barrage jammers were in external pods on five stations: one centerline and four underwing.

The main change in standard production aircraft was replacement of ALQ-53 by ALQ-86, with a narrower frequency range (0.5 to 10.75 GHz). Frequency coverage could be extended by adding antennas and tuners in a centerline pod. Two pods on wing stations could accommodate five additional tuners that could duplicate the most important frequency ranges. The prominent tail pod contained a broadband omnidirectional antenna and a sensitive rotating directional one. As in many other ESM systems, the "omni" was used to measure frequency and pulse characteristics. The operator had side-by-side panoramic and analysis screens. As with ALQ-53, the panoramic screen showed five traces representing signals on five selected bands, with frequency indicated from left to right. Traces could be expanded to show frequency more precisely, and a tuner could be

An A-7E carries a dummy B43 nuclear bomb. Carrier squadrons had to maintain nuclear readiness despite their grueling nonnuclear mission during the Vietnam War. This one was from CVW-15. Note the small radome, part of ALQ-126, on the upper lip of the air intake, above the jet intake's danger marking. EMIL BUEHLER LIBRARY, NATIONAL MUSEUM OF NAVAL AVIATION

stopped to measure frequency. The analysis screen could show modulation, pulse rate, pulse width, direction of arrival, and polarization (it could be photographed using a camera integrated with it). The operator could use headphones to help determine modulation (AM, FM, data link, interrupted continuous wave, or pulse), scan type (conical, raster, circular, etc.), and the approximate pulse rate. Audio presentations could be recorded on tape. A simplified receiving-antenna arrangement, using broader-band antennas, made it possible to provide six underwing pylons for jamming pods, which could be either the original ALQ-31s, ALQ-76s, or chaff dispensers.

The planned computer was not available in time. Detection—that is, the decision that the system was picking up a valid target radar—fell entirely to the operator, watching the system scan over its frequency range every two seconds. Typically, the operator listened for a characteristic pulse rate, integrated the system, and directed jammers on the basis of what he could see on his screens. In the dense electromagnetic environment that North Vietnam often presented, he could not do all that rapidly enough. Worse, as the EA-6A entered service the Soviets introduced tunable radars.

Two A-6As became prototype EA-6As. The first flew on 16 April 1963. Another five were converted as development aircraft, including one used for R&D (the NEA-6A). Six more A-6As were converted into operational EA-6As, and another fifteen were built as such. By the time the EA-6A was about to become operational, the Navy was using the EKA-3B, which was too large for most Marine airfields. The Marines were apparently given the EA-6A to replace their EF-10Bs while the Navy waited for the more sophisticated EA-6B.[31]

Even before the outbreak of war it was understood that self-defense jammers would not be enough. The EA-6A, whose prototype is show here completing carrier suitability trials on board the carrier *Kitty Hawk*, was essentially an A-6A with a standoff jamming system superimposed. Drop tanks and self-powered jamming pods could be hung from its pylons. The most obvious sign of its function was the big tail-top radome (sometimes said to be walnut shaped) characteristic of all EW A-6s. It housed receiving antennas. The jammer pod is an ALQ-31 (note the air turbine powering it). The outboard jamming pods did not last in service, and this outboard position was often occupied by chaff pods, some of which (ALE-31) resembled streamlined bombs. The inner pylons and the centerline carry three-hundred-gallon tanks. U.S. Naval Institute Photo Archive

This EA-6A of VMAQ-4 was photographed at El Toro on 20 August 1990, with jamming pods underwing. The rectangular antenna under the lip of the air intake was associated with the ALQ-92 jammer intended to disrupt enemy fighter-control links. The dome atop the airplane was for an automatic radio direction finder. *THE HOOK*

The Navy began work on a new, more integrated ECM system in 1963. Grumman began its own studies at about the same time, presumably in the knowledge that a new version of the EA-6 would soon be wanted. Without the planned central computer the EA-6A was a conglomeration of separately developed parts integrated by a human operator. This time, the system would be integrated by a central computer. It would automatically detect and log radar signals despite the density of the signal environment, splitting the mass of signals it detected into those from different enemy radars, based on measured signal characteristics. It also had to decide which emitters were important to jam, because much of what it detected would not be. To do that it correlated signals from different radars (differentiated by their signal characteristics) with location—because, at least on land, radar function could often be deduced from location. A somewhat analogous ability to detect signals automatically (and register them in a digital memory) was making it possible to produce digital track-while-scan radars like the AWG-9 in the Tomcat. Grumman pointed out that an airplane flying at four hundred knots at 20,000 feet would encounter about ten distinct emitters each second; the rate would double at 40,000 feet, because the airplane would see much farther. Accordingly, the new system automatically generated the jamming waveforms, to be amplified and transmitted by the devices in the pods, which in turn were powered by air turbines. Each pod contained an exciter (which generated signals of desired characteristics) flanked by two transmitters, with directional antennas underneath.

A formal requirement was issued in June 1965 for the new system, the ALQ-99. Cutler-Hammer received the development contract in August 1966. Although the -99 was more automated than the ALQ-86, it needed three operators rather than one. The A-6 fuselage was stretched to accommodate two more seats. The resulting aircraft was the EA-6B Prowler, its three prototypes being converted EA-6As.

Typically, EA-6Bs flew with two drop tanks and three pods (one on the centerline). The EA-6B was originally intended to neutralize the acquisition radars on which Soviet-type air-defense systems relied. It was designed to handle four bands that together accounted for 96 to 98 percent of the electronic threat in Vietnam. The natural next step was to jam defensive communications, without which it would be

The EKA-3B was the Navy's alternative to the EA-6A, while the EA-6B was being developed (the KA-6D tanker was a parallel development). This EKA-3B of VAQ-130 flew from the carrier *Franklin D. Roosevelt*, August 1969. The obvious additions were the canoe (visible forward of the refueling-hose housing) and four blister radomes, plus a radome atop the vertical fin. Note also the refueling probe at the after end of the canoe. The foremost blade antenna was for the automatic carrier landing system; the larger blade abaft it was for the ALQ-92 fighter-control link jammer. The canoe contained E/F-band antennas of the ALT-27 system. The blisters contained horizontally polarized antennas (also part of ALQ-92) to deal with lower-frequency North Vietnamese early-warning radars such as Knife Rest. The dome atop the tail housed an I-band ALR-28 antenna, which could detect gun-control and air-intercept radars. Aircraft had their ALQ-92s removed beginning in late 1972, the blisters being eliminated. This airplane was converted into a KA-3B in June 1967 and into an EKA-3B in January 1969. It was scrapped in 1973. These aircraft were initially called TACOS (Tanker/Communications or Strike), although the bomb bay was soon filled with jammers. U.S. NAVAL INSTITUTE PHOTO ARCHIVE

An EKA-3B of VAQ-125 from the carrier *Hancock*. Beginning with this airplane, all A3D-2s were delivered with provision for tanking, using a tank in the bomb bay. This airplane was converted into a KA-3B in June 1967 and into an EKA-3B a year later. It continued to operate in that role until stored in 1982; it returned to service, as a tanker, in 1985. U.S. NAVAL INSTITUTE PHOTO ARCHIVE

impossible to direct interceptors. During the late 1960s a fleet-air-defense mission was added: jamming the radars of the heavy missile bombers on which Soviet naval aviation relied. EA-6Bs could also provide early warning, since they could detect the approach of both Soviet missile-armed bombers and surface warships. About 1985 a requirement was added for capability to neutralize enemy radars physically using antiradar missiles (HARM, AGM-88). Meanwhile, EA-6B capability was increased by a series of standardized upgrades.[32]

The EA-6B became operational in Southeast Asia in June 1972, in time for the Linebacker operations.

Integrated Air Attack

As the SA-2 became less effective, the North Vietnamese had to emphasize the fighter element of their integrated system. For the Americans, fighter-versus-fighter combat experience had an important indirect impact on later attack-airplane development: sentiment, particularly in the Air Force, came to favor a small, agile fighter as an alternative to what pilots considered their current overlarge and oversophisticated aircraft. This pressure ultimately created the F-16 and the F/A-18. Yet it seems that the great lesson should have been that only a sophisticated command-and-control system, making maximum use of signals intelligence, could create a tactical picture so clear that long-range missiles could be used effectively.[33] Without it, pilots had to operate within extremely restrictive rules of engagement. During

the 1965–68 Rolling Thunder campaign, a pilot coming onto the tail of a North Vietnamese fighter might find himself having to go alongside to make sure that it was hostile, then fall back to fire. There were enough cases of mistaken identity that these rules seemed essential.

The initial solution was to trigger enemy IFF transponders, initially using QRC-248, devised as a result of continuing surveillance of Cuba. First tested against Soviet-supplied IFF off Florida in March and April 1965, it successfully interrogated the IFF of a Cuban airplane at a range of about two hundred nautical miles. In December 1966 an EC-121D (College Eye) equipped with QRC-248 picked up MiGs at a range of 175 miles when they were interrogated by North Vietnamese radars. A restriction on using this device (for fear of compromise) was lifted in July 1967, with dramatic results. Cruisers on PIRAZ duty were fitted with QRC-248 interrogators; they transmitted identifications to Navy fighters. Air Force and Navy EC-121s flying offshore also had QRC-248s.

In August 1967, the Air Force placed communications-intelligence operators, a QFC-248, and an ELINT display showing Fan Song radars on board an EC-121K—Rivet Top. Rivet Top control was credited with thirteen of twenty MiG kills by the Air Force between August 1967 and the end of Rolling Thunder.

The North Vietnamese seem to have been aware that their IFF emissions could betray them; two MiG-21s that shot down two Phantoms on 23 August

1967 achieved surprise by flying in ground clutter under the control of ground operators and keeping their IFF transponders silent (pilots were not amused to learn later that NSA had watched the North Vietnamese practice the new tactic).

An alternative was to place the interrogator on board fighters, which could then identify targets by themselves. In December 1971, the Air Force introduced Combat Tree (APX-80), which could trigger and detect enemy fighter IFF. In September 1972, the Air Force ascribed about seventeen of its last twenty air-to-air victories to Combat Tree. The North Vietnamese now used their IFF only when absolutely necessary during an interception.[34] A further attempted alternative was noncooperative identification, which relied on the physical characteristics of enemy aircraft.[35]

Yet another possibility was to jam enemy fighter-control communications. NSA generally vetoed the idea, because it was using these signals to help create an integrated picture of the air situation over North Vietnam (below).[36] Because the integrated system contained such highly classified elements, fighter pilots generally did not realize either how they worked or how crucial they had been—or the extent to which their successors might be crucial in future wars.

By 1965 U.S. Navy electronic-reconnaissance units had learned how to capture enemy search-radar pictures, matching the ground clutter in those pictures with known terrain. This technique was accurate to within about a tenth of a mile. By early 1967 about 26 percent of North Vietnamese early-warning and ground-control-interception radars had been located this way. Loss of these radars would make it much more difficult for the SA-2's own control radars to stay off the air. By 1967 other location techniques were about to enter service, using time of arrival.[37] There was also interest in jamming and deceiving the HF radio links between long-range radars and missile sites: because the SA-2s were so mobile, they could not rely on landlines.

Offshore monitoring airplanes could detect SA-2 radars and get their bearings; by maneuvering, the aircraft could make bearings change in such a way that the radars could be located. The slower the airplane, the longer such locating took; the North Viet-

namese learned to keep emissions as short as possible. As of 1967 U.S. ELINT aircraft typically localized SA-2 batteries within one to ten nautical miles. That was not good enough, because the sites were camouflaged and difficult to see in jungle. Moreover, strike aircraft generally did not navigate precisely enough to not need to be cued, either by ground-control radar (MSQ-77) or by an airborne control plane (E-2A), or else supported by photo reconnaissance with landmarks.

The most important development was integration of available information to create an air picture covering all of North Vietnam, showing fighter activity. No American radars covered all of North Vietnam; however, NSA could interpret and exploit North Vietnamese communications to track aircraft throughout that country.[38] The great achievement was to make NSA's highly classified data usable tactically—that is, without compromising sources. The Navy led, probably because its outlook was so different from the Air Force's.[39]

The Navy had spent decades controlling aircraft from ships, extending that control far beyond the horizon using early-warning and control aircraft. Ships and control aircraft far outside North Vietnam could safely and securely maintain an air picture. Ordering interception or evasion on the basis of that picture would not betray how it had been assembled, and whatever was shot down would carry no tell-tale equipment the North Vietnamese could capture. In its Vietnam War history, NSA made this point.[40] The Air Force, however, had very a different orientation. Where the Navy emphasized a network approach in which the aircraft controller exercised considerable control over pilots and hence could exploit all available information, the Air Force was pilot centered. It was not particularly interested in tactical air control using data links. That is why the Air Force installed IFF spoofers (Combat Tree) on board fighters, whereas the Navy put equivalent devices on control aircraft outside North Vietnamese airspace.

Beginning in 1965, an Air Force signals-intelligence detachment at Da Nang converted intercepted data into what appeared to be U.S. radar data. To maintain the illusion that it came from American radars, the system was limited to what airplanes flying at 40,000 feet 235 nautical miles from Da Nang could see—about half the distance to Hanoi. At first the conversion was done manually, hence slowly: warnings might take twelve to thirty minutes to reach pilots. They were often ignored, because they came over a heavily loaded guard channel.

Deficiencies became obvious when a group of U.S. aircraft strayed into Chinese airspace. Unfortunately, the usual link, a Navy EC-121M (Big Look) over the Tonkin Gulf, was not present. Attempts to route messages via Red Crown, the cruiser in the gulf, failed. The Chinese scrambled fighters, one of which was shot down; Beijing protested vigorously and threatened to widen the war.

A new control center (Iron Horse) was set up at Da Nang, and the system was computerized. At least as importantly, restrictions on the use of signals intelligence were relaxed, so that tracks were reported all the way north. Inputs also came from Navy and the Air Force electronic-intelligence aircraft, particularly EC-121s, operating offshore.

When attacks over North Vietnam resumed in 1972, the Navy made full use of SIGINT integrated into its air picture, but the Air Force did not; it suffered badly when the rebuilt North Vietnamese interceptor force showed much greater proficiency. The Air Force relied on Disco, EC-135Ks orbiting over Laos and the Gulf of Tonkin, out of reach of North Vietnamese fighters. Their special transmitters could trigger North Vietnamese IFF and thus positively identify air contacts as hostile. Unfortunately, neither Red Crown nor Disco could see low-flying MiGs west of Hanoi, which was a critical staging area for American attack missions. NSA U-2s (Olympic Torch) orbiting at high altitude over Laos and the Tonkin Gulf could pick up North Vietnamese air-control voice links, but the Air Force, unlike the Navy, had not created an all-source tactical air-control center. The commander of the Seventh Air Force, Gen. John W. Vogt Jr., told his service's Chief of Staff that he had to use SIGINT the way the Navy did—not only to evade North Vietnamese fighters but to shoot them down. He was given a new control center (Teaball) at Nakon Phanom in Thailand, which com-municated with pilots via a KC-135 relay airplane (Luzon). In effect Teaball demonstrated what the Navy had already learned: only all-source intelligence could create the air picture pilots needed.

Controllers at Teaball could see both the Air Force / Navy tactical picture and North Vietnamese data. They could even determine and transmit relative levels of MiG pilot experience. Air Force fighters could now use their long-range missiles freely. The exchange rate changed from about 0.5 to 1 (i.e., 2 U.S. aircraft lost per MiG killed) to 4 to 1 in September 1972. Between 29 July and the end of the war, American aircraft shot down thirty MiGs at the cost of ten of their own aircraft.[41]

In effect, integrated American attack defeated integrated North Vietnamese air defense.

Of the two air services, the Navy seems to have learned far more than the Air Force. An Air Force officer later ruefully recalled that during El Dorado Canyon, the 1986 attack on Libya, the Navy used an Air Force RC-135 for current intelligence but Air Force F-111s that had flown eight hours from airfields in the United Kingdom were not updated at all.

Outcome: Low Observables

Perhaps the key lesson of Rolling Thunder was that future attack airplanes should be far less visible to the enemy.[42] The best-known attempt at low visibility was the development of quiet helicopters. It appears that in about 1970 the Office of Naval Research, among others, sought to develop a night-attack airplane that would be both much less visible to radars and much quieter, hence much less susceptible to anti-aircraft fire. It would also have a reduced IR signature, to defeat short-range missiles like the SA-7. ONR seems to have attacked the problem in-house between June 1971 and October 1973. Its program may have been code-named (open sources are unclear) "Flying Banana." Rockwell declined a sole-source contract, but both Boeing and McDonnell Douglas apparently did receive contracts. McDonnell Douglas, whose airplane was designated "Model 226," was attractive to ONR because it already had a radar evaluation range with which it could test shapes and materials. Model 226 had a "blended wing" (a hybrid between a conventional design and a "flying wing," to avoid corner reflectors) and a V-tail, with air intakes in the wing roots. A high-aspect-ratio wing was selected as the best compromise to reduce aerodynamic noise. RAM (radar-absorbing material) was applied to the inlets and exhaust ducts. Thrust was provided by a two large-diameter fans "tip-driven" by gas from the core of a turbofan (tip-drive made it possible to run the fans at a low speed, which is inherently quiet). Like later stealthy airplanes, this one had an internal bomb bay (two 500-pound bombs); it could also carry bombs externally, up to a total of 12,000 pounds. The final version, 226-458, was offered to ONR in 1974. It would have had a maximum speed of 445 knots and, reportedly, a quiet speed of 115 knots. No such airplane was ever built; by this time DARPA was supporting Lockheed's antiradar Have Blue, which became the F-117.

Weapons for Limited War

China Lake seems to have taken the initiative in reviving conventional-weapon development. According to a January 1958 paper issued by CNO, "The Navy of the 1970 Era," by 1970 carriers should be optimized for limited war.[43] That July OpNav issued Operational Requirement CA-13501 for new free-fall weapons for use by high-performance aircraft.[44] By this time China Lake was already proposing non-nuclear weapons, to the point that in the fall of 1958 China Lake's advisory board commended its efforts in that direction. China Lake submitted a preliminary technical development plan in January 1959, postulating that weapons had to be usable from low altitude at relatively low speed. In June 1959 its planning group laid out a ten-year survey of requirements. Weapons were given "eye" names, the idea being that they were free-fall weapons to be delivered by eye.[45] The first were Hawkeye (later renamed Rockeye), Bigeye, and Gladeye.[46]

For Vietnam, the most important in the series was probably Snakeye, a means of slowing a Mk 81 (250-pound) or Mk 82 (500-pound) bomb dropped at low altitude and relatively low speed so that it

would not explode directly under the attacking airplane. Snakeye entered production in 1964. It became the weapon of choice for close air support, because it could be dropped from as low as four hundred feet in a ten-degree dive (slightly higher for Mk 82). Because it was somewhat unreliable, peacetime training was often conducted using twenty- or thirty-degree dives.[47] China Lake also developed a series of cluster-bomb dispensers (Rockeye, Sadeye, and Gladeye), area weapons for attacking troops and vehicles.[48]

Ground fire ruined the one precision attack technique the Navy had, dive-bombing. Typically, aircraft had to attack en masse to ensure both that some managed to deliver their bombs and that some of the bombs got close enough to destroy a point target. Some targets in North Vietnam were notorious for the vast number of unsuccessful sorties made against them, an example being the Thanh Hoa bridge near Hanoi. Bullpup, the only existing standoff missile (introduced in 1955), was dangerous to use (chapter 7); as was its custom, China Lake soon after developed an alternative, which became the Walleye television-guided missile. Past television-guided missiles had continuously transmitted their pictures back to a controlling airplane, the operator sending back commands. China Lake's new idea was to have the operator lock the missile onto its target, after which the airplane could leave.[49]

Like China Lake's Sidewinder air-to-air missile, Walleye would keep its sensor pointed at the target as it maneuvered. The camera was gyrostabilized, and the guidance was analogous to that of Sidewinder. China Lake developed the technique of locking the camera onto a prominent feature of the target image, initially a sharply defined edge. Since the aim point could slide along that edge, an alternative was an area of high contrast around the aim point.

In January 1957 two China Lake engineers wrote a memo describing the television-guided missile. China Lake envisaged a weapon with a 750-pound warhead, whose range when dropped from 30,000 feet would be sixteen miles. A January 1958 feasibility test led to a China Lake development program. A prototype was successfully guided on 27 November

The EA-6B Prowler replaced the EKA-3Bs as a tactical jammer; the KA-6D replaced it as a tanker. As in the EA-6A, the design concept of the -6B was to use self-powered underwing pods as jammers, powered by air turbines. The computer system in the airplane supplied the waveform that the jammers amplified. The EA-6B retained the tail radome of the -6A. There was also a self-defensive system like the one in the A-6 (note the glove radome visible on the leading edge of the starboard wing). ICAP airplanes like this could be distinguished by their ALQ-126B self-protective jammers. Unlike A-6Bs, the ALQ-126B in this airplane includes an omni antenna at the base of its refueling probe. In the tail, the transmitter and receiver antennas of the ALQ-126 were interchanged, because putting the transmitter high in the tail would have interfered with the receivers there (the ALQ-126 receiver is the "beer can" projecting from the tail radome). Reportedly the self-protection jammer was not well liked, because it tended to interfere with the main jamming system. ALQ-99 jamming pods can be seen under the wings and the belly, their air turbines spinning. XCAP and ICAP airplanes all had the strake visible above the body, leading to the vertical tail. The big blade antenna under the nose was for a communications (fighter-control) jammer. This airplane was from CVW-1. U.S. NAVAL INSTITUTE PHOTO ARCHIVE

An ICAP EA-6B of VAQ-128 in low-visibility paint. The big blade abaft the cockpits was for VHF radio, with the shorter UHF antenna abaft it. The low dome is for the automatic radio direction finder (ADF). The spine carries an HF antenna. The two bulges on the vertical tail are part of the ALQ-99 receiving system, for lower-frequency signals. The small antenna just forward of the big blade under the nose is for TACAN. The prominent air scoop visible under the forward end of the spine was for ALQ-92 cooling. U.S. NAVAL INSTITUTE PHOTO ARCHIVE

1962. By that time an operational requirement had already been issued (in August). In April 1963 China Lake submitted a technical development plan to BuWeps. Production engineering was farmed out to the Navy's avionics facility in Indianapolis and fuzing to the Naval Ordnance Laboratory at Corona, California. China Lake developed the warhead. Walleye is generally described as the first American precision standoff weapon.

Walleye entered production (by Martin, as the AGM-62A) in 1966. The first live warhead shot was made in April 1966. Walleye was declared ready for fleet use on 12 January 1967. It could be carried at Mach 1.9 at 35,000 feet and could glide about ten nautical miles. It had an 825-pound linear shaped-charge warhead. Its only defect was that even though, because it was unpowered, Walley could devote most of its weight to its warhead (825 pounds out of

1,125), the warhead was relatively small. By the time production ended in 1970, 4,531 had been made.

Initial reports were enthusiastic: of the first seven Walleyes fired in combat, at least six hit (the seventh was a probable). Early use in Vietnam was considered the missile's operational test: on 12 May 1967 Walleyes destroyed the Hanoi power station. The first Walleye unit, VA-212 (flying Intruders), made thirty-nine hits out of more than forty-three missiles launched, and five of the six Intruder pilots safely completed their tours (the sixth was shot down in a raid on the Bac Giang power plant and died in prison). As of March 1967, Commander in Chief U.S. Pacific Command expected to expend six hundred per month. Production would hardly meet his expected need for 4,800 missiles by the end of the year. However, in 1968 OSD pointed out that although television-guided missiles could hit within ten to fifteen feet of any high-contrast target, they had severe operational limits. A pilot had to point the missile camera by placing his pipper (in effect a cursor) on the target as he rolled in. Like those in other airborne televisions, the missile camera offered very poor resolution and suffered considerable jitter due both to the flexing of the airframe and to electronic noise. The worse the picture, the longer the pilot had to keep his airplane pointed vulnerably at the target while he designated the spot or edge the television would track—typically going into a long, shallow dive in a straight line. Statistics from Vietnam indicated that for these reasons average slant range at release was only 8,900 feet. Aircraft flying Walleye attacks took four times as many antiaircraft hits as in iron-bomb attacks, and the missile generally could not bring down steel bridges, which "iron" (conventional, unguided, air-dropped) bombs could. On this basis Walleye was actually only about twice as effective. On the other hand, it offered enough precision to be used in sensitive areas, that is, without risking civilian casualties. Hence its use against the Paul Doumer and Thanh Hoa bridges in 1967. Because of its inherent limitations, not the projected six hundred but only ten to fifteen Walleye I's were used each month. Navy pilots credited the missile with a 78 percent hit rate (Air Force pilots made 49 percent hits).

To solve the warhead-size problem, Hughes converted 1,481 missiles to Walleye IIs, with 2,000-pound linear shaped-charge warheads (total weight 2,400 pounds). Another 529 Walleye IIs were built. An extended-range Walleye had larger wings.

In parallel with the Walleye requirement, in 1962 the Navy approved a requirement for a longer-range television missile that would be locked onto its target after launch. Typically, a pilot locked the missile onto a prominent reference point near the target, then steered it onto the actual target (using the data link and the television in the nose of the missile). Alternatively, he could steer the missile directly onto the target. Control could also be handed over to another airplane.

This weapon was named Condor (ASM-N-11, later AGM-53). Proposals were received in 1964, and Rockwell received a contract in July 1966. With a solid-fuel rocket, Condor had a range of about sixty nautical miles (a turbojet would have given it a hundred). Compared to Walleye, it used much more of its weight for propulsion, so that at 2,130 pounds (to fit standard pylons) it carried only a 630-pound linear shaped-charge warhead. The missile first flew on 31 March 1970 and was reportedly very successful. The contract for pilot production (50 missiles) was signed on 28 June 1974, at which time all Navy attack and ASW aircraft were to have been armed with it; full-scale production (250 missiles) was planned for FY76. But Condor was abandoned as too expensive, its data link not reliable enough, and its warhead too small.

The most important legacy of the Condor program was probably its two-way data link, which was applied in modified form to Walleye to extend its effective range. Funding for an "extended-range datalink" (ERDL) Walleye was approved in February 1971 and an initial production run carried out between November 1972 and May 1973. Three ERDL Walleyes were dropped in combat in Vietnam, and five more were tested at China Lake; in these eight shots it made seven hits, always against targets invisible to the pilot at the time of launch. Many Walleyes were converted to ERDLs.

Laser guidance was an important alternative. The Army first tested it in 1962 as a way of guiding antitank missiles. After contracting with Texas Instruments to adapt a Shrike, however, it lost interest. Some staff members of its Missile Command shared their work with the Air Force's deputy for limited war in that service's Aeronautical Systems Division. The Air Force let contracts for competitive prototypes in November 1965; Texas Instruments won again, with a very simple, aerodynamically stabilized seeker head, bang-bang guidance, and tail control.[50] The first fifty seeker kits were made under a May 1968 contract. Requirements were a CEP of no more than twenty-five feet, at least 80 percent reliability, delivery in either a dive or level flight, and deployment no later than June 1968. The bombs were designated "Paveway." The Air Force estimated that although the computer in an improved A-7 improved target-kill performance by a factor of five, for laser-guidance the factor was two hundred. These estimates were apparently borne out both in Vietnam and in the Middle East. Paveway became the favored American unpowered precision weapon, typically an Mk 80 series bomb with an inexpensive adaptation kit.

The expensive part of the system was the designator, which tracked the target (to keep the bomb headed to it) while the bomb fell. It employed much the same lock-on technique as the Walleye seeker. The Air Force bought a large Philco-Ford designator, Pave Knife, sized for its F-111. It could maintain its lock as an airplane turned away from the target. Initially the Navy used a much simpler, handheld designator on Phantoms. It was deployed in June 1972; 372 modified Mk 83 bombs were dropped. Some A-6s were fitted with Pave Knife from November 1972 on; they dropped 33 Mk 83 and 21 Mk 84 bombs. The Air Force dropped about 25,000 laser-guided bombs.

These bombs were spectacularly successful. Of over 10,500 laser-guided bombs dropped by the Air Force between February 1972 and May 1973, over 5,100 were direct hits and another 4,000 had CEPs of twenty-five feet, based on pilot estimates. On average, two hits were needed to destroy a target. For example, a raid by twenty aircraft dropped six of the eleven spans of the Lang Giai bridge on a northeast rail line, twenty nautical miles from the Chinese border, without any losses.[51] Postwar, laser guidance was adopted for the Maverick air-to-surface missile, which in effect superseded Bullpup and Walleye.

Laser-guided bombs generally could not be salvoed, because the debris thrown up by the first bomb would obscure the target. Thus, many photographs of the effect of laser-guided-bomb attacks show direct hits plus craters far from it. The current solution is GPS guidance. Because the bomb looks up toward GPS satellites for guidance, it is not affected by debris. GPS guidance also avoids the major limitation of the tracking device that points the laser, that it needs considerable contrast or a sharp edge on its target. A current issue is whether the link to the satellites can be jammed (there are antijam GPS receivers).

The Marines developed an alternative means of guaranteeing accuracy, the Angle-Rate Bombing System. It was the ultimate development of techniques like dive- and loft bombing, the motion of the airplane being used to impart the desired motion to an unguided bomb. To do that the system had to know the range to the target and the motion of the airplane. The system could either track a ground target designated by a ground-based laser or track the target optically. The laser-spot tracker was important, because it guaranteed that the bomb would hit a target that Marines on the ground wanted hit. The tracker measured the angle to the target and the rate of change of that angle (hence the name), solving for range on the basis of airplane altitude and true air speed. As in other systems, the appropriate course was displayed on a head-up display, and the computer released bombs automatically. Typical target acquisition range was eight nautical miles for a ten-foot target with 10 percent contrast. Under similar conditions, the airplane could also detect the spot of a laser designator from another source about a thousand yards from the target. ARBS (Hughes ASB-19) was developed for the A-4M; tests began in the summer of 1976. The concept may already have been tested during the Vietnam War. Later it was employed in the Marines' AV-8B Harrier. The concept has been incorporated in many later airplanes, such as the F/A-18 Hornet.

An F/A-18 carrying two Harpoons underwing illustrates the return to an emphasis on warfare at sea. U.S. NAVAL INSTITUTE PHOTO ARCHIVE

12

WAR AT SEA

Even before the Vietnam War had ended, the U.S. Navy began to face a very different problem. Beginning in 1967 the Soviets maintained the Fifth Eskadra, a force of surface warships, in the Mediterranean. Until that time the main Soviet threat to U.S. carriers had been missile-carrying bombers, supported to some extent by submarines. Missile-armed surface ships, already in position to attack, could achieve a level of surprise bombers flying out of the Black Sea could not. The U.S. Navy had not faced significant ship-to-ship warfare at sea since 1945. The only major American advantage was that the eastern Mediterranean offered radar conditions unusually favorable to the carriers.[1]

The carriers represented the main tactical nuclear firepower available to NATO's Southern Command.[2] Yet the threat they faced was so severe that, from the late 1960s on, it was often suggested that at the outset of a war the carriers should not be risked within range of Soviet antiship missiles on ships. They might initially withdraw to the western Mediterranean or even the South Atlantic until Soviet warships could be rolled back.[3]

In the 1970s the Navy lacked surface-ship antiship weapons; its warships had been designed to deal with the two previous Soviet naval threats, submarines and missile-carrying bombers. Modern Soviet warships, moreover, seemed to bristle with layer after layer of antiaircraft weapons, because the main threat they faced was U.S. naval aircraft.[4] The United States began to develop the Harpoon antiship missile, but its small warhead was unlikely to sink a large Soviet warship.[5]

Vietnam experience offered a helpful insight. An accidental Shrike attack on the missile cruiser *Worden* in 1972 knocked the ship about so badly that she had to be withdrawn for several weeks of repairs (the Australian destroyer *Hobart* had a similar experience). The postwar HARM antiradar missile came to be seen as a valuable precursor weapon, one that would open the way for strike aircraft armed mainly with bombs. It took time for the fleet to develop tactics that exploited this new possibility. In the late 1970s the standard tactic was for A-6s to approach their target at minimum altitude, to avoid surface-to-air missiles. A-7s would split off to left and right and then pop up to loft bombs into the target ship. Both types of aircraft had to fly virtually over the target ship to hit it with their unguided bombs.[6] Given the sheer quantity of Soviet antiaircraft systems, pilots considered these tactics, which required them to fly straight at their targets, suicidal.

In the early 1980s the A-6s had new inertial navigation systems, FLIRs, laser target designators, and standoff weapons in the form of HARM and the longer-range Harpoon antiship missile. Typically, A-7s carried HARMs, and A-6s carried Harpoons. At least one squadron (VA-34) tried an attack in which the A-7s dropped HARMs from high altitude and considerable range while A-6s flew in low to launch Harpoons. EA-6Bs would jam enemy radars. It was soon obvious that Soviet ships could not carry enough weight to armor their radars against HARMs, and the *Worden* incident showed that HARM hits at the outset would at least temporarily put their air defenses out of action. Harpoons arriving shortly after the HARMs would create more chaos. Harpoon, because it was radar guided, would hit whether or not the Soviets turned off their radars, so its existence forced them to keep radars on and thus to attract the HARMs. Other attack aircraft would exploit that chaos by dropping substantial bomb loads to sink the ships—which would be practicable once enemy air defenses had been breached. In addition to iron bombs, carrier aircraft were armed with bomblets (Rockeyes) to spread disabling damage around ships' decks and perhaps ignite the big antiship missiles many Soviet warships carried on their upper decks. Rockeye was used effectively against a Libyan missile boat during the 1986 attack on that country. Aircraft also had laser-guided bombs and, in the 1980s, short-range Skipper laser-guided powered bombs.

VSTOL

Through the 1970s, the large-deck carrier force was under attack from within the Navy. In 1970 Adm. Elmo Zumwalt Jr. became Chief of Naval Operations. Unlike his predecessors, he was a surface sailor. He was enormously impressed by the big new missile-armed Soviet surface fleet.[7] Zumwalt feared that a U.S. Navy as dependent as it then was on a few large, expensive carriers could too easily be brought down. At the same time, the U.S. surface fleet was aging badly. Appointing Zumwalt, President Nixon promised that savings gained by mass decommissionings would go into new ships. Zumwalt sought inexpensive ones that could be built in numbers. That meant vertical takeoff, since without it the minimum size of the carrier would be set by catapults and arresting arrangements. The VSTOL Sea Control Ship (SCS) became a fixture of Zumwalt's tour as CNO.

Although in theory it would complement the large-deck carriers, many in the carrier air community suspected that it was intended to replace them. Even though a SCS would have far less capability than a conventional carrier, it seemed likely that politicians would equate the two. Admiral Zumwalt held a VSTOL fighter competition in 1971.[8] The Navy VTOL died in the next decade, partly as a result of the 1980 Sea-Based Air Master Study.[9]

The Post-Vietnam Fighter/Attack Airplane

Meanwhile the carrier Navy found itself developing and buying an attack airplane it did not want, largely due to congressional and OSD pressure connected with the supposed air lessons of the Vietnam War: the F/A-18 Hornet. Toward the end of the war the only question was whether the A-7 should be replaced by something with higher performance. A 1970 study showed that adding an afterburner to the A-7 would cost too much range.[10]

At this time the only Navy carrier fighter was the F-14, but by 1971 Grumman was in trouble and desperately wanted to raise its price.[11] OSD would accept the higher price only if the Navy adopted a "high/low" fighter mix, in what the Navy initially called a Fighter Modernization Program. OSD assumed that the F-14 was so expensive because of its Phoenix air-to-air weapon system; it ordered the Navy to examine a navalized F-15 as an alternative.

Despite Navy studies showing that this alternative would fail, OSD persisted. In June 1973 it reviewed the Navy's Fighter Modernization Program.[12] OSD sought something less expensive than F-14s to replace aging F-4s. To avoid delay and development effort, it limited alternatives to a modified F4J (soon dropped), the F-15, and a Phoenix-less F-14 (which could later be upgraded with that weapon). OSD envisaged a competitive fly-off between the latter two aircraft. Presumably a factor in its preference for the F-15 was that the F-14 was underpowered. Plans had originally called for a new engine in the sixty-eighth and later aircraft (to be designated F-14B) and for new all-weather-attack avionics in a later version (F-14C). In 1973 it seemed unlikely that the new engine or the avionics would become available.

Another factor was a strong perception that fighter experience in Vietnam had proven that U.S. fighters were too large and sophisticated to defeat more numerous, and more nimble, Soviet-bloc aircraft like the MiGs they had fought there. In effect, that paralleled Admiral Zumwalt's argument against big expensive carriers. An internal Navy lightweight-fighter proposal died about 1970 but seems to have had considerable impact in the Air Force.[13]

As the Vietnam War wound down, it seemed less and less likely that the United States would soon engage elsewhere in the Third World. OSD focused on Europe. Its air analysts envisaged hundreds of fighters caught up in a melee over Central Europe. Whatever the Vietnam War had proved about the virtues of a tactical picture built from all-source information, it was difficult to imagine such a picture (implying detailed fighter control) applying to such a massive air battle. That justified a focus on unsophisticated high-performance fighters.

The melee argument did not apply nearly so strongly to naval aircraft. The mobility of naval forces enables them to achieve local superiority, because an enemy cannot spread his force over the entire coastline a navy can strike. The wider the area within which a carrier can strike, the better the chance that its aircraft can enjoy superiority at that point, because the enemy cannot quickly mass everywhere within the threatened arc. That argument particularly applied to the very numerous but short-range Soviet interceptors.

The Navy tried to meet OSD's expressed need by reviving VFAX, which in theory might replace both the F-4 and the A-7. NavAir issued a "presolicitation" (i.e., precompetition) notice on 6 June 1974.[14] As of late August 1974, a request for proposals (RFP) was to have been issued on 1 September 1974, with a one-month turnaround. The project seems to have been designated "TS-169A." Like the previous VFAX, this one would be a single-seater suitable for modification into a two-seater for training and for some specialized missions, and its fighter armament would be Sparrow air-to-air missiles.

The two main missions were fighter escort (at least 400-nautical-mile radius using only internal fuel, with a goal of 450 nautical miles) and a strike mission with four Mk 83 bombs (radius 550 nautical miles).[15] Desired CAP loiter time was three hours, and desired deck-launched-interceptor radius was 200 nautical miles. The other missions were interdiction (carrying two Sparrows and two idewinders plus four Mk 83 bombs and two 600-gallon drop tanks) and close air support (the same missiles plus twelve Mk 82s and the two drop tanks). Desired interdiction radius, including sea-level dash in and out of the target, was 600 nautical miles. Apparently there was no long-range strike mission. The inclusion of the two Sparrows in the interdiction mission reflected the assumption that the airplane would be able to use the same radar for air-to-air and air-to-ground combat.

The operational requirement did not indicate a desired maximum speed. Experience in Vietnam showed that very high maximum speed was not useful but that supersonic cruising speed (supercruise) was desirable. Presumably, it would significantly reduce exposure to air defenses. However, it seems that supercruise would not be feasible for some considerable time to come.[16]

The airplane would have a multimission radar capable of ground mapping (for attack) and air-to-air search and track, manual terrain avoidance/following, and laser-spot tracking and following. A new requirement was a passive sensor (presumably infrared) that could detect a small missile boat (typified by the Soviet Project 183R "Komar," which carried two SSN-2 Styx antiship missiles on a length of about eighty-four feet) at sufficient range for the pilot to identify and attack it on the first pass at night. The airplane would have a projected map-display system. VFAX would have reliable inertial navigation set by the new standard carrier data link before flight.

Usable load factor at subsonic speed should be at least $7g$ (goal $7.5g$). Specific excess power at Mach 0.9 at 10,000 feet with maximum thrust should be 750 feet/second (goal 850). Acceleration time from Mach 0.8 to Mach 1.6 at 80,000 feet with maximum thrust should be no more than 110 seconds (eighty desired). Combat ceiling with intermediate thrust should be 45,000 feet (50,000 desired). Freefall-weapon delivery accuracy should be eight mils (six desired). There should be a substantial improvement in maintainability and reliability over existing aircraft, but figures were not given. Grumman papers indicate that NavAir demanded a high "ready rate" (78 percent) and a maintenance effort only a third that of the current F4J.

Desired takeoff weight was less than 30,000 pounds. The deck spot factor should be 1.1 (1.0 desired).

The operational requirement did not mention a VSTOL version, but the presolicitation notice did, because of the need for a Sea Control Ship fighter and also to replace the Marines' Harriers at minimum cost. NavAir wanted contractors to consider one versus two engines, dual versus single cockpits, and all-weather versus non-all-weather capability. Average flyaway cost should be limited to $4.5 million over a four-hundred-airplane program (the threshold was $6 million in 1975 dollars).

The airplane would enter service in 1981 (the Grumman proposal gave a date of 1982, which probably applied to an earlier requirement document).

Designs were offered by Convair, General Dynamics, Grumman, McDonnell, North American, Northrop, and Vought (LTV).[17] For VFAX Convair and North American offered much the same designs they had offered for the Sea Control Ship project but without the VSTOL feature.

By this time OSD had been besieged by a fighter "mafia" pressing for extremely agile, low-cost air-superiority fighters.[18] The Air Force was persuaded to conduct a fly-off between two prototypes, which became the General Dynamics YF-16 and the Northrop YF-17. In 1974 the House Armed Services Committee, which saw itself as the promoter of military reform, zeroed the Navy's request for VFAX money for the FY76 budget, arguing that it should be forced to participate in the Air Force Lightweight Fighter (LWF) program. Its Senate counterpart only considerably reduced the Navy's request but also pressed the Navy to see whether it could join the

Air Force program. Compromise budget language appeared to require the Navy to buy a version of whichever airplane won the Air Force contest. Ironically, the Air Force did not want the LWF at all, describing it as an experiment. It was Secretary of Defense James R. Schlesinger who made sure the FY75 budget provided money to missionize the LWF. It was renamed the Air Combat Fighter.

The F/A-18

VFAX died; the Navy agreed to participate in the Air Force's competition. The Navy ran its own phase, teaming each company with an experienced Navy contractor, which would modify its airplane to meet carrier and other naval requirements. General Dynamics Fort Worth was teamed with Vought; Northrop was teamed with McDonnell Douglas.[19]

The Navy did not suspend competition when the Air Force chose General Dynamics. Its requirements included a larger radar, Sparrow missiles, air-to-ground capability, and more fuel, as well as carrier compatibility. There would be limited commonality with the Air Force's airplane. The Navy successfully argued that congressional language bound it simply to buy a version of *either* of the two candidate lightweight fighters. In May 1975 it announced that it had chosen a version of the F-17. It was a twin-engine airplane, which the Navy preferred, had better carrier compatibility and mechanical backup to its fly-by-wire controls (a survivability issue).[20] The Navy requirements seem to have been much those set earlier for VFAX.[21] On this basis the Navy considered some of the characteristics of the Vought version of the F-16 unacceptable. It seems fair to say that the Navy's preference for twin engines was really a preference for greater range, since a twin-engine airplane could always shut down one engine while cruising on the other.

For a time the survival of the program was in question. First the Navy had to beat off a protest by Vought and General Dynamics, who argued that the congressional language in fact required selection of a navalized F-16. Also, OSD pointed out that the cost of development plus production would leave a net cost about that of the F-14. On this basis in

The F/A-18 Hornet offered combined fighter and attack capability in the post-Vietnam era but at a considerable cost in range and performance. This F/A-18C is shown over the Central Command area, 28 September 2020. The F/A-18C was externally identical to the -18A but had important internal improvements: avionics for better night and all-weather capability; provision for the AIM-120 AMRAAM air-to-air missile, the AGM-65F Maverick with IR seeker, and the Harpoon antiship missile; a common nose to support the reconnaissance mission; independent left and right fuel systems; provision for the laser-spot tracker; and a data storage set. It had considerably greater computer power and additional data buses. U.S. AIR FORCE PHOTO BY SENIOR AIRMAN DUNCAN C. BEVAN

1977 Under Secretary of the Navy R. James Woolsey Jr. suggested that the Navy use only F-14s and A-7s and the Marines only F-14s and AV-8s.[22] The F/A-18 program was saved by much lower estimated operating cost—for five hundred aircraft, about $150 million less each year.

The F/A-18 was the first fully digital Navy airplane. Its system was built around two AYK-14 computers (one tactical, one navigational), with digital avionics and weapon buses. By one account, it had the first of all truly modern "glass" cockpits (chapter 10), and a fully quad-redundant (a master and three backups) fly-by-wire system. Like the F-14, it had an inherently dual-purpose radar; its digital system was inherently dual-purpose as well. Initially the Navy announced that it planned to buy two versions of the same airframe, an F-18 fighter and an A-18 attack bomber. However, the avionics package conceived for VFAX was presumably well advanced by this time. The two separate airplanes may have been announced as a way of claiming that the Navy was buying a lightweight fighter, not a more sophisticated multirole airplane. The two designations were soon merged as the F/A-18 Hornet.

Once the F/A-18 was in squadron service, pilots were impressed by its performance and also by the glass-cockpit displays that made single-seat multimission operation possible.[23] The Marines too supported the project; they had no interest in the A-7. A FLIR pod with associated laser designator and range finder was developed concurrently with the F/A-18. In June 1981, the Defense Systems Acquisition and Research Council recommended full-rate production as a fighter; in December it recommended full production as an attack bomber. However, operational evaluation revealed problems. The Hornet's range was less than desired, its cycle time fell short, and its fuel reserve for returning to a carrier with maximum bomb load was inadequate; also, it required excessive wind over deck to launch from some older carriers.[24] The counterargument was that with precision weapons, heavy loads were no longer likely to be necessary. The F/A-18 did prove very reliable (one hundred sorties without maintenance) with satisfactory bombing accuracy and radar and engine operation. Pilots liked its fighter performance, which would improve its survivability.

The Maritime Strategy

Meanwhile, Navy strategy was changing to emphasize the importance of carrier-based strike. After entering office in 1977, the new Jimmy Carter administration asked for a statement of likely military performance, emphasizing the need to beat off a Soviet attack in Central Europe. That in turn implied that the important Navy mission was defense of reinforcement shipping in the Atlantic. The overall conclusion was that at best the United States and NATO could make a Soviet victory in Europe expensive; in any case, the Soviets would conquer considerable territory in Germany. Given limited funds, the Carter administration wanted to concentrate on improving the position in Germany. The Navy saw a very different solution to the weakness the administration identified.

The Navy's view was expressed in several studies, most notably Seaplan 2000. Instead of the defensive posture implicit in the administration's thinking, it envisaged an offensive using carriers. It (and also, it happened, the more land-oriented OSD) saw revolutionary possibilities in developing digital technology that the Soviets could not match. It might be possible to achieve "maritime air supremacy" in areas previously denied to carriers. Initially it seemed that carrier strikes could destroy so much of the Soviet fleet that the U.S. fleet could be freed to attack the flanks of any Soviet advance in Europe, slowing or stopping the Soviet attack and even threatening sensitive areas in the Soviet rear.

One participant in the Seaplan 2000 study was John Lehman, who would become the very activist Secretary of the Navy in the next (Reagan) administration. As a naval reservist, Lehman was an A-6 bombardier/navigator, so he had a clear idea of what naval strike aircraft could do. He was also an experienced political scientist, well attuned to policy issues

An F/A-18C launches from USS *Abraham Lincoln*, 21 September 2002. The shift to precision munitions like the laser-guided Mk 81 bomb, visible here, made a considerable reduction in carrying capacity acceptable. Note the two 300-gallon drop tanks, needed because the Hornet had limited fuel capacity. U.S. NAVY

An F/A-18C of VFA-137 in a 1994 exercise at "Strike U" (what was then the Naval Strike Warfare Center at Naval Air Station Fallon, Nev.), carries a laser-guided bomb on the left wing and a FLIR pod incorporating a laser detector/ranger in the left cheek position. The pod made it possible for the airplane to mark a target and also to detect a target marked by another laser. U.S. NAVAL INSTITUTE PHOTO ARCHIVE

and to how Congress worked (the subject of his PhD thesis). Seaplan 2000 argued that the mobility of naval strike forces could be used to force the Soviets out of their preferred way of attacking Europe.

The first Carter-era senior naval commander to press for an offensively oriented naval strategy was Adm. Thomas B. Hayward, who in 1978 commanded the Pacific Fleet. It had been assumed that in the event of a NATO war his ships would swing west to reinforce the fleet fighting the hot war in Europe. Hayward pointed out that they could do something far more useful: they could tie down the very large Soviet force already deployed facing China. Otherwise, the Soviets would feel free to move these forces, at least their air components, to add to an offensive in Europe. Hayward's fleet could even make attacks that would seem to support a Chinese seizure of

Siberia, territory the Chinese claimed the Russians had seized illicitly in centuries past. Moreover, Pacific ships swinging west probably would not arrive in the Atlantic before the Soviets had largely won the ground war. Hayward's key weapon would be the naval strike aircraft on board his carriers. Other fleet commanders felt encouraged to examine the pressure their ships could exert.

Ronald Reagan was elected in 1980 on a platform promising to rebuild American defense in the face of an aggressive Soviet Union. He soon realized that the Soviets themselves were in deep economic trouble, so much so that the United States might be able to force them to give up the Cold War altogether. This was not an entirely popular view in Washington, and it might be argued that much of the bureaucracy never accepted it. Reagan's first budget director resigned, convinced that the defense buildup was the beginning of endless insupportable deficits. Undoubtedly Reagan felt that heavy spending might cause the Soviets to collapse and end the Cold War—and the deficit—altogether.

Naval forces had a special role. They alone could demonstrate a forward U.S. posture—which would stretch Soviet resources—without incurring great risks. During a period of tension they, but not ground forces, could position themselves to improve the military balance—for example, by reinforcing northern Norway. Lehman campaigned to become Reagan's Secretary of the Navy, doubtless making exactly this point. He was well aware that Soviet naval outposts, particularly the massive base structure of the Kola Peninsula, were potential pressure points. In the past, NATO exercises in the north had been avoided as provocative. Now provocation was welcome, because it could force the Soviets to spend resources they did not have. After the Cold War, a former Soviet flag officer told Dr. Lehman that after a major exercise the Soviet navy had demanded that its resources be tripled to secure the Kola. That was impossible, and Lehman was convinced that it was the major reason that Mikhail Gorbachev realized that he had to back off from the Cold War. Reagan administration policies further made it impossible for Gorbachev to do as his predecessors had in the face of such problems, which was to warm relations with the West while facing down any internal dissent. Reagan demanded that external relaxation had to be accompanied by internal relaxation. That proved fatal to the regime.

Dr. Lehman is best known for his success in building up the Navy, maintaining fifteen active carriers in place of the twelve accepted by Carter. As Secretary of the Navy, he led the creation of an explicit Maritime Strategy. Lehman's initial briefing simply merged the outline war plans of his senior commanders. All of them understood the use of sea power in much the same way. Attack aircraft would be central in the new strategy. As it developed, the crucial role of the Kola became clear. It turned out to be not only a lucrative naval target area but one of the few naval assets that the land-oriented Soviet

leadership valued. To the Soviets, the key to victory was the nuclear balance. During an initial nonnuclear phase of a major war they hoped to hunt down Western strategic submarines and at the same time to preserve their own strategic submarines in sanctuaries ("bastions") they could protect—mainly north of Norway and the Sea of Okhotsk in the Far East. This very Soviet view of naval warfare and assets had in the past been suggested in the West, but in the early 1980s a very secret (as yet unrevealed) intelligence source confirmed the idea. The Kola Peninsula dominated the approaches to the most important Soviet bastion.

In a big European war, carrier attack aircraft operating on the flanks of a Soviet offensive could help slow it—but only if the carriers could survive. The main threats were Soviet land-based missile bombers (based in the Kola) and submarines. Putting pressure on the northern bastion would keep many Soviet attack submarines home, defending strategic submarines in the bastion. These submarines would not be attacking NATO shipping that was keeping the NATO army on the Central Front alive. However, the fleet of Soviet long-range bombers could be even more destructive than the submarines.

Of all the NATO navies, only that of the United States was air oriented. It saw Soviet long-range missile bombers as the worst threat NATO shipping faced; no NATO surface ship could shoot down the bombers before they launched their own missiles. American maritime strategists became interested in the way in which the Soviets could be forced to expose the bombers to the fighters on board strike carriers. How well such a ploy would work depended on the effective range of the carrier attack aircraft, as well as on how well the carrier's fighters could deal with the bombers launched against them.[25] As the long-range carrier bomber, the A-6 became the critical element in the strategy. To drive the point home, Lehman arranged for carrier-based Intruders to make thousand-mile sorties toward Soviet airspace.

The Navy's role demanded a high degree of readiness. Dr. Lehman's largely unheralded achievement was to move the Washington part of the Navy into a prewar state of mind—the Navy at sea was

The "maritime strategy" of the 1980s made the ability of an A-6 to deliver a nuclear weapon at maximum range vitally important: the threat to vital Soviet interests—such as the ballistic-missile-submarine sanctuary in the Barents Sea—would draw out Soviet naval bombers so that they could be destroyed. A-6Es flew thousand-mile practice strike flights specifically to dramatize this capability. This A-6E (converted from an A-6A) carries a shape representing a B61 bomb. Note the large drop tank on the centerline. The big air scoop near the tail marks this as a CAINS conversion, with an inertial-navigation set on board the airplane, initialized by the carrier. CAINS aircraft all had provision for later installation of the TRAM turret, not yet installed in this airplane. Associated with the intake was the small exhaust visible lower down the fuselage. The object under the wingtip is for the APR-25 warning receiver, which replaced the earlier ALR-15 threat warner (this airframe change was dated 30 April 1968). For a time at least, some aircraft had both.
EMIL BUEHLER LIBRARY, NATIONAL MUSEUM OF NAVAL AVIATION

The prospect of dynamic war at sea emphasized the carriers' need for organic signals-intelligence support, which EA-3Bs like this one could provide. These dedicated electronic-reconnaissance aircraft were flown by specialist VQ squadrons. The reconnaissance crew of four sat facing consoles on the left side of the airplane, lit by two windows on the other side. Reconnaissance equipment was in the canoe visible below the airplane and often elsewhere as well. This VAQ-2 airplane is shown off the Spanish coast in July 1980. The two SIGINT squadrons often adopted fictitious tail codes. Note that no bureau number can be seen. EA-3s were the last operational Skyraiders, some serving as late as the Gulf War. U.S. NAVAL INSTITUTE PHOTO ARCHIVE

already there. Dr. Lehman was well aware from personal experience in Intruders that this crucial airplane was aging. It was not fully digital, hence not very flexible and not amenable to incremental upgrades. Also, many existing A-6s had served grueling tours in Vietnam. What should replace it? Given the orientation toward readiness to fight and the present importance of those thousand-mile sorties, Lehman needed a near-term solution. His choices were a radically upgraded and digitized A-6 (the A-6F) and a stretched, two-seat F/A-18.[26] Range was so important that he had to choose the A-6F.

The Navy experimented with different compositions of carrier air wings. One was the "all-Grumman wing," without any A-7s but with more A-6s. The big A-6 offered range, greater carrying capacity, and all-weather capacity. Fewer A-6s could deliver more weapons. It turned out, however, that the A-7 could deliver its bombs more precisely, because its displays were easier to use to put bombs on target. The two airplanes had comparable systems, the difference being the displays. Hence the decision to modernize the A-6 with new displays, so that it could fully exploit the accuracy of the weapon system.

Dr. Lehman also had to look farther ahead to new technology the United States was developing: stealth, composites, and new engines. These advances might not be available until the 1990s, and to maintain pressure on the Soviets—to win the Cold War—the Navy had to remain entirely credible in the interim. A stretched F/A-18 probably would not be ready much before a stealthy attack bomber was. There was also some question as to how the composite structure of a stealthy airplane would stand up on board a carrier. Shredded or damaged composites could be dangerous to handle and difficult to patch.

Around 1985 it seemed that a jump in airframe technology was about five years off, so it seemed that the Navy would be wise to start a program in 1990 for an airplane that might enter service about 2000. Meanwhile the Navy sought to gain experience by contributing to the avionics and ECM being developed for the Air Force's stealthy Advanced Technology Fighter, which would become the F-22.

A Navy "blue-ribbon panel" (with membership expert and distinguished enough to speak with independence) supported a program to upgrade the A-6 as an interim step and begin a new multirole aircraft (VMX) in five or six years. By this time the Navy had had considerable unhappy experience with multimission aircraft; it was impossible for the same airplane to be a good dogfighter and a good bomber. The F-4 had been a good interceptor and a good bomber but not a good dogfighter; the F-14 was a good fighter and a bad bomber.

Inevitably, by about 1987 the administration no longer saw the situation so urgently. Mikhail Gorbachev was beginning to soften on many issues and, as importantly, was being forced to liberalize at home. The administration's strategy had never been very well understood within the government itself, and efforts were made to reduce what seemed to be excessive defense spending. Lehman resigned in April 1987, and his interim A-6F was soon abandoned in favor of what seemed to be a far better, if longer-term, stealthy attack bomber, the A-12. In retrospect, the A-12 seems to have been the unobtainable "best" that killed the "good enough" A-6F, which could have served for decades more.

Modernizing the A-6

It was essential both to threaten the Soviets with the longest possible strike range *and* to deal with massed Soviet missile-bomber attacks. The digital technology that provided the Hornet with a combination of air-to-air and air-to-ground capability could provide an Intruder with useful air-to-air capability using new AMRAAM missiles (carried in a third underwing rack). Ground-mapping improved the capability to deliver unguided bombs. The new APQ-173 had twice the acquisition and tracking range of the APQ-148 of the A-6E. It offered synthetic- and inverse-synthetic-aperture modes, the latter important for identifying ships at sea. Nonradiating GPS replaced the earlier Doppler radar-navigation system. A modernized airplane would use the same new AYK-14 computer as the F-14D, F/A-18, and AV-8B upgrades then planned. The airplane would be fully digital,

Grumman's upgraded A-6F Intruder II looked like an A-6E but in fact was heavily redesigned. Note the dual air scoops near the tail. The ALQ-126 defensive ECM system was replaced by ALQ-126B. Note its omni antenna at the base of the refueling probe, as in EA-6Bs. Not visible is the planned towed ALE-50 decoy. Note also the extra (outboard) wing station for an air-to-air missile, such as AMRAAM (AIM-120). Its presence reflected the idea that all carrier aircraft should be able to contribute to the Outer Air Battle against Soviet naval bombers.
U.S. NAVAL INSTITUTE PHOTO ARCHIVE

with a cockpit having five multifunction and head-up displays instead of the earlier optical sight. The J52s would be replaced by F404-GE-400D turbofans, with 15 percent more thrust. There were new defensive countermeasures, including the ALQ-165 jammer then being adopted for other naval aircraft, and a towed decoy to handle monopulse radars. The airplane had new wings, essentially EA-6B wings made out of composites. They were modified for increased lift, including a change to the fillet/slat to extend the slat seven inches outboard. The fuel system was redesigned to reduce its vulnerability, and an auxiliary power unit was added. The modernized A-6 offered greater payload/range than the alternative stretched F/A-18, and the Navy preferred it to the F/A-18; the Marines preferred the stretched F/A-18 (they received the two-seat F/A-18D).

Development of this Intruder II was ordered in 1984, and five full-scale development aircraft built in 1986–87.[27] The aerodynamic and engine prototype flew on 25 August 1987. By that time the program was in jeopardy, as the Navy was being forced to cut its budget. It was probably also relevant that Dr. Lehman, its champion, was no longer Secretary of the Navy. The FY87 Defense Authorization bill ordered the Navy to drop the A-6F in favor of the stealthy ATA (see below), although the FY88 appropriations bill included money for it.

In the fall of 1988 a less ambitious A-6G program was announced. It would employ some of the new avionics (such as the APQ-173), the new composite wings, and J52 engines. At the end of 1988, OSD decided not to fund the program; Grumman and its congressional allies failed to revive it.

Stealth as a Black Program

The story of the stealthy A-12 Avenger is linked with the Reagan administration's program of "black" (specially classified) programs. This classification seems to have been adopted specifically as part of the campaign to destroy the Soviet Union by over-stressing its weak economy.[28] To the extent that a black program was not compromised, it could suddenly be disclosed, as a kind of surprise attack.

The Soviets would feel compelled to counter such a surprise with a crash program. Given the rigidity of their planned economy, any such program would be disproportionately expensive. Further, the effort to penetrate a black program would soak up scarce Soviet hard cash used to subvert westerners—the higher the degree of security, the greater the cost of penetration. Splitting a category of programs, such as stealthy aircraft, into separate compartments would particularly increase the cost of penetration and the chance of surprise. Some black programs apparently involved projects that were known to be physically impossible. The administration understood that the Soviet leadership believed the Americans could do anything technological and that their politicians often brushed off claims by Soviet scientists and engineers that the intelligence they saw described impossibilities.

Access restrictions applied to those in the Defense Department who would normally track programs and their costs. The situation was undoubtedly complicated by the degree of deception (against the Soviets) inherent in the programs. In the past, only intelligence projects had been treated like Reagan-era "black" ones, the justification being that disclosure would destroy invaluable sources, such as broken codes or human agents.

Once a black program began, it was difficult for the few decision makers who did have access to monitor, let alone others. Limiting access suppressed potential critics. To pilots, low-observability reversed the widely held belief that battlefields had become far too lethal for them to survive in. The central, actual justification for making programs "black"—to destroy the Soviet economy—seems to have been forgotten after President Reagan left office, and quite possibly before that. The use of black programs has continued as a way of concealing exotic technology and, some cynics may suggest, of shielding programs from serious review.

Low-observable (largely antiradar but also anti-IR to some extent) technology was certainly real, but prior to the Reagan administration it was not highly classified.[29]

In the late 1970s three companies offered alternative approaches to antiradar stealth. Lockheed exploited work by a Russian, Pyotr Ufimstev, that made it possible to calculate the radar cross section of a plane at any set angle to the radar. An airplane could be built up of small triangular planes and its total radar cross section calculated by adding up those of the different planes. On this basis, Lockheed was able to minimize the cross section at fire-control-radar frequencies. Its approach did very little good at the lower frequencies used by long-range search radars.[30] Northrop and General Dynamics both offered flying wings. Although it lost the initial stealth competition, the latter continued to refine its design, which was a nearly pure delta.[31] In September 1981 General Dynamics' stealth designer briefed the General Purpose Forces director in OSD's Program Analysis and Evaluation (PA&E) branch. He was interested in a naval version, and the company thought that a contract might follow.

On 29 April 1982 Dr. Lehman and the commander of the Naval Air Systems Command visited Lockheed and saw the company's F-117 light stealth bomber.[32] In June 1983 Deputy Secretary of Defense Paul Thayer attacked Lehman's aircraft budget, which included the A-6 (the A-6F) and the F-14 (the F-14D). The Navy, he urged, should develop a new stealth bomber, at a total cost of about $3 billion. DDR&E pressed the Navy to join the Air Force stealth program.

Lehman created a Blue Ribbon Oversight Committee on Strike Aircraft to review the proposed A-6F program and also to consider the possibility of adopting a new all-weather attack aircraft. The alternatives were stealth, supercruise, a stretched F/A-18, and an upgraded A-6. On 8 December the panel reported that a shift to stealth would be premature: the Air Force was only beginning work on the B-2. Viable alternatives were a navalized F-117, an upgraded A-6, and a stealth program maintained in "advanced development" status to keep advancing that technology. The upgraded F/A-18 would be inferior to the A-6. The new strike airplane was tentatively designated VMX (it was also the Advanced Technology Aircraft).

Companies were invited to offer low-observability proposals. Requirements issued in September 1983 included susceptibility goals: a combination of signature (including visual), countermeasures, and tactics. Up to 7 percent larger than an A-6E, the airplane was to be supersonic and air-to-air capable, have twice the range of an A-6E, and possess the STOL capability the Marines wanted. The main constraints were a 1.5 spotting factor and a maximum eighty-two-foot wing span; the airplane also had to use an approved engine. It seemed unlikely that it could be ready until the end of the 1990s. That made the F-14D/A-6F program an essential interim step. A minority thought VMX could enter service as early as 1993. Dr. Lehman concluded that he should add a new stealth program.

Lehman later said that OSD had hyped the Soviet air-defense system to promote its favored stealth technology. The Blue Ribbon report included an annex on Soviet air defenses that pointed to gaps in them. When the U.S. Navy raided Tripoli in 1986, it was more heavily defended than any Soviet city, yet the attack was carried off without any losses. The logic of black programs suggests that OSD's main motive may have been to force the Soviets to spend much more on their national air defense and hence to help bankrupt them. OSD stealth advocates may also have been looking ahead to new generations of Soviet radars and air-defense missiles. It is often said today that stealthy aircraft like Joint Strike Fighter are needed to defeat "double-digit" antiaircraft missiles that did not figure in analyses conducted in 1983. Dr. Lehman considered the stealthy airplane a limited special-purpose type, a precursor to clear a path for more conventional attackers. That is much the way the Navy now describes its version of JSF.[33]

By 1984, when the Defense Department developed its long-range program including a new Navy attack bomber, the B-2 program had slipped several years from an early prediction that it would enter

The best that killed the good enough: an artist's impression of the unfortunate A-12 Avenger II shows some of its stealth features. The dark area closest to the centerline is one of its two air intakes. Angled louvers in the intake kept hostile radar from seeing the compressor of the jet engine. The wing covered the engine exhaust. The darker area farther out along the wing covered a conformal radar antenna. The tandem cockpit was adopted after Secretary of the Navy John Lehman left office; as an A-6 bomb/nav, he much favored a side-by-side arrangement. *THE HOOK*

service in 1987. Money was put into the FY85 and FY86 budgets. A formal program memorandum issued on 23 February 1984 called for an initial operational capability in 1995. Lehman wanted to keep the airplane in a research-and-development status, entering service late in the 1990s.[34] He also wanted to limit cost by making maximum use of technology developed for other programs, such as the B-2. (Later he said that the new Navy airplane should enter service five years after the B-2.)[35]

On 5 April 1984 the Navy's Special Program Review Group, responsible for black programs, approved the Advanced Technology Aircraft. OSD in effect had pushed Lehman and the Navy into developing a stealthy attack bomber on an accelerated schedule. It apparently helped by grossly underestimating cost.[36]

Black programming made the Navy's preferred cost-control technique, competition, complex. Competitors would have to combine experience with low-observables, carrier operation, and system integration. To do that, the Navy decided that companies would form teams to compete for "concept formulation" and then for the demonstration/validation phase—a fly-off. Lehman's fear was always that one team would bid low and buy in, then raise its price dramatically knowing only its entry could go into production. To avoid that possibility, he wanted the teams to compete again for "full-scale engineering development" and also for successive production batches. The leading stealth companies—General Dynamics, McDonnell Douglas, Lockheed, and Northrop—were told to team either with each other or with the other major companies: Grumman, Boeing, Rockwell, LTV, and Fairchild.

Both General Dynamics and Northrop were offering flying wings. In November 1984 General Dynamics agreed to team with McDonnell Douglas, combining its stealth experience with the other company's carrier experience. This was much the way Northrop had teamed with McDonnell Douglas in the F/A-18 program a few years before. This time Northrop picked Grumman as its teammate; Vought became a major subcontractor.[37] Lockheed did not want a partner, and it decided not to bid.

The Navy signed a concept-formulation contract with the General Dynamics–McDonnell Douglas team on 28 November 1984. A year later the team offered three alternative designs, having evaluated twenty subsonic and supersonic configurations. Two were flying wings optimized for attack. The third (Configuration 403) was supersonic, with swing-wings and a V-tail. Maximum speed would have been Mach 1.5. The flying wings were preferred because they were inherently more stealthy; the swing-wing aircraft would have relied more heavily on radar-absorbing material.[38] Ultimately the team offered a pure delta wing.[39] It used flush inlets under the leading edge, with exhausts under the trailing edge ramp. The Northrop/Grumman team offered a scaled-down B-2, a flying wing with a sawtooth trailing edge. Intakes and exhausts were on the upper surface of its wing, as in the B-2. Both airplanes combined stealthy shape with considerable amounts of radar-absorbing composite material, which might or might not be part of the airplane structure. Neither had a vertical tail—but no one had landed a tailless airplane on a carrier.

Both teams expected to use large amounts of composite materials but had little experience with them. NavAir seems not to have had any idea of the difficulty Northrop was having with the composite wing of the B-2. Without composites, the stealthy carrier bomber would be badly overweight.

Northrop's experience with the B-2 project had already taught it how risky the project was. Its president said that neither team had sufficient net worth to absorb the losses it might well incur under the Navy's preferred fixed-price contract. Northrop was used to the Air Force's cost-plus-fixed-fee contracts. Painfully conscious himself of the risks of escalating Defense costs, Dr. Lehman did not want such a contract. It seems arguable that "blackness" had kept his advisors from understanding just how much would go into a stealthy airplane—even contract terms were closely guarded.[40] Although Northrop's price was high and many of its features did not meet the requirement, NavAir decided to keep both teams in the project, presumably because these airplanes involved new technology that might have unpredictable twists. There was some suspicion too that General Dynamics–McDonnell was investing its own money to buy into the program. Even so, it had exceeded the Navy's cost cap—which was raised to keep the program alive. Not surprisingly, the Navy also tried to force both contractors to offer lower prices as "best-and-final offers" of the current stage of the program.

Both teams submitted best-and-final offers on 15 September 1987. General Dynamics–McDonnell significantly improved their entry's performance; Northrop lost performance, presumably because it had made its estimates more realistic. It also offered less stealth.[41] The difference in performance was enough to make General Dynamics–McDonnell the clear winner. The new airplane was formally ordered as the A-12 Avenger II in January 1988, with first flight scheduled for twenty-nine months later, in June 1990. This was much like the schedules of conventional aircraft. Stealthy aircraft (F-117, B-2, and F-22) typically took much longer.

By 1988 the A-12 was no longer a "silver bullet" (one-time game changer). The A-6F having been discarded, it would replace A-6s on a one-for-one basis. Without it, the Navy would no longer be able to execute deep air strikes. Expected strike radius was 785 nautical miles, with a load of two generic air-to-surface missiles (total 4,500 pounds) and two air-to-air missiles (1,000 pounds); gross weight would be 63,970 pounds. In the antisurface mission, combat radius would be 815 nautical miles carrying two HARMs (1,600 pounds), two Harpoons (2,320 pounds), and two air-to-air missiles (1,000 pounds), at a gross weight of 63,702 pounds. For close air support (560 nautical miles) the load would be sixteen Mk 82 Snakeyes (563 pounds each) and two air-to-air missiles (70,146 pounds). In the tanker role, the airplane would loiter for 1.5 hours 200 nautical miles from a carrier, with 13,750 pounds of fuel on board, plus two air-to-air missiles and two 400-gallon drop tanks (74,103 pounds).[42] Other contract requirements were a maximum speed of 540 knots, a maneuver load of 6.5 *gs*,

specific excess power of 180 feet per second, a single-engine rate of climb of 200 feet per minute, an approach speed of 132 knots, a specific range of 0.1240 nautical miles per pound, and a cruise ceiling of 45,000 feet.

Compared to the A-6, the A-12 offered both greater range and payload. In common with other aircraft of its generation, it also promised lower ownership cost: twice the reliability and half the maintenance man-hours per flight hour of the A-6. By the late 1980s it was a joint project, meant to replace not only the A-6 but also the Air Force's F-111. Total envisaged production was huge: 858 for the Navy (including 104 for the Marines) and 400 for the Air Force.

No one seems to have understood how risky the project was. In 1984 the only stealthy airplane in existence was the F-117, whose technology was radically different from the wideband stealth that Northrop incorporated in the B-2 and that General Dynamics thought it understood from its work on flying wings. In any case, Northrop was finding it very difficult to produce large composite structures for the B-2, and General Dynamics had not yet tried. Even if Northrop and the Air Force were willing to share the unfortunate parts of their experience, very few people in NavAir or in the two companies building the A-12 were cleared to ask about it or use it.[43] General Dynamics and McDonnell Douglas were following a structural approach very different from that of the B-2, a conventional aluminum structure with a composite skin rather than a massive composite skin supported by a specially shaped supporting structure. There were also necessary design changes. For example the wing had to be thickened to carry more fuel, and that affected the way the engine inlets reflected radar signals.

Through 1988–90 the estimated unit cost of the A-12 kept rising, to between $140 and $180 million. Yet to fill out the fleet's long-range attack squadrons, unit cost had to be held to something like $50 million. That was what Dr. Lehman had understood several years before, when he had described the A-12 as a silver bullet (see note 33), to be bought in very limited numbers. In November 1990, the two contractors realized that as Northrop had already realized, the fixed-price contract was deadly. They expected to overrun the projected development cost by $1 billion or more, and they faced further losses as projected production was reduced. Their investment of more than $2 billion had bought only a full-scale mock-up and a few parts. NavAir understood and was willing to restructure the full-scale development and production contracts.

The end of the Cold War killed the A-12. In 1990 Secretary of Defense Richard Cheney reviewed expensive major aircraft programs in the expectation that Congress would demand deep cuts: the Air Force's F-22 and B-2, the Navy's A-12, and the Air Force's C-17 transport. Each involved major new technology. At least one had to die. It was difficult to kill any of the Air Force programs, as prototypes were already flying. As for the A-12, although the production schedule required that much of it already exist by now, it did not. Worse, Cheney had been assured, he said, that the A-12 was on time and on budget.[44] He said that he had been lied to, and in his fury he canceled the A-12 program in January 1991. It turned out that he did so in apparent violation of the contract terms. The companies involved sued, and in the process a remarkable amount of documentation became public, making it possible to trace the program despite its "blackness."

That same month, Secretary Cheney directed the Navy to begin work on a successor called "A/X."[45] The Air Force joined the program, and the follow-on A/F-X was intended to replace fighters as well as attack aircraft. For the first-day mission (no external weapons, for minimum observability) it was to have an unrefueled radius of seven hundred nautical miles (including a low-altitude dash into and out from the target) with four 1,000-pound bombs and two air-to-air missiles. Five teams were formed: Grumman/Lockheed/Boeing; Rockwell/Lockheed; McDonnell Douglas / LTV; General Dynamics–McDonnell Douglas / Northrop; and Lockheed / Boeing / General Dynamics. All five received study contracts at the end of 1991, for reports to be

submitted in September 1992. Reported proposals included a navalized F-22 (NATF) and a Boeing variable-sweep airplane intended to have better loiter and carrier suitability. A wind-tunnel model showed twin vertical tails, canted outward. A Lockheed sketch showed a twin-engine swing-wing airplane with inlets under its wing roots and twin tails canted outward for stealth. The two-man crew sat in tandem under a bubble canopy.

Because A/F-X would not be in service before about 2006, there was interest in an interim attack airplane. The two alternatives were modified F-14s and F/A-18s.

The Marines and VTOL

The Marines first ordered the Hawker Siddeley Harrier in 1969, and it entered service in 1971. Unlike other VSTOLS, it used a single engine for both lift and cruise rather than separate ones. That insured against the usual VTOL problem, that separate lift engines might not all work together. As a turbofan, the Pegasus engine provided two sources of lift: air taken directly from the fan, for the fore end, and exhaust gas for the after end. Outlets for the fan stream and the jets were linked mechanically.

Because its development in the United Kingdom had been funded partly by the U.S. Mutual Weapons Development Program, the U.S. services were invited to test the initial version, the Kestrel, in 1966. The Marines were interested enough to evaluate a production Harrier at Royal Aircraft Establishment Farnborough in 1968. The Marines bought the Harrier as the AV-8A, the attack designation, *A*, indicating its role. It was a simple airplane, with virtually no avionics, but it offered the Marines what they wanted: an attack airplane that could operate close enough to their troops to be their artillery.

The Harrier had its detractors. It was more vulnerable and carried less than the Skyhawk it replaced, and it was no more sophisticated electronically.

Hawker delivered 102 AV-8As and eight TAV-8As (plus eleven AV-8As and two TAV-8As for the Spanish navy). By this time Douglas had been negotiating for a production license in the event that Congress required the Marine aircraft to be made in the United States. The first of three Marine Harrier squadrons, VMA-531, began working up in April 1971.

Soon the Marines and Hawker Siddeley (which became part of BAE in 1977) were considering advanced versions of the Harrier, beginning with one tentatively designated AV-16 (i.e., twice as good as an AV-8). It proved too expensive, so a simpler "AV-8+" was designed.[46] The object was to more than double mission performance while improving reliability and maintainability. In 1975 the Marines wanted the advanced VSTOL airplane to enter squadron service in 1982. Unlike the AV-8A, it would provide a useful capability at sea as well as ashore. The project, which was to incorporate American technology, was initiated in May 1975. It combined a modified inlet and larger auxiliary air doors for increased lift power, as well as lift improvement devices (LIDs) and a new, larger wing with a higher aspect ratio (using leading-edge extensions) and a supercritical airfoil section.[47] Much of it was built of composites. The new wing would increase lift through positive air circulation via slotted flaps and drooped ailerons. The forward exhaust nozzles were modified. New elliptical air intakes improved internal airflow. Later the rear fuselage was lengthened, the forward fuselage modified with a bubble canopy, and leading-edge wing-root extensions added. Although the airplane would use the existing version of the Pegasus engine, it would be designed so that the next-generation engine could be installed later.

The AV-8+ became the AV-8B Harrier II, a unique joint U.S.-British program in which McDonnell and British Aerospace each provided components to the other's assembly line. The British equivalent to the AV-8B was the Harrier GR.5. The American project was formally approved on 27 July 1976. Planned first flight was in December 1978. The first YAV-8B development airplane, a modified AV-8A, flew on 9 November 1978, six weeks ahead of schedule. The airplane had seven weapon stations (versus four in an AV-8A), four of which could take drop tanks (as against two in an AV-8A).[48] Its modern weapons-delivery avionics (ARBS) would improve accuracy

by a factor of three. It is not clear whether the AV-8+ initially was to have had the radically new avionics that featured in its AV-8B production version: a central tactical system integrated by the same AYK-14 computer that drove the F/A-18 system. The combination of a central computer and a data bus made it relatively easy to add new equipment, such as the APG-65 radar. Thus in 1985 the advanced Harrier was incapable of delivering important missiles such as HARM and Walleye or even two-thousand-pound Mk 84 bombs—but it would gain those capabilities through software changes.

This airplane, the smallest U.S. tactical aircraft of its time, was too limited in space and weight to add a second AYK-14. That became less important as computer components became much more compact and upgrades became easier. A new commercial-off-the-shelf (COTS) core "open system" was announced in October 1996; the aircraft first flew in May 1998. The airplane also ended up with a programmable multipurpose head-down display (for a glass cockpit) in addition to the HUD of the AV-8A.

In July 1977, in connection with the attempt to develop a VTOL naval air force, NavAir was asked for a feasibility study of an AV-8B as an alternative to the F/A-18. It reported in September 1977; although the F/A-18 was selected for development, the upgraded AV-8B+ project survived, and McDonnell Douglas submitted a report in 1979.[49] To carry out the F/A-18 mission, the "AV-8B+" needed a radar, an autopilot, and self-contained navigation. Since it would probably operate without the benefit of cover by conventional fighters, it also needed an all-weather missile, in this case the self-guided AMRAAM. Despite the additions, McDonnell Douglas tried to keep performance close to that of the AV-8B. The new version had an uprated Pegasus 11F-35 engine (1,500 pounds more thrust), with a redesigned fan fed by a refined inlet, and more fuel. To accommodate more tankage, its fuselage was extended fore and aft (from the forty-five feet seven inches of the Harrier to forty-six feet three inches). It was given a second multifunction display to provide the pilot with a radar picture. The planned radar was the

The first production AV-8B airplane shows its enlarged canopy and new wing. U.S. NAVAL INSTITUTE PHOTO ARCHIVE

same APG-65 that equipped the F/A-18; the new airplane could carry the same range of weapons, including Sparrow and AMRAAM and nuclear weapons. Additional electrical power was provided to support the additional avionics. The AV-8B+ was stressed for ski-jump and rocket-assisted takeoffs. Nothing came of this project at the time, but the evolving AV-8B was given the lengthened fuselage, presumably to give it the potential for an upgrade. In 1982 Dr. Lehman tried to promote the AV-8B+ despite a lack of interest within the DCNO (Air) organization.

To keep the AV-8B itself alive, the Marines had to beat off formidable opposition. CNO received the Marines' requirement document (for a VSTOL airplane) in October 1975 but never formally approved it. In 1977 the AV-8B collided, as noted above, with the F/A-18 program. There was considerable pressure to increase the number of F/A-18s, so late in 1977 the secretary of defense decided that the whole light-attack force (including the Marines) should buy them. At this time the Marines operated five squadrons of A-4Ms and three of AV-8As; complete replacement by AV-8Bs would require 336 aircraft. The Marines could have new AV-8s only if they could prove that they needed a VSTOL airplane on a cost/effectiveness basis. They wanted any new attack airplane to be capable of operating from small ships

(helicopter carriers) and from small, austere shore bases. This qualification ruled out anything but a VSTOL airplane. Only the AV-8B offered the desired increased range and payload.

The Marines completed an analysis in October 1977 showing that life-cycle costs of the F/A-18 and AV-8B forces were equal and that relative effectiveness depended very much on the scenario. In the Marines' scenario, the AV-8B was substantially more effective, thanks to its basing flexibility. The assistant secretary of defense for PA&E argued that the Marines had not taken into account the logistical burden of supporting dispersed AV-8B bases. The Marines rebutted his arguments. The secretary of defense ordered a fly-off. The Marines pointed out that the existing prototype lacked full equipment. No realistic fly-off could be carried out until the airplane had undergone full-scale development—which would cost considerably more.

The Marines convinced Congress to appropriate money for the AV-8B, but the secretary of defense resisted. The Under Secretary of Defense for Research and Engineering refused to permit the Navy to obligate $108 million of the $123 million in full-scale-development funds that had been appropriated for FY79, and the secretary of defense decided not to request any FY80 money for the AV-8B—actions that would have killed the AV-8B program. They did not stand: four full-scale development aircraft were ordered in April 1979. Before the prototype flew on 5 November 1981 the decision to put the airplane into production had been announced (on 24 August).

The AV-8B had a longer "ferry" range (maximum fuel, minimum equipment) than the F/A-18. Its maximum speed (clean) was about half that of the F/A-18, Mach 0.92 rather than 1.8. With drop tanks, the two were comparable. Since it was larger, when carrying similar bomb loads the F/A-18 outranged the AV-8B on internal fuel. Empty weight was just over half that of an F/A-18, 12,750 versus 21,830 pounds (takeoff weights were, respectively, 29,750 and 51,900 pounds). Maximum ordnance loads with internal fuel were, respectively, 8,260 and 19,000 pounds.

The first 161 AV-8Bs were clear-air day strike aircraft, with an ARBS (ASB-19) in the nose, an inertial navigation system (ASN-130), and a radar warner (ALR-67). There was provision for a podded jammer. The airplane was fitted "for but not with" the Maverick missile and was wired for nuclear weapons. The Marines emphasized the airplane's ability to take off in a very short distance (STO), giving it a roll that increased its payload: maximum STO takeoff weight was 26 percent greater than that of the AV-8A; the new version could lift twice the weapons load or take the same load twice as far.

The eighty-seventh airplane became the prototype Harrier II night-attack airplane, with a FLIR in its nose, an upgraded HUD, a digital moving map, two-color cockpit displays, and cockpit lighting compatible with night-vision goggles. The airplane also had larger leading-edge root extensions (for greater agility) and four upward-firing chaff/flare dispensers added to its after fuselage. The first of sixty-five production aircraft (aircraft 167–232) was delivered in September 1989. Flight tests using the upgraded F402-RR-408 engine (three thousand pounds more thrust, greater durability) began in 1990. The engine caused some problems, but after they were corrected the -408 engine was retrofitted into all of the night-attack aircraft and installed in later Harrier IIs.

In September 1990 the United States, Spain, and Italy agreed to support integration of the APG-65 radar into what was now known as the "Harrier II+"—essentially what had been offered a decade

The first flight of the AV-8B+, its nose radar making it an effective air-to-air fighter comparable to the last version of the British Sea Harrier. U.S. NAVAL INSTITUTE PHOTO ARCHIVE

earlier. It had the -408 engine. The prototype (the 205th airplane) flew on 22 September 1992. This change applied to forty-two new aircraft (production numbers 233 to 262) and meant the remanufacture of others originally built as day-attack aircraft (numbers 263 to 336). (The last of the Harrier II+ type were delivered in 2003.) In the late 1990s the Marines integrated a new FLIR, the AAQ-28 Litening II pod, with the aircraft; the pod included a laser designator. Tests began in 2000. Pods were later provided with an air-to-ground link enabling aircraft to

pass what they saw to ground units. AV-8Bs were also fitted with GPS receivers, which enabled them to employ GPS-guided JDAM Joint Direct Attack Munition) bombs.

The first operational AV-8B squadron attained initial operational capability in August 1985; its aircraft replaced the Marines' A-4Ms. Unlike Skyhawks, they could be deployed on board large amphibious ships, beginning with VMA-331 on board USS *Belleau Wood* in 1987. During the 1991 Iraq War (Desert Storm) the Marines deployed Harriers on board the amphibious assault ship *Nassau*.

The last HMM-266 Harrier lands on board the large-deck amphibious ship *Nassau* (LHA 4) after participating in the first strikes during Operation Allied Force against Serbian forces. The advent of the Harrier provided the Marines with an integral bomber that could accompany their seaborne forces ashore. U.S. NAVAL INSTITUTE PHOTO ARCHIVE

Of the 102 AV-8As delivered, 47 were modernized as AV-8Cs, incorporating as much of the AV-8B as possible. The "decision coordinating paper" authorizing the program was issued 24 April 1980 (the first airplane had been operationally evaluated in November 1979). The LIDs developed for the AV-8B were used. Instead of ARBS, the modified airplane had a weapon delivery system based on a continuously computed impact point (CCIP), a mechanism that had been considered as an alternative to ARBS for the -8B. Other features were secure voice radio, formation lights (to assist aircraft flying closely in limited visibility), a digital-interface / weapon-arming computer, a radar warning receiver (with antennas at the wingtips and tail warning in the bullet fairing aft), an integral flare and chaff dispenser, and onboard oxygen generation. Estimated CEP of a Mk 82 bomb delivered using CCIP at 545 knots at five hundred feet was 150 feet.

The electronic revolution in action: an EA-18G Growler of VAQ-139 from USS *Nimitz*. The two-man crew does the job of the four people in an EA-6B, and the electronically scanned antenna in the nose functions as both a sensor and jammer. The airplane carries EA-6B-type jamming pods under the centerline and under each wing. That leaves enough space for the fuel tank shown and for missiles. The wingtip pods are permanent. USAF PHOTO BY SSGT JUSTIN PARSONS

13

AFTER THE COLD WAR

The end of the Cold War brought a new era of crises. Without East/West tension to control them, many existing local problems erupted. This was nothing like the Cold War. Deep strike by aircraft no longer seemed vital. Crises would typically involve a shallow strip of land bordering the sea: the great majority of the Third World lives in such geographical settings. The short-range F/A-18, often derided as the "lawn dart," seemed to be good enough. It was far less expensive to maintain than the longer-legged A-6 and A-7, which were retired.[1] Deep strike could be left to Tomahawk cruise missiles. The death of the A-12 was therefore less painful than it might otherwise have been.

Existing naval aircraft had generally been designed to last for eight thousand flight hours. Given a normal peacetime rate of fifty hours per month and expected major overhauls, that gave them lives of about thirty years—meaning that the flood of aircraft built during the Reagan administration might not need replacement until about 2012. The massive aircraft industry built up during the Cold War could not survive until then, so Defense Secretary Les Aspin pressed companies to shed capacity by merging. Six did so: Lockheed with Martin, Northrop with Grumman, and Boeing with McDonnell Douglas. Chance Vought, as such, disappeared. Aspin was not interested in new programs, so A/X and then A/F-X were essentially study efforts.

Aspin did not take into account the many crises that the end of the Cold War engendered. Usage rates were much higher, so replacement (or complete reconstruction) was needed sooner.

Weapons

The 1991 Kuwait war demonstrated that air attack had changed radically. In Iraq, Saddam Hussein had built an integrated air-defense system. The war began with a coordinated strike against the whole system, including its Baghdad headquarters and its outlying interceptor-control stations as well. Surveillance radars were destroyed (the first shots of the war were from attack helicopters against early-warning radars in western Iraq). Stealth made it possible for F-117s to destroy Iraqi air-defense control centers at the outset. Tomahawks contributed. With long-range radars, headquarters, and ground intercept sites wiped out, Iraqi fighters found it difficult to work. American and coalition aircraft could attack at high altitude, where they were safe from the antiaircraft fire that had been so destructive in Vietnam. They suffered very few losses. One carrier air group (from *Saratoga*) employed the low-level attack tactics of earlier years and suffered typical Vietnam-era losses. Throughout the war, Iraqi guns and short-range, IR-guided missiles remained active—but aircraft generally did not have to risk them, absent the long-range missiles that would have forced them down.

The U.S. force stimulated Iraqi radar-controlled systems by firing large numbers of decoys, then followed up with HARMs that shut down the radars. Apparently, the Iraqis had not invested heavily in spares or in repair personnel, so HARM effects lasted. The chief ground-attack weapon was the laser-guided bomb, supplemented by a few television weapons (such as Walleye) and imaging IR weapons (such as the Navy SLAM, a version of Harpoon).

This was the first war in which the GPS navigation system featured, but its full potential was unrealized. Tomahawk became the first U.S. weapon to use GPS guidance, against targets in Serbia in 1995. Then GPS was embodied in a bomb guidance kit comparable to that used in laser-guided bombs; low-rate production began in April 1997. Production of these bombs became a major factor in the timing of the 2003 attack against Iraq.

For an attack bomber, GPS guidance makes an enormous difference. Using laser or television guidance, the airplane has to approach its target. It can attack from outside the range of point-defense systems, but an attack outside the range of an area-defense system requires considerable sophistication and preplanning—and long-range missiles. The attack requires high visibility, since the bomber must see the target and the descending bomb has to be able to detect laser radiation the target reflects. Presumably the laser on board the airplane is itself detectable, and an alerted enemy may track it or even use it as an aim point. Finally, debris thrown up by the bomb precludes further hits.

A GPS-guided bomb or missile flies to a point in space designated either by the airplane system or before takeoff. The attacker releases the bomb and flies off. It does not matter how many bombs are sent on their way at the same time. Bombs can, moreover, be dropped from higher altitude and therefore from greater range. These weapons change the way carrier firepower is evaluated, in terms of the number of targets the ship can hit each day. When it took a mass sortie to hit even a single target, capacity was defined by how many deck-load strikes the carrier could launch per day—two or three at most. With the advent of guided bombs, the number of targets might increase to about fifty—up to one hit per sortie. With GPS, the criterion becomes targets hit per sortie. The new capacity motivated the redesign of the flight deck in the current *Gerald R. Ford* class. In these ships the emphasis is on how quickly attack aircraft can be reloaded and turned around, not so much on how many are on board.

The Bombcat

The F-14 multimode radar was inherently suited to attack, and its initial software included an attack function. That was ruled out, apparently because the bombs had to be carried under the fuselage; airflow in the tunnel between the two engines tended to suck bombs up. Grumman began working on a strike version of the F-14D once its team lost the ATA competition in 1987. It offered an unsolicited proposal, Quickstrike (Block IV upgrade). That added air-to-ground modes to the APG-71 radar.[2] The two glove pylons would carry off-the-shelf pods: one with a terrain-following radar and a wide-angle FLIR, the other with a steerable FLIR coupled to a laser target designator. The cockpit would be modified with a new head-up display and a color moving-map display and for compatibility with night-vision goggles. Although Congress initially approved Quickstrike development, the Navy rejected the idea, having decided that a developed Hornet (Super Hornet) would be both less expensive and more survivable.

In 1988 Grumman proposed a Super Tomcat 21.[3] New engines (GE F110-429s) and additional fuel in a beefed-up airframe would give it roughly the high-low-low-high mission radius of the A-6 (550 nautical miles) with eight thousand pounds of ordnance. The airframe would have been modified to reduce radar cross section, presumably from the ahead aspect. Additional fuel would have been stored in a thickened and widened glove area. Pylon stations used only for drop tanks in existing F-14s would have been modified to carry weapons, and some of the sensors developed for the A-12 would have been

The F-14D was modified as an effective long-range bomber.
U.S. NAVAL INSTITUTE PHOTO ARCHIVE

incorporated, such as the Night Owl II FLIR. Existing F-14s could be rebuilt to Super Tomcat 21 configuration. It did not help that in November 1989 Secretary of Defense Cheney included termination of the F-14D in a list of defense cuts. Production ended in 1994.

The Navy began looking for a long-range strike bomber as it faced the retirement of the A-6; the last Intruder squadron began its last deployment in June 1996 (ending in December). The projected replacement, the Super Hornet, did not fly in production form until December 1998, and it did not deploy until 2002. In the meantime, the F-14, with its six-hundred-nautical-mile unrefueled mission radius, was the only viable candidate.[4] VX-4 tested bombing with Mk 83 and Mk 84 bombs late in 1987, and in July 1992 the F-14 was certified to drop general-purpose bombs.[5] In September 1995, F-14s dropped laser-guided bombs on Bosnia, but other aircraft had to designate the targets.

Meanwhile the Atlantic Fighter Wing began a crash program to provide F-14s with laser designation, using a version of the Air Force LANTIRN (Low-Altitude Navigation and Targeting for Night) pod.[6] The aft cockpit was fitted with a hand-controller and panel, and the FLIR imagery was piped into the display. A squadron of modified F-14Bs (VF-103) deployed on board *Enterprise* in June 1996. Capability was extended to GPS-guided JDAMs, and LANTIRN designation altitude increased from 25,000 to 40,000 feet. Modified aircraft were called "Bombcats." They fought during Operation Allied Force (Kosovo, 1999) and during the 2003 Gulf War. They were retired in 2006.

The Super Hornet

McDonnell Douglas first offered the Navy an enlarged version of the Hornet about 1984. This supersonic replacement for the aging A-6 was rejected in favor of the stealthy A-12. In 1988, however, the Navy and McDonnell Douglas conducted a joint Hornet 2000 program. Among new requirements was that the Hornet follow-on be able to bring expensive unexpended weapons back on board a carrier. Other requirements were a degree of stealth and increased volume for avionics.

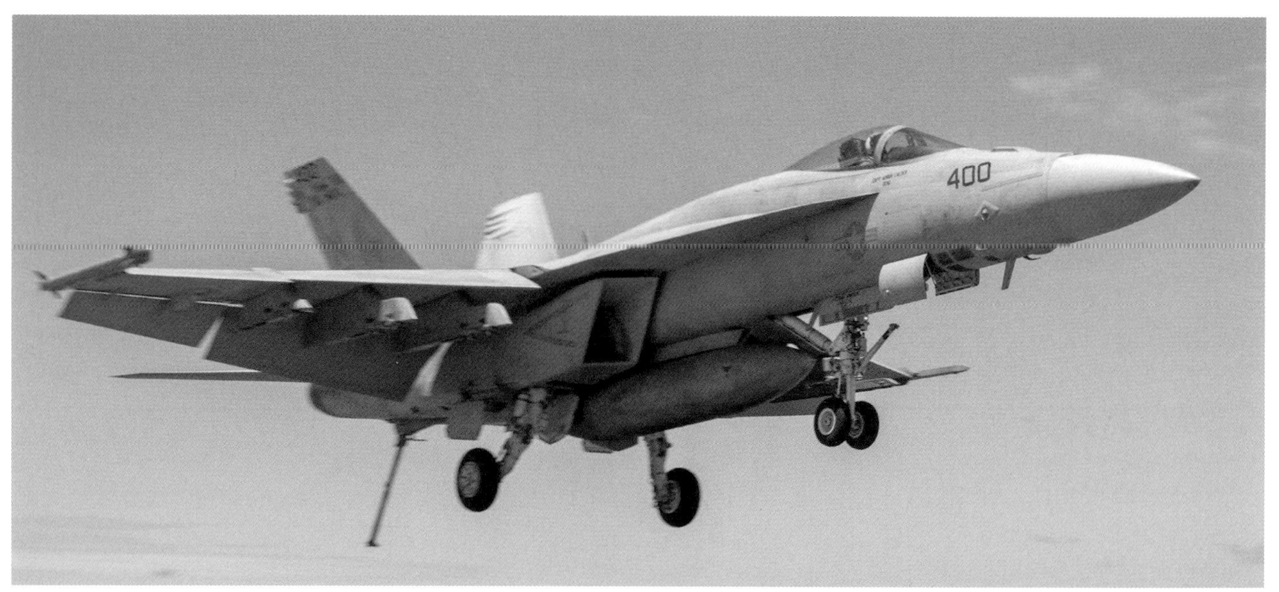

An F/A-18E of VFA-195 landing on board USS *Ronald Reagan*, 12 August 2020. U.S. NAVY BY PETTY OFFICER 2ND CLASS SAMANTHA JETZER

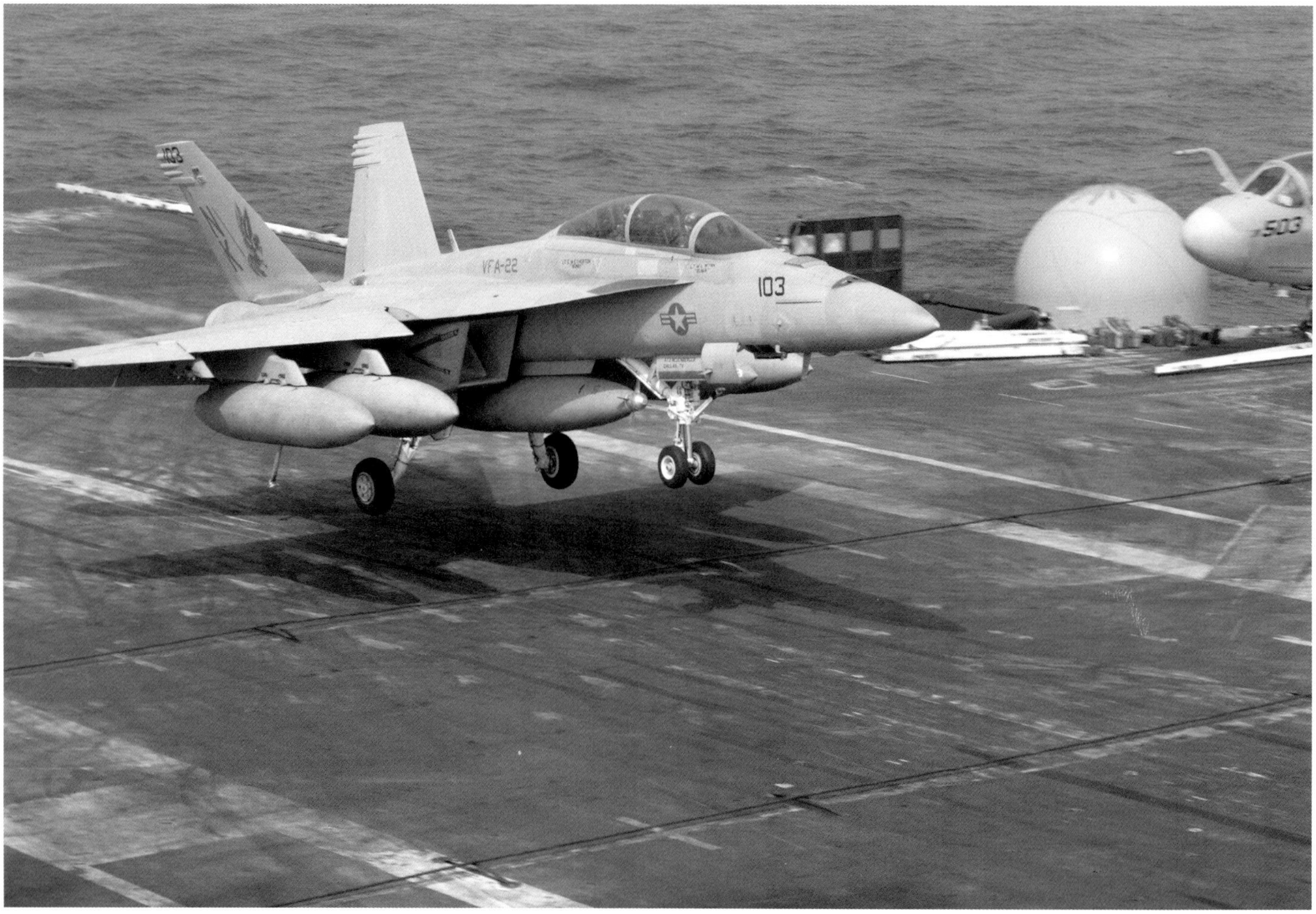

A two-seat F/A-18F of VFA-22 lands on USS *Ronald Reagan*, July 2009. Many Super Hornets are used as tankers, because they have much greater endurance than earlier Hornets. This practice is presumably ending as F/A-18Fs in the fleet are all replaced by F-35Cs. This airplane is acting as a buddy tanker; note the small propeller (turbine) that powers the pump in the centerline tank (fuel is also fed through the other four tanks). U.S. NAVY

The group's obvious option was an enlarged F/A-18, using much the same avionics as the F/A-18C/D, built around a new mission computer. The hope was that by avoiding major avionics costs and by using much the same aerodynamics (albeit scaled up), development could be quick and relatively inexpensive. McDonnell Douglas offered a range of possible upgrades.[7]

Option I was the FY88 baseline (F/A-18C/D) with planned avionics upgrades, a survivability improvement, and more powerful engines. All other upgrades involved increased fuel, more-powerful engines, an active-array radar, and the new Integrated EW System.

Option II was the least elaborate redesign, with a raised hump (for more fuel) and a stiffened wing.

Option III added a larger increased-chord wing to Option II.

Option IIIA added fuselage "plugs" to the larger wing.

Option IIIB had the raised hump (dorsal) plus the larger wing.

Option IIIC, like IIIA, had the fuselage plugs and the larger wing but not the hump.

Option IV, the most radical, had fuselage plugs and a new cranked-arrow (a variation of the compound delta) wing plus canards.

The Navy chose something close to Option IIIC, using a new F414 engine descended from the F404 of the F/A-18 and the F412 of the canceled A-12. It had 36 percent more thrust than the F404. Fuselage plugs added four feet to length, and the wing was 25 percent larger, with a thicker cross section and less twist. It carried two more pylons. The leading-edge extensions were reshaped for better performance at high angles of attack. The Super Hornet carried a quarter more fuel.

Measures to reduce radar cross section are evident in the forward aspect the airplane would present as it approached a defended target. The obvious sign was the redesigned engine inlets, which are shaped internally to block the radar's view of the engine compressor, a major contributor to radar cross section. Some low-observable materials were used and the landing-gear doors and access panels realigned. Reportedly radar cross section was reduced by 90 percent over the F/A-18C/D, although the new airplane is substantially larger.[8]

The pilot could land on a carrier with 9,000 rather than 5,500 pounds of fuel and weapons; typically, 4,000 pounds of fuel was the minimum for final approach.

As built, the Super Hornet lacked an active-array radar, but that was a preplanned improvement in Block 2. The Super Hornet incorporated a towed decoy (initially ALE-50, then the fiber-optic ALE-55) intended specifically to defeat monopulse radars.

The Super Hornet was about the size of a Phantom, so it was still much smaller than an F-14; its gross takeoff weight was 66,000 rather than 73,000 pounds. Its mission radius with two 330-gallon drop tanks and four Mk 83 (1,000-pound) bombs plus two Sidewinders was four hundred nautical miles, compared to 460 for a similarly armed F-14 with two 280-gallon tanks. The F-14 was faster, but it was more expensive to operate.[9]

The Navy described the Super Hornet (F/A-18E/F) as a simple derivative of the earlier Hornets. On that basis OSD approved engineering and manufacturing development in May 1992, without the usual prototyping phase. McDonnell Douglas received the contract in July. The prototype flew in November 1995, a month ahead of schedule. The single-seat version is the F/A-18E, the two-seater the F/A-18F.

The new wing now presented real problems.[10] In March 1996 it was found that one wing would abruptly lose lift in high-speed, high-*g* maneuvers. The aircraft would suddenly roll out of a hard turn. A similar problem had surfaced during tests of the original F/A-18, solved by deleting the snag ("dogtooth") that McDonnell Douglas had added to the original Northrop design. The wing snag (meant to reduce "flutter") was restored in the Super Hornet. Attempts to solve the aerodynamic problem by modifying control software failed. Finally in 1998 it was solved by placing a porous skin over the wing-fold area to minimize the differential (right/left) loss of lift at high angles of attack. Also, more generally, the Super Hornet was publicly criticized for acceleration and maneuverability inferior to those of the F/A-18C/D. Despite the ongoing aerodynamic problem, initial low-rate production was approved in March 1997.

Operational evaluation began in May 1999 and was completed in November, the month the first readiness squadron received its first airplanes. In service, Super Hornets have often been used as tankers rather than strikers, thanks to their greater stores-carrying capacity. That mission became essential when the last Lockheed S-3s were retired in 2007. This diversion helps explain why the first unmanned carrier aircraft, Boeing's MQ-25 Stingray, is to be a tanker.

Once the expensive Joint Strike Fighter was being built, Boeing (which had bought McDonnell Douglas in 1997) proposed an Advanced Super Hornet, with some stealth enhancement, as a much less expensive alternative. The Navy appears to see JSF as a "day one" silver bullet, Super Hornets making up required numbers for strikes once enemy air defenses have been attacked.

Once the Super Hornet had an active-array radar, another option opened. Each element of such a radar can receive or radiate separately, under computer control. Boeing pointed out that the radar could function as a high-powered jammer. It also argued that the four aircrew of an EA-6B were no longer needed, that a two-man cockpit could suffice. The usual wingtip Sidewinders could be replaced by jamming pods. The resulting EA-18G Growler has replaced the EA-6B Prowler. An order for 140 Growlers helped reduce the unit cost of Super Hornets.

Super Hornets are built in "blocks." The first 178 aircraft (both -18E and -18F) constitute Block I (lots 21–25). They have mechanically scanned APG-73 radars and systems similar to those of late-production F-18C/Ds. Block II (first in service 2006) had a new forward fuselage with the Raytheon APG-79 active-array radar, the ALQ-124 Integrated Defensive Electronic Countermeasures suite, Joint Helmet-Mounted Cueing System, ASQ-228 ATFLIR (multispectral targeting pod), Link 16 connectivity, and new weapons. As with all Hornets, capability is defined in part by software; after production, Block IIs were given ASQ-34 long-range IR search and track sensors, for

Digital flexibility made it possible to adapt the F/A-18F as jamming airplane. This EA-18G Growler of VAQ-139 lands on board USS *Nimitz* in the Indian Ocean, 6 October 2020. The missile outboard is an antiradar HARM. U.S. NAVY

example. Block III has five engineering changes: new displays (part of an advanced cockpit), conformal (i.e., to the aircraft's profile) fuel tanks rather than drop tanks, networking capability (beyond Link 16), signature improvements, and a nine-thousand-hour design life instead of the six thousand hours programmed for earlier versions. The new cockpit uses eighteen-inch touch screens rather than the previous five-inch screens; they save considerable weight. The conformal fuel tanks were conceived to reduce drag when the very draggy ALQ-249 jammer was added to the EA-18. Apparently their substantially reduced radar cross section was an unexpected bonus; CNO considered them the most significant Block III improvement. The networking is associated with an EA-18 program called Distributed Targeting Process–Networked, using a new waveform (Tactical Targeting Networking Technology) that supersedes Link 16, enormously increasing its capacity. An additional processor is added to handle the additional load. Much of the Block III capacity can be retrofitted into Block II (but not Block I) aircraft under a projected service-life modification program, including treatments that reduce radar and other signatures.

JSF

Under the Clinton administration, A/F-X was a "paper program," unlikely to produce near-term hardware.[11] DARPA was already conducting a program to develop a supersonic STOVL strike fighter (SSF). It was based on NASA-funded studies conducted between 1980 and 1987 to lead to an AV-8B successor.[12] In the fall of 1986 DARPA awarded the Lockheed Martin Skunk Works (otherwise Advanced Development Programs) a nine-month contract for an exploratory study of a supersonic STOVL for the Marines. It would replace both the F/A-18 and the AV-8B. To achieve VSTOL performance the airplane needed a thrust-to-weight ratio of about 1.2, compared to the 0.6 of the F/A-18.

DARPA awarded the Skunk Works a follow-on contract for the STOVL in January 1988; McDonnell Douglas and General Dynamics received parallel contracts.

Unlike A/F-X, this project involved a new kind of capability, hence a prototype (at least a large-scale model) to test it. It became a candidate for a next-generation bomber.

The design missions were close air support, combat air patrol, and deck-launched intercept. The only explicit requirement was that empty weight be less than 24,000 pounds, which was 5 percent more than that of an F/A-18C; DARPA used weight to control cost. In the fall of 1989 DARPA had the three contractors present their concepts to NavAir. All received follow-on contracts to refine their designs and investigate the use of stealth in a 1990 naval environment.

Supersonic performance demanded twice the engine thrust of earlier aircraft. Half of it had to be moved forward. Lockheed engineers realized that the best way to do that was to use a shaft to drive a fan, which offered maximum mass flow. The after vertical thrust would be provided by vectoring. For roll control, engine bypass air could be ducted to the wingtips.[13] For horizontal flight, the fan was declutched and the doors through which it ingested and exhausted air were closed. For fore-and-aft balance, the engine exhaust could be directed downward by swiveling the nozzle. Since the fan could be decoupled from the engine, the engine could be sized for conventional flight. The thrust produced by the fan would more than balance the additional weight involved. Too, the lower exhaust temperature and pressure involved would make for a more benign ground environment during hover. One advantage of Lockheed's approach was that when the fan was designed out, the space vacated was ideally located for additional fuel. The lift fan was also far more efficient than a lift jet. On the other hand, it was technically risky, because so much power had to be handled by the clutch and the gearbox (which turned the axis of rotation ninety degrees). Allison developed and qualified the fan only barely in time. DARPA seems to have found Lockheed's approach very attractive.

Lockheed's SSF had a large semidelta wing (with tapered trailing edges) with active canards to reduce buffeting. A more conventional highly swept wing would have produced unstable pitch at even moderate angles of attack. The canards were deflected when the center of lift moved aft during supersonic flight. The airplane had two vertical tail fins, canted outboard at a large angle to minimize radar signature. It was armed with two AMRAAMs and two Sidewinders.

During 1991 DARPA and the Skunk Works sought funding to bring SSF to maturity. Assistant Secretary of the Navy for Research, Development, and Acquisition Gerry Cann asked the Naval Research Advisory Committee (NRAC) to assess SSF feasibility. In April 1992 the Deputy Air Force Chief of Staff for Requirements at Air Combat Command visited the Skunk Works and was briefed on the SSF, though the Air Force had no need for a STOVL airplane. However, it was interested in an F-16 replacement. The Skunk Works pointed out that its STOVL could easily be transformed into a conventional airplane by replacing the lift fan with a fuel tank. It turned out that the fuel would weigh about as much as the fan and its driveshaft, giving both airplanes roughly the same midmission maneuverability. This possibility suggested that it might be possible to develop a "common strike fighter" for all three services. Lockheed Skunk Works saw this as a major opportunity. DARPA arranged a round of briefings, and OSD proposed the ideas to the service secretaries.

NRAC endorsed the SSF and suggested that the Navy work on it with the Air Force to produce a highly common multirole strike fighter. Congress appropriated $65 million for a joint STOVL/CTOL Strike Fighter Program. DARPA released a request for proposals in August 1992 calling for technology demonstrations of what was then called the "Common Affordable Lightweight Fighter" (CALF). It would use a fan driven by either a shaft or by ducted gas from its engine. In March 1993 the Skunk Works received a contract to develop the shaft-driven fan, while McDonnell Douglas received one for a gas-driven fan. Congress later (March 1994) appropriated somewhat less to study a lift/cruise-engine concept like that used by the Harrier.[14] Boeing agreed to match

the appropriation with its own funds and won this contract; Congress appropriated more money in 1995.[15]

In each case the contractor was required to design both operational and demonstrator aircraft, showing that the lift/cruise combination was feasible—for example, that ingesting hot exhaust gas would not be a problem. The emphasis changed from a fighter with some strike capability to a strike airplane with some fighter capability. The combination of stealth and long-range air-to-air missiles made it much less likely that aircraft would be called on to dogfight, so the two Sidewinders were eliminated and the airplanes given an internal bay large enough for two two-thousand-pound bombs and two AMRAAMs. That increased frontal area and wave drag. The Skunk Works reduced its canards, based on data from the F-22 program (later they were eliminated altogether and the airplane given a horizontal tail). Its propulsion system, built out of components of existing engines, operated successfully in February 1995 and was then installed in a full-size airframe model.

In February 1993, while all this was happening, the Defense Department began a Bottom-Up Review of current plans, an important objective being to rationalize the five ongoing aircraft development programs: the Air Force F-22 and MRF (Multi-Role Fighter), the Navy F/A-18E/F and A/F-X, and DARPA's CALF. The Air Force and Navy jointly proposed developing a highly common fighter based on DARPA's SSF, which they called the "Joint Attack Fighter." The Navy planned a conventional airplane, but the Marines, supported by the Air Force, argued for the DARPA STOVL fighter.

In September 1993 the review recommended canceling both the MRF and A/F-X and developing the Joint Attack Fighter to replace the AV-8, F-16, and F/A-18 when they retired beginning in 2010. The new program was called "Joint Advanced Strike Technology" (JAST); its first director was the former Deputy Air Force Chief of Staff for Requirements at Air Combat Command, who had attended the DARPA briefing in 1992. The first concept-exploration contracts were awarded in May 1994. Initially JAST did not include a Marine STOVL variant. However, in October 1994 Congress directed that the focus of JAST be the Marines' STOVL version, from which other variants would be derived.[16] OSD, for its part, appears to have initially seen JAST as a paper project rather than a likely airplane.

At the end of the DARPA demonstration phase, McDonnell Douglas decided to abandon its gas-driven lift fan for a shaft-driven one. When Lockheed Martin refused to give the necessary permission, McDonnell Douglas shifted to a lift engine. It did not conduct a large-scale demonstration. Boeing had demonstrated its lift-plus-cruise-engine arrangement only in the hover, whereas Lockheed Martin had demonstrated both engine cycles (lift and cruise).

Formal development of a Joint Strike Fighter was approved in February 1996, a request for proposals being released the following month. Lockheed Martin planned to build one prototype as a STOVL, the other as the Air Force version. The latter could be modified into the Navy version by replacing wing flaps and slats. In November Boeing and Lockheed Martin were chosen to build concept demonstrators. The Marines rejected the McDonnell Douglas proposal, on the ground that the airplane would need far greater maintenance effort because it would have separate lift and cruise engines. They were also aware that the Soviet Yak-38, which used separate engines, had been extremely unreliable in service. (As noted above, McDonnell Douglas soon merged with Boeing.) Northrop Grumman joined the Lockheed Martin team, as did BAE. Because wind-tunnel tests validated the fan/engine combination, Lockheed Martin decided to concentrate on carrier handling qualities with its first prototype. The second would fly first as an Air Force prototype and then be converted to STOVL. The conventional X-35A flew on 24 October 2000 and was converted into the STOVL X-35B. On 20 July 2001 it demonstrated the full mission capability: a short takeoff followed by a supersonic flight, hover, and then a vertical landing. The naval X-35C flew on 16 December 2001.

On 26 October 2001, the program office announced that Lockheed Martin had won the competition; the win was credited to the airplane's lift-fan system. The fighter version of the X-35 became the F-35.[17] Compared to the X-35A/B, the follow-on F-35A/B has slightly greater wingspan (for better maneuverability and range) and slightly enlarged rudder and horizontal tails. The weapon-bay doors on the STOVL version were designed to open during vertical landing to capture the lift jet air reflected by

The next generation: a VFA-147 F-35C lands, 15 September 2020. U.S. NAVY BY PETTY OFFICER 3RD CLASS HAYDEN SMITH

A folded F-35C of VFA-147 on board USS *Carl Vinson*, 13 September 2020. It wears the new stealth finish associated with radar-absorbing material. U.S. NAVY BY SEAMAN KATLYN HRUSKA

the ground (the "fountain") and to counter "suck-down" in ground effect. Weights added included a gun, strength (maneuver limit raised from 7.5 to 9 *g*s), and pylons for external weapons, for a total of more than 3,000 pounds by January 2004. Lockheed Martin asked its employees to find ways of reducing weight and managed to claw back 2,700 pounds. The first production airplane, an Air Force F-35B, was rolled out on 19 February 2006 and flown on 15 December 2006. The first F-35B (STOVL) was rolled out on 18 December 2007, flying on 11 June 2008.

The F-35 was intended as a simple, inexpensive airplane. The attempt to limit cost focused on what could be estimated: platform characteristics, such as performance and stealth. Successive program managers tried in effect to recover capability by adding sophisticated systems and software supporting them. It was impossible to trade off software features to control the price of the airplane. Because of that,

program managers seem to have treated software as cost-free, which is very far from reality. For a time it seemed unlikely that the mass of software required could possibly be written in time. On the other hand, software is unlike hardware: reproduction and installation really are virtually without cost. The more airplanes built, therefore, the lower the cost per software installation. Against that, major block upgrades require new generations of software, hence massive new investments.

The best-known enhancements are an ELINT capacity and the pilot's ability to "see through" the airplane in any direction using his helmet display. The unusually elaborate ESM antennas cover a wide frequency range and offer high accuracy. They and the inboard electronics were intended to provide a pilot with unusually comprehensive situational awareness, based on a stored emitter database. Functions included separating out pulse trains so that the system could try to identify each radar the airplane

Taking off on 22 July 2020, this Marine Corps F-35B shows its vertical-launch system: the tailpipe tilted down, the big flap lifted over the opening for the fan, and the auxiliary opening abaft it.
U.S. NAVY BY MASS COMMUNICATIONS SPECIALIST SEAMAN WESLEY RICHARDSON

An F-35B of VMFA-211 launches from the large-deck amphibious ship *Essex* in the Indian Ocean, 22 July 2020.
U.S. NAVY BY MASS COMMUNICATIONS SPECIALIST 1ST CLASS RAWAD MADANAT

detected. Normally this sort of capability would be provided only on electronic reconnaissance airplanes such as the EA-6B.

The ability to see through the airplane was more obviously spectacular. Cameras under the airplane provide images that the computer fuses into a picture extending all the way around the airplane. The pilot sees the image on his helmet display.

Given its expensive software, F-35 was described as the most expensive military aircraft in history, at about $150 million each. This unit price has declined considerably as production has increased with substantial foreign sales. In 2020 it is claimed that an F-35 costs less than a modernized F-15.

Aircraft Data Tables

Data are in imperial units (feet, inches, pounds). Dimensions are mostly feet and inches (ft-in), a few of them feet and decimal, to two places. Note that guns in turrets are *not* included in lengths. Speeds are all in miles per hour (mph), and distances are in statute miles. Speeds are indicated as at a stated altitude, so that 250/15,000 means 250 mph at 15,000 ft. In range and combat radius, the figure after diagonal is the speed in mph for that range. Note that combat radius was not calculated until well into World War II. Rate of climb is in feet per minute or altitude reached in a given number of minutes. Roll means the distance the airplane must travel on the ground before taking off; the figures show how it is affected by the strength of wind flowing over a flight deck (wind over deck, WOD). The great unassisted-takeoff distances for jets show why catapults have been so important.

Engine ratings are in horsepower for piston engines and in pounds of static thrust for jets; at 375 mph, one pound of thrust equals one horsepower, but above that the equivalent horsepower increases in proportion to the speed. For jets, takeoff rating is with afterburner, normal rating without.

All gross weights are takeoff weights.

Performance is given for the airplane with a full bomb load or with the bomb load indicated, e.g., 275/torpedo means 275 mph carrying a torpedo (this does not apply to unofficial data, for which loads are uncertain). Stall speeds are at landing weight, so that they approximate carrier approach speeds. For piston-engine aircraft, the service ceiling is the altitude at which the rate of climb is reduced to 100 ft/min. For U.S. jets, the service ceiling quoted is the combat ceiling, at which the rate of climb is reduced to 500 ft/min.

Unless otherwise indicated, figures are from official tables. For the period from 1931 through 1939, they are taken from Bureau of Aeronautics summaries of aircraft data. World War II data are from official Airplane Characteristics and Performance (ACP) reports, for postwar, from Standard Aircraft Characteristics Charts (SACCs). These papers offer characteristics at various loads.

A

Torpedo Bombers

DESIGNATION	DT-2[a]	CS-1[b]	SC-2[c]
NAME	N/A	N/A	N/A
ENGINE	Liberty 12A	T-2	T-3
RATING			
NORMAL	450	525	585
TAKEOFF	–	–	–
SPAN	50-0	56-6	56-6⅞
FOLDED	19-1	N/A[d]	N/A
WING AREA	707	856	856
LENGTH	34-1⅝	38-5⅜	37-8¾
HEIGHT	13-7¼	15-2½	14-8
WEIGHT			
EMPTY	3,737	4,690	5,007
GROSS	6,503	7,908	8,422
ARMAMENT			
GUNS	1x0.30	1x0.30	1x0.30
BOMBS	torpedo	torpedo	torpedo
MAXIMUM SPEED			
LOW	103.5/SL	101.5/SL	103/SL
HIGH	N/A	N/A	N/A
CLIMB RATE AT SL	5,000/10 mins.	5,000/12.2 mins.	N/A
SERVICE CEILING	9,600	9,100	7,470
RANGE	208	452	336/torpedo
COMBAT RADIUS	N/A	N/A	N/A
ROLL			
0 WOD	N/A	N/A	N/A
25 kts. WOD	N/A	N/A	N/A
STALL SPEED	49	51	53

DESIGNATION	TB-1	T2D-1	T3M-2
NAME	N/A	N/A	N/A
ENGINE	1A-2500	R-1750(2)	3A-2500
RATING			
NORMAL	770/SL	525	710
TAKEOFF	–	–	–
SPAN	55-0	57-0	56-7
FOLDED	21-8	26-0	22-9
WING AREA	868	886	883
LENGTH	40-10	42-0	41-4
HEIGHT	14-4	15-11	15-2
WEIGHT			
EMPTY	6,298	6,011	5,814
GROSS	10,015[e]	10,523	9,503
ARMAMENT			
GUNS	1x0.30	2x0.30	1x0.30
BOMBS	torpedo	torpedo	torpedo
MAXIMUM SPEED			
LOW	106/SL	123.5/SL	109.4/SL
HIGH	N/A	120/5,000	78/8,000
CLIMB RATE AT SL	4770/10 mins.*	5,000/6.8 mins.	5,000/16.8 mins.
SERVICE CEILING	9,700*	12,600	7,900
RANGE	850	442	634
COMBAT RADIUS	N/A	N/A	N/A
ROLL			
0 WOD	N/A	481	550
25 kts. WOD	N/A	138	220[f]
STALL SPEED	58*	58	55.1

DESIGNATION	T4M-1	XT6M-1	XT3D-2
NAME	N/A	N/A	N/A
ENGINE	R-1690-24	R-1860	R-1830-54
RATING			
NORMAL	525	575	800/SL
TAKEOFF	N/A	N/A	N/A
SPAN	53-0	42-3	50-0
FOLDED	21-4	N/A	19-1
WING AREA	656	502	636
LENGTH	35-7	33-8	35-6
HEIGHT	14-9	13-10	14-0
WEIGHT			
EMPTY	3,931	3,500	4,885
GROSS	7,387	6,866	8,759

ARMAMENT			
GUNS	1x0.30	2x0.30[g]	1x0.30
BOMBS	torpedo	torpedo	torpedo
MAXIMUM SPEED			
LOW	114/SL	130.7/SL	153.2/SL
HIGH	106/5,000	116/10,000	161/5,000
CLIMB RATE AT SL	5,000/14 mins.	5,000/8.8 mins.	5,000/6.6 mins.
SERVICE CEILING	10,150	11,600	15,700
RANGE	363	323	778
COMBAT RADIUS	N/A	N/A	N/A
ROLL			
0 WOD	674	N/A	600
25 kts. WOD	185	N/A	185
STALL SPEED	56.7	N/A	60.4

DESIGNATION	TBD-1	TBF-1C[h]	TBY-2[i]
NAME	Devastator	Avenger	Seawolf
ENGINE	R-1830-64	R-2600-8	R-2800-22
RATING			
NORMAL	850/8,000	1,450/12,000	1,600/16,400
TAKEOFF	900	1,700	2,100
SPAN	50-0	54-2	56-11
FOLDED	25-8 ½	19-0	27-6
WING AREA	422	490	440
LENGTH	35-0	40-0	39-2
HEIGHT	15-1	16-5	15-6
WEIGHT			
EMPTY	5,627	10,555	11,366
GROSS	9,359	16,412	17,491
ARMAMENT			
GUNS	2x0.30	3x0.50, 1x0.30[j]	3x0.50[k]
BOMBS	torpedo	torpedo	torpedo
MAXIMUM SPEED			
LOW	192/SL	249	292
HIGH	207/8,000	257/12,000	312/17,700
CLIMB RATE AT SL	840/min.	1,430/min.	1,770/min.
SERVICE CEILING	19,400	22,400	29,400
RANGE	460	1105/153	1179/179
COMBAT RADIUS	N/A	225	224
ROLL			
0 WOD	840	1,071	900
25 kts. WOD	328	435	401
STALL SPEED	66.8	76.9	80.4

DESIGNATION	XTB2D-1[l]	XTB2F-1[m]	XTB3F-1[n]
NAME	Skypirate	N/A	N/A
ENGINE	XR-4360-8	R-2800-22(2)	R-2800-34W[o]
RATING			
NORMAL	2,400/13,500	1,600/16,000	1,850/15,500[p]
TAKEOFF	3,000	2,100	2,100
SPAN	70-0	74-0	60-0
FOLDED	36-0	33-0	24-5
WING AREA	605	777	549
LENGTH	46-0	52-0	42-1
HEIGHT	22-7	21-2	13-2
WEIGHT			
EMPTY	18,405	22,227	13,370
GROSS	28,545	35,720	19,285
ARMAMENT			
GUNS	7x0.50	7x0.50	2x20 mm
BOMBS	2 torpedoes	2 torpedoes	1 torpedo
MAXIMUM SPEED			
LOW	292/SL	300	251
HIGH	312/15,600	322/17,900	290/19,900
CLIMB RATE AT SL	1,390/min.	1,710/min.	1,460/min.
SERVICE CEILING	24,500	25,200	28,600
RANGE	1,250/162	2,041/189	1,180/148
COMBAT RADIUS	247	454	175
ROLL			
0 WOD	1,019	1,032	532
25 kts. WOD	485	518	214
STALL SPEED	86	91.7	78.6

[*]   See note e.

[a]   DT-2 data from BuAer sheet prepared to trade data with the RAF, 1924; BuAer secret correspondence, NARA.

[b]   Data from Ray Wagner, *American Combat Planes of the 20th Century* (Reno, Nev.: Jack Bacon, 2004). This was the fourth edition of his massive book. He used official reports, but his data are not nearly as detailed as those in the ACPs and SACCs. Peter M. Bowers, *Curtiss Aircraft 1907–1947* (London: Putnam, 1979) gives a maximum speed of 105 mph, cruising speed 84 mph, range 570 miles, climb rate 450 ft/min., and service ceiling 7,200 ft.

[c]   Data from Wagner, *American Combat Planes*.

[d]   Wings folded, but folded width is not known.

[e]   Data are mainly from Wagner, *American Combat Planes*. Some data, indicated by asterisk, are from the sheet of estimated data for Design 35 dated 20 June 1925.

Weight with torpedo is given as 10,265 lb. Initial rate of climb was 615 ft/min. The engine for these calculations was rated at 730 hp. Estimated speed was 112 mph.

[f] In 20 knots of wind, from a BuAer paper comparing the T3M-1 to the F8C-2.

[g] Two flexible guns in rear cockpit.

[h] Data dated 1 July 1943.

[i] Data dated 1 September 1944.

[j] Two 0.50s in nose, one in turret, 0.30 in ventral (tunnel) position.

[k] Two wing guns plus turret gun; the ventral gun was omitted in production aircraft.

[l] Data issued 1 February 1944 but reissued unchanged 1 January 1947.

[m] Data issued 28 September 1943; this is not the final version with 75-mm cannon.

[n] Data issued 12 July 1945.

[o] Plus Westinghouse 19XB jet rated at 1,600-lb thrust at sea level. Takeoff runs are with the jet engine (never actually used). Originally the piston engine was to have been an R-3350-24W, but that engine was unavailable and was changed to an R-2800; the TB3F-2 was to have had the more powerful engine.

[p] This is the combat power rating (2,380 hp at sea level). Equivalent military rating was 1,700 hp at 16,600 ft (2,100 at sea level).

B

Dive-Bombers

DESIGNATION	XT5M-1	F8C-4	BM-2
NAME	N/A	Helldiver	N/A
ENGINE	R-1690-22	R-1340-88	R-1690-44
RATING			
NORMAL	525	450	600/6,000 ft
TAKEOFF	N/A	N/A	N/A
SPAN	41-0	32-0	41-0
FOLDED	N/A	N/A	N/A
WING AREA	417	308	436
LENGTH	28-4	25-11	28-9
HEIGHT	12-4	10-10	12-4
WEIGHT			
EMPTY	3,084	2,513	3,563
GROSS	5,693	3,783	6,131
ARMAMENT			
GUNS	2x0.30	3x0.30[a]	2x0.30
BOMBS	1,000 lb	500 lb	1,000 lb
MAXIMUM SPEED			
LOW	133.9/SL	137/SL	N/A
HIGH	129/5,000 ft	134/5,000 ft	145/6,000 ft
CLIMB RATE AT SL	5,000/7.8 mins.	5,000/6 mins.	5,000/7.3 mins.
SERVICE CEILING	13,250	15,000	15,200
RANGE	442	455	415
COMBAT RADIUS	N/A	N/A	N/A
ROLL			
0 WOD	620	434	580
25 kts WOD	190	130	183
STALL SPEED	61.6	59.2	61.4

DESIGNATION	F4B-1	XF6B-1	BFC-2
NAME	N/A	N/A	N/A
ENGINE	R-1340-8	R-1535-44	R-1820-78
RATING			
NORMAL	500/6,000 ft	625/5,500 ft	700/4,000 ft
TAKEOFF	–	–	–
SPAN	30-0	28-6	31-6
FOLDED	N/A	N/A	N/A
WING AREA	227.5	252	262
LENGTH	20-1	22-2	25-0
HEIGHT	9-4	10-6	10-7
WEIGHT			
EMPTY	1,950	2,668	3,111
GROSS	2,750	4,104	4,712
ARMAMENT			
GUNS	2x0.30	2x0.30	2x0.30
BOMBS	500 lb	500 lb	500 lb
MAXIMUM SPEED			
LOW	N/A	N/A	197.9/4,000 ft
HIGH	176/6,000 ft	193.1/5,500 ft	177/15,000 ft
CLIMB RATE AT SL	5,000/2.9 mins.	5,000/4.2 mins.	5,000/2.6 mins.
SERVICE CEILING	27,700	22,300	20,700
RANGE	371	394	560
COMBAT RADIUS	N/A	N/A	N/A
ROLL			
0 WOD	280	606	645
25 kts WOD	83	224	243
STALL SPEED	58.8	68.2	69.2

DESIGNATION	BF2C-1	SU-1	XSE-1
NAME	N/A	Corsair	N/A
ENGINE	R-1820-04	R-1690-40	R-1820
RATING			
NORMAL	700/8,000 ft	600	600/8,000 ft
TAKEOFF	730	–	–
SPAN	31-6	36-0	49-9
FOLDED	N/A	N/A	N/A
WING AREA	262	325.6	411
LENGTH	23-0	26-3	29-8
HEIGHT	10-10	11-6	12-2½
WEIGHT			
EMPTY	3,329	2,984	3,985
GROSS	5,086	4,502	6,298

ARMAMENT			
GUNS	2x0.30	3x0.30[c]	3x0.30[d]
BOMBS	500 lb	None	None
MAXIMUM SPEED			
LOW	N/A	170.4/SL	173/SL
HIGH	215/8,000 ft	148/15,000 ft	179.4/8,000 ft
CLIMB RATE AT SL	5,000/2.6 mins.	5,000/4 mins.	5,000/4.9 mins.
SERVICE CEILING	22,700	19,900	22,600
RANGE	700	654	1394
COMBAT RADIUS	N/A	N/A	N/A
ROLL			
0 WOD	818	433	790
25 kts WOD	331	133	272
STALL SPEED	73	60.5	64.9

DESIGNATION	**XBY-1**	**BG-1**	**XB2G-1**
NAME	N/A	N/A	N/A
ENGINE	R-1820-78	R-1535-82	R-1535-82
RATING			
NORMAL	600/8,000 ft	700/8,900 ft	700/8,900 ft
TAKEOFF	N/A	750	N/A
SPAN	50-0	36-0	36-0
FOLDED	N/A	N/A	N/A
WING AREA	361.5	384	384
LENGTH	33-8	28-9	28-9
HEIGHT	10-6	11-0	11-9
WEIGHT			
EMPTY	3,834	3,853	4,248
GROSS	6,607	6,297	6,802
ARMAMENT			
GUNS	1x0.30	2x0.30	1x0.50, 1x0.30
BOMBS	1,000 lb	1,000 lb	1,000 lb
MAXIMUM SPEED			
LOW	154/SL	171	N/A
HIGH	170/8,000	188/8,700 ft	197/8,900 ft
CLIMB RATE AT SL	7,600/10 mins.	5,000/5.5 mins.	5,000/6.5 mins.
SERVICE CEILING	20,500	20,100	19,500
RANGE	628	549	582
COMBAT RADIUS	N/A	N/A	N/A
ROLL			
0 WOD	930	580	670
25 kts WOD	338	208	235
STALL SPEED	67.6	67.1	61.0

DESIGNATION	SF-1	SBU-1	SBC-4
NAME	N/A	N/A	Helldiver
ENGINE	R-1820-84	R-1535-82	R-1830-34
RATING			
NORMAL	725/3,000 ft	700/8,900 ft	750/15,000 ft
TAKEOFF	N/A	750	950
SPAN	34-6	33-6	34-0
FOLDED	N/A	N/A	N/A
WING AREA	310	327	317
LENGTH	24-11	27-9	27-6
HEIGHT	11-1	12-0	13-2
WEIGHT			
EMPTY	3,254	3,563	4,847
GROSS	5,067	5,533	6,907
ARMAMENT			
GUNS	2x0.30	1x0.30,1x0.50	2x0.30[e]
BOMBS	2x116 lb	500 lb	500 lb
MAXIMUM SPEED			
LOW	199/SL	174/SL	213/SL
HIGH	206.4/3,000 ft	198.5/8,900 ft	220/9,800 ft
CLIMB RATE AT SL	1,650/min.	1,180/min.	1,590/min.
SERVICE CEILING	23,100	23,300	26,100
RANGE	880	531	1,070/175
COMBAT RADIUS	N/A	N/A	N/A
ROLL			
0 WOD	495	549	823
25 kts WOD	184	204	378
STALL SPEED	64.6	66.4	70.6[f]

DESIGNATION	SB2U-1[g]	BT-1	SBD-3
NAME	Vindicator	N/A	Dauntless
ENGINE	R-1535-96	R-1535-94	R-1820-52
RATING			
NORMAL	750/9000	750/9500 ft	800/16,000 ft
TAKEOFF	825	825	1000
SPAN	42-0	41-6	41-6
FOLDED	16-0	N/A	N/A
WING AREA	305	319	325
LENGTH	34-0	31-8	32-8
HEIGHT	10-3	9-11	13-7
WEIGHT			
EMPTY	4,676	4,314	6,345
GROSS	7,278	7,087	9,407

ARMAMENT			
GUNS	1x0.50,1x0.30[h]	1x0.50, 1x0.30	2x0.50,2x0.30
BOMBS	1,000 lb	l,000 lb	l,000 lb
MAXIMUM SPEED			
LOW	216/SL	196/SL	224/SL
HIGH	234/9,500	213/9,500	241/16,000
CLIMB RATE AT SL	880/min.	1,080/min.	950/min.
SERVICE CEILING	22,900	23,700	22,800
RANGE	635	555/135	1205/152
COMBAT RADIUS	N/A	N/A	N/A
ROLL			
0 WOD	811	688	1,250
25 kts WOD	338	272	580
STALL SPEED	71.1	68.2	81.8

DESIGNATION	SBN-1[i]	SB2C-4	SB2A-4[j]
NAME	N/A	Helldiver	Bermuda
ENGINE	R-1820-38	R-2600-20	R-2600-8A
RATING			
NORMAL	750/15,200	1,450/15,000	1,450/12,000
TAKEOFF	950	1,900	1,700
SPAN	39-0	49-9	47-0
FOLDED	N/A	22-5	21-5
WING AREA	259	422	379
LENGTH	27-8	36-8	39-2
HEIGHT	12-5	13-2	15-5
WEIGHT			
EMPTY	4,503	10,547	9,785
GROSS	6,759	15,189	13,811
ARMAMENT			
GUNS	1x0.50, 1x0.30	2x20mm, 2x0.30	2x0.50, 2x0.30[k]
BOMBS	500 lb	l,000 lb	l,000 lb
MAXIMUM SPEED			
LOW	226/SL	261/SL	248/SL
HIGH	248/15,200	285/16,400	268/13,000
CLIMB RATE AT SL	1,730/min.	1,430/min.	1,590/min.
SERVICE CEILING	26,800	27,000	23,800
RANGE	1,015/117	1,165/149	1,095/161
COMBAT RADIUS	N/A	305	N/A
ROLL			
0 WOD	654	925	1,060
25 kts WOD	280	412	485
STALL SPEED	70.7	86.6	89.7

DESIGNATION	XSB2D-1[l]	XBTC-1[m]	XBT2C-1[n]
NAME	Destroyer	N/A	N/A
ENGINE	R-3350-14	R-4360-8A	R-3350-24
RATING			
NORMAL	1,900/14,000	2,400/13,600	1,900/14,800
TAKEOFF	2,300	3,000	2,500
SPAN	45-0	50-0	47-7
FOLDED	20-8	22-6	22-6½
WING AREA	375	406	416
LENGTH	38-7	38-7	38-8
HEIGHT	16-11	16-8	16-5
WEIGHT			
EMPTY	12,458	13,947	12,268
GROSS	18,168	19,067	17,118
ARMAMENT			
GUNS	2x20 mm, 2x0.50	4x20 mm	2x20 mm
BOMBS	1,000 lb[o]	1,000 lb	1,000 lb
MAXIMUM SPEED			
LOW	311/SL	323	297
HIGH	338/16,100	365/16,000	330/17,000
CLIMB RATE AT SL	1,460/min.	1,820/min.	1,890/min.
SERVICE CEILING	22,800	26,300	26,300
RANGE	1,480/185	1,140/185	1,435
COMBAT RADIUS	N/A	155	283
ROLL			
0 WOD	1,090	737	778
25 kts WOD	545	341	361
STALL	89.4	84.7	92.7

[a] Two forward-firing, one flexible.

[b] Performance without bomb load.

[c] One fixed, two flexible. Could be provided with racks to carry four light bombs underwing, not counted in these data. The 1933 description in Jane's included only one flexible gun.

[d] Two forward-firing, one flexible.

[e] Nose and flexible guns; nose gun could be 0.50.

[f] For SBC-3, which should be similar.

[g] Data from 1 October 1939 BuAer sheet.

[h] One 0.30 or 0.50 in right wing, provision for same in left wing, 0.30 flexible gun.

[i] Data sheet dated 11 November 1942.

[j] Data dated 1 March 1943.

[k] Two fixed 0.50 in nose, two 0.30 in wings, two 0.30 in dorsal mounting.

[l] Data from 1 December 1942 sheet.

[m] Estimates dated 1 August 1944.

[n] Estimates dated 1 August 1945. Gross weight (not empty weight) includes APS-4 radar "bomb."

[o] Bomb-bay capacity: 2x500 or 1,000- or 1,600-lb bombs; wings could carry 2x100-lb or 2x325-lb depth bomb.

C
Single-Seat Bombers

DESIGNATION	BTD-1[a]	XBTK-1	AM-1
NAME	N/A	N/A	Mauler
ENGINE	R-3350-14	R-2800-34W	R-4360-4
RATING			
NORMAL	1,900/14,000	1,850/16,000	2,400/16,200
TAKEOFF	2,300	2,100	3,000
SPAN	45-0	48-8	50-0
FOLDED	–	21-0	23-4.75
WING AREA	373	380	496
LENGTH	38-7	38-11	41-2
HEIGHT	13-7	15-8	16-10
WEIGHT			
EMPTY	12,900	9,959	14,510
GROSS	18,140	13,871	23,386[b]
ARMAMENT			
GUNS	2x20 mm	2x20 mm	4x20 mm
BOMBS	1,000 lb	1,000 lb	2,000 lb
MAXIMUM SPEED			
LOW	319/SL	274/SL	277/SL
HIGH	344/16,100	318/20,200	313/17,700
CLIMB RATE AT SL	1,650/min.	2,220/min.	1,490/min.
SERVICE CEILING	23,600	31,000	24,100
RANGE	1,395/188	1,105/149	1,800
COMBAT RADIUS	270	220	679
ROLL			
0 WOD	1,092	541	960
25 kts. WOD	549	228	468
STALL SPEED	91.5	84.5	106

DESIGNATION	AD-4[c]	XA2D-1[d]
NAME	Skyraider	Skyshark
ENGINE	R-3350-26WA	XT40-A-6
RATING		
NORMAL	2,300/SL	5,035[e]
TAKEOFF	2,700	5,035
SPAN	50-0	50-0
FOLDED	23-10½	20-6
WING AREA	400	400
LENGTH	39-3	41-4
HEIGHT	15-8	16-11
WEIGHT		
EMPTY	11,712	14,139
GROSS	21,483	23,944
ARMAMENT		
GUNS	4x20 mm	4x20 mm
BOMBS	2,000 lb[f]	2,000 lb[g]
MAXIMUM SPEED		
LOW	N/A	451/SL[h]
HIGH	301/18,800	427/28,500
CLIMB RATE AT SL	1,480/min.	4,120/min.
SERVICE CEILING	22,800	39,500
RANGE	1,277/231	909/335
COMBAT RADIUS	598/228	426/331
ROLL		
0 WOD	1,400	1,150
25 kts. WOD	710	660
STALL SPEED	101	113

[a]   Estimates dated 17 December 1943.

[b]   Gross weight with one 2,000-lb bomb and two wing drop tanks; 21,410 lb with one 2,000-lb bomb and one APS-4 radar "bomb."

[c]   Data dated 1 November 1952.

[d]   Data dated 1 October 1950.

[e]   And 830-lb thrust.

[f]   One 2,000-lb bomb plus twelve 5-in HVARs (high-velocity aircraft rockets) and two 150-gal drop tanks. Maximum bomb capacity for shipboard is 6,500 lb (9,000 ashore).

[g]   Two 1,000-lb bombs, eight 5-in rockets, one 150-gal drop tank.

[h]   In combat condition, 19,160 lb; maximum at 28,200 ft was 494.5 mph; combat speed was 455.4 mph at 1,500 ft.

D

Strategic Bombers

DESIGNATION	P2V-3C[a]	AJ-1(A-2)[b]	XA2J-1[c]
NAME	Neptune	Savage	N/A
ENGINE	R-3350-26W (2)	R-2800-44W (2)[d]	T40-A-6 (2)
RATING			
NORMAL	2,700	2,300/30,000	5,035/14,300[e]
TAKEOFF	3,200[f]	4,600	N/A
SPAN	100-0	71-5	71-6
FOLDED	N/A	49-6½	46-0
WING AREA	1,000	836	850
LENGTH	76-10	63-1	70-2
HEIGHT	28-1	21-5	22-6
WEIGHT			
EMPTY	33,504	30,776	32,169
GROSS	74,100[g]	50,963	58,000
ARMAMENT			
GUNS	2x20 mm	None	2x20 mm
BOMBS	10,000 lb	Mk 15[h]	10,500 lb
MAXIMUM SPEED			
LOW	356.5./12,900[i]	357/SL	446/SL[j]
HIGH	320/30,000[k]	339/26,000	451/24,000
CLIMB RATE AT SL	N/A[l]	970/min.	3260/min.[m]
SERVICE CEILING	N/A	33,700	37,600
RANGE	4,198	1,731/270	1,760/400
COMBAT RADIUS	2,099/178[n]	828	978/400
ROLL			
0 WOD	N/A[o]	1,420[p]	1,190
25 kts WOD	N/A	795	645
STALL SPEED	103.5[q]	119.6	102.4

DESIGNATION	A-3B(CLE)[r]	A-5A
NAME	Skywarrior	Vigilante
ENGINE	J57-P-10 (2)	J79-GE-8(2)
RATING		
NORMAL	10,500	10,900
TAKEOFF	N/A	17,000
SPAN	72-6	53-0
FOLDED	49-2	42-0
WING AREA	812	700
LENGTH	75-8	76-6
HEIGHT	22-9	19-5
WEIGHT		
EMPTY	37,077	32,714
GROSS	73,000[s]	55,160
ARMAMENT		
GUNS	None	None
BOMBS	2x2,050 lb	Mk 28
MAXIMUM SPEED		
LOW	643/SL	805/SL[t]
HIGH	590/35,000	687/20,000[u]
CLIMB RATE AT SL	5,440/min.	8,000/min.
SERVICE CEILING	41,500	41,400[v]
RANGE	2,610/501	2,013/560
COMBAT RADIUS	1,323/501	788/560
ROLL		
0 WOD	3,680	1,800
25 kts WOD	2,360	4,050
STALL SPEED	142.6	154.7

[a]   Data based on RG 72: BuAer correspondence concerning the P2V-3C project, 1948–1950; it did not include a standard characteristics sheet. This file includes graphs showing estimated JATO takeoff data.

[b]   Characteristics dated 30 June 1957.

[c]   Data from Standard Characteristics Chart compiled prior to first flight (date unreadable, presumably 1950). Later data show wing area of 836 sq ft.

[d]   Plus J33-A-10 jet engine, 4,600-lb thrust.

[e]   Plus 1,225-lb thrust (turbojet engine).

[f]   Combat rating. Takeoff rating is the "normal" one.

[g]   For data sheet, maximum weight is 58,000 lb. The normal maximum allowable gross weight was 62,000 lb, and as of March 1950 VC-5 had typically taken off from carriers 12,100 lb heavier. For VC-5, assumed landing gross weight was 60,000 lb.

[h]   And two 300-gal. tip tanks (3,600 lb of fuel in addition to 7,302 lb internal). Tip tanks not included for SL speed.

[i]   At 74,000 lb, maximum permissible smooth air speed was 240 kts. (276 mph).

[j]   At combat weight of 42,258 lb, after expending fuel; combat speed was 404 mph at 1,500 ft.

[k]   However, it was estimated that the airplane could make 414 mph for sixteen minutes using chemical supercharging (nitrous oxide). Estimates were dated 3 August 1948, at which time estimated combat radius was 2,570 miles.

[l]   At 58,000 lb, service ceiling was 24,400 ft; maximum speed at sea level was 258 kts. (296.7 mph); maximum speed was 279 kts. (320.9 mph) at 9,200 ft.

[m]   At combat weight, 5,815 ft/min.

[n]   Radius, not combat radius—i.e., half of range. Original estimate was 2,235 nm (2,570 miles). It included a combat run over the target at 20,000 ft, transits out and back at 1,500 ft.

[o]   At 58,000 lb, takeoff runs were 1,690 and 910 feet at, respectively, 0 and 25 knots of wind.

[p]   Takeoff runs with all engines; with piston only, 2,200 and 1,300 ft.

[q]   Estimated takeoff speed at 74,000 lb. Power-off stall speed of a P2V-3 at 58,000 lb was 80.8 kts. (92.9 mph).

[r]   A-3B(CLE) sheet dated July 1967; the cover photo shows the airplane without tail guns.

[s]   Listed as limit for carrier takeoff, high-altitude attack mission.

[t]   Given as combat speed for a low-level attack, with afterburner. No sea-level speed is given for the fully loaded airplane.

[u]   Presumably without afterburner; combat data assume afterburning.

[v]   In combat (47,530 lb), rate of climb at sea level with full afterburners on was 33,900 ft/min. and maximum speed at sea level was 700 kts. (805 mph). Combat speed at combat altitude for a high-altitude attack was 1,090 kts. (1,254 mph) at 54,000 ft. Data given here are at takeoff weight for this mission. Stall speed was 109 kts. (125.4 mph).

E
Light Jet Bombers

DESIGNATION	A-4E[a]	A-6E[b]	A-7E[c]
NAME	Skyshark	Intruder	Corsair II
ENGINE	J52-P-6A	J52-P-88(2)	TF41-A-2
RATING			
NORMAL	8,500	9,300	15,000
TAKEOFF	N/A	N/A	N/A
SPAN	27-6	53-0	38-9
FOLDED	N/A	25-4	23-9
WING AREA	260	528.9	375
LENGTH	41-4[d]	54-9	46-1½
HEIGHT	15-0	16-2	16-1
WEIGHT			
EMPTY	9,624	26,456	18,546
GROSS	18,311	54,719	34,781
ARMAMENT			
GUNS	2x20 mm	none	1x20 mm
BOMBS	Mk 28	B43[e]	12 Mk 81[f]
MAXIMUM SPEED			
LOW	635/SL	605/SL	645/7,500[g]
HIGH	N/A	N/A	N/A
CLIMB RATE AT SL	7,100	5,330	8,630/min.
SERVICE CEILING	38,800	37,200	34,750
RANGE	965/498	2,818/474	1,342/508
COMBAT RADIUS	200	877/393	497/463
ROLL			
0 WOD	2,660	3,890	3,670
25 kts. WOD	1,880	2,850	N/A
STALL SPEED	121	103.5	162.4

DESIGNATION	AV-8B[h]	F/A-18C	F/A-18E[i]
NAME	Harrier II	Hornet[j]	Super Hornet
ENGINE	F402-RR-406	F404-GE-402(2)	F414-GE-400(2)
RATING			
NORMAL	18,720[k]	11,000	–
TAKEOFF	21,550	17,750	22,000
SPAN	30.33	40-4	42-10[l]
FOLDED	N/A	27-6	32-8
WING AREA	230	400	500
LENGTH	46.33	56-0	60-2
HEIGHT	11.65	15-4	16-0
WEIGHT			
EMPTY	12,835	23,832	30,564
GROSS	31,000	37,708	47,881[m]
ARMAMENT			
GUNS	1x25 mm	1x20 mm	1x20 mm
BOMBS	6 Mk 82[n]	4 Mk 83[o]	4 Mk 83
MAXIMUM SPEED			
LOW	571.6/10,000	N/A	N/A
HIGH	N/A	1,360/36,000	1,280/36,000[p]
CLIMB RATE AT SL	7,800/min.	50,000	N/A
SERVICE CEILING	27,500	50,000	45,500
RANGE	1,135/438	N/A	N/A
COMBAT RADIUS	584/460	424/530[q]	598
ROLL			
0 WOD	1,530	N/A	N/A
25 kts. WOD	1,030	N/A	N/A
STALL SPEED	187	154[r]	N/A

DESIGNATION	A-12[s]	F-14D[t]	F-35B
NAME	Avenger II	Tomcat	Lightning II
ENGINE	F412-GE-D52	F110-GE-400	F135-PW-600
RATING			
NORMAL	13,000	11,800	28,000
TAKEOFF	N/A	26,950	43,000
SPAN	70-3.2	64-1.5	35
FOLDED	36-3.2	38-2.4	N/A
WING AREA	1,317	565	460
LENGTH	37-3	61-10.6	51.2
HEIGHT	11-3.4	16-0	14.3
WEIGHT			
EMPTY	37,327	41.353	32,472
GROSS	63,970[h]	72,646	60,000 max.

ARMAMENT			
GUNS	none	1x20 mm	1x25 mm
BOMBS	2x2,250 lb[u]	missiles	N/A[v]
MAXIMUM SPEED			
LOW	653[w]	684/10,000	N/A
HIGH	N/A	N/A	Mach 1.6
CLIMB RATE AT SL	N/A[x]	10,800	N/A
SERVICE CEILING	43,100	38,600	50,000
RANGE	N/A	1410/482	over 1,000
COMBAT RADIUS	965	173/482	580
ROLL			
0 WOD	N/A	3,200	N/A
25 kts WOD	N/A	2,230	N/A
STALL SPEED	151.8	151.8	N/A

DESIGNATION	**F-35C**
NAME	Lightning II
ENGINE	F135-PW-100
RATING	
NORMAL	28,000
TAKEOFF	43,000
SPAN	35
FOLDED	N/A
WING AREA	668
LENGTH	51-6
HEIGHT	14.4
WEIGHT	
EMPTY	34,581
GROSS	70,000
ARMAMENT	
GUNS	1x25 mm
BOMBS	N/A
MAXIMUM SPEED	
LOW	N/A
HIGH	Mach 1.6
CLIMB RATE AT SL	–
SERVICE CEILING	50,000
RANGE	N/A[y]
COMBAT RADIUS	N/A
ROLL	
0 WOD	N/A
25 kts WOD	N/A
STALL SPEED	N/A

[a] Data dated 1 July 1967. No high-altitude speed was listed; this was nuclear attack at sea level.

[b] A-6E with TRAM, data dated November 1979.

[c] Data dated April 1972.

[d] Not including refueling problem.

[e] B43 nuclear bomb plus four 300-gal tanks. High-low-high mission. Combat weight, with tanks dropped, was 45,907 lbs. At that weight speed at sea level was 644 mph, and rate of climb was 7,670 ft/min.

[f] Snakeyes.

[g] In combat condition (30,767 lbs), maximum sea-level speed is 644 mph, and rate of climb is 8,630 ft/min.

[h] Data dated October 1986.

[i] Data from Wagner and other sources. Wagner gives maximum weight as 63,500 lbs, but NATOPS gives 66,000 as maximum for catapulting.

[j] Data from Wagner; gross weight is probably as a fighter. Combat radius is with four Mk 83 (1,000-lb) bombs (probably also two wingtip Sidewinders and pods). Maximum weight 51,900 lbs. Cruising speed is probably with bombs.

[k] Combat rating (10 minutes).

[l] Without missiles; span with wingtip missiles is 44ft10in.

[m] Weight from Wagner, but Navy figure for maximum for attack mission is 65,980 lbs (for F/A-18C, approx. 51,884 lbs. Wagner weights are for fighter missions.

[n] High-Low-High mission with gun pod and two 300-gal tanks. Bombs are Snak-eyes. Combat weight (tanks dropped) is 25,710 lbs; maximum speed is 567 mph at SL. Combat ceiling is 31,650 ft.

[o] F/A-18C can carry up to 13,700 lbs of external ordnance, compared to 17,750 for F/A-18E.

[p] Speeds are equivalent to Mach 1.7 for F/A-18C and to Mach 1.6 for F/A-18E, but other (probably Navy) figures are more than Mach 1.8 in level flight for both types.

[q] A Navy vu-graph shows high-low-high combat radius with four 1,000-lb bombs: 369 nm for the F/A-18C (lot XIX) carrying three 330-gal drop tanks, vs. 520 nm for the F/A-18E carrying three 400-gal drop tanks.

[r] Approach speed to land, not stall speed. Minimum wind over deck for recovery is 19 kts (for launch, 35 kts). For F/A-18E these figures are, respectively, 15 and 30 kts. F/A-18C takeoff run is less than 1,400 ft.

[s] Contract requirements from Stevenson, $5 Billion Misunderstanding, 147 and 306, latter giving NavAir projections when the program was killed. Weights are those specified at the Full-Scale Development stage.

[t] Data for fighter role, July 1985, with maximum load: 4 Phoenix, 2 Sparrow, 2 Sidewinder, 2x280-gal drop tanks. With tanks dropped, combat weight is 64,172 lbs. Combat speed is 1,210 mph at 35,000 ft, and rate of climb is 14,800 ft/min. at 35,000 ft. Rate of climb at sea level is 34,000 ft/min. Maximum speed at sea level is 796 mph.

u   Internal capacity, ten Mk 83 (1,000 lb each) in two bays, or four Mk 84 (2,000 lb each) or two TSSAM. The radius was for a strike mission on which the airplane carried two generic air to surface missiles (total 4,500 lbs) and two air to air missiles (1,000 lbs).

v   Maximum weapons payload 15,000 lb; 18,000 for F-35C. Four internal stations and six external stations on wings; capacity 5,700 lbs internally and 15,000 lbs externally, but note somewhat lower total limits.

w   Compared to initial requirements of 903 miles and 621 mph (785 nm and 540 kts).

x   Internet sources give a rate of climb of 5,000 ft/min. and a substantially lower maximum speed than the guaranteed figure in Stevenson. Stevenson does not give a climb figure, only the change in climb rate due to added weight.

y   For the similar F-35A, combat range on internal fuel is 769 miles. The weights quoted for both versions are maximum takeoff weight, presumably with external stores. The takeoff weight associated with the 769-mile combat radius of the F-35A is 49,540 lbs (maximum 70,000 lbs).

Notes

Chapter 1. The Fleet's Aircraft

1. This account of BuAer organization is based largely on the fourth volume of the World War II–era administrative history of the Bureau of Aeronautics, *Determination of Military Characteristics,* which is undated but covers the period between 1939 and 1947. It includes the development of torpedo bombers as a case in point.

2. *Determination of Military Characteristics,* 43.

3. Some of the insights about changes in BuAer procurement methods are from a 20 July 1971 essay on past practices by Lee M. Pearson, the NAVAIR historian of the time, following up an earlier discussion; unnumbered memo from AIR-06B to AIR-320, Record Group [hereafter RG] 343, National Archives and Records Administration, College Park, Md. [hereafter NARA II]. In Pearson's view, development was typically driven more by the potential offered by new technology than by perceived tactical changes, as tactics and the value of trade-offs kept shifting. This book suggests otherwise, but it may be that the tactical or strategic rationales were often conceived to justify new technology already desired. Pearson also pointed to the perceived urgency of technology development (hence airplane development) once the Korean War broke out and it seemed that World War III was entirely possible. In his view the C-5 transport of the 1960s was "the first victim of national disillusionment with the philosophy that a good airplane is worth whatever it costs."

4. For details, see my *The Fifty-Year War: Conflict and Strategy in the Cold War* (Annapolis, Md.: Naval Institute Press, 2000). Mechanization and new technology greatly increased the cost of maintaining armies during and after World War II; hence the dramatic decrease in the number of formations, such as divisions, in Western armies. During the 1960 campaign, one of John F. Kennedy's closest advisors was Lt. Gen. Maxwell Taylor, who had been involved in the Army's attempt to force President Eisenhower to rebuild the Army. As a senator, Kennedy had been much concerned with the developing war in Vietnam.

5. Alain C. Einthoven and Wayne K. Smith, *How Much Is Enough? Shaping the Defense Program 1961–1969* (New York: Harper & Row, 1971).

6. Unrestricted submarine warfare, which might also have starved out Japan, was ruled out on the ground that the United States had gone to war in 1917 because the Germans had mounted exactly such an operation. The Germans' sin had been to sink not only British but also American ships: a U-boat captain could not always tell the difference. U.S. submarines in the Far East would surely sink neutrals. Since the British had the world's largest merchant fleet, that would mean British ships—which might not have meant much had the U.S. strategists realized that the British would be on the same side.

7. The study, dated 28 May 1944, is filed with the papers of the OpNav Strategic Plans Division,

RG 38, NARA II. The Marines later got more: in December 1944 plans called for eight Marine assault escort carriers and sixteen Marine escort carrier air groups (for 100 percent reserves). By that time, six groups had been formed. Ten Marine fighter squadrons went on board fleet carriers to relieve a shortage of carrier fighters facing kamikazes. In December 1944 it was proposed that the Marines provide the aircraft for the four existing large *Sangamon*-class escort carriers and also for the first twelve of the *Commencement Bay* class (*Commencement Bay* would be used for training). The CNO, Fleet Admiral King approved: embarking Marine squadrons for close support of Marines on the ground was "a step in the right direction." It was essential that the Marine airmen be integral with the Marines on the ground: "Otherwise Marine squadrons become just other Navy squadrons wearing the uniform of the Marines." In May 1945 the Marines had four carrier air groups, on board the escort carriers *Block Island* (CVE 106), *Gilbert Islands* (CVE 107), *Vella Gulf* (CVE 111), and *Cape Gloucester* (CVE 109), of which the first three saw combat. *Block Island* fought at Okinawa. Marine aircraft (all Corsair fighter-bombers) used this type of carrier again during the Korean War. The ships were thereafter retired, since they could not accommodate later heavier support aircraft. Marine combat aircraft were embarked on board Navy attack carriers.

8. The measure of wing length is "aspect ratio," the square of span divided by wing area. The greater the aspect ratio, the lower the induced drag of the wing.

9. BuAer memo on Naval Aviation contributions to the aircraft industry, 21 January 1929, entry 63, 602-0 pt. 3, RG 72, NARA). The earliest aircraft in this book used liquid-cooled piston engines. During World War I the U.S. government sponsored development of the 400-hp Liberty (V-1650 in later terms). Its postwar successor was the 585-hp Wright Tornado (T-3) (V-1950 in later terms), which was more efficient. Not counting its radiator, it weighed 1,000 pounds compared to 845 for the Liberty: it produced 0.585 hp per pound, compared to 0.473 hp per pound.

10. U.S. piston engines were designated by their displacement in cubic inches, the volume swept out by their pistons, as in the radial *(R)* R-2800, a radial with that displacement (most liquid-cooled engines used vee-cylinder arrangements, hence had *V* designations). Versions of an engine were indicated by "dash numbers," such as R-1820-6. Power per cubic inch increased dramatically. The V-12 Liberty used in the first U.S. naval attack airplane had displacement of 1,649 cubic inches and produced 400 hp. With about the same displacement, the Packard Merlin (V-1650) produced about five times as much power in 1945. Air-cooled engines produced somewhat less power per cubic inch, but they were lighter overall, which was what counted.

11. Graham White, *Allied Piston Engines of World War II* (Warrendale, Pa.: SAE, 1995), 222, considers the R-2800 the most significant U.S. World War II aircraft engine. According to White, Pratt & Whitney originally planned a 2,600-cubic-inch engine but decided to enlarge the engine when it heard that Wright was also developing an R-2600. Work began in August 1936, and the decision to enlarge it was made in March 1937; the engine was ready for production, rated at 1,850 hp, in 1940.

12. White, *Allied Piston Engines*, 4–5.

Chapter 2. Carriers

1. U.S. General Board files (420-7) include Goodall's 22 August 1918 memo on carrier characteristics:

> Air fighting has become a feature of Naval Operations, and the tactical movements of a fleet before an engagement opens will most probably be governed by information obtained from air scouts loosely observing the movements, disposition, and strength of the enemy. A series of fights between opposing aircraft will most likely be a preliminary to a fleet action. A fleet should therefore be attended by reconnaissance machines and fighting machines[;] . . . the attendance of aircraft on the fleet necessitates one or more vessels fitted for the special purpose of carrying seaplanes or airplanes with their attendant personnel and equipment, and providing good flying-off and landing platforms for them.

He went on to describe the carriers under construction when he had left England late in 1917: *Argus*, *Furious*, and the unnamed carrier that became *Hermes*. Goodall estimated that the carrier should displace about 22,000 tons and be about eight hundred feet long. An elaborate report on aircraft in the British Grand Fleet dated 18 November 1918 (File 9106) is in the General Board 449 file for 1919. It was originally submitted by the Commander in Chief (CinC) Atlantic Fleet to the CNO. Grand Fleet staff officers considered reconnaissance aircraft most important, followed by fighters, and torpedo bombers. American officers added spotting, to guarantee long gun range; fighters were wanted to defend the spotters and to protect fleet torpedo planes from enemy fighters. Cdr W. S. Pye was specially detailed by the fleet CinC to write a study of aviation policy, dated 14 March 1919, General Board 449 file. In his 2 April 1919 endorsement, the fleet commander wrote that the fleet had to have carriers, because shore-based aircraft were "likely to be unavailable when needed." He specifically rejected a proposal to design but not build a carrier; it was essential to have carriers to support the development of seagoing wheeled aircraft, the ones with the highest performance.

2. The Washington Treaty allowed each signatory a fixed total tonnage in several categories, including carriers. Tonnage for the major sea powers was set in a ratio of 5:5:3:1.75:1.75 for, respectively, the United States, the United Kingdom, Japan, France, and Italy. New carriers could not be replaced until they reached the same twenty-year lifetimes as specified for battleships. However, existing carriers (including *Langley*) were classed as experimental and could be replaced at any time.

3. Typically their 33,000-ton displacement entries carried an asterisk indicating that they had additional air and torpedo protection, its extent not disclosed. When the Secretary of the Navy asked in 1925 whether the ships could be cut back to 33,000 tons, he was told that the result would be unacceptable. Note that the treaty established a new "standard tonnage" that was not the tonnage previously used. It omitted fuel and reserve feed water (the previous "normal" tonnage typically included two-thirds fuel). Under the treaty, no

limit was placed on ships displacing ten thousand tons or less. Only Japan availed itself of this provision, with the carrier *Ryujo*. BuAer was anxious to add carriers to the U.S. Navy, so it pressed for ten-thousand-tonners (which might be conversions of the first U.S. heavy cruisers), to no avail.

4. BuAer estimated that each of the two big carriers could stow ninety assembled airplanes in its hangar but could place only sixty on its flight deck. It could also stow thirty disassembled airplanes in a storeroom. That seemed to be far more than any other navy could fit on board its converted capital ships, but the British already had more carriers that could accommodate additional aircraft.

5. General Board 420-5 serial 1362.

6. In this game, Reeves had not only fighters on board his carrier but also fighters on board tankers, an idea that BuAer was then promoting. His battleships carried fighters too.

7. In 1931 the Royal Navy noticed that the U.S. Navy was using arresting gear and deck parks to operate many more aircraft. The Air Ministry rejected larger carrier air groups, however, because it was concerned that an imminent (abortive) League of Nations arms-control conference would limit total numbers of combat aircraft, costing it the land-based aircraft it valued. The Royal Navy seems to have forgotten deck parking until its carriers operated the U.S. Navy during World War II. The 1931 incident is described in an *Eagle* "Cover," or official British design record, National Maritime Museum, Greenwich, U.K. [hereafter NMM].

8. The LSO (see also chapter 1) predated the new system of operation. It seems to have been inspired by early practice on board the prototype carrier *Langley*. Pilots trying to land discovered that they could judge their approaches by the reactions of the ship's air officer, who soon began using hand signals to, for example, signal the "cut," the moment when the pilot should shut off his engine. An American observer in China was surprised by how slowly a prewar British carrier, without any LSOs, landed its aircraft.

9. The first U.S. naval aviator, Eugene Ely, landed his airplane in 1911 into a series of cross-deck wires ("pendants") weighed down with sandbags, which stopped him. Arresting gear was incorporated into

the carrier *Langley* and into all later carriers. In the 1920s the Royal Navy abandoned arresting gear as an unnecessary complication in carriers whose pilots landed by eye.

10. The idea of the deck park seems to have predated the barrier. Apparently a barrier was improvised after a pilot landing on *Langley* almost crashed into twelve fighters parked forward.

11. Prestrike reconnaissance of shore targets seems not to have figured in prewar thinking; until about 1941 BuAer concentrated on photography for mapping. The F4F-7 (or -7P) demonstrated that a fighter pilot could operate a camera remotely. Atlantic Fleet F4F-4s equipped with cameras supported the invasion of North Africa. The main wartime carrier photo-reconnaissance aircraft were Hellcats (F6F-3P and -5P).

12. BuAer *Confidential Bulletin*, November 1952, 50–51. These folders were joint products of the Office of Naval Intelligence and the Air Force Directorate of Intelligence. They were created to meet a recognized need for better and more comprehensive air target intelligence than had been available during World War II, as well as from a realization that to be useful, intelligence had to be in a readily usable form, immediately available should the Cold War turn hot. That the folders did not include mobile enemy forces was probably not a serious problem until the Soviets deployed mobile surface-to-air missiles in the 1960s. Target folders were the basis of planned nuclear strikes to the end of the Cold War.

13. A Development Characteristic (CA-03501) for a reconnaissance system was issued in September 1953. A 2 November 1953 memo from Commander Naval Forces Atlantic argued that existing photographic aircraft lacked the performance to survive without air superiority; also, their capability was too limited. Commander in Chief Pacific Fleet (CinCPacFlt) wrote (7 September 1954) that the growing need for photo intelligence demanded that such aircraft keep abreast of carrier aviation's offensive potential. On 30 September 1955 Commander Naval Air Force Pacific Fleet strongly recommended developing a new photo airplane from the ground up. At this time another Development Characteristic (CA-11502) covered electro-photo reconnaissance equipment and its ground support equipment, and an Operational Requirement (IO-13501) had been issued for airborne mapping and reconnaissance equipment and supporting ground equipment. There does not seem to have been any explicit requirement for a companion electronic-intelligence system.

14. Desired capabilities included forward-looking photography, stabilized vertical photography, horizon-to-horizon photography, and side oblique photography ("trimetrogon"), all operating simultaneously (the pilot would have a viewfinder); night photography with artificial illumination; automatic camera control for all stations; a side-looking high-resolution radar with photographic recording; radarscope photography; automatically recorded electronic data covering the spectrum from 50 to 40,000 MHz (frequency to within 20 kHz, pulse repetition rate, pulse width and modulation, antenna scan rate and pattern, polarization, and location); low-light-level television (soon crossed out); and correlation and recording of aerological data. Data had to be marked with accurate time, orientation, and location data supplied by the airplane—that required accurate (preferably nonradiating) navigation. The covering memo mentioned atomic radiation detectors, but the tabulation of requirements did not. The airplane was to be capable of in-flight refueling. Versions of CA-03501 beginning in 1958 are in the summary of the BuWeps evaluation report in NAVAIR evaluation division file, carton 7, marked VFX Studies '67, VFAX, and studies 1960, RG 84, entry UD-22 441, NARA II.

15. The Navy wanted an unrefueled radius of two thousand nautical miles, so that the new airplane would considerably exceed the thousand-mile radius then standard for the nuclear-attack role in its design competitions. A two-thousand-nautical-mile radius was consistent with a radius of only five hundred miles if the airplane flew at supersonic speeds over the target area. The Air Force wanted a thousand-nautical-mile reconnaissance radius. Secretary of Defense McNamara tried but failed to force the Air Force to buy the Vigilante. He considered it by far the best U.S. reconnaissance airplane. The Marines bought a version of the Phantom (RF-4B); the Air Force bought the RF-4C.

16. The file includes an undated comparison between the "development characteristic," an unofficial Marine Corps requirement, and the Vigilante and Phantom reconnaissance versions. The original requirement called for all photographs to be reduced to transmittable form and transmitted, but the Marines were willing to reduce that to selected photographs. The Navy expected advanced aircraft to be able to transmit side-looking airborne radar (SLAR) data on a delayed basis. Where the requirement called for a viewfinder, the A3J would provide a television for its back-seater ("the guy in the back seat," a naval flight officer or intelligence specialist responsible for navigation, here the reconnaissance systems, and the like). ELINT data would be transmitted (albeit on a delayed basis) in advanced aircraft. The SLAR specified for both the Vigilante and the Phantom was APQ-56, but with fifteen-foot antennas in the Vigilante and ten-foot antennas in the Phantom. Neither airplane would meet the requirement for a missile-defense system. Because the Marines did not use long-range attack aircraft, they were willing to accept a combat radius exceeding that of their light attack aircraft, which would be significantly less than what the Navy wanted.

17. The Vigilante had multiple cameras with an electronic strobe for night work, an infrared (IR) camera (AAS-21), an electronic intelligence-collection system (ALQ-61), and a SLAR (APD-7). The SLAR benefited from the airplane's ability to fly a straight path using its inertial navigator. The system was controlled by the back-seat "reconnaissance-attack navigator" (RAN), typically a naval intelligence officer. The ASB-12 system of the A-5 was retained, for navigation and camera targeting. According to Alfred Price, *The History of U.S. Electronic Warfare* ([Alexandria, Va.]: Association of Old Crows, 1984–2000), 2: 229, AIL Systems Inc. won the ALQ-61 competition with a scaled-down version of its Air Force ASD-1. As such it had unusually sensitive wideband receivers backed by a computer that sorted incoming pulses into "trains" from which parameters such as pulse repetition frequency could be computed (pulse-rate hoppers could be identified by their consistent direction, pulse width, and amplitude). Emitters could be localized from the way in which direction varied as the airplane

flew. The system automatically selected pulses that seemed to come from new radars for manual analysis. The Air Force system was associated with an automated Air Force intelligence sorting and analysis system, GIRDHS. Price describes IOIC as a scaled-down GIRDHS. According to Price, the *ALQ* designation was deliberately chosen to hide the ELINT role of the ALQ-61, for fear that the National Security Agency would take it over. It was declared a "passive ECM" device for fleet tactical support to suggest a completely bogus jamming role. The comment on neglect of the ELINT system during the Vietnam War is from Price, *U.S. Electronic Warfare*, 3:59–60, describing a visit to *Enterprise*, probably in 1967. However, RA-5Cs routinely flew missions up and down Vietnam to find SA-2 and other radars, following the "blue" and "black" tracks, the former off the coast and the latter over Laos. To get accurate bearings the airplanes had to fly straight tracks; Robert R. Powell, *Vigilante: 1200 Hours Flying the Ultimate U.S. Navy Reconnaissance Aircraft* (Forrest Lake, Minn.: Speciality, 2019), 62.

18. The June 1938 tactical orders (Commander Aircraft Battle Force *Current Tactical Orders and Doctrine, U.S. Fleet Aircraft*, vol. 1, *Carrier Aircraft*, USF 74 Vol I, June 1938, RG 38, NARA II) advocated early breakup of an attack group into attack units as a way of reducing the chance of early detection and aerial counterattack. Squadrons attacked together to saturate defenses and to ensure hits against maneuvering targets. Dive-bombers were expected to strike at maximum speed, directly from the approach formation, without maneuvering and preferably on the approach course. Prewar tactical instructions discussed the pros and cons of "divided" and "stream" attacks. Alternative approaches were those at high altitude, at high speed (for maximum surprise), zigzagging (to overcome enemy antiaircraft, not to be used until after discovery, because it slowed the attack), or a divided attack.

> In delivering dive bombing attacks, units attacking the same target should endeavor to vary the angle of approach in such manner as to avoid enfilade fire. Divided or three-way attacks tend to hamper enemy return

fire, but this consideration should not out-weigh the principles of speed and surprise, and does not justify additional exposure to enemy antiaircraft fire nor jeopardy of concealment while maneuvering to attain positions for divided attack. If complete surprise be effected, divided attack tends further to harass enemy antiaircraft control and fire. If not . . . [it] permits additional antiaircraft batteries to be brought in action, and is of doubtful value.

Much the same language appears in the March 1941 edition of USF 74 Vol I. It adds that "once a heavy bombing group is within striking distance of the enemy, it is extremely difficult for enemy fighters to stop their attack unless they be present in overwhelming numbers. It is essential, however, that the bombers maintain a type formation which can be rapidly closed for defense and that is readily maneuverable." The best cruising formation in the face of fighter opposition was a six-plane division in two sections, the second section taking position under and slightly astern of the first. Divisions or squadrons not engaged were to press home the attack, to deliver the maximum possible number of bombs on target.

> There is little to fear from fighters once the dive is started. Our speed-retarding devices cause enemy fighters to overshoot so rapidly that effective gunnery is most difficult. After bomb release, do not run away from an enemy fighter. To do so will merely present a no-deflection shot. When the fighter starts his approach, the best maneuver is to turn towards him, striving at all times to give the free gunner the best possible target and steady gun platform. . . . Experience indicates that two scout bombers can effectively counter the attack of one fighter by maintaining a lateral separation of 200 to 300 yds [which] permits the unengaged bomber to close the fighter if he should start an attack. . . . Should fighter opposition be met, close formation should be maintained until just before the push over.

The same volume warned that unsupported torpedo bomber attacks would most likely fail—as happened at Midway. Unfortunately, no current carrier tactical orders for 1942–43 seem to have survived.

19. This was written *after* the June 1944 "Turkey Shoot," in which the U.S. fleet failed to destroy Japanese carriers but managed to shoot down almost the whole of the attacking Japanese force. This emphasis seems to explain the orders given Admiral Halsey before Leyte Gulf to destroy the Japanese carriers if an opportunity arose or could be created, even at the expense of uncovering the U.S. invasion force.

20. The 1974 Naval Warfare Publication on strike warfare, declassified in 1982, mentions TOT attacks but does not discuss them. It discusses Alfa strikes and cyclical attacks in greater detail.

21. The General Board 420–7 file includes a series of papers headed by an 8 May 1939 memo from CinC U.S. Fleet to the General Board on recommendations for the next carrier (*Essex*) based on operating experience with *Yorktown* and *Enterprise*. Commander Battle Force wrote that "the whole mission of the ship depends upon the mental and physical alertness of the pilots who, before departure, will frequently be required to spend several hours in an alert condition. The ready rooms must be so located, arranged, and equipped, as to impose upon the pilots as little discomfort as possible during this standby period. Considering the necessity for the pilots wearing heavy clothing (due to sudden calls to take off and attain high altitude) air conditioning of the ready rooms is completely justified." Commander Carrier Division [CarDiv] 2, in USS *Yorktown*, 3 March 1939, recommended centralizing all squadron activities on the gallery deck: squadron offices and ready rooms outboard, port and starboard; gear lockers, storerooms, armories, ammo-belting rooms, bombsight stowage, and radio storerooms inboard. Each squadron would have its own assigned ready room, with twenty-five seats, egress to the nearest flight-deck catwalk, adequate ventilation, and indirect lighting. As described by the Bureau of Ships on 7 August 1940, the *Essex* design showed four ready rooms with necessary service rooms adjacent, two just aft of the midships (not yet deck-edge) elevator and two forward of the after elevator; Ready Room

No. 1 had twenty-seven seats to accommodate the enlarged fighter squadron then planned. All would be air-conditioned.

22. According to the *Carrier Tactical Orders* (USF 77) of June 1938, which were probably typical, the preflight briefing had to include: "(a) The general tactical situation including own and enemy forces; (b) The mission; (c) Special orders for the flight; (d) Detailed instructions regarding the intelligence and communication plan, including recognition signals; (e) Point Option; (f) Required time of return to carrier or base, based on existing rules as to fuel reserve; (g) Time and position (with respect to Point Option) of expected rendezvous with the carrier; (h) Bearing and distance to nearest land if in the vicinity; and (i) Latest meteorological data (airplanes in the air would be informed of changes if necessary)."

23. *Current Tactical Orders, Aircraft*, USF 12, 12 September 1934.

24. USF 12 (1935 edition) includes instructions for carriers operating together. It cautions that carriers working together should never become separated by more than three miles.

25. A copy of the report of an air navigation board for the Battle Force written in August 1936 is in RG 313: U.S. Fleet Battle Force, box 2 (A2 to A2–2), NARA.

26. *Carrier Tactical Orders*, USF 77, February 1943.

27. BuAer, *Confidential Bulletin* 1-51 (March 1951), 49–50, Naval Aviation History Division, Naval History and Heritage Command, Washington, D.C. [hereafter NHHC]. In the first system test (September 1951) the device was flown down a straight railroad track between Petersburg and Suffolk, Va.; highway crossings were used as visual endpoints. In November that year the device flew from Washington to the West Indies and back to Raleigh, N.C. Over water it detected ocean currents, but their effects could easily be canceled out.

28. BuAer, *Confidential Bulletin* (November 1952).

29. BuAer *Confidential Bulletin* (May 1953), 18–20, reporting on an exhibition of nine new navigational devices at Naval Air Warfare Development Center Johnsville (so called, actually in Warminster, Pa.) for a conference on radar ground stabilization and ground position indicator devices.

Chapter 3. Torpedo Bombers

1. Adm. Bradley Fiske proposed (and patented) a torpedo bomber before 1914. BuOrd forwarded a sketch of a launching arrangement in a 20 December 1916 letter to the Naval Gun Factory, in Washington, D.C. In July 1917 CNO asked BuOrd to develop an eight-hundred-pound torpedo specifically for aircraft, capable of withstanding a drop from thirty to forty feet. A dummy torpedo was dropped unsuccessfully (due to vibration) in August 1917. C&R and the OpNav aviation unit argued that so small a torpedo was worthless. In BuOrd's view, that delayed torpedo plane development. In September 1917 BuOrd forwarded drawings of a 2,000-pound torpedo; meanwhile the Bliss Torpedo Factory built an eight-inch torpedo for Admiral Fiske. CNO, Admiral Benson, pointed out that existing seaplanes could lift only 450 to 600 pounds; it would take a 400-pound warhead to do serious damage. It seemed unlikely that any near-term development could much affect the war. The Navy would concentrate on aircraft for scouting, training, and attacking U-boats. With the appearance of the F-5 seaplane in 1918, the Navy finally had an airplane capable of lifting a 1,000-pound torpedo under each wing. In July, OpNav (Aviation) proposed experiments. Meanwhile, on 3 July BuOrd received a telegram from Adm. William S. Sims in London describing British interest in torpedo-carrying airplanes to operate from *Furious*-class carriers. He wanted a torpedo-carrying seaplane developed. Finally, on 23 November 1918, a seaplane carried a 400-pound dummy torpedo under one wing. Further tests used torpedoes weighing up to 1,000 pounds. In January 1919 Sims cabled that the British considered that they had successfully developed aerial torpedo attack. BuOrd chronology of its work on torpedo bombing before the Armistice dated 28 December 1945 in Aviation History Division, NHHC.

2. *L* indicated a Liberty engine, and there were additional outboard wing panels to add lift. The Curtiss R was originally essentially an enlarged JN trainer, which was also used for antisubmarine scouting. The same type of airplane was also widely used by the Army and by the Royal Naval Air Service. The initial Navy R-3 had increased span to carry the weight of its floats (the Army used the

wheeled R-4). R-6 was an improved R-3 with dihedral (upward-slanting wings, for stability) and a Curtis 200-hp VXX engine. All had the pilot in the rear cockpit and the observer in front, an arrangement reversed in the final R-9 version.

3. An August 1919 C&R aircraft performance chart showed the R-6-L with a Mk D torpedo. Type D was a short version of the eighteen-inch Mk VII (Mk D or Mk VIID) weighing 1,036 pounds and 11 feet 7¾ inches rather than 17 feet ¾ inches long. The warhead weighed two hundred pounds. It was capable of two thousand yards at thirty knots. A total of thirty-six were made, apparently specifically for the initial drop tests using R-6Ls. BuAer considered its range inadequate, citing wartime German aerial torpedo attacks against British merchant ships, which had failed because the airplanes came under defensive fire. Account of early experiments from BuOrd letter to Commander Aircraft Squadrons Pacific Fleet, 37589/548-(Ma3)-0, not dated but probably October 1922, in entry 63, file 204-5, RG 72, NARA I.

4. BuOrd letter, 9 November 1921. BuAer urgently wanted thirty torpedoes for training, and BuOrd offered to modify between two and three hundred pending policy decisions. Its four-thousand-yard (at thirty-two knots) torpedo carried a 260-pound charge. The alternative was a modified seven-thousand-yard (twenty-seven-knot) torpedo with a 245-pound charge; entry 63, folder 204-5, RG 72, NARA I. In 1920, about five hundred new Mk VII torpedoes (eighteen inches by 5.2 meters) were available for use by aircraft. Mod 4 carried a 314-pound charge, with a range of three thousand yards at thirty-three knots. Mod 5 was rated at four thousand yards at thirty-five to thirty-eight knots. These torpedoes weighed about 2,040 pounds. BuOrd rejected twenty-one-inch torpedoes as too heavy for a carrier-based torpedo bomber. In any case, the twenty-one-inch Mk X (2,215 pounds) carried only eighty pounds more TNT and ran only a knot faster more at three thousand yards. No surplus twenty-one-inch torpedoes were available.

5. BuAer letter 12 February 1923, endorsement on request for aircraft torpedoes, entry 63, file 204-5, RG 72, NARA I.

6. "Torpedo Bombers: The Prewar Years," *BuAer Confidential Bulletin*, October 1949. The article was written by Lee M. Pearson, who was preparing a wartime technical history of BuAer. It was one of a series on the development of various types of combat aircraft through the end of World War II.

7. There was also a Marine Expeditionary Plane, to equip Marine advanced base units: a land-and-water plane for reconnaissance, mapping, and offensive operations against troops and material. The Marines were currently using the existing DH-4B; the DT would probably be its successor. Neither corresponded to the 1920 requirements.

8. As listed in the 1920 Navy Department Annual Report. Two Martin MBTs were on order, but they were not considered suitable for shipboard use. Handwritten notes on early torpedo bombers in NHHC Aviation History Division files suggest that the CT came first and that BuAer initially favored it. There was also a bid from the Thomas Pigeon Sales Company, but the firm soon dropped out. Stout requested wind tunnel tests on 30 October 1920; copies of Curtis reports went to the wind tunnel at the Washington Navy Yard on 8 November 1920. Late in 1921 the BuAer material division was preparing specifications for a wheeled ST-1 and a floatplane ST-2, both to be powered by the new Packard 1A-1237 engines. Expected maximum speed was 105 mph (slightly over the desired ninety knots [103.5 mph]), with the desired three-hour endurance fully loaded; General Board 449 file, 1921. ST first flew on 25 April 1922. Cancellation put Stout out of business. He was later responsible for the Ford trimotor airliner.

9. William F. Trimble, *Jerome C. Hunsaker and the Rise of American Aeronautics* (Washington, D.C.: Smithsonian, 2002), 73. Douglas returned to Washington to redesign his Cloudster as the DT.

10. René J. Francillon, *McDonnell Douglas Aircraft since 1920*, 2nd ed. (Annapolis, Md.: Naval Institute Press, 1988), 1:55, quotes the Acting Chief of the Army Air Service as proposing purchase of a DT-2 (to be powered by a Packard 2025) to arrive "at some definite conclusions as to the advantages and disadvantages of a single engine bombardment airplane, similar in performance and weight carrying to twin-engined airplanes in use."

11. Francillon, 45. The report on initial performance trials of the DT-1 (Airplane A-6031) was dated 14 December 1921. In addition to U.S. Navy aircraft, Norway bought one DT-2B and built another eight under license, and Peru bought four DTBs (similar to DT-2Bs) powered by 650-hp Wright Typhoon engines. The DT-2B had the nose radiator and tail surfaces developed for the Douglas World Cruiser derivative of the DT.

12. Fokker's FT was a single-engine twin-float monoplane built in the Netherlands, initially as his T.II torpedo bomber (there was no T.I). It might be considered equivalent to German World War I floatplane torpedo bombers, although it was a fresh design. Although it was relatively large, with a wingspan of sixty-five feet one inch, it was not much heavier than a DT (7,306 pounds gross versus 7,293 pounds for a DT-2) and had the same single Liberty engine. The initial request to authorize a contract with Fokker is dated 28 December 1920. Acting chief of C&R Captain Stocker wrote that "it would appear that the performance guaranteed exceeds what responsible American constructors have so far been willing to promise for a single engine seaplane but is considered possible to realize by careful design. The new features involved are a steel fuselage and veneer-covered wings which will be new to this country although such constructions have been successfully made by Fokker in the past." He quoted a span of sixty-five feet and a length of fifty feet or less, a gross weight of 7,500 pounds, and a maximum speed of ninety knots. The fuselage would have a steel-tube structure, with a V-shaped bottom aft to deflect the torpedo splash. A requisition for three Fokkers was issued in January 1921, the purchase having been approved on 7 January. The initial specification was dated 20 December 1920. At that time span was given as seventy-two feet and length as fifty-one feet eight inches; all-up weight was 7,500 pounds, and estimated speed at sea level was 100–105 mph fully loaded. Span was later reduced to sixty-five feet and weight to 7,300 pounds, speed at sea level being 105 mph. The Navy would supply their Liberty engines. A 7 June 1923 trials report is in file O-ZA-39, RG 72, NARA: the airplane handled very nicely in the air, was easily controllable and pleasant to fly, but it was very difficult to handle on the water and was not considered satisfactory. Its load-carrying capacity was considered poor, and the bombardier would have had very poor visibility. A May 1923 letter in this file rejected a proposal by Mr. Noorduyn (presumably from Fokker) that the FT be redesigned in such a way as to save two hundred kilograms of structural weight and reengined with a Napier model. BuAer was unwilling to have torpedo-plane development carried out for it abroad, its experience having shown that close cooperation would be quite impossible. It would be equally impossible to redesign the airplane in the United States, since the entire Fokker design staff was in Holland. There were also specific objections. The use of built-up wooden wing spars was considered unsafe for tropical service and also too heavy, as heavy fittings were required to give sufficient strength at the joints. The veneer covering was unsatisfactory for tropical service and did not permit periodic inspection, drying out, and revarnishing of the internal construction of the wing. The deep-chord monoplane design entailed both a longer fuselage and greater span than desired. "In general, the monoplane type is handicapped for storage aboard ship." The file on the Swift (O-ZA-40) begins with a 24 March 1921 reference to an earlier price quote and a January 1921 ONI report on the airplane. The attaché reported that the British Air Ministry was satisfied with the design and had ordered three more. On 25 April C&R and SE sent a joint letter to the Secretary of the Navy recommending purchase: the Swift was the latest and best British torpedo plane, with a full speed of ninety-eight knots at an all-up weight of 6,000 pounds, with fuel for four hours at full power carrying a 1,650-pound torpedo. Necessary funds ($100,000) had already been set aside. The purchase was authorized on 2 May. In contrast to the Fokker, which was tested only at Anacostia (in the District of Columbia, across the Anacostia River from downtown Washington) and at Norfolk, Va., the two Swifts were sent to the Battle Fleet for service trials, to establish general serviceability and suitability of the airplane and of its power plant. The Swift was comparable to the DT. The report was dated 22 August 1924. Measured maximum speed (light) was 95.5 mph (eighty-three

knots), well below the ninety-eight-knot claim. Fully loaded, maximum speed was 90.2 mph (seventy-eight knots). The oleo (pneumatic/hydraulic) shock absorber lived up to its advertising. The engine was accessible and easy to start but occasionally overheated and lost oil pressure. Takeoff compared favorably with a DT under full service load. The airplane handled well under all conditions except in bad air. In very rough air, under full load, it was difficult to handle. With sufficient altitude, it readily righted itself, but near the ground "the pilot is kept busy." However, it seemed that maneuverability under full load was better than that of a DT-2. On the ground, however, it was much more difficult to taxi with or across a strong wind. Features worth using in future designs were the method of wing folding and its compactness of stowage and of the exhaust heater. The Swift performed well enough that Naval Air Station San Diego wanted one of the two converted into a photo plane. BuAer rejected that as too expensive. The Swift did not enter British service, but the Dart, derived from it, did.

13. For years, Douglas stationery carried the legend "First around the World." The World Cruiser had sixty-two-gallon tanks in the lower wing roots, but the upper-center section tank held sixty rather than eighty gallons. In addition it had a 150-gallon tank behind the engine firewall, 160 gallons under the pilot's seat, and 150 gallons under the observer's seat, for a total of 644 gallons (maximum still-air range 2,200 miles). It did not have the longer wings proposed for the long-range version of the DT-6.

14. File O-G-28 (in the series of ex-C&R files taken over by BuAer), RG 72, includes the rejection of a 28 July 1923 offer by Douglas to modify the DT into what it called the DT-6 (-5 crossed out) three-purpose airplane as an alternative to the CS, with a narrower nose and a bombardier's position satisfying the "rigid requirements of the air service as to vision." Crew would be three rather than the two of the torpedo bomber, a bombardier ("bomber") being added. The scout would have a pilot, an assistant pilot, and a gunner. This design provided a seat alongside the pilot's seat to accommodate a reserve pilot for long flights. A detail specification was enclosed. Provision was made for nonstop flights of up to 2,800 miles. Full-load weight as a torpedo bomber would be 7,527 pounds (8,289 pounds as a seaplane). Estimated performance as a torpedo bomber was a maximum speed of 108 mph (ninety-four knots), cruising speed 90 mph, stall speed 55 mph, ceiling 9,500 ft, duration at full power 3.5 hours, and range at full power 378 miles (540 cruising). As a scout, full-power range (i.e., one-way) was 825 miles (1,130 cruising). With full tanks (long-range option), full-power distance was 1,750 miles and cruising distance 2,400 miles. (It is not clear how this range can be squared with Douglas' reference to 2,800 miles.) Span was fifty-six feet ten inches, overall length thirty-nine feet ten inches. The long-range version would have a span of seventy-three feet eight inches. The torpedo bomber had two sixty-gallon tanks built into the side walk wings and one eighty-gallon tank built into the upper-center wing section. As a scout, the torpedo bomber could carry an external 140-gallon "torpedo tank" in place of the torpedo. For long range, a 290-gallon tank would be added in the bomb bay, a 270-gallon tank in the fuselage aft of the bomb bay, and a fifty-gallon tank in the fuselage below the pilot's floor, for a total of 810 gallons of gasoline plus sixty gallons of oil (compared to eleven in the torpedo bomber and twenty-two in the scout). As in the DT, the rear gun was a flexibly mounted Lewis light machine gun.

15. The NARA RG 72 correspondence file (O-G-29) indicates that SDWs were initially to have been designated "DT-5" (that designation was reused for a reengined DT-2). SDW-1s had the pilot's cockpit moved back to just ahead of the gunner's, to facilitate the use of dual controls when fitted. The relief pilot in the second cockpit had to sit facing forward rather than (as a gunner) aft, and he needed navigational instruments and a chart board. The pilot had to be able to wake him up as needed. These aircraft had nose rather than wing radiators (the increased weight of the latter more than offset their reduced drag), and the same Wright T-3 (initially T-2) engines as the CS-2. A deep fairing covered the fuel tank that replaced the torpedo. To maintain balance, as much as possible of the fuselage gasoline had to be moved to the upper wing. There were five new fuel tanks: a seventy-two-gallon rectangular tank with a circular top;

a 330-gallon rectangular with a cylindrical top; a thirty-seven-gallon semicylindrical tank; a 160-gallon rectangular tank; and a 160-gallon cylindrical tank with hemispherical ends. At least two of the SDW-1s had dual controls, although the gunner's cockpit was considered too cramped when they were fitted. Comparative performance trials were ordered in May 1924. Note that the three aircraft were initially sent to Chance Vought (presumably Wright's subcontractor) for modification. Wright's quote was dated 25 August 1923.

16. There was apparently no DT-3. DT-4, to test the Wright T-2, was converted into DT-5 by substituting a geared T-2. The Wright P-1 (prototype Cyclone radial) was tested in DT-6.

17. Memo, 4 March 1922, in file OA-45 (in the series of C&R files taken up by BuAer), RG 72, NARA. OG-7 in this series begins with a 14 November 1921 memo from the Material Division to the Plans Division enclosing two blueprints (unfortunately not in the file copy) of Design Study No. 1 of a scout. Prints of Design Study 5 were forwarded on 6 December 1921; No. 6 followed on 7 December. No. 4 followed on 9 December and No. 7 on the 14th. A torpedo-bomber study followed on 20 December. It seems to have been No. 8 in the series. Studies 9 through 13 were of an observation plane (VO). A memo dated September 1924 refers to the airplane as a "Class VS [i.e., scout] Torpedo Plane."

18. This idea was embodied in the air section of the *U.S. Naval Policy* approved by the Secretary of the Navy on 1 December 1922. Development of heavier-than-air craft was to concentrate on spotting planes and on torpedo, bomb, and scout planes that could operate from ships. The three latter functions should be combined in one plane.

19. This seems to be based on a statement made by Navy witnesses during 1925 congressional hearings on aircraft procurement that CS was a Navy design. They dated it from 1922, which is when Design 8 was produced. The Navy file on CS does not indicate its design origin.

20. Letter from Commanding Officer [hereafter CO] USS *Langley*, 12 June 1925, correspondence file VSC1/F1-1 (Design CS-1), RG 72, NARA. No report of landing trials is appended.

21. Conference summary in file OG-24, vol. 1, NARA [hereafter OG-24, vol. 1]. The conference was chaired by the BuAer chief, Rear Admiral Moffett, and included Capt. Emory S. Land, the bureau's senior technical expert. The conference met on 26 March 1924.

22. According to the 10 July 1924 report of the Board of Inspection and Survey (in file OG-24 vol. 1), CS-1 handled well in horizontal flight but required continuous attention of the pilot at all times: "Owing to the sluggish action of the ailerons and difficulty in turning in the air, it is advisable to have 500 ft altitude in making turns." With a weight equivalent to a thousand-pound bomb, service ceiling was 4,025 feet. In climbs the engine tended to drop revolutions, for an undetermined reason.

23. Lt. Cdr. Marc Mitscher, assigned to BuAer, said that no single-engine airplane could be a satisfactory bomber, because it could not meet the requirements set by the 1923 Fire Control Regulations; it could not have sufficient visibility: "The bombing man [bombardier] must be out in front." BuAer had already tried a small nacelle under the wing, in an airplane that crashed. Cdr. Henry B. Cecil thought it would be best to put the bombing (i.e., bombardier) position in the bottom of the fuselage under the engine, giving ninety-degree visibility from the vertical forward. That could be done by changing the lines of the fuselage; a similar modification could be made to a DT.

24. N. L. Snaples Jr., "Institutionalizing Aircraft Procurement in the U.S. Navy 1919–1925," PhD dissertation, Texas A&M University, August 1999, 251. Snaples accepts Curtiss' claim that it had been forced to spend heavily to adjust to all-metal construction, which was new to it, but other authors describe the CS as a conventional wooden airplane, the unusual feature of which was the greater span of its lower wing.

25. Memo on cost of CS-1 planes, 21 February 1924, file OG-24 vol. 1, referring to a Curtiss proposal dated 14 February to build six, eight, or ten CS-1s as an experimental order. They would have been a follow-on to the initial six, ordered in June 1922. BuAer rejected the company's proposed 100 percent overhead, 10 percent contingency charge, and 15 percent profit, the latter on the ground that it

did not reflect the current state of the trade or the amount of competition. At a conference about 1 March, Curtiss representatives were told that their proposal had been rejected and asked to return with a new offer. The file includes orders for comparative trials between the CS-1 and SDW-1; that suggests that the point of the SDW-1 project was to create a possible alternative to CS-1 and thus force Curtiss to cut its price. Nothing in the file makes this point explicitly.

26. According to a 21 October 1925 BuAer memo to the Secretary of the Navy on aircraft procurement (which defended the Martin purchases) in VV vol. 3, RG 72, NARA, at the time of the first contract (for SC-1s) Martin was

> instructed to develop many features in the plane and to change all metal parts from heat-treated alloy to mild steel; to accomplish this, as well as other modifications and improvements, the Martin Company found it necessary to prepare a complete stress analysis and also to prepare all drawings from which these airplanes are constructed[;] . . . no drawings of any description from the Curtiss Company were used. This indicates that the type of plane, though originally a Curtiss design, is now to all intents and purposes a Martin design. During the past 18 months the Martin Company has done practically no other business than develop an organization capable of producing SC type airplanes up to the high standard of quality desired, and has built up a complete outfit of special tools and jigs for the purpose of manufacturing all parts with interchangeability. . . . [F]urther airplanes produced by the Martin Company from the tools and methods employed will necessarily give to the Naval Service standard parts, standard equipment, and standard practices so far as upkeep, maintenance, and repairs are concerned. This in itself is a very important economy. For this company to have a continuity of orders would undoubtedly mean marked saving in any future Government business.

This applied to twenty-four more SCs, which became the T3M-1s. This memo was drafted by Capt. (later rear admiral) E. S. Land and signed by Admiral Moffett. It dealt with an ongoing investigation of the shift from Curtiss to Martin, Curtiss charging that Martin was undercharging for an inferior airplane.

27. According to a 12 June 1926 report by Commander Aircraft Squadrons Scouting Fleet, the SC was a great advance over the PT and DT. However, as a bomber (with a thousand-pound bomb) it could not reach the desired ceiling of ten to twelve thousand feet (it was limited to about six thousand). As a scout it suffered from poor visibility (which was to be corrected in future designs), cramped space for its navigator, and weakness in a moderate or rough sea. This report, derived from that of a Scouting Fleet board, was considered particularly valuable because it described an airplane that had been in service for some time.

28. Memo dated 16 July 1924, OG-24 vol. 1.

29. The BuAer file on this negotiation is in OG-47. Curtiss' detail specification, which is not in the file, appears to have been submitted in July 1924. The project was discussed in detail at a 13 August 1924 conference in BuAer, with a Curtiss representative present. It was decided that if a contract for forty aircraft were let on a competitive basis, it would not include the conversion of a CS-1 as prototype; that would be the subject of a subsequent contract for a "flying mock-up." The S2C designation was never official, so it was used later by BuAer for a completely different airplane.

30. Correspondence in VSC5/F1-1 (Design SC2), RG 72, NARA, responding to a 7 May 1926 report by VT-1. All SC-2s in VT-1 had had their compression ratios changed, reducing power to 525 hp. VT-1 quoted the Board of Inspection and Survey: although the SC-1 with a T-3 engine did not fulfill the required performance guarantees, the board recommended that it be accepted. However, since the SC-1 was neither a satisfactory bomber nor torpedo plane, the board recommended buying no more. According to VT-1,

> When loaded with a torpedo, the maximum gas that can be carried and still get the plane into the air is about 145 gallons. The plane in flight is so heavily loaded that it requires practically full throttle to keep

[it] in the air and altitude is gained very slowly as the gas load is lightened. Operating under these conditions, the planes have an endurance range of approximately three hours, or about 240 miles. . . . Prior to starting gunnery I ran a series of full load tests with the SC-2 and succeeded in getting the plane into the air only after a water run of about ten miles. This was done in the inner bay in smooth water. It would be quite disastrous to attempt a takeoff in the open sea with a full load. While the plane can be forced into the air by bouncing off a swell, the pick up in air speed is so sluggish that the plane drops before it can attain flying speed. A series of poundings such as these would wreck the plane. In time of war it would be necessary to operate with full military load, with full assurance that all planes could take the air. This is beyond the performance of the SC-2 except under most favorable conditions with a strong surface wind to aid in the take off.

Service ceiling was five thousand feet as a seaplane and seven thousand as a land plane, but bombing altitude needed to be at least between eight and ten thousand feet. The side radiators blocked the pilot's vision so badly that he could not make a proper approach to the target. Lying prone in the tail, the bombardier looked down through a narrow slot, and his lateral vision was blocked by the floats. Some sort of harness was needed to direct the pilot across the target. The VT-1 commander thought a bomber should have multiple engines with sufficient reserve power to reach sufficient bombing altitude. The arrangement of the pilot and bombardier should be that in the new Douglas bomber (T2D).

31. Chief BuAer Letter from Bureau chief to CO Naval Air Station Hampton Roads, 24 October 1924, in connection with a request that the Board of Officers who had evaluated the DT-2 continue their work, in RG 72 file OA-47, in series of C&R files carried over by BuAer. In the FY26 (mid-1926) fleet annual report, Commander Air Squadrons Battle Fleet, Captain Reeves, complained that the CS found it difficult to rise from the water with a full load. It was slow and cumbersome and generally unsatisfactory. Despite its supposed versatility, Reeves did not consider it a real bomber; it was useful mainly to develop tactics. The possibilities for improved engine performance were (a) over-compression; (b) supercharging; (c) gearing; and (d) a combination of gearing with either over-compression or supercharging. The simplest was to increase the compression ratio, using a higher-octane fuel to overcome knocking. Supercharging would require development. Gearing was attractive because it allowed an engine to drive a more efficient propeller at lower speed; against that, it added weight. However, it was too late to modify engines on the existing contract. The larger the airplane, the smaller the relative penalty involved. By this time supercharging had already been tried in a DT-2; BuAer wanted to try it in a CS. The first test of gearing would be the DT-5 powered by a geared T-2 engine. The next would be a CS-3 seaplane powered by a geared T-3, which was being reconditioned at the Naval Aircraft Factory. In both cases the gear ratio was 5:3. A CS-2 at Anacostia had a T-3 engine with a 6:1 compression ratio. Planned tests included the use of a CS-2 and a CS-3 with two compression ratios, 6:1 and 6.5:1. In addition, CS-2s and -3s would be tested with both available superchargers, mechanical (Roots type) and turbo-driven (Moss type).

32. The SC-3 designation seems to have been avoided because of possible confusion with the CS-3. Although the post-1926 BuAer correspondence files include several folders which seem to refer to SC-4 and SC-5, their contents do not. No folder refers to SC-8. SC-9 was the follow-on T3M-1. SC-6 (designation assigned in June 1925) was an SC-1 powered by a 730-hp Packard 1A-2500 engine. SC-7 was a CS-1 with a supercharged and geared Wright T-3 engine. Both were described in a June 1925 BuAer memo as two—rather than three—seaters. BuAer was also interested in installing geared Packard 1A-1500s in either a CS-2 or an SC-2. This engine was rated at 575 hp and geared down 2:1 from its 2500 RPM. It had already been installed in PN-9 seaplanes.

33. Memo dated 15 July 1924 in RG 72 VV vol. 2, NARA, listing types of aircraft in service and under design. It is not clear which design this was.

34. Memo in RG 72 VV vol. 2, NARA, cover memo dated 5 January 1925, "Torpedo, Bombing, and Scout Planes—Experimental Development." It recommended that the two types be developed as soon as possible, and that the Naval Aircraft Factory be assigned Design 36. That could be done without assigning further funds (experimental funds for FY25 had already been assigned), and the factory had experience with large twin-engine flying boats. A table gave some details. There were two twin-engine designs, No. 30 (two liquid-cooled 850-hp Packard 1500s) and No. 36 (two air-cooled 800-hp engines); No. 35 had a single 730-hp Packard 2500. Gross weights were, respectively, 10,800 pounds, 9,910 pounds, and 9,660 pounds; maximum speeds were 115, 116.5, and 112.5 mph (all stalled at 55.5 mph). Cruising ranges were 433, 472, and 485 miles. Spans were 59, 57, and 55 feet. All could carry a 1,000-pound bomb and 1,300 pounds of gas. Data were for seaplane versions.

35. There was no T2M as such. Martin's SC-1 is sometimes listed as T2M-1, T2M-2 being its SC-2, but these designations seem never to have been used. The TM designation would have corresponded to the twin-engine Martin MT land plane ordered on 30 September 1919. The T3M-1 was initially designated SC-9 and then SM-1, to indicate its scouting role.

36. 12 March 1927 memo was from the BuAer carrier desk to the assistant bureau chief. In folder "Aircraft, Design, Procurement & Production, Pre–WW II" in the Aircraft archive, Naval Historical Center (now NHHC). The paper is coded "Aer-M-111-HBJ CV2&3." It is marked in pencil "info from Mr. Abang, formerly of SI [ship installations], for use in Meade Report Book." The Meade Report was the "ten year history of BuAer" produced in 1945.

37. The BuAer design chief preferred twin-engine aircraft. He agreed with the Naval War College that twin-engine aircraft were tactically superior despite their lower performance. Some of the tactical disadvantages of the single-engine airplane had been reduced by improving the position of the bombardier, but the single-engine airplane was still inferior in the protection provided by its machine guns. It was no surprise that two of the three experimental VT-S were twin-engine. The question of single versus twin engine, he argued, should be decided by tests, ideally using competitive development squadrons of each type. This idea was killed by early operating experience with the T2D.

38. The T4M tail had to be jacked up for its wing to be folded without striking the deck. Great Lakes' new mechanism allowed folding while the airplane was on its wheels, but it added drag that cost 6 mph.

39. This airplane was originally ordered in July 1930 from Detroit Aircraft Corporation as TE-1, but it received a Great Lakes designation because it was built under a Great Lakes subcontract. A TG-1 reengined with the TG-2 engine achieved 116 mph. Plans to replace the engines of all the TG-1s were canceled in 1932; money was too short.

40. Memo dated 11 January 1927, VV vol. 3, RG 72, NARA.

41. "The Employment of Aviation in Naval Warfare" numbered 2005-9, box 74, RG 4, Naval War College Archives [hereafter NWCA]. Its cover is dated September 1937, but the mimeographed paper inside is marked "10-26-36," so the lecture was presumably first given the previous year.

42. General Board 449, 1924–25, paper dated 28 September 1925, GB 449 serial 1295. RG 80, NARA.

43. General Board 449, 1926–29 file; memorandum for the president of the Naval War College dated 20 February 1926, RG 80, NARA.

44. The Naval War College claimed that a recent British Boulton-Paul twin-engine bomber demonstrated that a twin-engine airplane need not be much larger or less maneuverable than a single-engine one. Unfortunately, the T2D would soon show that twin engines were a major problem on board a carrier. According to Lee M. Pearson, "Torpedo Bombers: The Prewar Years" in *BuAer Confidential Bulletin* for October 1949, the initial P-2 engines were too unreliable. The R-1750s that replaced them generated unequal torque at low landing speeds, pulling the airplane sideways just when it should be settling in the "groove" for landing. The aircraft were all assigned to shore torpedo squadrons at Coco Solo and at Pearl Harbor.

45. General Board 449, 31 January 1927, in 1926–29 file, BuAer Aer-M1-A A1-3 F1, NARA. The head of the design section, Cdr. E. E. Wilson, was among

the Taylor Board witnesses. So were Capt. Holden C. Richardson, Constructor Corps (CC), head of the Material Division of BuAer, and Cdr. W. H. White Jr., head of the BuAer Planning Division.

46. Sheets marked April, May, and June 1927 in attack aircraft file, Aviation History Division files, NHHC. The designs were No. 64 (direct-drive air-cooled engine, large upper and small lower wing, with reduced fuselage, fuel consumption, and other items; narrow fuselage would not accommodate the usual three five-hundred-pound bombs); No. 65 (Design 64 with equal-span- and- chord wings and no "stagger," i.e., the upper wing not set forward of the lower one); No. 66 (a monoplane like the earlier Fokker FT but with an air-cooled engine); and No. 70, a bomber-torpedo plane without scout provisions, powered by a Hornet engine, with wings folding to a span of eighteen feet so that it could use either elevator. (No. 70's wings had both sweep-back and stagger and so could be folded with the tail skid on the ground; the design would have under-wing bomb racks, and the gunner would have two upper guns and one lower); data from preliminary design sheets (there is no sheet for No. 70). A previous VT-VS had been Design 51 of 16 March 1926, powered by a Packard 2500 engine and thus broadly comparable to the T3M. Gross weight as a land plane was 9,050 pounds, and maximum speed was 124.3 mph (stall speed 57 mph). Service ceiling was 13,200 feet, and cruising range was 555 miles. Span was not given, but wing area was 848 sq ft, compared to 883 for a T3M-1.

47. The design section had studied this possibility in terms of adapting the existing Curtiss Falcon with a Pratt & Whitney Hornet engine. It appeared that it could readily be adapted as a light bomber with a five-hundred-pound bomb load and a maximum speed of over 150 mph. Its other characteristics would make it a good carrier observation airplane (spotter) and a fair two-seat fighter after it had dropped its bombs. To the design section, this study showed that it would be better to combine the scouting and two-seat fighter functions with a light bomber of this type to replace the amphibious aircraft currently on board carriers, designing the heavy attack aircraft primarily for torpedo attack, with a purely secondary heavy-bomber role. The Falcon became the F8C Helldiver.

48. The questionnaire and the conclusions of the Taylor Board are in General Board 449 serial 1353, in 1926–29 file, NARA.

49. GB 449 serial 1388, dated 12 December 1929, in 1926–29 file, NARA.

50. When he reviewed the General Board paper, Assistant Secretary of the Navy for Aeronautics Edward P. Warner wrote that "there is a possibility of engaging the enemy with bombing planes at a range far beyond that of gunfire, and the consequent possibility that actions may be conducted with the heavy airplane and its bombs as the initial and final, if not the only, weapon employed." The battles of the Coral Sea and Midway, the first in which carrier aircraft would in fact fight far out of sight of their carriers, were almost fourteen years away.

51. Marine Corps letter in General Board package, 17 December 1928, titled "Marine Corps Aviation a part of Naval Aviation," RG 80, NARA.

52. This airplane seems to have originated as a BuAer design. BuAer ordered prototypes with two alternative engines: T5M-1 with an R-1690 Hornet, and T2N-1 from the Naval Aircraft Factory with an R-1750. Both flew in March 1930, the Martin version being selected for production. It was Martin's, not BuAer's, Design 77. The Martin contract was dated 18 June 1928.

53. The Secretary of the Navy approved the board's proposal on 5 December 1928.

54. Typed sheet on torpedo bomber progress, clearly copied from internal BuAer files, in pre-1945 attack aircraft file, Aviation History Division, NHHC. It is included in the preliminary design sheets but not dated; the aircraft had a gross weight of 6,300 pounds (land plane only, not a seaplane) and was powered by a 550-hp Wright engine; wing area was indicated as 495 sq ft, but span was not given. Stall speed was 60 mph and maximum speed 135.3; initial climb rate was 940 ft/min. (service ceiling 15,700 ft). No endurance figures were given. This may have been the beginning of the T5M/BM/T2N.

55. An undated (probably summer or fall 1929) BuAer memo compared the T6M with the T4M and T5M. The chief of the design section remarked that "the general trend downward in the size of aircraft carriers emphasizes the importance of reduction in

size of airplanes." He meant USS *Ranger*; the shift back up in size represented by the *Yorktown*s was about a year away. Thus the memo compared the storage space needed by the three aircraft: 24,993 cubic feet for a T4M; 13,972 for T5M; and 18,159 for T6M. T5M had the smallest height, which mattered because aircraft were often triced up from the overhead of the hangar. The memo compared actual performance for the T4M with estimates for T6M and T5M, in that order. T4M maximum speed was 114 mph, compared to 123.4 for T6M (which had to be large enough to handle the long torpedo), and 138.1 for a modified T5M. Stall speeds, which were important for carrier landings, were, respectively, 56.3, 64.8, and 65.3 mph. The effect of smaller size showed most dramatically in climbing performance: altitude after ten minutes was 3,650, 5,360, and 6,330 ft, respectively. Cruising endurance was, respectively, 4.28, 4.42, and 4.60 hours, which translated into cruising ranges of 300, 388, and 409 miles—showing that T5M would be a superior scout.

56. "Dive Bombers—General Information," memo, 25 January 1930, VB file, RG 72, NARA. According to the cover sheet, this idea originated with Cdr. (later admiral) R. K. Turner.

57. The 8 May 1930 sheet describing BuAer's Design 97 is marked "XT3D-1." The design was powered by a geared supercharged Hornet engine (560 net hp). Loads were a single thousand-pound bomb, or two five-hundred-pound bombs, or a torpedo. Maximum speed at six thousand feet (to which the engine was supercharged) was about 130 mph. The file includes different versions developed in March 1930: the 97-A, with an R-1690 in a Townsend Ring; 97-B, with an R-1820; and 97-C, with a supercharged 575-hp R-1820. Unfortunately, the surviving BuAer design sheet gives performance but not dimensional data. Note that Design 97-A (5 March) was a Class VB bomber, meaning a heavy dive-bomber. The summary sheet shows both the basic 6,628-pound version and a smaller, 6,300-pound, version of 97-A, maximum speeds being 124.6 and 127 mph. Design 97-B (8 March) was another VB, somewhat heavier (7,200 pounds, and 7,277 for Design 97-C); maximum speed was 124.2 mph. On a 28 March sheet, 97-C is described as a bomber (not VB) with alternative loads of a torpedo, a thousand-pound bomb, and two five-hundred-pound bombs. The XT3D-2 was credited with a maximum speed of 153.1 mph at 6,000 feet and a service ceiling of 15,700 feet. With its greater fuel capacity (225 rather than 160 gallons) it had a greater range at economical speed (778 miles on 180 gallons) and a longer endurance at 75 percent speed (6.22 hours). It had the same SGR-1830 engine (geared 3:2).

58. The descriptive sheet for Design 98 is dated 20 May 1930. As a pure bomber, it did not have to lift the weight of a torpedo. The basic design employed an R-1690, as in the T4M, and weighed 6,647 pounds at takeoff, compared to 8,071 for a loaded T4M. Maximum speed would be 123.3 mph compared to 114 for the T4M. Range for this non-scout was 409 miles at cruising speed; endurance would be 4.7 hours at 90 mph. An alternative version was powered by an R-1750 supercharged to six thousand feet. For the bomber role, it would carry a 1,000-pound bomb. According to a 15 October 1931 memo from the Airplane Design Section recommending cancellation, empty and gross weights were, respectively, 3,670 and 6,898 pounds; span was forty-eight feet (no length was given). Useful load as a bomber carrying a 989-pound bomb was 3,028 pounds. Maximum speed at six thousand feet was expected to be 129.1 mph, and stall speed with the bomb was 81 mph (Naval Aircraft Factory said 63 mph, but that was probably without the bomb). Service ceiling was 13,500 feet. Endurance at maximum speed was 2.59 hours (4.72 hours at 90 mph). Calculations assumed engine output of 525 hp at six thousand feet. Construction was authorized on 7 June 1930, and the mock-up was inspected on 1 October. BuAer ordered cancellation on 31 October 1931: performance was judged inferior, and the airplane would be obsolete before it could be completed. Any conclusion as to the value of a 130-mph bomber carrying a thousand-pound bomb could be drawn from experience with the XT3D-1, which would be delivered within a week (the XBN-1 was not expected until April 1932). Although the XT3D-1 was somewhat larger, it had much the same performance.

59. That is, firing into the mass of the enemy (on land, "the brown") in the expectation of getting at least some hits. This was often proposed for submarines and destroyers firing torpedoes from long range, the idea being that ships in the enemy formation filled most of its total length and would find evasion difficult.

60. The BuAer Plans Division summarized General Board and BuOrd action in a 31 October 1929 memo contained in Secret papers, RG 72, NARA. It is coded Aero-D-158-OL S-75. (S-75 was the filing code for torpedoes.) BuOrd had asked for specific limits on overall length, diameter, and weight, as well as on maximum launch speed and minimum launch altitude, for recommendations as to whether the torpedo gyro should be capable of being set externally, and any other required features. OpNav did not want to set specific limits on so complex a project, but the weight was not to exceed the 1,737 pounds of the existing aerial torpedo. Maximum launch speed was set at 145 knots and the torpedo was to withstand a drop from forty feet. It was considered desirable to shorten the torpedo to fourteen feet and also to improve its streamlining, exploiting its better length-to-maximum-width, or "fineness," ratio (7 rather than the previous 12). The General Board suggested a launch speed of a hundred knots and a fifty-foot drop. In a 29 January 1929 memo to CinC U.S. Fleet, BuAer stated that a fourteen-foot torpedo (168 inches) could hardly have a range beyond four thousand yards, compared to six thousand for the Mk VII-2A, which was seventeen feet nine inches long. If the fleet wanted greater range, it would have to accept a much larger torpedo bomber. A development plan for the short torpedo would be sent to BuOrd if the fleet found the reduced range acceptable. According to a cover memo, War Plans rightly doubted the effectiveness of current aerial torpedoes against capital ships; BuAer itself doubted that torpedo bombers without fighter cover could penetrate enemy air defenses. It might be forced to accept a twin-engine VT carrying a heavier torpedo. That idea would recur later. It turned out that BuOrd had been working since 1922 on a new aircraft torpedo to replace the interim Mk VII. The goal of a new project (G-6) begun in February

1925 was a 21-by-216-in torpedo launchable at 140 knots (forty-foot altitude) with a 350-pound warhead and a range of four thousand yards at thirty-five knots. This project was stopped in 1926 in favor of modifying the Mk VII-2A torpedo so that it could be launched at 125 knots at fifty feet. (According to the 1929–30–31 edition of the BuOrd *Torpedo Record*, chap. 7, instructions to this end were issued in October 1928.) G-160 was a new project for a short, streamlined aircraft torpedo. A sketch was offered to BuAer in February 1930. Work was suspended in October 1930 because of the omission of torpedo stowage on the new *Ranger*, but BuOrd expected to resume in FY32 (beginning 1 July 1931).

61. In February 1930 BuOrd proposed to replace normal air with enriched air (40 percent oxygen). BuOrd considered this its first opportunity to try the enriched-air technique it had been developing since 1922. The Naval Research Laboratory (NRL) was working independently on oxygen fuel; it recommended using pure oxygen rather than a mixture of oxygen and compressed air, which it considered very dangerous. The main problem would be keeping the oxygen away from any contact with lubricating oil. Unknown to the U.S. Navy, the Japanese were developing torpedoes with high oxygen content. Also unknown to the U.S. Navy, the British had used high-oxygen-content torpedoes on board some cruisers but abandoned it because of problems encountered in lightning storms. In February 1930 BuOrd wrote its Torpedo Station, which developed torpedoes, that while it was very interested in oxygen, the new aerial torpedo should be treated as a separate project. BuAer liked neither oxygen nor enriched air. Other options were an aluminum body (BuAer feared corrosion) and a stronger air flask, which would take up less internal space. Torpedo stability in the water was complicated by the new short shape desired. NRL papers are in BuAer Confidential series F41-9 vol. 1, RG 72, NARA II.

62. Letter to BuOrd from Commander Aircraft Battle Force, date not entirely legible but 1931 (a later reference indicates that it was dated 5 September), in BuAer Confidential file F41-9 vol. 1, RG 72, NARA II.

63. BuOrd *Torpedo Record* 1933–1934, 171. This seems to have been the final edition, as NARA II has a complete set up to this point. The initial Mk 13 had a rated range of 4,500 yards at about 30 knots. With a stronger air flask, Mod 1 could run 6,000 yards at 33.5 knots. This longer range seemed excessive, so Mod 2 traded air-flask volume for more explosive (six hundred pounds). It also had a more powerful turbine (170 rather than 93–98 shp) and was credited with 4,000 yards at 40 knots. At this time the British were satisfied with 1,800 yards at 40 knots or 3,000 at 27. The initial Mod 0 was 164 inches long and weighed about 1,950 pounds. Mod 1 was 161 inches long (1,930 pounds). Mod 2 was the same length as Mod 1 but weighed 2,125 to 2,150 pounds (Mods 0 and 1 had voids in their noses to maintain balance).

64. BuOrd *Torpedo Record* 1932, 127.

65. Memo, 27 November 1933, in VVS vol. 4, RG 72, NARA.

66. Material to Plans, memo, undated but with the typed date of "7-19-33" on the bottom of the last page, VB file, RG 72, NARA.

67. Dated 17 October 1933, in VB file, RG 72, NARA.

68. The Douglas proposal (as revised) is dated 4 January 1934, entry 28, in RG 72. The power plant was an 800 bhp (at 6,500 ft) radial engine driving a two-position, three-blade, controllable-pitch propeller (the engine was not otherwise specified but weighed 1,172 pounds). The Mk 13 torpedo was to be carried so that it would be ten degrees down by the nose when the airplane was in level flight. Its release mechanism served as the attachment points for two five-hundred-pound bombs abreast, with another attachment point on the after part of the torpedo carriage for a third five-hundred-pound bomb. Gun armament was one fixed and one flexible 0.30-calibre machine gun. Dimensions given were a span of fifty feet (twenty-four feet ten inches folded), length thirty-five feet, and height over tail 13 feet 7½ inches. Crew was a pilot, assistant pilot, and radio operator / rear gunner. Douglas claimed that when carrying three five-hundred-pound bombs, the airplane would make 217 mph at 6,500 ft, landing at 60.1 mph. Rate of climb would be 1,440 ft/min. at 6,500 ft, and service ceiling would be 24,400 ft. Takeoff run in a 30-mph wind would be 290 ft, and range at three-quarters speed would be 556 miles. With a torpedo, maximum speed would be 224 mph; speed empty would be 234 mph. Takeoff distance could be reduced to 190 feet by using a continuously variable–pitch propeller, under development but not yet available. Gross weight was 7,210 pounds without weapons, or 7,785 with two five-hundred-pound bombs, or 8,500 with a torpedo.

69. Initially 114 were ordered, fifteen more being added in 1938, after the airplane was operational. One tested in 1939 as a floatplane was designated TBD-1A. An abortive version offered in 1940 to the Dutch used a 1,200-hp R-1820.

70. Letter to BuOrd, 21 June 1941, in F41-9 vol. 4, RG 72, NARA II. Commander Battle Force (Admiral Pye) endorsed Halsey's idea and added the successful attack against the British cruiser *Liverpool*. He cautioned that the percentage of failed torpedo attacks was unknown. CinC Pacific Fleet (Adm. H. E. Kimmel) also agreed. On an August 1941 route sheet Kimmel pointed out that the big new VSBs could each carry *two* Mk 13s externally with only very minor effect on them. "Perhaps this can be extended to the XSB2C [Helldiver]." It seemed that the VSB and VTB types were approaching each other.

71. BuOrd disliked the drogue, on the theory that the problem was either pilot error (in altitude, attitude, or speed) or in the torpedo itself. However, it made twenty-four drogues and shipped them to Hampton Roads, in Virginia. BuOrd to Naval Torpedo Station 375, letter, 8 September 1922, 37589/544 (A5)-0-MES, entry 63 file 204-5, RG 72, NARA II. This file incudes a 5 July 1922 letter from the Inspector of Ordnance at the Newport Torpedo Station describing the drogue fitting for a Mk VII torpedo.

72. BuOrd memo, 26 October 1926, Confidential BuAer Correspondence, F41–9 vol. 1, RG 72, NARA II.

73. BuAer to CNO, 4 May 1929, Confidential F41-9 vol. 1, RG 72, NARA II.

74. The British had analogous "monoplane air tails" (MATs). It is not clear whether BuOrd was aware of them while developing its boards.

75. Barrett Tillman, *TBD Devastator Units of the U.S. Navy* (Oxford, U.K.: Osprey, 2000), 34, quotes a noncommissioned naval aviation pilot, Thomas F. Cheek, describing the first live torpedo drop he ever made. After a run in within limits (ninety knots, a hundred feet), the torpedo dropped "clean" and ran true for the first 100–150 yards, then veered sixty degrees to port and ran off over the surface. A second drop behaved similarly, but this time the torpedo veered slightly to starboard, toward the tug towing the target. Cheek wrote that according to the "scuttlebutt" at Pearl Harbor these experiences were the rule, not the exception. The fleet exercised torpedo attacks only rarely; each pilot made a live drop only once a year, and there were no practices (as there were later) using concrete dummy torpedoes. Cheek made no drop at all in 1939, then two dry runs followed by a live drop in 1940. Simulated runs were carried out during fleet problems and local maneuvers. Tillman (page 45) reports that during 1940 fleet gunnery exercises, four of the ten Mk 13s launched sank, and five ran erratically. Tests showed that torpedoes veered to port on entering the water. Their wooden stabilizers often broke, as did their propellers. Depth-keeping was erratic. In 1941 it was claimed that Mk 13 was 80 percent reliable when dropped from 140 feet, but fleet squadrons limited themselves to a maximum speed of 110 knots and drop altitudes of 50 feet or less. The ninety-five known drops in eleven squadron-strength attacks between February and June 1942 made only ten hits.

76. BuOrd *Confidential Bulletin* 1-43, 25 March 1943.

77. Transcript of 7 January 1941 conference (in BuOrd Torpedo Section), F41-9 vol. 3, RG 72, NARA II.

78. F41–9 vol. 5, RG 72, NARA II, notice issued by BuOrd 14 April 1942.

79. The origin of the "pickle barrel" is not clear from the BuAer torpedo file (F41-9), but the file includes a request for ninety "drag rings" dated December 1943. At this time the rings were only just entering production.

80. BuOrd *Confidential Bulletin* 1-45, of 31 March 1945, explains the functions of the three modifications. The July 1944 BuAer *Confidential Bulletin* associated the new fittings with a torpedo rack that could be fitted to the Helldiver. According to

this article, the new tail cost about half a knot of speed (to thirty-three knots). Depth and deflection control were so improved that during an experimental run, with depth set to 5 feet, the torpedo could be seen to follow the rise and fall of the ground swell. Tests indicated that 80 to 85 percent of torpedoes entering the water at angles of between ten and twenty degrees would recover in about fifty feet of water. By this time torpedoes had been launched at altitudes of 1,800 feet at three hundred knots.

81. BuOrd *Confidential Bulletin* 1-44, 30 April 1944.

82. Mod 1 war shots were not to be dropped above three hundred feet, to avoid throwing their turbines out of alignment after release but before hitting the water. For other versions, in fairly deep water normal drop speed was 210–230 knots between three and eight hundred feet, or slightly higher at 230–250 knots. For shallow water, recommended speed was 225–245 knots at 300–330 feet.

83. BuOrd *Confidential Bulletin* 3-45, 30 September 1945.

84. N. J. M. Campbell, *Naval Weapons of World War Two* (London: Conway Maritime, 1985), 159, reports an initial launch limitation of fifty feet and 110 knots and a late-war capability of 2,400 feet and 410 knots.

85. The first Navy bombsight (1918) was the British Wimperis plus a pilot director, the result being designated the Pilot Directing Bombsight Mk III. According to Albert L. Pardini, *The Legendary Norden Bombsight* (Atglen, Pa.: Schiffer, 1999), three thousand were ordered from International Signal in April 1918 but canceled shortly after the Armistice. By this time the Navy had approached Sperry for a stabilized bombsight, but he had been unable to achieve the degree of accuracy desired. The British used the Wimperis in postwar experiments against the radio-controlled target ship HMS *Agamemnon* in which 114 bombs were dropped, without any hitting. In January 1920 BuOrd invited gyro expert Carl L. Norden, a private consulting engineer who had left Sperry to set up his own business in 1915, to improve the Wimperis. He integrated a stabilized telescope and a speed- and course-setting mechanism into it but

said that an entirely fresh design was needed. After testing the stabilized sight in July 1920, BuOrd hired Norden to develop a sight that would measure aircraft speed over the ground by timing. It became the Navy's Mk XI (the Navy and the Army used the same series of mark numbers, so for the Navy Mk XI came after Mk III). A production order was placed in 1927. Tests showed an average error of 190 feet, compared to 286 for the Wimperis. By May 1928 it appeared that every salvo dropped using the Mk XI sight would straddle a target, whereas with other sights only one in four would. However, the sight was too difficult to maintain, and it could not be used below 2,500 feet. BuOrd wanted something better, so in 1929 Norden began work on Mk XV, the famous "Norden bombsight." It was first tested in 1933. Although designed specifically for high-altitude bombing, it was easily adapted to low altitude, outperforming the simpler Mk III. Series production began in 1935.

86. The 1 June 1936 edition of "Aircraft Gunnery Notes" for the Battle Force (USF 49) described horizontal bombing in detail. Attackers would try to establish the long axis of their pattern at right angles to the long axis of the target. Using Mk XV bombsights, the mean error for bombs dropped in line should be about 100 feet. The mean range error should be fourteen mils (thousandths of range), the mean deflection error twenty mils. For bombing from 10,000 feet, these errors would translate into 140 and 200 feet, respectively. With bombs in line, 100 feet apart, attacking a battleship-size target, three airplanes should have a 41 percent probability of making at least one hit; nine should have an 83.8 percent probability. Calculations included the effect of target maneuvering. Prewar air doctrine laid out tactics: bombers would approach their targets at 12,000 to 13,000 feet and bomb from 10,000. Ideally, bombers would attack out of the sun, from ahead or astern. They would do best if supported by dive-bombers. Nine-plane divisions would attack shore targets or ships at anchor; divisions of six would attack heavy ships when the threat to them would be enemy aircraft rather than

antiaircraft fire; and a three-plane division when antiaircraft fire was as serious a threat as enemy aircraft. Bombing records showed that the minimum effective formation was probably six aircraft; nine was the largest that could improve hitting without suffering too many misses.

87. It was claimed that a bombardier with the U.S. Norden bombsight could hit a "pickle barrel" from 20,000 feet or above. (Norden formed a "pickle barrel club.") In prewar tests unopposed bombers managed to drop half their bombs within a hundred yards or so of a fixed target, but that was hardly good enough to hit a moving ship. Royal Air Force studies showed that only 7 percent of all U.S. Army Air Forces bombs (dropped mainly by Norden sights) fell within a thousand feet of their targets; William Wolf, *U.S. Aerial Armament in World War II: The Ultimate Look*, vol. 2, *Bombs, Bombsights, and Bombing* (Atglen, Pa.: Schiffer, 2010). As late as 1944, the Bureau of Personnel training manual *Aircraft Fire Control* claimed that "with the help of American-built bombsights, U.S. Navy bombers can practically split the sides of a floating beer barrel from 30,000 feet up." During the experimental bombing of the old armored cruiser *Pittsburgh* in September 1931, the Mk XV made four hits out of eight from five thousand feet, far better than the Mk XI. In October 1932 the Naval Proving Ground, Dahlgren, Va., claimed that in good weather the accuracy of Mk XV was eighteen mils in range and 13.5 in deflection between altitudes of 4,000 and 14,700 feet. Even better was claimed for a production model (8.5 and 8.0 mils, respectively). Mk XV was officially adopted in 1933, and it also became the standard Army bombsight. It imposed altitude and speed limits based on the optical angles and bomb trail. In 1941, for example, the Norden company envisaged limits of 1,500- to 30,000-foot altitudes and speeds of up to 250 mph (the Army Air Forces wanted a higher speed and minimum altitude).

88. They thought that horizontal bombers would fly too high for antiaircraft guns to be effective. This is evident in remarks included in the official 1940 account of U.S. Navy gunnery practices.

89. From COMINCH, *Secret Information Bulletin No.1: Battle Experience from Pearl Harbor to Midway, December 1941 to June 1942 (including Makin Island Raid 17–18 August 1942)*, former NHHC documents in NARA II.

Chapter 4. Dive-Bombers

1. Peter C. Smith, *Dive Bomber!* (Annapolis, Md.: Naval Institute Press, 1982), 8–9. The concept was described in a British magazine as early as December 1914. British pilots described dive-bombing attacks during World War I. However, it appears that the main World War I method of ground attack was strafing with machine guns, supplemented by bombs dropped at low altitude (but not generally in a dive). Smith points out that under later definitions many attacks would have been classed as "glide bombing" rather than dive-bombing, the latter involving a dive at an angle of seventy degrees or more. The first formal tests of dive-bombing were conducted at the RAF Armament Experimental Station at Orfordness early in 1918 (pp. 13–16) using a Camel and an SE 5. Pilots sighted the target using their Aldis air-to-air sights. The sight was considered dangerous, because it gave the pilot no sense of how close he was to the ground. An alternative vertical diving technique proved effective, the pilot looking along the top of the sight. Despite great accuracy, dive-bombing was rejected as too unsafe for average pilots.

2. The Army's used war-built DH-4s conducted what would later be called glide bombing, diving at no more than a forty-five-degree angle (their wings would not have withstood steeper dives). They used underwing racks for small bombs. After completing the course at Kelly Field, Rowell joined the Army's 1st Pursuit Group. There he met Lt. (later brigadier general) George P. Tourtellot, who told him about wartime British dive-bombing and about his own dive-bombing attack against the ex-German target battleship *Ostfriesland* during the exercises off the Virginia Capes in 1921. Tourtellot had hit the ship but considered the technique too dangerous in the face of antiaircraft fire.

3. Thomas Wildenberg, *Destined for Glory: Dive Bombing, Midway, and the Evolution of Carrier Airpower* (Annapolis, Md.: Naval Institute Press, 1998), 10. Wildenberg argues that Wagner was able to make near-vertical dives because his new F6C Hawk fighters were particularly rugged, with reinforced-metal leading edges to their wings. However, Smith describes earlier near-vertical dives.

4. According to Lee M. Pearson, "Dive Bombers: The Prewar Years," *BuAer Confidential Bulletin* (July 1949), only the UOs could be fitted to carry light bombs (presumably underwing) for the 1926 tests. Pearson reports that once the tests proved successful, wing racks were installed on board existing standard fighters: VE-7s, F6Cs, and FBs. Aircraft dived at forty-five-degree angles.

5. Initially pilots were ordered to zoom up immediately after releasing their bombs. It turned out that some bombs did not actually leave until after the aircraft had turned, ruining accuracy. BuAer then proposed that pilots continue on for another 200 feet, so they had to release at higher altitude, typically 1,000 or 1,500 feet. There was also some fear that if the pilot turned up immediately after releasing a bomb, it might hit his tail surfaces as he turned up.

6. Report dated 18 October 1927 in BuAer Secret A5 file, NARA II. Trials began on 11 October and were completed on 24 October.

7. Comments on a paper on the employment of two-seat fighters, which were expected to maneuver less strenuously than single-seaters; Material Division, memo to Plans Division, 11 February 1927, correspondence VF vol. 1, RG 72, NARA. This in turn was a comment on a paper by the bureau chief, 26 January 1927, on two-seat fighters, which Admiral Moffett considered essential.

8. Early in 1928 BuAer had access to the Falcon because the Marines had bought two prototypes (XF8C-1s) and four production aircraft to meet a 1927 Marine Corps specification for a multipurpose airplane combining fighter, dive-bomber, and observation roles—most likely mainly observation and light bomber. In this form it carried light bombs underwing. These aircraft were derived from the Army's O-12 version but had bomb racks and the extra wing guns of the Army A-3 attack version. The Navy applied a fighter designation because there was no specific attack designation.

The Marines redesignated their F8C-1s as OC-1s. The F8C-2 was a fresh design, but the F8C-3 (soon redesignated OC-2) was a modified F8C-1. These airplanes were descended from an airplane entered in a 1924 Army contest for an observation plane powered by the Liberty engine. It lost out to Douglas' O-2, but the Army opened a new contest in 1925 for aircraft powered by the Packard 1A-1500. Curtiss won (the engine was changed to its D-12 Conqueror), and the airplane was built as the O-1. Unusual features were the aluminum-tubing fuselage frame and a wing moved forward for maximum pilot access and visibility, the outer wing panels being swept back nine degrees to maintain balance. These wings were characteristic of the whole series.

9. The bureau chief's July 1929 memo to the assistant secretary included some performance figures. The F4B-1 had demonstrated a maximum speed of 168.8 mph, compared to an estimated 149 mph for the F8C-2 prototype. It could climb to 5,000 feet in 2.5 minutes, compared to 3.2 for the F8C-2. In ten minutes it could climb to 15,300 feet, compared to 12,500. It also had greater range: 466 miles at cruising speed versus 332.

10. BuAer memorandum of conference, 14 December 1928, attended by Bureau and Curtiss representatives, in BuAer Correspondence, VF vol. 2, RG 72, NARA. At this time Curtiss expected its new F8C-2 prototype to be ready by May 1929. It was to be a "dog plane" (presumably a developmental prototype) toward a two-seat fighter/dive-bomber, with a view toward future production. In parallel Curtiss would offer a very similar airplane as the O2C (to speed production); it would be a "dog ship" for observation and scouting planes with interchangeable wings, struts, wires, and tail surfaces, with the same strength factor as the fighter. A third plane would be the full prototype F8C-2, with complete ordnance equipment and other features required for dive-bombing. It would be ready for production in case the F8C-2 prototype was accepted for production; Curtiss said that it would be ready by September 1929. Curtiss asked for a contract change (for the strengthened tail) on the F8C and a new contract covering the other two aircraft. The Marines' O2C-1 was originally designated

F8C-5, an F8C without carrier fittings. Sixty-three were built. Curtiss modified two Helldivers with 575-hp R-1820 Cyclone engines instead of the earlier 450-hp R-1340 Wasps. The Navy bought one of them as the XF8C-8 and two others as O2C-2s. After one of them crashed, Curtiss replaced it with a heavily modified airplane designated XF10C-1. It had a 650-hp R-1820-E engine, single-strut undercarriage, and a new tail.

11. The O2U impressed some because it seemed to be faster than the F8C, but that was partly because it did not carry many of the same loads: oxygen gear, floatation gear, gun mounted with ammunition drums, two sets of gunsights, a hook, and a hoisting sling. Too, the F8C, but not the O2U, could be spun, stalled, and maneuvered violently.

12. There seem to have been problems. A 10 May 1929 summary of the two-seat fighter situation stated that it did not look as though the F2U would arrive before the end of the fiscal year, after which the relevant funding would lapse.

13. CO VF-1 of Battle Fleet, memo to Commander Carrier Divisions, 15 January 1931, in BuAer Correspondence, VF vol. 2, RG 72, NARA. The pilots attributed low performance to the fact that the airplane had been designed for fighter-type maneuvering (hence for a large safety factor) and also to the sheer size of its rear cockpit, which had controls and instruments. There were also complaints of air sickness by gunners facing aft. Note that the high stress limits were associated with the secondary dive-bombing role. By this time improvements in single-seat fighters had rendered the F8C-4 obsolete as a fighter; surviving aircraft were converted into O2C-1s for the Marines and for Pensacola.

14. I have not found the papers that launched the project in 1928. It may correspond to an undated Design 75 (ca 1928), a wheeled airplane powered by a 550-hp Wright engine, weighing 6,300 pounds loaded, with a wing area of 495 square feet. Maximum speed was 135.3 mph, somewhat slower than that actually achieved. Service ceiling was 15,700 feet. The very abbreviated sheet describing this design does not indicate dimensions, a weapon load, or empty weight. In 1931 BuAer described the T5M as a response to fleet requests; Chief of Bureau to Assistant Secretary of the Navy for Aeronautics, 12 January 1931, in VB file, RG 72, NARA.

15. BuAer Correspondence, VF vol. 2, RG 72, NARA, includes a proposed technical note, dated 27 November 1929, "VF and VB Types of Airplanes: Precautions in Maneuvering."

16. Its R-1340 Wasp engine was reliable—but only at low speed. The fleet did not even use its full reliable endurance, which was 5¼ to 5½ hours at low power. Typically O2Us formed a line ahead of the scouting line, immediately ahead of the battleships, which in turn were advancing at about twenty knots. At reduced power the O2U could press ahead at seventy knots. When its time in the air was limited to three hours for safety, it could position itself only fifty-eight nautical miles ahead of the scouting line (or seventy with four-hour endurance). That had been the situation in Fleet Problems 10 and 11. This was hardly good enough: fast, long-range scouts were needed. Unfortunately, high-powered engines, such as the R-1690s in the torpedo bombers, were not yet reliable enough. Comments on the O2U are from the 1930 Annual Report of CinC US Fleet, microfilm publication of the annual reports, 1921–1941, NARA.

17. BuAer VS vol. 1 file, NARA.

18. VS vol. 1 file.

19. The SC designation had already been used. The S2C was an adapted Army attack plane, bought to test its novel arrangement of flaps and slats. It was not considered for carrier use.

20. Memo of conference, 13 March 1933, BuAer Correspondence, VS vol. 1, RG 72, NARA. Contemporary documents do not seem to explain why these aircraft were needed.

21. BuAer chief, Rear Adm. E. J. King, memo to Secretary of the Navy, 2 October 1933, BuAer Correspondence, VS vol. 1, RG 72, NARA.

22. Cdr. R. K. Turner, Plans Division, memo to chief of BuAer, 14 August 1931, VB file, RG72, NARA.

23. BuAer memo for file describing meeting held 24 August 1931 in assistant chief's office to decide policy on buying experimental dive-bombers; NHHC Aviation History Division. The meeting was chaired by the assistant chief, Capt. A. B. Cook.

24. Compared to the F4B-3 and -4, the F6B-1 had 10 percent greater wing area, more power, and fully cantilevered main-gear legs, which they did not, to clear a centerline bomb. It lacked the high fuselage structure the F4B had abaft the cockpit, which seems to have created unwanted turbulence at high speed. The BuAer R&D history also lists Martin's abortive XFM-1 (its Model 127; Model 129 was the BM-2). The designation was later used for the Eastern Motors (GM) version of the Grumman Wildcat (F4F).

25. According to a BuAer history file in NHHC Naval Aviation History, the F11C was derived from BuAer Design 120. The earliest Design 120 sheet is dated 19 February 1932; it shows a variety of increasingly heavy aircraft powered by R-1535s. Another sheet (28 March) compares Design 120 versions with alternative projected ("theoretical") engines and with various wing areas. The fighter version with an R-1535 is credited with a speed of 213.4 mph. The final sheet describing Design 120 is dated 8 April 1932; it is listed as a fighter (3,946 pounds) or a bomber (4,100 pounds, with two 116-pound bombs), powered by an R-1535 rated at 685 hp, with a wing area of 240 sq ft. Maximum speed (as a fighter) was 202.8 mph (199.3 as a bomber). It seems likeliest that Design 120 was a yardstick against which to compare Curtiss' proposal rather than the origin of the F11C/BFC. Note that the bomber version carried two 116-pound bombs underwing rather than a 500-pounder under the centerline. It is not clear from surviving documents when the single 500-pound bomb was adopted; the two 116-pound bombs survived as an alternative load.

26. Head of Materiel Division, memo, 1 January 1932, VV vol. 5, RG 72, NARA. Several fighter projects had lower priorities, hence would be pursued only if there was sufficient money.

27. Unknowable due to excellent Japanese security, although some Japanese aircraft were seen during the 1932 attacks on Shanghai.

28. Confidential Correspondence of Secretary of the Navy and CNO, 9 December 1933, RG 80, NARA: folder VS to VV. It lists pure fighters in proposed future carrier aircraft complements. In 1931 the Navy's F4B was competitive with current European service aircraft, but European air forces were developing higher-performance interceptors.

29. Export sales of this Hawk III were one demonstrator, twenty-four to Thailand, 102 to China, one to Turkey, and ten to Argentina. The Chinese aircraft,

delivered between March 1936 and June 1938, saw combat against the Japanese. Export sales of Hawk II (BFC-1): nine to Bolivia, four to Chile (plus aircraft built under license), fifty to China, twenty-six (floatplanes) to Colombia, four to Cuba, two to Germany (in 1933), twelve to Thailand, and nineteen to Turkey.

30. Marine Corps Headquarters to BuAer, 9 September 1933, Aviation History Division, NHHC. Like the Navy, the Marines wanted an airplane that could deliver a thousand-pound bomb. They wanted about the same speed BuAer would request the following year—180 kts (207 mph)—but they were willing to accept shorter range (250 nm, 287.5 miles) and lower ceiling (15,000 ft).

31. Tactics described by Smith, *Dive Bomber!*, 107.

32. Memo, 6 May 1935, BuAer Correspondence, F41 vol. 2, F41-8(377), NARA II. The object of penetrating thick deck armor is from the BuOrd administrative history compiled at the end of World War II, a copy of which is in the Navy Library rare book collection.

33. According to a December 1941 BuAer memo, if released from a sufficient height, the 1,600-pound bomb could penetrate five to six inches of armor rather than the usual three to four. Although all current dive-bombers could carry the bomb and drop it horizontally, dive-bombing required that the trapeze be modified. That could not be done at an overhaul base.

34. The Norden is shown in an undated photostat drawing in BuAer Correspondence, F41 vol. 3, NARA II.

35. The September 1935 handbook for this device is ORD 625, RG 74, NARA II.

36. When properly set up, the Norden sight worked about 80 percent of the time. However, at dive angles of forty-five degrees or less, the cowling obscured the target. Between forty-five and sixty degrees, the rate of drift was too high to permit satisfactory adjustment before some maneuver threw the target behind the cowling. There had to be a better way to apply corrections to the attitude of the diving airplane. Gyros had to be locked and unlocked automatically. The sight calculated altitude on the basis of air pressure but registered too slowly, owing to the length of the tube from wing to sight.

37. Dahlgren apparently tested both Mk 16 and Mk 17 during the summer of 1943. According to the postwar BuOrd history, the Mk 16/17 project was abandoned during 1944. Mk 17 was a simplified Aiming Sight Mk 3 (Bombsight Mk 17). It did not provide an automatic means to correct for lateral deflection caused by drift. Mk 17 did not enter production. Dahlgren apparently also tested the British DBS-1 dive-bombing sight in 1944 but does not seem to have considered it a viable alternative. The DBS-1 was a new approach to the problem, using a separate sighting head (with gyro) and a remote computer (bomb releaser). The sighting head had an illuminated reticle like that of the standard American illuminated gunsight (Mk 8). As in the Norden sight, the gyro tracked the target. Relative wind could be taken into account. The computer allowed for bomb ballistics; according to BuOrd many pilots believed that the greatest source of error was the curved path the bomb followed. The computer/releaser could be installed on board dive-bombers equipped with Mk 8 sights. This combination would be effective between 2,000 and 3,000 feet. With target tracking, that might increase to 4,500. With the remote computer, it might be possible to attack from 6,000 or 7,000 feet. Correspondence 23 Jan 44 (vol. 1944-1), NARA II, describes the desired test in an SBD-3, but this and other 1944 files do not describe results. BuOrd had received a descriptive pamphlet dated June 1942 but apparently preferred to continue with the Norden project.

38. For example, *BuAer Bulletin 1943-2* announced that Mk 8 sights would be installed in the SBD-3 and -4 (airplane Changes 128 and 14). All SB2C Helldivers, apart from prototypes, had Mk 8 sights.

39. *Current Tactical Orders and Doctrine, U.S. Fleet Aircraft*, vol. 1, *Carrier Aircraft*, USF 74B, November 1944, 4–8.

40. These three ships were attacked because they would have been employed as a floating battery in the final defense of Japan against invasion.

41. A high-wing monoplane powered by an R-1830 (Design 112), a sesquiplane (a biplane with one wing less than half as long as the other) powered by an R-1830 (Design 113), and a biplane powered by a liquid-cooled SGV-1800 (Design 115). Design 113

weighed 4,412 pounds (gross); maximum speed in that form was 199.8 mph. With the R-1535, that rose to 206.4 mph. Range at economical speed was 504 miles, and takeoff run with 30-mph wind over the deck was 135 ft. With the R-1535, range at most economical speed was 474 miles, and speed at critical altitude was 218.5 mph. Performance was calculated for fighter and bomber loads, the latter with two 116-pound bombs underwing. The design was dated 24 November 1931; a revised version was dated 11 February 1932. Design 115 was dated 7 December 1931. Gross weight was 4,946 pounds; there was also a scaled-up version (5,013 pounds). Maximum speed was 225.6 mph at sea level. No revised version appears in the package of design studies. Although Design 112 (and the Curtiss and Douglas monoplanes) exceeded the desired wingspan (forty-two to forty-four feet), they were considered acceptable because more high-wing monoplanes than biplanes could be parked in a given space. Design 113 exceeded the thirty-five-foot limit for a biplane by a foot. The Curtiss XF10C-2 exceeded the 65-mph stall-speed limit, but that could be cured by reducing the fuel load to a hundred gallons and the range to six hundred miles. Grumman's FF-1 also exceeded the stall-speed limit, but with reduced fuel its range would still be more than 550 miles. Data for the Douglas design were based on the Army's XO-31, but it seemed that a smaller airplane (span forty feet) could be faster (215–220 mph).

42. There was a parallel study of "special" (high-performance) designs with higher stall speed (70 mph) and possibly reduced useful load. Designs were: Lockheed monoplane with R-1830; Design 111 with R-1830 (similar to 112 but no radio or life raft); Design 114 sesquiplane with SGV-1800 engine; Design 116 biplane with GV-1570 (liquid-cooled) engine, similar to 115 but no radio or life raft; Design 117 monoplane with GV-1570 engine (similar to 111); and Design 118 monoplane with SGV-1800 engine and no life raft. Design 118 was the most promising, with an estimated speed of over 240 mph. Lockheed's design probably could not meet the stall-speed requirement with landing gear down, and there were possible objections as to vision, takeoff, communication, and maneuverability.

43. Plans, memo to Design, 5 June 1933, referring to an 18 May memo to chief of BuAer, VB file, RG 72, NARA. Grumman's proposed dive-bomber version of its two-seat FF-1 fighter (SBF-1) was not adopted.

44. According to the final report on service trials of the XFD-1 (18 June 1933–14 August 1934) in the pre–World War II attack airplane file in Aviation History Division, NHHC, this airplane was designed to carry two 116-pound bombs, not the 500-pounder of published accounts.

45. Memo, 5 June 1933, BuAer Correspondence, VS vol. 1, RG 72, NARA.

46. Request dated 19 March 1934 in Aviation History Division files, NHHC.

47. Evaluation dated 15 June 1934, Aviation History Division, NHHC.

48. VB-1B, comments to Commander Aircraft Battle Force, 2 April 1934, Aviation History Division, NHHC.

49. These windows were recommended in the annual Reports of Gunnery Exercises (which included bombing). According to a 20 March 1936 memo from the BuAer Armament Section to the Airplane Design Section (in NHHC's Aviation History Division files), squadron personnel had already installed these windows in the standard heavy and light dive-bombers: BMs and F4B-4s. "Dive-bomber pilots say that any window, however small, is of great value."

50. Memo, 22 May 1935, Aviation History Division, NHHC. This memo included data

	BT-1	Great Lakes 66-A	Consolidated mod XB3Y-1
Gross weight (lb)	6,198	6,310	6,250
Maximum speed (mph)	212.9	209.5	210 (at 8,900 ft)
Stall speed (flaps) (mph)	62.1	60.4	63.5
Service ceiling (ft)	24,900	21,200	22,000
Range at 75% max. spd (mi)	538	500	N/A
Scout range (mi)	1,106	1,042	N/A
Takeoff (30 mph wind) (ft)	237	204	N/A
Terminal vel. in dive (mph)	350	325	250
Option price (27 aircraft)	$24,800	$19,000	N/A

51. Memo, 27 September 1935, in Aviation History Division, NHHC.

52. Memo to Bureau Chief, 7 April 1936, in Correspondence, VB file, RG 72, NARA.

53. Evaluation data from a file of prewar competition figures in RG 72 (formerly RG 402) entry

P7, NARA II, which includes the draft evaluation report. Both XBT-1 and XSB2U-1 used the same 750-hp R-1535-88 engine. As a bomber, the XSB2U-1 was considerably faster at 9,000 feet (237.3 versus 216.5 mph), and it had greater range (660 versus 594 miles), but it had a much longer take-off run (384 versus 290 feet) and a higher stall speed (69.4 versus 67.8 mph). As a VSB, with a 500-pound bomb, the same SB2U-1 had a maximum speed of 242.6 mph, a range of 611 miles, and a 298-foot takeoff run (stall speed 66.2 mph). The SBA-1 had a more powerful R-1820-22 engine developing 845 hp (1,000 for takeoff) rather than the 750/825 hp of the others.

54. The 54 SB2U-1s were followed by fifty-eight SB2U-2s ordered in January 1938. The main modification was to make the bomb easily replaceable with a drop tank. There were also two light underwing bomb racks. The SB2U-3 was a much longer-range version for the Marines, meeting a requirement for what was called an "expeditionary airplane." Chief of BuAer to CNO, 17 January 1938, in BuAer Confidential Correspondence, VV file, RG 72, NARA II. It referred to a confidential CNO letter, Op-12Y-MG, (SC) F1-2 serial 365, of 13 November 1937. OpNav War Plans wanted a convertible seaplane/landplane so that offensive strength could be built up rapidly at an advanced base even if there was no shipping space to get airplanes there. Ideally the airplane should also be carrier capable. Range as a seaplane (without armament) was to be two thousand nautical miles. Chance Vought offered a modified SB2U. Curtiss offered a twin-engine airplane based on its Army A-18, which was rejected because it failed to meet proposed requirements, had many dubious features, and had no carrier fittings. According to a BuAer note, neither contractor understood the significance of the floats. Vought was told that float operation would be rare (it was for delivery purposes). The company inferred that any floats at all had been specified "merely to comply with some law, and that that law also required twin floats." Curtiss initially offered an SBC with a single float "and went away with the idea that what was wanted was a twin-float job *plus* twin engines." Its airplane was far too large and heavy for carrier operation, whereas Vought offered carrier features if desired. Also, the Curtiss airplane could not dive-bomb. Twin floats had been abandoned in Navy floatplanes because rugged ones were too heavy. Yet the Marines required a floatplane which could dive-bomb. The SB2U-3 offered increased forward-firing firepower (four 0.30s or two 0.50s), because the Marines expected to use it for close air support, which included strafing. In addition to the usual trapeze, it had three additional racks for small bombs. It was powered by an uprated R-1535-102 engine (fifty-seven ordered September 1939). In addition to the American aircraft, the French ordered twenty in February 1939 as the V-156-F1, then another twenty in May as the -F2, and fifty as -F3s in March 1940. This last order, undeliverable because of the fall of France, was transferred to the British under the designation V-156-B1 (Chesapeake, in British service). The French aircraft fought during the Battle of France in 1940. The only American combat use of the SB2U was by Marine aircraft at Midway.

55. The contract was shifted because Brewster was fully occupied with building the more urgently needed F2A-1 fighter. The design had to be revised in 1937 with an upgraded engine in a revised cowling, a three-bladed propeller, and a revised canopy. In this form it made 263 mph, making it the fastest VSB yet tested. The first SBN-1s were not delivered until 1940 or the last until 1942, by which time the SBN was entirely obsolete.

56. As reported in a 2 April 1936 report, "Status of Class Desk 'B' Projects." These roughly monthly status reports from January 1936 through June 1945 are in box: "NAVAIR Aircraft: Aircraft, Attack VA Status Reports (Mostly WW II)," Aviation History Division, NHHC.

57. According to Heinemann, the idea came from NACA aerodynamicist Charles Helms, who had been sent out to Northrop. He described the effect of the holes as the "alternating vortex theory." Edward H. Heinemann and Rosario Rausa, *Ed Heinemann: Combat Aircraft Designer* (Annapolis, Md.: Naval Institute Press, 1980), 35–36.

58. Paper, 20 June 1939, entry 28: proposals, box 62 (with Douglas proposal for 1939 VSB competition), RG 72, NARA.

59. These reports led to installation of shields, such as gun tubs, on board U.S. warships. The new *Fletcher*-class destroyers also had light protective plating over their machinery, a very unusual feature for a destroyer.

60. Of existing dive-bombers, the single-gun SB2U-1 and -2 were upgraded to two forward-firing 0.50s.

61. In mid-1943 the fleet began installing drop tanks on the left wings of SBD-5s; Douglas was asked to install drop tanks under both wings on its next SBD-5 contract.

Chapter 5. Dive-Bombers of World War II

1. In a 23 July 1934 memo Plans Division, for example, wanted "to build a scout diving bomber of an entirely new design using two small engines instead of one larger one as in previous types, with a view to obtaining greater performance through the use of increased power in smaller engine units." The preliminary designers had just produced the twin-engine Design 132 and the single-engine Design 133, the latter apparently a template for the 1934 torpedo-bomber competition. It would make 175 mph. Design 132 offered 180 mph (fifty-foot span, 8,900 pounds—which was high at the time). It was ordered from the Naval Aircraft Factory—a 23 November 1934 BuAer paper refers to the design of its wing—but was not completed. At that time the airplane was expected to make 176 mph at 12,000 feet on 1,000 bhp with a gross weight of 5,780 pounds. Wing sections (including a complete center section) and a mock-up were approved. In 1938, two really powerful engines were envisaged: the 1,500-bhp R-2600-4 and the 1,700-bhp R-2800. The next step down was the 1,200-bhp R-1820-G231. Power cost weight: a R-2600 weighed 1,930 pounds "dry" (with no lubricants or fuel, even in fuel lines) compared to 1,300 pounds for the R-1820-G231—1.3 pounds per bhp for the R-2600 compared to 1.08 for the R-1820, which was a far more mature engine. Output of the more powerful engines would, it was presumed, probably rise quickly, soon making them more efficient in terms of weight per unit power.

2. The data sheet, which unfortunately does not include speed, is dated 25 February 1938. Time to 15,000 feet was 8.9 minutes. There do not seem to be any contemporary data sheets for other dive-bombers powered by lower-powered engines.

3. Memo from Plans Division to Chief of BuAer, 31 March 1938, copy in Aviation History files, NHHC.

4. Plans emphasized the need for strafing. It also argued that really fast scouts could force combat on large enemy bombers. Both required more firepower: at least two fixed 20-mm wing cannon (or 0.50-calibre machine guns) instead of the earlier alternate 0.50 and 0.30-calibre weapons. Soon Plans also called for a 0.50-calibre fixed synchronized gun in place of the earlier 0.30 (the SBD was similarly upgraded).

5. Versions of the specification are in the BuAer Confidential VSB file vol. 1, Correspondence 1922–44, RG 72, NARA II..

6. This reasoning is not explicit in the relevant BuAer file, but it seems implicit in the analysis of the relationship between weight and engines.

7. BuAer liked it: the yoke would give a pilot greater leverage in moving the ailerons and also allow him to use both hands for control when pulling out of a dive or carrying out other violent maneuvers.

8. The systems were retraction and extension of landing gear, operation of landing flaps, operation of double split flaps, folding and spreading wings, opening and closing bomb-bay doors, operation of bomb displacing gear (trapeze) and brake, charging of Madsen (23-mm) wing cannon, locking of automatic slots, and possible operation of a flexible-gun turret.

9. The next highest, Vought B, was rated only 65. The other rated designs, in order of preference, were Douglas No. 5, Vought A, Vultee, Douglas No. 2 (tricycle landing gear), and Consolidated LB-13 (tricycle landing gear). Vultee used the 1,400-hp R-2180, an engine soon discarded. Nearly all the others used R-1830s. Of the rejected designs, Consolidated's LB-13 was the most exotic. Its one 150-bhp Allison V-1710 was buried in the fuselage abaft a pilot sitting far forward, over a long driveshaft. The bomb was carried internally. A 37-mm cannon could fire through the propeller hub. Gross weight would be 10,750 pounds with a thousand-pound bomb and full fuel load. The sketch showed tricycle landing gear, but that could be modified with conventional gear. The

company estimated that it would make 287 mph carrying a five-hundred-pound bomb, or 285 mph with a thousand-pound bomb; range with a bomb would be about 1,000 miles, 1,570 without, as a scout. Douglas offered its DB-200 as an SBD successor with much the same configuration (it could have tricycle landing gear). It could be powered by either an R-1820 or R-1830 with two-speed supercharger offering, respectively, 1,000 and 1,050 bhp at various altitudes from 4,500 to 17,500 feet. Unlike the SBD, this airplane would have folding wings (span forty-seven feet extended, compared to forty-one feet six inches for an SBD). It had a longer greenhouse and a deep belly for the internal bomb bay and heavier gun armament now wanted. Performance was little better than that of the SBD. With a 1,000-pound bomb on board (10,732 pounds) it would make 230 mph at sea level and 271 mph at 17,000 feet; service ceiling would be 27,200 feet, and time to reach 5,000 feet would be 2.9 minutes. Takeoff run against a twenty-five-knot wind over the deck would be 255 feet. Range at economical speed would be a thousand miles. Vought's V-169A was powered by a supercharged R-1830; gross weight was about 10,400 pounds. With a 500-pound bomb, maximum speed would be 262 mph, far short of what Curtiss and Brewster offered. Vultee's VSB-50 limited its folded width by using double-folding wings. Flaps would limit diving speed. Limiting weight with a 1,000-pound bomb was 10,750 pounds. With an engine rated at 1,150 hp at 7,000 feet (1,060 at sea level) Vultee claimed a range of 411 miles at 210 mph, a maximum speed of 232 mph at sea level, and a service ceiling of 26,900 feet. Takeoff run with twenty-five knots wind over deck would be 241 feet. This was not remotely competitive with the Curtiss design. The 30 December 1938 memo on selection from the Engineering Division did not describe any of the lesser designs or explain their ranking; data on failed proposals from the proposal file, entry 28 (which is erroneously marked as containing only rejected proposals), RG 72, NARA. Performance data from proposals are manufacturers' estimates, *not* estimates by BuAer. Of the designs submitted, only Vought's was *not* included in some way in entry 28, RG 72, at NARA.

10. According to the cover sheet for the 18 January 1939 memo by the chief of BuAer ordering the contract award.

11. With wing flaps the diving speed of the SB2U-1 was just under 300 mph, and with trailing-edge flaps the SBC-3 dived at about 290 mph. Memo, 19 January 1939, memo, Confidential Correspondence 1922–44, VSB vol. 1 file, RG 72, NARA II.

12. Harold Andrews, *The Curtiss SB2C-1 Helldiver*, Aircraft Profile 124 (Windsor, U.K.: Profile, 1971). Andrews was the NAVAIR historian.

13. Rejected alternatives were extending the leading edge slats to full span and drooping the ailerons. The brakes-open diving speed was also higher than desired, but nothing could be done beyond providing the brakes with variable opening angles. Andrews, *Curtiss SB2C-1 Helldiver.*

14. The Marine expeditionary version (SB2C-2) was analogous to the SB2U-3, with a range of 2,600 miles. It would have four 0.50-calibre wing guns and one 0.50-calibre flexible gun, and it would have racks for fifteen 30-pound demolition bombs. Like the SB2U-3, it would have an alternative float landing gear so that it could be ferried to a lagoon.

15. Mod III included deletion of the 0.50 turret in favor of a twin free 0.30 (as in a Dauntless), as well as structural strengthening, some change in the control system (a "bobweight," a weight that applies a nose-down force on the elevator), a nonretractable tail wheel, and a self-sealing fuselage fuel tank. Andrews, *Curtiss SB2C-1 Helldiver.*

16. The -1A version differed from the naval Helldiver mainly in having electrically operated flaps, a turret with a single 0.50, no folding wings or provision for catapulting, an Army radio installation, no search radar, and additional armor. They were not considered satisfactory as procured, so in 1944 BuAer organized a modification program, the main center for which was Roosevelt Field at Mineola, Long Island. The main items were to modify turrets or replace them with CH-150-2s and to revise the radio installation to Navy standards. As of early March 1944, 140 aircraft were ready. Some were allocated to Marine squadrons for training.

17. Surviving papers do not describe how draft requirements evolved during the run-up to the January 1941 circular, so it is not clear just when BuAer became interested in tricycle arrangements. The concession in stall speed was presumably offered because it would be easier to land a tricycle airplane on a carrier. Note that tricycle designs offered in 1938 were not favored.

18. The monthly progress reports by the VSB Desk (Naval Aviation History, NHHC) show how engine problems delayed the SB2D program. First flight was originally scheduled for about 30 June 1942, but as of November 1941 it had been pushed to 15 August 1942. Even then 2,200- rather than 2,300-hp engines would have to be used for early flights. By early March 1942, the first flight had been set back to 19 September 1942. According to the 28 April 1942 VSB Desk progress report, delivery date had been pushed back from 19 September to 30 December 1942, because of a "lack of engineering talent due to other Douglas projects," probably the Army A-26. Of the delay, six weeks was due to the engine, one to revised stress analysis, and four to engineering changes required when the original four-bladed propeller was replaced by a three-bladed one to simplify production. The prototype actually flew in April 1943.

19. By October 1943 two aircraft were involved, the first (Bureau Number [BuNo] 04962) to have a Westinghouse 19A turbojet and the fourth BTD (BuNo 04964) the 19B. A 19A was shipped to El Segundo early in January 1944; at that time the 19B was not expected to arrive before April. Flight tests were to begin in March 1944. Tests were delayed because the thrust bearing of the jet engine failed in ground tests. That was expected to delay the program by six weeks, but it is not clear that any dual-power flights were made. Arrival of the 19B was delayed, first to early June 1944, then to 1 October (as of 1 September 1944). Then it slipped to December 1944. The whole first airplane was ordered stricken by OpNav, because of the obsolescence and defective design of the engine and the uneconomical rework required to adapt the first BTD-2 to carry the 19B engine. The second airplane awaited installation of the 19B, after which it was to be tested by NACA. It seems unlikely that this airplane ever flew with its jet engine. It is possible that the 19A airplane flew in May 1944; Peter C. Smith, *Douglas AD Skyraider* (Rambury, U.K.: Crowood, 1999), 18. That, however, is not reflected in the monthly VSB progress reports.

20. Date from the VSB desk progress report. This flight ended after five minutes because the landing gear refused to retract.

21. The 1 May 1944 VSB desk report gave details. The current airplane weighed 18,154 pounds and had a load factor of 6.47 g at 17,000 pounds. Combat radius was 304 miles (460 gallons), and takeoff run an unacceptable 558 feet. Adopting the Curtiss propeller reduced weight to 18,138 pounds and the takeoff run to 517 feet and increased combat radius to 312 miles. The new higher-powered version of the R-3350 would reduce takeoff roll to 480 feet. Increasing wingspan would further reduce takeoff to 452 feet, but the load factor would drop to 5.6 g. Extending the ailerons and increasing the load (strength) factor to 6.0 (the spar caps in the wing structure would be good for 7.5) would increase weight to 18,147 pounds (111 pounds could be saved by detail improvements) but reduce takeoff run to 447 feet. Eliminating the bomb bay and consolidating fuel in one forward tank would cut weight to 17,828 pounds (347-mile radius, 420-foot takeoff). Changing the wing-joint forging would increase strength to 7.5 g without much weight increase. Given all these changes, Douglas proposed a new fuselage and a new engine installation with water injection. The pilot would be raised six inches, and the wings would be lightened by using 758T alloy. This airplane would weigh 17,260 pounds, its load factor would be 7.5 g at 16,500 pounds, radius would be 360 miles, takeoff run 376 feet. As of late May 1944, it was proposed that the 161st airplane be changed about March 1945 with increased wing strength and as many other items as possible. They would include the extended wingtips and ailerons, external bomb racks, torpedo quick-change provision, a load factor of 7.5 g at 17,000 pounds, a single fuselage fuel tank, space for water injection if desired, and at least 300 pounds saving in gross weight. This version would be designated BTD-3. It was never built.

Chapter 6. Torpedo Bombers of World War II

1. Memo to Assistant Chief, 12 February 1937, in BuAer VTB vol. 1, Confidential file 1922–1944, NARA II.

2. At best an 11,000 pound single-engine VTB could make 235 mph. The twin-engine Design 145 could achieve 270 mph on much the same weight as a single-engine bomber.

3. Handwritten cover note on a memo for the chief, 8 May 1937; he had not discussed this with anyone.

4. An undated penciled calculation sheet attached to the package of 1937 design papers (in the BuAer correspondence file) seems to have been the basis for the Plans Division conclusions. It shows two airplanes powered by 1,150-hp, liquid-cooled V-1710 engines:

	Single	Twin
Span (ft)	52.5	54.1
Gross weight (lb)	9,861	10,488
Wing area (sq ft)	476.6	507
Stall speed (mph)	65.3	65.3
Max. speed (mph)	235	267 (at 25,000 ft; 286 at 1½ minute rating)
Service ceiling (ft)	25,000	33,300
Time to 5,000 ft (min.)	5.5	4.9
Time to 15,000 ft (min.)	17.4	14.6
Takeoff (25 kts. WOD) (ft)	235	320
Max. range (mi)	770	798

There were also estimates for the R-1830-64 (900 hp at takeoff), XR-2600 (1,500 hp), XR-3350 (1,800 hp), and R-1830-66 (1,050 hp). Maximum speeds at rated altitudes were all below 240 mph. Design 145 used two XR-1535-92 engines. Engineering Division forwarded its performance estimate to Plans on 9 February 1937. A calculation sheet dated 21 January 1937 shows it with two 825-hp engines and a gross weight of 11,225 pounds (span not given, area 449 square feet). At the rated altitude of 8,000 feet, maximum speed was 275.8 mph, stall speed was 63.8 mph, service ceiling was 26,500 feet. This sheet does not show time to climb or deck roll.

5. According to a 17 February 1967 letter NAVAIR history office wrote answering questions by Dutch aviation historian Jack W. A. van Dongon; TBF file, Aviation History Division, NHHC. Design 183 dimensions were span fifty-eight feet six inches (four feet three inches more on each wingtip than for the TBD), folded span twenty-nine feet three inches, length thirty-nine feet, and height 13 feet 8¼ inches; normal weight with a torpedo was 11,958 pounds (overload 12,658 pounds). Using the engine rated at 1,700 hp for takeoff and 1,340 hp at 22,000 feet, maximum speed at 22,000 feet was 310 mph and service ceiling was 33,800 feet. Range with normal load was a thousand miles (1,500 when overloaded). The takeoff run with twenty-five knots of wind over the deck was 287 feet, and stalling speed was 69.8 mph. A spotting study on *Lexington* showed that twenty could be accommodated in the same space as eighteen TBDs; this study was described in a 12 January 1939 BuAer memo. Unfortunately, the correspondence file does not include any sketch of Design 183.

6. The stall requirement forced down maximum speed, because it demanded a larger and hence "draggier" wing. It had been written around an unusually severe landing condition: an emergency landing soon after takeoff. Specifying stall speed with half fuel would result in much better performance.

7. This circular is in VTB vol. 1, BuAer Confidential series for 1922–1944, RG 72, NARA II.

8. Engineering Division to Chief of Bureau, 21 October 1939, in BuAer Confidential series for 1922–1944, VTB vol. 2, RG 72, NARA II.

9. Chief of Bureau via Class Desk "O" to Heads of Engineering, Plans, and Material Divisions, 17 October 1939, in BuAer Confidential series for 1922–1944, VTB vol. 2, RG 72, NARA II.

10. As described in a 6 December 1939 memo from Engineering Division to the bureau chief. The two wing 0.30s were replaced by a single synchronized gun (there was insufficient space for a 0.50, which in other airplanes was an alternative). A single tunnel gun was added abaft the bombsight, covering a sector from five degrees above to fifty-five degrees below the horizontal and thirty degrees athwartship on both sides. To make space, the top cowl line (i.e., of the canopy) was raised four inches and carried forward to include the pilot, giving him improved vision. The bomber's window was moved aft slightly, making it possible for the Automatic Flight Control (AFC, part of the Norden system)

to be carried permanently. Maximum speed fell from 321 to 308 mph at critical altitude, and service ceiling was slightly reduced.

11. Bureau Chief, memo, 1 August 1940, BuAer VTB vol. 2, Confidential 1922–1944 files, RG 72, NARA II. Vought's unit price did not include $500,000 for planned plant expansion.

12. TBF procurement in 1942 amounted to 645 airplanes in two groups (beginning with 325) plus two prototypes, one of which was the XTBF-2. Procurement in 1943 amounted to 739 TBF-1s in two lots (435 and 304), 446 TBF-1Cs, and two TBF-3 prototypes. In 1943 there were also 458 for foreign (presumably British) customers: sixty TBF-1Bs, eighty TBF-1s, 255 TBF-1Cs, and a second batch of sixty-three TBF-1Cs. That totaled 1,832 for the Navy and 458 for others. The first contract with General Motors' Eastern Aircraft Division was dated 23 March 1942; the first airplane was delivered in November 1942, using parts Grumman supplied. Grumman delivered a total of 2,311 TBF-1s. Eastern delivered 2,882 TBM-1s and 4,664 TBM-3s. Deliveries to the United Kingdom totaled 395 TBF-1s, 334 TBM-1s, and 192 TBM-3s. In addition, sixty-three TBF-1s were delivered to New Zealand. Figures from a BuAer letter to VS-38, which thought it was the last Pacific TBM-3 squadron (it was not), 13 August 1953. In addition to operational aircraft, a few Avengers were used for tests not related to later versions. Records in the NARA II VTB file mention one airplane converted to a single-seater by the Naval Aircraft Factory and being evaluated at Anacostia in September 1942. Another was given in July 1942 to Wright Aeronautical as a test bed for their R-3350 engine.

13. Summary of the current naval aircraft production program produced for Representative Carl Vinson in connection with a review of the TBY program, Correspondence file VTB 1945, 3 August 1945, RG 72, NARA II.

14. Design 183 was worked out both as a two-seat glide bomber and as a three-place horizontal bomber. In the former condition, carrying a torpedo and normal fuel, it would weigh 11,964 pounds and would have a maximum speed at critical altitude of 310 mph and a stall speed of 69.8 mph. With a 1,000-pound bomb, it would weigh 11,057 pounds and have a speed of 311.2 mph at critical altitude. Takeoff runs (twenty-five knots wind over deck) would be, respectively, 287 and 234 feet. With overload fuel for 1,500 miles, weight would be 12,664 pounds, speed 308.8 mph, and deck run 331 feet. The three-place airplane would weigh 11,958 pounds with a torpedo and normal fuel (310 mph, deck run 287 feet) or 11,507 pounds with a 1,000-pound bomb (310.8 mph, 260-ft deck run).

15. Letter to BuAer, 27 November 1941, Correspondence file, VTB vol. 2, RG 72, NARA II. The two endorsements came after the outbreak of war (Halsey's was dated 2 January 1942).

16. VT-8 reported that its pilots, after an initial dive, in order to reduce the length and speed of the final dive, aimed and then released bombs at one to three thousand feet. Longitudinal control was usually satisfactory, but lateral gradually worsened as speed increased. As aircraft "set" in their dives it became difficult or impossible to correct aim except in a longitudinal direction. The steeper the dive and the lower the release altitude, the less lead required. Pilots were warned that altimeters might overstate altitude by about three hundred feet, because they could not keep up with rapid change. It was vital to hold the airplane in its dive for about two seconds after the bomb was released. The following February VT-12 reported that the lower the release point the greater the accuracy; a TBF could pull out within eight hundred to a thousand feet from a fifty-degree glide at three hundred knots without exceeding limits or causing blackout or dimout. In the margin of the BuAer copy of these instructions, someone wrote, "I should say not! When you pull TBF hard enough to black out, it will probably be permanent blackout!" Steep dives badly stressed the airplane, not designed as a dive-bomber. After some were lost while diving, BuAer decided that they should be fitted with recording accelerometers. In March 1943 the VTB Design Section compared design date with the stresses incurred in glide bombing. As designed, at a gross weight of 14,500 pounds the TBF could stand a maximum

acceleration of 4 *g*s, and its maximum airspeed was 313 knots. It actually demonstrated 315 knots at 14,229 pounds and 3.9 *g*s. The actual maximum load on a 13,000-pound airplane diving at 304 knots was 4.5 *g*s, and the average of twenty-six available records of training dives was 5.51 *g*s (290 knots, 13,338 pounds). In service glide bombing the maximum load was 4.5 *g*s at 13,000 pounds and 304 knots, With four 500-pound bombs and two-thirds fuel a TBF-1 would weigh about 15,200 pounds, 18 percent over the load limit (weight times acceleration). Not surprisingly, squadrons were reporting wing failures, some minor and some catastrophic. Pilots opened bomb-bay doors and extended their landing gear to hold down diving speed. The question was whether to strengthen the airplane, for glide bombing, diving torpedo approaches, and maneuvering to use the fixed gun.

17. TO (Technical Order) 37-43 issued on 31 May 43 modified an earlier TO, 84-42, and listed permissible speeds and accelerations. Flaps were not to be used above 130 knots IAS, and the full throw of ailerons was not permissible over two hundred; at higher speeds deflection was limited to the force required at two hundred knots IAS. Maximum speed to extend landing gear was two hundred knots, and it might be impossible to retract gear completely at over 150. With wheels down, bomb bay open, and minimum power to keep the engine cleared, a dive angle of forty to forty-five degrees could be maintained without exceeding the restricted speed. Airplanes were to be thoroughly inspected after glide bombing.

18. Grumman was asked for a prototype installation of the Martin 250 CE 4 twin 0.50-calibre turret, despite the increase of four hundred pounds involved. This project is mentioned in the 1 September 1942 BuAer memo on the status of VTB projects in Correspondence, VTB vol. 2, RG 72, NARA II. The test airplane was BuNo 01746; according to René Francillon, *Grumman Aircraft since 1929* (Annapolis, Md.: Naval Institute Press, 1989), 168, the turret was planned for Grumman's proposed G-56, a derivative of the Avenger to be powered by an R-2800 engine. Francillon reports that the volume of fire was less than that of the

single turret and that firing had to be stopped more often to avoid hitting the tail surfaces. Grumman's G-56 performance calculation dated 12 November 1942 is in the BuAer proposal file. Normal design gross weight was given as 16,436 pounds (overload 17,305 pounds). The engine was an R-2800-27 producing 2,000 hp at takeoff rather than the 1,700 hp of the R-2600-8. Grumman compared G-56 to the two-hundredth TBF-1 (14,697 pounds gross weight, 199 versus 245 gallons of fuel). Maximum sea-level speed would have been 265 rather than 253 mph, and at critical altitude it would have been 313 mph at 15,500 feet compared to 265 mph at 10,750 mph. Stall speed would have been the same (75.3 mph). Initial rate of climb at sea level would have been 2,215 rather than 1,610 feet per minute. Range would have been 1,000 miles (1,310 with overload fuel of 334 gallons and overload ammunition) rather than 990. The R-2800 project seems to have been revived in June 1944 under the designation XTBF-4; in this case the engine would have been a 2-speed C-series R-2800-22. The proposal file includes weight and balance but not projected performance. Grumman described the airplane as a TBF-1C with the new engine. Loaded weight would have been 16,319 pounds.

19. In the Helldiver, the C subdesignation was associated with cannon, but in the Avenger letters seems to have been in sequence of modification. There was no -1A. The -1B designation referred to aircraft for Britain. The version with ASD/APS-3 radar was the -1D. The -1E had the ASH/APS-4 radar; it seems originally to have been informally designated -1H. There were also informal designations: a -1J winterized version, and -1L with a searchlight. The informal designations appear in Francillon, *Grumman*, 169, but not in either the RG 72 correspondence series or in the NHHC file.

20. BuAer files include a June 1943 report from the escort carrier *Bogue*. Its aircraft were finding that U-boats, with more powerful antiaircraft guns, were trying to shoot back when caught on the surface. The change was formally ordered on 26 July 1943. A BuAer engineer doubted the TBF was particularly suited to fixed guns, which imposed a

serious weight penalty, but BuAer could not resist so many fleet requests. Wing modification included reinforcement. Because production of new wing panels was insufficient, existing wings had to be adapted in the field.

21. Takeoff run with twenty-five knots of wind over deck was 315 feet at 17,500 pounds, compared to 260 for the TBM-1C. Handling notes and figures from Patuxent River TED No. PER 2139 dated 17 August 1944, TBF file, NHHC.

22. In August 1944 the first ten available aircraft were ordered diverted to Quonset for installation of ASD-1 radars, to be followed by another five in September, and possibly more.

23. Initially some escort carriers used SBD Dauntlesses, which were being produced in numbers beyond those needed in the Pacific.

24. William T. Y'Blood, *Hunter-Killer: U.S. Escort Carriers in the Battle of the Atlantic* (Annapolis, Md.: Naval Institute Press, 1983). The shift to offensive ASW exploited code breaking. This practice was contentious, the British fearing that the Germans might realize that their signals had been compromised. The Germans' failure to do so remains a major puzzle of the Battle of the Atlantic.

25. CVE Aircraft Anti-Submarine Warfare Operational Instructions, 18 August 1943, envisaged only day operations. A copy of this draft guidance is in Tenth Fleet papers in the file "A/S Warfare by Aircraft," RG 38, NARA II.

26. Mk 24 was called FIDO. Photographs of it in Avenger bomb bays show the same combination of box tail and "pickle barrel" that characterized heavy torpedoes on Avengers in 1944. The sonobuoys were not, as later, a means of search; they were a way to relocate a U-boat that had just dived. They were sometimes dropped in patterns. Y'Blood, 147, mentions MAD (magnetic anomaly detection), as used during a March 1944 search by Avengers from USS *Block Island*. This seems to have been one of the earliest such employment by Avengers.

27. As summarized in the postwar U.S. analysis of the Battle of the Atlantic in NDRC Division 6, *A Summary of Antisubmarine Operations in World War II*, OEG 53 (Washington, D.C., 1946), 3:48–49.

OEG 53 omits all mention of signals intelligence. The revival of the open-ocean battle in October was due in part to German gross overestimation of the success of the anti-escort homing torpedo.

28. U-boats could sustain high speed only on the surface. Previously they had dived only when in danger from, e.g., aircraft. Until codes were changed in June 1943, the U-boat offensive was supported by German code breaking, which enabled packs of U-boats to maneuver into position. Later, the German U-boat command tried unsuccessfully to substitute traffic analysis and reconnaissance by long-range aircraft based in France. The new tactics are described in the December 1943 issue of the *U.S. Fleet Anti-Submarine Bulletin*. This issue also summarizes developments in airborne searchlights.

29. Y'Blood, 137. He credits the first suggestion for CVE night operations to USS *Santee* in August 1943, with further thoughts that October. Presumably the key development was the revelation in November 1943 that U-boats were now operating submerged during the day. *Croatan* had VC-1 on board. The 7 March 1944 register of U.S. naval aircraft locations shows eleven TBF-1Cs and no TBF-1Ds, so the radar attack was executed using ASB radar. This list shows no TBF-1Ds at all on board escort carriers at the time; they do not appear in quantity (in official lists, at least) in the Atlantic escort carriers until the 30 July 1944 list, in which all Avengers on board *Guadalcanal* are TBF-1Ds. *Solomons* had two -1Ds and ten -1Cs (in April and May it had been the only Atlantic CVE credited with such aircraft, and no CVE was listed as having them in March). At the beginning of October, Atlantic CVE air wings typically included three to five TBM-1Ds alongside TBM-1Cs. In December, Atlantic CVEs had six TBM-1D or -3Ds, half the total number of Avengers (*Card* had eight). Barrett Tillman, *Avenger at War* (London: Ian Allen, 1979), credits Capt. Dan Gallery (commanding USS *Guadalcanal*) with the idea of maintaining Avengers continuously aloft during the night. He began night patrols on 7 April 1944. He kept four Avengers continuously airborne; with four-hour on-station time, he needed eight in all.

30. Y'Blood, 143, credits *Card* with this innovation. Two TBFs were stripped, and two 58-gallon wing

tanks and a 275-gallon bomb-bay tank added. With weight reduced by 451 pounds, endurance was sixteen hours. Another source claims fourteen-hour endurance. Actual flights by the *Card* Owls lasted up to thirteen hours.

31. The *U.S. Fleet Anti-Submarine Bulletin* for December 1943 describes initial efforts, noting that the standard patrol-plane light was too massive to fit an Avenger bomb bay. General Electric was developing a lightweight (100-pound) light on a retractable mounting. A prototype was expected in six to eight weeks. A list of aircraft searchlights in the July 1944 issue featured two retractable eighteen-inch GE searchlights in an L-8 series with different outputs. Aircraft with searchlights were designated TBM-1L or -3L (aircraft with temporary searchlights were not specially designated). Lists of the locations of Navy aircraft show no -1Ls.

32. This sight was described in the February 1945 issue of the *U.S. Fleet Anti-Submarine Bulletin*. The periscope extended up from the greenhouse of the airplane and had a streamlined upper end. BuAer Correspondence Confidential file F36-6(2) vol. 3, RG 72, opened January 1944, NARA II, includes a 2 August 1944 report on the searchlight installation using a periscopic sight. This file includes a 10 August 1944 report that three (presumably the first operational) TBM-1 searchlight airplanes had been delivered (BuNos 46278, 46279, and 46282), with the remaining three due by 24 August. These aircraft had L8H lights; additional deliveries would depend on receipt of parts.

33. According to Y'Blood, 239, the first escort carrier so equipped was *Bogue*, which in mid-September 1944 deployed with four searchlight-equipped TBM-1Ds. According to Y'Blood, on 9 July Admiral King had instructed Atlantic Fleet to assign four searchlight Avengers to each of the next three escort carriers to deploy. Training was to be expedited. Weight compensation seems to have been rocket rails and rockets.

34. This project is described, with photographs, in BuAer Confidential file F36-5 vol. 1 for 1944, in BuAer Confidential Files 1922–1944, box 656, RG 72, NARA II. The project was also described in the *U.S. Fleet Anti-Submarine Bulletin*. The pilot had a sensor that could measure the brightness of the sky; the output of the bulbs was adjusted accordingly. The idea was also tested on an Army B-24. This was not Project Yehudi (for "the man who wasn't there"), which employed lighting to eliminate the shadows that otherwise made it possible to see airplanes at a distance.

35. The XTBF-3 flew on 20 June 1943. As upgraded in 1945, the -20 engine produced 1,900 hp for take-off. The XTBF-3 was later modified with a British Halford H-1B turbojet (planned for the XF15C) in its bomb bay. In this form it flew in December 1944.

36. Report dated 29 December 1944 referring to the period 25 September–17 October 1944, Correspondence folder VTBM3/A4, RG 72, NARA II. The squadron had previously operated TBM-1s and -1Cs.

37. As reported in the February 1945 BuAer *Confidential Bulletin*.

38. According to Price, *U.S. Electronic Warfare*, vol. 1, 203–207, in the fall of 1944 each carrier air group was given equipment for five aircraft that had the APR-1 receiver and the APT-1 and APQ-2 jammers. Two aircraft had APR-2 and -5 receivers. According to Navy doctrine, aircraft accompanying a first strike would detect and locate enemy radars; the second strike would include at least two jamming aircraft. In the spring of 1945 more electronic warfare (EW) aircraft were allocated: each Avenger squadron was to include five aircraft carrying APT-1 and -2 jammers. By this time Avengers had both intercept receivers (APR-1) and direction finders (APR-11). Avengers typically carried chaff. Jammers were typically preset to frequencies determined in the initial operations. APR-1 covered 100–950 MHz, which was extended to 40–3,300 MHz (Japanese sets typically operated in the metric band, about 200 MHz). APR-2 was an automatic recording receiver (90 to 1,000 MHz). APR-5 covered centimetric radars (1,000 to 3,100 MHz). APT-1 ("Dina") covered 90–220 MHz (output twelve watts). APT-2 ("Carpet") covered 450–720 MHz (output five watts). Both were standard wartime sets produced in large quantities. The designations Price mentions suggest that at least some -3Qs were wartime conversions so designated at the end of the war. The April

1946 *Airborne Radar Countermeasures Operator's Manual* (RADTWELVE) describes a typical Avenger installation: it included an APR-1 receiver, an APA-38 panoramic adapter, APA-11 pulse analyzer, and an AS-242/A "snooper" direction-finding antenna. Aircraft had an AT-37/APT stub antenna for nondirectional search using the APR-1 (40–320 MHz), an AS-124/APR for 300–1,000 MHz (on the upper starboard bomb-bay door), an AS-125/APR HF cone (not in all ECM aircraft) for nondirectional general search with the APR-5A receiver (1,000–3,000 MHz) and two AS-65/APQ-2A stubs on the port and starboard sides of the bomb bay, primarily for use with the APQ-2A "Rug" jammer. Each carrier was furnished with between two and twenty-five of the following: APR-1 (APR-4) search receivers, APA-11 pulse analyzers, APR-5A microwave search receivers, APR-2 recording search receivers, APA-23 recorder attachments, APA-38 panoramic adapters, APT-1 jammers, APQ-2 jammers, and AM-14 and -18 jamming amplifiers.

39. The receiver was APR-1, -4, or -5A. The pulse analyzer was APA-11, which was used with an APA-38 panoramic scope; the jammer was an APQ-2. These details and the numbers cited are from an Internet article by Rick Morgan, "The Enigmatic TBM-3Q" accessible on Tommy Thomason's https://tailspintopics.blogspot.com, July 2015. Aircraft could be visually distinguished by their special antennas: a rotatable twin dipole (for direction finding) under the belly, a white radome on the port or starboard bomb-bay door (probably for the jammer), and a blade (omni) antenna roughly abreast the tail wheel. These aircraft were intended to work with strike units, so they retained their turrets and guns. The TBM-3Q first appeared in monthly naval aircraft lists in April 1946 as part of VT-81, typically integrated with TBM-3Es in the new attack squadrons on board *Essex*-class carriers. In 1948 they were replaced by ECM versions of the new AD-1 and AM-1 attack bombers, only one squadron remaining in 1949. Numbers of TBM-3Qs are not known but probably were in the high seventies.

40. By early November 1944 three night air groups were forming in the Hawaiian area and three replacement groups at Quonset. On 9 November 1944 the director of Military Requirements wrote to the directors of Engineering and Maintenance about the special configuration involved. It would be impossible to prototype the new version or tactically evaluate it before it entered combat. Equipment and tactics could not be standardized, but a list of equipment needed included the "sniffer" blind-bombing radar (APG-4), a tail warning radar (APS-13), a Pioneer flux-gate compass, and an air position indicator. The aircraft would already have provision for the APS-4 radar "bomb." Photographs suggest that, at least initially, VT(N)s used TBM-1Ds. For example, one from VT(N)-41 is shown in Tillman, *Avenger at War*, 113, on board the light carrier *Independence* on 10 October 1944. According to Tillman, Night Group 90 (USS *Enterprise*) had twenty-seven TBM-1Ds.

41. BuAer, *Determination of Military Characteristics*, 84.

42. Correspondence, VTBM4/F41 1945, RG 72, NARA II. As of July 1945 there was some interest in substituting the Emerson CH-150-1 hydraulic turret used in the TBY for the Grumman electric turret. Cuts in the TBY program made a decision urgent. The Emerson turret was lighter (175 pounds), easier to maintain, trained and elevated faster, had better visibility, and was unlikely to affect performance. The idea was killed because it would move the airplane's center of gravity, a hydraulic system was more vulnerable, and it would complicate logistics and training. The new turret had not been tested, and it would not add firepower. It also seemed foolish to adopt a new turret when there was considerable pressure to eliminate the turret altogether to save even more weight at a time when Japanese air opposition seemed to be evaporating.

43. According to a 13 November 1944 memo in the NHHC TBF file, VTB Design Branch asked the Experiments and Developments Branch to set up a TBM-3 weight-reduction project at NAMU Johnsville, using XTBF-3 BuNo 24341. The turret and all accessories would be removed, as well as the tunnel gun, all ASB radar equipment, and the radioman's seat in the tunnel compartment. A Douglas twin 0.30 mount (as in the SBD) would be installed on an adapter fitted to the turret ring. NAMU would make and install a suitable sliding

canopy over the gun mount. Necessary radar and radio controls would be installed in a new single station for a radioman-gunner forward of the gun. Radio and radar equipment not requiring adjustment in flight would be moved to the after end of the fuselage. The resulting airplane would be tested.

44. Data from BuAer *Confidential Bulletin*, May 1945.

45. TBY program summary 30 April 1945, Correspondence, 45 VTBY1/A4, RG 72, NARA II. Strengthening details were in the June 1945 BuAer *Confidential Bulletin*. See note 13 above for a history of the program written for Carl Vinson early in August 1945, after cancellation.

46.

	VTSB	TBF-2
1,600-lb bombs	2	1
1,000-lb bombs	2	1
500-lb bombs	4	4
Forward gun	1x0.50	1x0.30
Rear gun	2x0.50	1x0.50, 1x0.30
Armor	front and rear	rear only
Fuel (protected) (mi)	1,500	1,300
Radar search	ASD	ASB
Engine critical altitude (ft)	20,000	22,000
Gross weight (lb)	approx. 17,500	15,370
Range (mi)	1,000	1,000
Overload with torpedo (lb)	1,500	1,300
Vmax* plane critical (mph)	approx. 350 mph	293
Full-load power stall (mph)	approx. 75	75
Time to climb to 20,000 ft (mins.)	11	19.5
Service ceiling (ft)	above 30,000	28,000
Takeoff run in 25-kt. wind (ft)	approx. 350	350

* maximum speed.

47. Memo to Inspector of Naval Aircraft at Douglas, 9 February 1942, BuAer Confidential files 1922–1944, VTB vol. 2, RG 72, NARA II. BuAer wanted a preliminary design and a firm proposal with estimates of cost and delivery for two prototypes. The airplane would be protected against 0.30-calibre hits, but provision to resist 0.50s might later be wanted. The torpedo-plane function included glide bombing at angles up to fifty degrees. Although completely protected fuel for 1,500 miles was desired, BuAer would accept a design in which half the fuel was protected and the other half droppable. According to an attached cover sheet, BuAer estimated that this airplane could be built within 17,630 pounds "if the contractor exercises care in design and does not add any margin for guarantee purposes." The goal should be 17,500 pounds.

48. Memo, 27 January 1942, BuAer Confidential files 1922–1944, VTB vol. 2, RG 72, NARA II.

49. Heinemann, memo to Resident Inspector of Naval Aircraft at El Segundo, 12 March 1942, BuAer Confidential files 1922–1944, VTB vol. 2, RG 72, NARA II.

50. Memo of conference at BuAer, 16 April 1942, to evaluate Douglas VTSB designs; BuAer Confidential files 1922–1944, VTB vol. 2, RG 72, NARA II. Both twin-engine designs offered mediocre speed at sea level—266 and 288 mph—although they offered more at maximum altitude (337 mph at 27,000 feet and 317 at 20,000). The X-Wasp airplane offered 328 mph at sea level and 362 mph at 17,000 feet; the V-3420 airplane offered 330 mph at sea level and 425 mph at 25,000 feet. The twin-engine airplane with the two-speed supercharger offered about the same performance as the XTBU, which had been a 1939 winner, except for the very superior takeoff and stalling speeds of the XTBU. Heinemann estimated that the single-engine design was about 10 percent better than a twin-engine design. A single liquid-cooled Allison engine buried in the fuselage added too much weight, whether it drove a conventional propeller via a driveshaft or two wing propellers. Heinemann abandoned the two power-driven, remote-control turrets of the SB2D, because of their weight. Most of the BuAer evaluators preferred a twin turret atop the fuselage plus a handheld "tail stinger" (the practicability of a power-driven stinger would also be investigated). By this time horizontal bombing, which would be done from high altitude, was not as important as it had been, so it was impossible to justify a design that offered advantages only at high altitude at the expense of performance lower down. In contrast to conclusions offered a few months earlier, a Navy radar expert pointed out that a single-engine airplane with its radar antenna on the wing might have some advantages. This antenna might be able to cover the full 360 degrees, which would be impossible with the nose installation in a twin-engine airplane. An airplane approaching from astern and below might be picked up at three miles or beyond. At this time the XR-4360 was rated at 3,000 bhp for takeoff and at 1,500 feet on military (i.e., maximum) power and at 2,400 bhp at 13,500 feet. Normal rating was 2,200 bhp at 14,500 feet.

51. Highlights identified in a 6 November 1942 report by the head of the Engineering Branch were a maximum speed of 339 mph with one torpedo on board (355 mph after it was dropped); a range (protected tanks only) of 1,750 miles as a scout (2,700 with drop tanks); a range with one torpedo and wing tanks only of about 1,500 miles; speed brakes; racks allowing the airplane to carry two torpedoes (takeoff roll 363 feet with a thirty-knot wind and fuel for a range of 860 miles); provision for four bombs, each up to 1,600 pounds (extreme overload); two 0.50-calibre guns in an improved turret, a single tunnel gun with power boost, and two fixed 0.50s; and tricycle landing gear. The airplane could operate from an *Essex*-class carrier. Evaluation in the BuAer proposal file entry 1044F, RG 72, NARA II.

52. The Douglas proposal for this TB2D-1 (D-548) in the BuAer Proposals file (entry 1044F, RG 72, NARA II) is dated 25 September 1942. It gave a 377-foot takeoff run in a twenty-five-knot wind with one torpedo and contraprops.

53. Heinemann argued that a tricycle airplane did not need a barrier, since it would not tend to bounce over the arresting gear wires on landing. With two torpedoes, in the new configuration the airplane would take off at a gross weight of 24,505 pounds (compared to 25,405 in the earlier configuration), carrying 320 gallons of fuel in each case. Maximum speed would fall from 346 to 328 mph, but takeoff distance in a twenty-five-knot wind would be reduced from 520 to 502 feet (in both cases better than what BuAer wanted). Maximum range, assuming the torpedoes would be released 375 miles from takeoff, would be 1,013 rather than 993 miles. With a single torpedo, gross weight would be 22,500 rather than 23,400 pounds; maximum speed would be 342 rather than 348 mph. As a scout, gross weight would be 21,718 rather than 22,618 pounds (510 pounds of fuel in either case); speed would be 357 rather than 349 mph (takeoff 328 feet rather than 374). Maximum range would be 1,760 rather than 1,642 miles.

54. The connection is pointed out in the BuAer, *Determination of Military Characteristics*, 83.

55. The 14 April 1942 Grumman letter is in the TB2D folder in the BuAer proposal file, entry 1044F, RG 72, NARA II.

56. TB2F file, Naval Air History Division, NHHC. A typed note refers to conversations between Grumman and BuAer in April and May 1942. According to a 6 October 1942 note in the file (retyped in the Naval Air History file), "This and the TB2D are the only irons in the fire to provide an eventual replacement for the TBF-TBU." Another note, probably from a route sheet, is that "Grumman always pleads 'lack of engineering talent' but later pops up with something that takes such talent. This [memo, presumably] should smoke him out as to what he has been doing on twin engine VTSB."

57. Chief of BuAer to Inspector of Naval Aircraft at Bethpage, N.Y., 7 December 1942, referring to Grumman's 29 October 1942 letter; BuAer Confidential files 1922–1944, VTB vol. 2, RG 72, NARA II. According to the cover sheet, Grumman had proposed a version of the TBF using an R-2800 engine "but nothing on 2 engine design."

58. Grumman's outline specification (Grumman history archive, Bethpage, G-55 file) is dated 15 December 1942. Gross weight with a torpedo is given as 33,294 pounds (design weight 34,000 pounds). Principal dimensions were: span eighty feet (thirty-five folded) by length fifty-three feet by height over vertical tail twenty feet. The airplane would be stressed for 4 *g*s; data sheet dated 7 January 1943. The Design 55 drawing in this file, dated September 17, 1943, shows a high-wing airplane with both a dorsal turret and a belly turret, the two not quite in line with each other. There is a single tail. Bound in with this report is a data sheet on G-55C/G-55D, the former carrying its torpedoes internally. Gross weight for -55C was 29,490 pounds with one torpedo or 31,646 pounds with two. Takeoff run would have been 246 feet in a twenty-five-knot wind with one torpedo, and stall speed with power on would have been 66.1 mph. G-55D is described in a report dated 8 March 1943. There was also a G-55E (report dated 21 June 1943) with a gross weight of 29,621 pounds, capable of 338 mph at critical altitude (18,000 feet), with a maximum range of 1,225 miles at five thousand feet with all bombs retained (1,265 miles if bombs were

dropped at the halfway point). Takeoff roll in the standard twenty-five-knot wind would have been 294 feet, and stall speed was 71.1 mph with power on. This version had a twin upper turret but only a single lower gun, in a tunnel. All versions had tricycle landing gear.

59. Grumman letter to BuAer, 19 March 1943, in BuAer Confidential file 1922–1944, VTB vol. 2, RG 72, NARA II, referring to a 15 January 1943 conference at BuAer. The engines were two-speed R-2800-2SC13Cs (2,100 hp at takeoff and 1,600 hp at 16,000 feet). Space and weight sufficed for future installation of R-2800-SSC22G engines with two-stage superchargers (weight penalty 773 pounds).

60. Corresponding figures for G-55D were 274 feet for takeoff, 74.8 mph stalling speed, 315 mph at 30,500 feet, and a gross weight on landing of 24,499 pounds.

61. VTB Design Section to Director of Planning, memo, 26 July 1943, in BuAer Confidential file VTB vol. 2, RG 72, NARA II.

62. Comments from the various BuAer Desks are from the TB2F file in Naval Aviation History Division, NHHC.

63. Current VTBs spotted ten to twelve airplanes per hundred feet of flight deck, but for the TB2F that figure would be only five and a half per hundred. Given its twenty-foot-one-inch height over the tail, it could not fit in a hangar without using an unproven nose-leg extender to tip it down at the tail. Folded unloaded aircraft (fifty-two feet six inches by thirty-six) could fit the two centerline elevators when but not on the deck-edge elevator.

64. Note dated 22 May 1944. Slightly later the same officer wrote that the TB2F was "an obvious attempt to provide something comparable to the B25 and B26 which can be operated from a CVB [*Midway*]. The necessary lower landing speed and shorter takeoff make it actually larger wing area [*sic*] than either of B25 or B26." The development history seems to show that this was hardly the case.

65. Penciled note in TSF-1 file, BuAer Proposals file, entry 1044F, RG 72, NARA II. This decision was dated 13 July 1944.

66. The Grumman design report showed a span of fifty-nine feet four inches (thirty-two feet folded), length forty-six feet four inches, and height over vertical tail sixteen feet. Note how much shorter span its was than the TB2F's.

67. VA (general) file, Naval Aviation History Division, NHHC.

68. Estimated normal gross weight was 23,874 pounds. Maximum speed at sea level with war-emergency power would be 389 mph (423 mph at 19,500 feet). Takeoff roll (twenty-five-knot wind over deck) would be 351 feet (stall speed with power would be 81 mph). Even at an overload gross weight of 26,090 pounds, carrying extra fuel, the airplane would make 411 mph at 19,500 feet. The F7F-2 fighter was credited with a speed of 417 mph at 22,000 feet on military power and a takeoff roll of 375 feet at its normal gross weight of 21,645 pounds. For long-range scouting G-66A could use a droppable bomb-bay tank and underwing drop tanks.

69. This project is mentioned in a 21 July 1944 memo on the VTB program. It would be armed with two fixed guns (20-mm or the new 0.60-calibre) and one free gun (one 0.50 or a twin 0.30) and would have the same droppable APS-4 radar as the TBM-3E.

70. Gross takeoff weight was 15,400 pounds (10,342 empty). The takeoff run was given as 217 feet. Combat radius would have been 130 nautical miles with 240 gallons of fuel on board, or 445 nautical miles with 440 gals. Stall speed with power off was 74 mph. By way of comparison, a TBM-3 (gross weight 16,940 pounds) had a combat radius of 250 nautical miles with 335 gallons on board and a maximum sea-level speed of 254 mph. Takeoff distance of the TBM-1C with the same twenty-five knots of wind over deck would be far longer, 403 feet. Both airplanes were about the same size, with spans of about fifty-four feet and lengths about forty-one. On 30 August Grumman offered three versions: G-70 (R-2800-22), G-70A (R-3350), and G-70B (R-4360). The power of the big R-4360 did balance off its added weight, so that it offered 352 mph at sea level (338 at continuous military rather than emergency power). Data from XTB3F file in Grumman history archive.

71. First run in April 1944, it was never given a J-series designation; output was two thousand pounds of thrust.

72. Design 70D file in Grumman history archive 70C also had an I-20 jet engine. A calculation sheet is dated 12 September 1944. A Design G-70F file is dated 9 October 1944. An internal Grumman memo dated 28 February 1945 refers to the possibility of using the GE I-16 jet engine instead of the Westinghouse 19XB. The undated G-70C and -70D sheets almost certainly were produced during August 1944.

73. Grumman history archive, referring to a drawing in the Naval Aviation History Division, NHHC, files, for which there was no Grumman documentation. The drawing showed inlet ducts in the wing roots and a tailpipe extending straight aft.

74. From a transcript of this conference in BuAer, *Determination of Military Characteristics*, encl. G.

75. TBF file, Naval Aviation History Division, NHHC, which included a compiled history of the TB3F/AF.

76. BuAer Confidential, VPBJ vol. 3, RG 72, NARA II. The route sheet on the account of the catapulting of the PBJ-1H using a special bridle notes, "If the PBJ is to be catapulted in service, a new [bridle] installation will be used which eliminates the need for a Y-bridle." That suggests definite interest in routine operation from carriers. The paper was a 6 December 1944 account of a special launching bridle developed by the Naval Air Materiel Center (NAMC) in Philadelphia. It had been developed in response to a 24 October 1944 BuAer request.

77. Memo, 15 June 1942, in BuAer Confidential files 1922–1944, VTB vol. 2, RG 72, NARA II, describing the conversation between Lt. A. B. Metzger USN and Lockheed Vega. It refers to a decision taken at a 4 June conference attended by representatives of the Plans, Engineering, and Production divisions and also to a decision by the Planning Division. Metzger was sent to Vega, meeting its officers on 10 June "for the purpose of discussing and initiating a new VTB design project." His account of the meetings suggests that he was visiting at the company's behest. At this time, however, Vega was producing Army B-17s, and the Army resisted any new Vega project.

78. The undated Vega brochure is in entry P 1044F, box 91 (proposal file), RG 72, NARA II. In this brochure the R-4360 was rated at 2,300 hp (military rating) and the R-2000 at 1,150. In pencil, the R-4360 was given military ratings of 2,400 and then 3,000 hp; it was credited with 3,000 hp at sea level (low blower, high blower at altitude). The conventional design had a design gross weight of 20,790 pounds, and the others were heavier. Vega credited V-141A with 309 mph at a critical altitude of 13,500 feet and with a service ceiling of 28,000 feet. V-141B and -C were slower. In each case, stall speed was 65 mph. Takeoff run was 300 feet (275 for V-141C). Maximum range with a torpedo was a thousand miles. Alternatives to a single torpedo included two 1,600-pound armor-piercing (AP) bombs (four in -141C). Each version had a pair of forward-firing 0.50-calibre machine guns, a two-gun upper turret, and a flexible 0.50-calibre gun; V-141B had a second flexible gun, thanks to its exotic configuration. Each had a radar (ASD for -141A and -141B, ASF for -141C). Given the need to slow quickly to drop a torpedo, Vega offered forward-opening upper-wing-surface perforated flaps (air brakes). The designs using R-4360 engines had contraprops. A drawing of bomb-bay arrangement for the modified V-141A-4 is dated 24 September 1942.

Chapter 7. A New Kind of Attack Bomber

1. The key 15 December 1941 memo, signed by Murphy, is in BuAer Confidential file, VSB vol. 3, RG 72, NARA II. In a 26 January 1942 memo, John G. Crommelin Jr. (who was from a different division) proposed that Murphy be assigned to temporary duty at Douglas to support the redesign of the SB2D; BuAer Secret VSB file, RG 72, NARA II. Murphy's cover sheet in this file, dated 28 January, mentions current work on both a new single-seat VBT and a single-seat version of the SBD-3. Conversion of the SB2D was discussed at the 21–22 January meeting: "From a performance standpoint this airplane comes nearest to an existing two-seater design that could be altered to a single seater. The balance difficulties would appear to require redesign to the point where almost a new airplane would result." Such extensive redesign would be required that the consensus was to hold off action for the present; the Armament chief concurred. The GE remote-controlled turrets

planned for the SB2D were being developed with high priority; if they worked, it would be better to leave the SB2D alone. The record of the conference is in BuAer Director of Planning and the Head of Engineering, memo to the Bureau Chief, 29 January 1942, Confidential Correspondence 1922–1944 file, VTB vol. 2, RG 72, NARA II.

2. BuAer to Resident Inspector of Naval Aircraft at El Segundo, 6 February 1942, on desired improvements and a single-seat version of the SBD-3, Confidential series 1922–1944, VSB vol. 3, RG 72, NARA II. Two separate studies were wanted: one with a better engine (R-1820-40) and provision for a torpedo, the other a single-seat dive-bomber with a secondary torpedo-attack role. The engine would be the 1,200-hp R-1820-40 (normal power 1,000 hp up to 4,500 feet). Armament of the former would be one 500- or 1,000-pound bomb or one torpedo. Guns would be two synchronized 0.50s with provision for two wing guns. Vision over the nose would be improved to make a reflector sight more effective; the present four-degree down angle over the nose was considered inadequate for a dive-bomber, at least seven and a half degrees being wanted. According to the 5 August 1942 edition of the VSB status report, the torpedo-bomber idea did not appear to offer sufficient range to make it attractive. The added weight (thirty-eight pounds) was enough to make the airplane unattractive as a dive-bomber as well. The second, single-seat, project was not mentioned.

3. Chief of BuAer to Inspector of Naval Aircraft at Curtiss Columbus, 9 February 1942, BuAer Confidential series 1922–1944, VBT file, RG 72, NARA. Unfortunately the file does not include the enclosure, a set of outline specifications. This file does include both the initial Curtiss proposal and a revised one dated 13 July, in which speed at critical altitude was reduced to 378 mph and takeoff run to 229 feet.

4. BuAer rejected a Curtiss proposal using a pusher engine (it was then working on the Army P-55 fighter with a pusher) as too experimental.

5. As indicated in the monthly VSB Desk progress reports in Naval Aviation History Division, NHHC. The A-40 was not, as sometimes claimed, a version of the SB3C.

6. Chief of BuAer to BuAer Representative (BAR) Columbus, Ohio, 18 September 1944, in BuAer Confidential file, VBT vol. 1, RG 72, NARA II. The Curtiss proposal letter was dated 7 August 1944. The tail would have been lengthened by twenty inches, as in the proposed SB2C-6, and there would have been a bubble canopy. Maximum speed at critical altitude was given as 343 mph, stall speed without power as 90.3 mph, takeoff run as 377 feet, service ceiling as 26,800 feet, gross weight 17,596 pounds. Features listed in the 1 February 1945 BuAer bomber report were an R-3350-24 (BD) engine, 410 gallons of internal protected fuel, a twenty-inch-longer tail, a blown-bubble canopy that raised the pilot two inches (with strong consideration for a jump-seat for a radar operator), bullet-resistant glass integral with the windshield (used as a gunsight reflector), a revised bomb bay that could completely enclose a Mk 13 Mod3 torpedo, metal-covered control surfaces, and high-aspect-ratio horizontal and vertical tail surfaces.

7. VSB Progress Report, 1 February 1945, Naval Aviation History Division, NHHC.

8. Memo, BuAer Confidential file, VTB vol. 2, RG 72, NARA II.

9. Entry 1044F, RG 72, NARA II. Remarkably, this file does not include the Kaiser Fleetwings design BuAer actually bought.

10. Director Engineering Division to Assistant Chief BuAer, 6 January 1944, AM-1 Mauler file, Naval Aviation History Division, NHHC.

11. VA-17A received its first airplane on 1 October 1948. It deployed on board the *Essex*-class carrier *Leyte*. It had already been decided that the aircraft would be assigned only to *Midway*-class carriers (CNO letter, 7 June 1948, and Op-55 Quarterly Summary, 6 July 1948), and thereafter virtually all deployments were on *Midway*s. Pilots found the Mauler a good dive-bombing platform but did not like it nearly as much as the Skyraider. According to the AM-1 file in Naval Aviation History Division, NHHC, quoting the Op-55 Quarterly Summary of 19 January 1950, the Mauler had been accepted with operating restrictions that had not been lifted even after more than a year, and the fleet had reported unsatisfactory flying qualities.

12. Smith, *Douglas AD Skyraider,* 57.

13. E. H. Heinemann, "Douglas Skyraider," in Naval Aviation History Division, Skyraider file, NHHC. This manuscript account is undated, but since is identifies Heinemann as the chief engineer of the Douglas El Segundo branch, it must have been written before 1961. This paper includes examples of direct weight savings over the BTD: 270 pounds by simplifying the fuel system (one tank rather than five), 50 by adopting short carburetor and oil-cooler ducts instead of the cowl leading-edge type, 70 by adopting fuselage dive brakes, 50 by doubling pressure in the hydraulic system, 200 by elimination of the bomb bay, 40 by adopting a bubble canopy, 40 by using powder (explosive) rather than mechanical (trapeze) bomb-displacing gear, 100 by using a continuous-wing center section (which was possible when the bomb bay was eliminated), and 20 by using a continuous horizontal tail. There was a multiplier effect, as savings in direct weight also cut that of other items, such as structure. There were also other advantages. For example, the single fuel tank freed a pilot from having to switch from tank to tank; operational statistics showed that many airplanes lost power when that was done incorrectly. Heinemann pointed out that eliminating the bomb bay improved get-away performance at a slight speed penalty when approaching the target.

14. Naval Aviation History Division attack aircraft folder, NHHC; original in BuAer Confidential file, VB vol. 1, RG 72, NARA II.

15. Confidential file, VBT file for 1945–46, RG 72, NARA II. The year is not clearly shown, but the 1944 date seems likely because Heinemann was requesting information on gas turbines; he would presumably have been well aware of their characteristics in 1945.

16. Production figures from Francillon, *McDonnell Douglas Aircraft.*

17. Confidential file, VAD 1951, RG 72, NARA II. ASW conversion of -3Qs had been approved as a local change, but consideration was being given to such conversion of -3Q and -3N aircraft on the service level. This file includes a 3 April 1951 endorsement by ComAirPac (Commander Air Force Pacific Fleet) on a 22 March letter from VC-35 recommending conversion of AD-3Ns and -3Qs to ASW aircraft pending availability of AD-4NLs. A 31 May 1951 report from VC-23 (Atlantic Fleet) recommended conversions of AD-3Ns. Expected late delivery of -4NLs led to a fleet request for conversions of -3Ns. The sole single-package AD-3S was BuNo 122918. Equipment included a sonobuoy dispenser and an ARR-31 sonobuoy receiver, as well as an underwing APS-31 radar. BuAer Military Requirements decided not to pursue the project, because it would compromise the airplane in its intended missions "and injure the current production ASW program, by taking from them [sonobuoy] dispensers and receivers. Also to have a good ASW aircraft it should have a searchlight (which AD-3N do not have and requires major rework to get) and a periscope [for the searchlight operator]. At this time AD-4Ns should be in sufficient quantities to meet fleet requirements."

18. Confidential file, VAD 1951, RG 72 NARA II. According to the route sheet, eighteen AD-3Q were available. AirLant asked for a carrier-based, high-speed, high-altitude target aircraft to support shipboard gunnery training. All versions of the AD-4 were in critical supply; CNO therefore approved modification for this purpose of AD-3Qs, which became AD-3QUs (the *U* for Utility). Francillon, *McDonnell Douglas Aircraft,* 374, claims that twenty-three aircraft were ordered specifically as AD-3QUs but that the success of the Mk 22 target and the AD-2QU made them unnecessary.

19. The Confidential file, VAD4 1951, RG72, NARA II, includes a 4 December 1951 letter from ComAirPac, Adm. T. L. Sprague, to CNO via CinCPacFlt calling for more AD-4 production. Except for the AD-4, no attack aircraft was compatible with an unmodernized *Essex*-class carrier. ComAirPac wanted more AD squadrons formed; the F4U piston-engine fighter was being replaced by jets far less adapted to ground attack. The night-fighter situation was as bad: carriers still relied on night Corsairs (F4U-5N). ComAirPac suggested replacing F4U-5Ns with old Hellcats (F6F-5Ns), because all current production of the AD-4, "a preferable alternate for CV-9 night fighter use," was already committed. He wanted a new Pacific carrier air

group to be "stood up" using training-command Bearcats (F8F) and Hellcats (F6F)—that did not happen. All of this was in reaction to the fact that the aircraft "shopping lists" for 1951–53 showed more fighters, at the expense of attack aircraft. It was impossible to trade more AD-4s for F3Ds, because the bottleneck was R-3350 engines, not production capacity at El Segundo. The FY53 program ultimately increased AD-4s.

20. The *B* suffix presumably indicated provision for remote arming, as no carrier airplane was authorized to take off carrying an armed nuclear weapon, for fear of the consequences of a crash. In at least one case remote arming required a quite visible underbody "saddle."

21. The 26 November 1951 OpNav (CNO staff) notice describing the -4B lists as its only new capability the ability to carry 3,500 pounds on its centerline, that being the approximate weight of the new nuclear bomb. Primary missions were given as general-purpose day attack and close air support. Delivery was expected to begin in December 1951. Delivery of the 165 newly built aircraft was expected to begin in June 1952. The OpNav notice is in Confidential file, VA4D 1951, RG 72, NARA II. The same file includes a 10 February 1951 memo on tests of the compatibility of the AD-4 with the TX-7 external store (i.e., with the Mk 7 nuclear bomb). The syllabus of flights envisaged referred to both lightweight and heavyweight versions. The airplane and bomb were tested in dives at angles of up to sixty degrees. This syllabus did not include over-the-shoulder delivery (a variant of toss bombing in which the aircraft performs a loop, releases the weapon just past the vertical, and half-rolls away in a different direction to escape the blast), which apparently had not yet been invented. Francillon, *McDonnell Douglas Aircraft*, 376, associates the -4B designation with strengthening to permit over-the-shoulder bombing with nuclear weapons, but the development of the BuOrd Aircraft Rocket (BOAR) nuclear rocket was premised on the inability of the Skyraider to escape effectively from a nuclear explosion (see chapter 9, note 8).

22. In October 1950 VC-35 asked BuAer to approve adding sonobuoy receivers and dispensers to four AD-4Qs. BuAer agreed, provided the airplanes could easily be returned to their original configurations. ComAirPac asked for authority for further modifications should they be needed. In August 1951 BuAer pointed out that AD 3Q and 4Q were needed in air groups and that modifications might compromise them for that use. BuAer had proposed to modify AD-3N, -3Q, and -4Q but decided not to issue a service change. No dispensers were available unless diverted from the AD-4N program. AD-4Ns would fulfill the requirement; twenty-nine had already been delivered, and thirty-six more would be by November 1951. That made the shortage of carrier ASW aircraft temporary; only a few -3Q and -4Q were being used for this purpose. Modification of two more AD-3Ns was approved, but BuAer did not want to divert too much Douglas engineering manpower to this project. As soon as practicable, the aircraft modified locally were to be returned to their original configurations.

23. The extra wing guns may have originated with an 18 December 1950 letter from VA-55 cited in a 10 January 1951 letter to the chief of BuAer from the Commander Air Force Pacific Fleet. Confidential file, VA 1951, RG 72, NARA II.

24. According to a note in the BuAer Confidential file, VAD4 file 1951, RG 72, NARA II, in September 1951 CNO (meaning DCNO [Air]) was considering procuring three hundred AD-5s instead of AD-4/6s (apparently for the Marines). At this time Op-56 was recommending some changes to improve the AD-4's ability to attack at low altitude and to provide close air support. At a 15 August 1951 meeting with representatives of Op-52 and -55 BuAer called for additional quickly detachable armor and other changes so that a pilot could more quickly apply control (moving the stabilizer trim-tab control to the stick and the speed-brake control onto the throttle quadrant—which might have been impractical, as shown by tests in the prototype AD-5). A proposal to add an ejection seat was rejected because it would require major redesign. Nor could the pilot's seat be moved aft. The canopy could be made jettisonable. Strengthening the structure in the area of the centerline bomb rack could make it possible to carry 3,500-pound

special (nuclear) weapons; this structural change was being incorporated on the production line in 165 AD-4s (which became -4Bs). To carry all weapons on this rack, it would have to be moved forward fourteen inches, which would entail some production problems. The AD-5 and -6 could have more electronics: an Airborne Direction Finder (ADF), an air-support guidance system (for the Marines), an ultra-high-frequency (UHF) transceiver, a Mk X IFF, a YG beacon receiver, and a radar altimeter. An AD-4 would be limited to the Mk X IFF (identification friend or foe), YG receiver, and radio altimeter. Later AD-4s would be wired for an underwing APS-19B radar, but to speed up acceptances only one AD-6 in three would be fitted. Water injection would be deleted in AD-5 and -6, and they would not have the automatic cowl flap of the AD-4. The wing stubs were moved forward. With its larger cockpit and rearranged electronics, it could have more complete electronics, although the air-support guidance system and VOR (a navigational beacon) receiver would have to be installed after delivery.

25. Douglas was asked to submit a brief study of appropriate modifications. Naval speedletter, 13 July 1951, Confidential file, VAD6 1951, in RG 72, NARA II. This file has no other content.

26. Memo for CNO, 27 October 1961, Op-508/jj serial 22350P30, in Skyraider file, Naval Aviation History Division, NHHC, warned that unless Intruder procurement was stepped up, further extension would be needed. By the end of June 1962 all of the forty-eight AD-6s now in reserve stock (there were no reserve AD-7s) would be active, unless a fifth tour was authorized. With four tours, aircraft would begin to end their service lives in FY62. By June 1964 there would not be enough Skyraiders and Intruders to fill out squadrons. An extension to five tours (ten years) would provide enough aircraft through FY67. The Skyraider was so important in Vietnam, and the inventory so badly affected by attrition, that in May 1965 Secretary of the Navy Paul H. Nitze recommended reopening production; memo for the Secretary of Defense, 3 May 1965, Op-50B/-1, 08009 P 50, Skyraider file, Naval Aviation History Division, NHHC. Douglas wrote that it could begin producing AD-7s (A-1Js) six months after a go-ahead; two hundred aircraft would cost $138.1 million. Nitze pointed out that there were few choices but also that the Navy could not at that point fund the new aircraft; they would go mainly to South Vietnam and to the U.S. Air Force. The project might be paid for out of FY66 funds. Yet another possibility, laid out in a May 1965 Navy teletype message, was to buy back French Skyraiders. The French refused but transferred thirty of them to Cambodia.

27. The agreement that ended the French war in Vietnam prohibited outside powers from supplying weapons to the combatants. It was violated covertly by both the Chinese and the Americans. According to Smith, *Douglas AD Skyraider*, 151, the first six Skyraiders arrived in Saigon on 23 September 1960 to replace F8F Bearcats already there. A second batch of twenty-five AD-4s arrived in May 1961. Such aircraft had already been supplied to France in 1959, to fight in Algeria (forty AD-4NAs and fifty-three AD-4Ns). According to Smith, release of AD-6s still in U.S. Navy service was also authorized early in 1960. These aircraft were in short supply. A working-level conference between representatives of Op-03, Op-63, and OSD/ISA (security assistance) suggested instead modified T-28 trainers, which were available in sufficient numbers. However, in February 1963 two AD-6 (A-1H) were approved for transfer to South Vietnam. Another twenty-five were included in the FY63 Military Assistance Program (Op-50 letter dated 12 March 1963, Skyraider file, Naval Aviation History Division, NHHC). Another thirty were included in the FY64 Military Assistance Program (twenty-five came from the Fleet Reserve—twenty of these aircraft being reworked—and five others from the Litchfield Park aircraft storage depot). According to Smith, 154, in May 1964 fifty A-1Es (AD-5) and twenty-five A-1Hs (AD-6) were prepared at Alameda Naval Air Station; the two-seat A-1H was converted to have two sets of controls. The United States was not yet officially in the war. According to Smith, the presence of two sets of controls allowed it to be claimed that a U.S. Air Force pilot was merely observing the South Vietnamese pilot in the other seat. The final batch for South Vietnam was shipped in January 1966

on board the escort carrier *Core*; Smith does not give a figure. According to Smith, VNAF Skyraider transfers amounted to fifty-seven A-1Es, eight A-1Gs (EA-1), 121 A-1Hs, and ten A-1Js. His figures omit the initial transfers of AD-4s (A-1Ds). Many of the later aircraft were transferred from the Air Force. The Air Force received two ex-Navy AD-5s in 1962 for its Special Operations Forces, which, with its counterparts, the Kennedy administration hoped could covertly win the developing Southeast Asian war. In March 1963 the Air Force requested sixty AD-5s (A-1Es) for its Special Operations Forces. Later it also received AD-6/7 aircraft.

28. Copy in Skyshark folder, Naval Aviation History Division, NHHC. This folder includes a history dated 6 August 1957 of the A2D project, in a series of design histories of failed naval aircraft programs prepared for the Senate Preparedness Investigating Subcommittee chaired by Senator Lyndon B. Johnson.

29. Memo, 14 June 1945, Skyshark folder, Naval Aviation History Division, NHHC. Most work to date had addressed the problems of designing and building the gas turbines, leaving two other sets of problems: first, reduction gearing and arrangements of turbines, gearing, and shafting; and second, the design of airplanes around turbines, particularly with respect to arrangement, balance, stability, and control. Douglas El Segundo was already very interested and was "eager to get in on the ground floor."

30. Allison responded with the T40, its first turboprop, to a request for a 4,100-hp engine to work at medium altitudes (below 20,000 feet). A proposal was submitted in November 1944 and a Navy design contract awarded in December 1945. T40 consisted of two T38s (Allison 501s, 2,925 equivalent horsepower each) geared to one shaft; either could drive both contraprops when the other was declutched. The combination was rated at 5,100 equivalent shaft horsepower (eshp). Allison had geared V-1710 piston engines to long extension shafts for P-39 and P-63 fighters. In this case, shutting down one section could add as much as 20 percent to the range of the airplane. The T40 was chosen for the P5Y in 1947, then for the A2D, and

then for the A2J early in 1948. Heinemann's paper on the Skyshark shows other engine arrangements he investigated. One, a composite jet/turbine engine (developed but not built) used two 24C jets whose exhaust would drive a turbine connected to contraprops. Two side-by-side 24Cs showed promise but did not satisfy takeoff and endurance requirements; Skyshark folder, Naval Aviation History Division, NHHC. Heinemann used the General Electric TG-100 (T31) in his 1945 studies. GE soon abandoned it in favor of pure jets. An installation was built for a Skyraider, but the engine was not delivered. The other 1945 turboprop projects were Pratt & Whitney's T34 (flown in 1950) and the abortive Westinghouse T30 (a version of the J34 turbojet). The T34 did not drive contraprops, which was a problem in a single-engine carrier airplane, as it would have taken a very large propeller to absorb its 5,000 hp.

31. Lots I and II were both credited with a gross weight of 21,500 pounds with a 2,000-pound bomb and a stress factor of 7. Maximum sea-level speed was estimated at 424 mph for I and 427 for II. At 15,000 feet, II might attain 445 mph (no figure was given for I). Estimated takeoff distance with the normal load was 250 and 288 feet, respectively (426/483 feet with an overload of six hundred gallons in drop tanks). Stall speed was 91.8 mph with flaps down and power off at full weight. Combat radius would be 166 and 258 nautical miles, respectively (504 and 651 nautical miles with drop tanks). For II, the most economical cruising speed would be 200 to 270 mph at sea level, or 22 to 290 at ten thousand (no figures for I were given). These airplanes would have tricycle landing gear and a jump seat adjacent to the pilot for a radar operator. Fixed guns would be two 20 mm, and there would be four major bomb stations, each of which could carry one 2,000-pound bomb or one torpedo or one Tiny Tim rocket, plus twelve rocket stations (each to carry one five-inch High Velocity Aircraft Rocket or a 250-pound bomb). No details of Lot III were given. Buttler and Griffith, *American Secret Projects*, 218–21, provides sketches and some details of the D-557, -557A, -557B, and -557C. Heinemann's initial D-557 sketch was dated

25 January 1945. Its noticeable feature was a cockpit pushed forward almost to the contraprops so that the long, narrow turboprop engine could fit beneath it. Span was fifty-four feet ten and a half inches, length was forty-five feet, and takeoff weight was 21,350 pounds, including a 2,000-pound bomb. There were two wing 20-mm cannon. Maximum speed without bombs was 430 mph at sea level and 447 mph at 20,000 feet; rate of climb at sea level was 5,885 feet per minute. The drawing shows a radar pod on the starboard wingtip. The twin-engine D-557A was fifty-three feet eight inches by forty-five feet ten inches; carried 23,100 pounds of fuel; maximum sea-level speed would have been 401 mph (415 mph at 20,000 feet). D-557B returned to the D-557 configuration but had a faired-in canopy rather than a bubble. Dimensions were fifty-two feet by forty-five feet ten inches (21,825 pounds of fuel); maximum speed would have been 407 mph at sea level (419 mph at 20,000 feet). D-557C had a similar configuration, forty-six feet six inches by forty-one feet eight and a half inches (17,400 pounds), with a maximum speed of 389 mph at sea level and 394 mph at 20,000 feet. D-557B and -557C showed pronounced air scoops under their propellers, whereas D-557 had cheek intakes. Note that none of these aircraft approached Heinemann's vision of a fighter-bomber. D-557 had conventional three-point landing gear; the others all had tricycles.

32. NHHC's Aviation History Division file refers to submissions by several companies, but the only comparable airplane apparently offered at the right time was Vought's V-358 attack-fighter, which a Vought history dates to 1947. A brochure is in the BuAer proposal file in NARA II, according to which, at BuAer request, Chance Vought conducted a study of long-range carrier fighters with a radius of 1,200 nautical miles or less. At the same time it was making preliminary studies of carrier attack or strike aircraft and realized that it could offer a dual-purpose airplane, much as Heinemann had. It analogized the airplane to the dual-purpose Corsair. The airplane was powered by two 4,500 shp (2,770 shp at 30,000 feet

and 500 mph, plus 606 hp contributed by the jet) Pratt & Whitney PT-2 turboprops in separate nacelles. Space was provided under the canopy for a two-man crew, on the ground that a second crewman would become mandatory as electronics grew more complex. Armament was four thousand pounds of bombs (two bombs underwing) and two 20-mm cannon, and the airplane had a thousand pounds of electronics on board. The assumed attack mission began with a climb to five thousand feet, rendezvous, and then cruise out and back at 45,000 feet, with ten minutes of combat at maximum speed. Radius as a bomber was 425 nautical miles; takeoff weight was 28,520 pounds. Span was fifty-one feet one inch, length forty-six feet (wing area was 435 square feet). Maximum speed was 480 knots at sea level, 441 knots at 40,000 feet. Cruising speed was 350 knots.

33. Memo, 1 March 1971, A2D folder, Naval Aviation History Division, NHHC. The first was accepted in 1950 and flew at Edwards Air Force Base (the former Muroc Field) on 26 May 1950. It crashed there in December, never having flown anywhere else. The second was accepted 15 April 1953 and immediately went to Indianapolis for development work and then to NAMC Philadelphia.

34. One of the two power units failed, the compressor of the dead unit absorbing the complete power output of the remaining one as well as energy from the propeller. The pilot did not realize how much drag had been added and did not compensate. The second airplane incorporated instruments to alert the pilot to the failure of a power section. Later an automatic decoupler was added to separate the failed power section from the propeller drive.

35. According to the 1957 history, the first cancellation was sixty-two aircraft on 11 February 1952, after a delay in the first flight of the second prototype. Another forty-seven were canceled on 21 August 1952, after it became clear that no aircraft would be operational until mid-1954, when 124 should have been delivered. Another thirty-two were canceled on 7 October 1952, to release production facilities to the A3D-1. Funds were transferred to the A4D. In August 1953 the engine still could not pass its fifty-hour test, and AD-6s were urgently

needed. The program was now reviewed by the secretary of defense. On 22 September 1953 the remaining 180 production aircraft were canceled, the ten service-test aircraft surviving. They were wanted for a tactical evaluation of a turboprop airplane.

36. British visitors to BuAer and American aircraft companies in 1948 attributed the policy of carrying all stores externally to Douglas, which told them that to carry 10 percent of the gross weight of an airplane in bombs they would have to provide a bomb bay weighing 3 percent of gross weight, a bad bargain. "They have put a lot of work on an optimum design of a high-speed container," which must have been the predecessor of the Mk 80–series shape. It had a fineness ratio of 8.6 and was fitted with a tail; it weighed 2,500 pounds loaded and was fifteen feet long and twenty-one inches in diameter. It was designed to be satisfactory at Mach 0.85, the optimum wing/store gap being about 60 percent of the diameter of the store. The pylon was 0.2 calibres thick, with a 10 percent thickness ratio. The nacelle drag was only a fifth that of a standard two-thousand-pound bomb. With the store fitted to a wing, the drag coefficient began to increase at Mach 0.8; minimum buffeting speed was Mach 0.84, a characteristic of the wing rather than of the store; "Notes on U.S. Naval Aircraft for the Naval Aircraft research Committee," 12 January 1949, PRO DSIR23/17889, The National Archives, Kew, London, U.K. The group visited BuAer, Grumman, Chance Vought, Douglas, North American, and McDonnell during October–November 1948.

37. The proposal (Boeing Model 400-2-2) in the BuAer proposal file, entry 1044F, RG 72, NARA II, is dated 12 January 1945. Drawings showed a two-seater with a large, framed, glasshouse-type canopy. A drawing showed a torpedo, two drop tanks, two rockets, and a radar "bomb" under the starboard wing. The proposal mentioned a 10 November 1944 BuAer letter asking Boeing to substitute a single-stage, variable-speed, single-rotation engine for the two-stage, variable-speed, dual-rotation (contraprop) engine, move the pilot forward for better vision over the engine, add a second crew member to operate radio and radar, provide for external mounting of a Mk 13-3 torpedo, an APS-4 radar "bomb" under the starboard wing, and external loads: rockets, gun packages, and drop tanks. Standard radios were to be added, including a homing receiver, and there should be provision for the APG-17 bombing radar. Boeing offered everything except the engine change, considered the existing one essential for performance even at low altitude. That also precluded moving the pilot forward. In the file is a penciled note of a 6 February 1945 conference; no further action was to be taken to modify the XF8B until after evaluation at Patuxent River. Another sheet compares the original XF8B, the modified version, the current single-seat dive-bombers (BTM, BT2D, and BK-1), and the F6F-5E. In this chart the gross weight of the XF8B-1 is given as 21,730 pounds, the current modified version as 22,084 pounds. The next-heaviest was the XBTM-1, 21,069 pounds. The original XF8B-1 was credited with a combat radius of 520 nautical miles and a take off run of 569 feet. The modified version had a 300-nautical-mile combat radius and would take off in 630 feet, which was entirely unacceptable. By way of comparison, the XBTM-1 offered a combat radius of 285 nautical miles but could take off in 428 feet. Maximum speed at sea level in combat condition (torpedo, tanks, and radar dropped) was 340 mph (XF8B-1) or 337 mph (Modified F8B), compared to 340 mph for the XBTM-1. At 15,000 feet, speeds were 385, 382, and 367 mph, respectively. In two hundred feet of an *Essex*-class deck, fifteen Boeing fighters could be spotted, compared to twenty XBTM-1s or XBT2D-1s, or twenty-four XBK-1s, or thirty Hellcats. Also in this file is a 31 January 1945 BuAer estimate of performance with various external loads: a thousand-pound bomb and an ASH (APS-4) radar; a torpedo and a radar; and two Tiny Tim rockets and a torpedo. In each case takeoff run was unacceptable: respectively, 538, 630, and 777 feet. Combat radius would be 469, 300, and 300 nautical miles, respectively.

38. Tony Buttler and Alan Griffith, *American Secret Projects: Fighters, Bombers, and Attack Aircraft 1937–1945* (Manchester, U.K.: Crecy, 2015), 177–78. This design is not in the BuAer proposal file,

suggesting that it was never formally submitted. The drawing is marked "land-based fighter," but in the Vought design list it is "Navy VF design." Span was forty-eight feet (wing area 365 square feet) and length thirty-eight feet seven and a half inches. Gross weight as a fighter (no torpedoes) was 13,733 pounds (overload 14,368 pounds). The drawing shows an odd armament arrangement in which two of the six 20-mm cannon fired through the propeller disc. Buttler and Griffith found no performance data, and they do not give a date. However, the detailed French history of Vought aircraft dates the project to 1942.

39. Engines were two J46-WE-2 rated at 6,100 pounds of thrust with afterburner and 4,080 without. Gross weight was 34,000 pounds (26,500 pounds in combat). Span was forty-seven feet and length fifty-one feet, so the airplane was just larger than an elevator when not folded. Folding the outer wing sections reduced width to twenty-five feet. An inboard profile shows two bombs in tandem in a high-speed rotary bomb bay, which Martin advocated. In addition to the single 2,000-pound bomb, the bay could carry two 1,000-pound or four 500-pound bombs, or other weapons. In combat, Martin estimated maximum speed at sea level as 688 knots; BuAer estimated 643. Estimated combat radius was, in BuAer's view, 395 rather than 524 nautical miles, and average cruising speed 478 rather than 505 knots: "There appears to be optimism on the part of Martin in estimating airplane drag." The BuAer evaluation is dated November 1949; the May date is from Tony Buttler, *American Secret Projects: Bombers, Attack, and Anti-Submarine Aircraft 1945 to 1974* (Hersham, U.K.: Midland, 2010), 110–11.

40. BuAer Confidential series, VA2D1 1951 file, RG 72, NARA II. The operational requirement is in the Skyshark file at NHHC's Naval Aviation History Division. There is no indication that any design competition was called. An account dated 28 February of the conference is in the BuAer Confidential series, VA 1951 file, RG 72, NARA II.

41. Entry P7, RG 72 (formerly RG 402), NARA II. This airplane was comparable in size to an AJ: span was 61.85 feet, length 54.92 feet, and height 21.16 feet. Normal gross weight would have been 58,280 pounds, and useful load was greater than that of the 1946 heavy attack bombers, 20,367 pounds including 6,000 pounds of bombs. Gun armament would have been four 20-mm cannon. Maximum speed at 24,000 feet with military power would have been 500 knots. Takeoff distance in a 25-knot wind over deck was given as 570 feet at 58,280 pounds, stalling speed with power off as 93 knots. A brochure of 1 October 1951 gave combat radius as 390 nautical miles at 430 knots or 560 nautical miles at 413 knots (with two wing-tip tanks).

42. CinCLantFlt endorsement, 20 August 1952, in BuAer Confidential series, VA 1952 file, RG 72, NARA II. The fleet commander had not seen anything to show that the recommendations were no longer valid, so he affirmed them.

43. Dates are from the relevant card in NHHC's Naval Aviation History Division folder for "Aircraft, Design Competition Summaries (R&D)." The envisaged system was OS-134, the missile being OS-133. The circular distributing the outline specification was dated 28 August 1953. OS-134 documentation mentioned a solid-fuel motor and gave data on the 250-pound bomb. It also included the clause about a target at ten-thousand-foot altitude; entry 1048, "Proposal Files" for Airplanes, Helicopters, and Missiles, 1934–1961, box 632, RG 72, NARA II. A provision that the missile be usable against targets at altitudes of up to ten thousand feet suggests that something more than an air-to-surface weapon was envisaged, but nothing came of it.

44. Bell had developed the Tarzon guided bomb and the secure-command system for Regulus. It had also developed the Shrike (not the later antiradar missile), Rascal, and Meteor missiles. Chance Vought was developing Regulus. Convair had developed a version of the Lark antiaircraft missile, as well as a missile of the V-2 type (which became Atlas); it was working on Terrier and had developed missile radars. Douglas was developing the Sparrow air-to-air missiles as well as the Nike and Honest John. Fairchild had produced a version of Lark. Hughes had done missile studies and was responsible for air-to-air-missile fire-control systems. Martin had developed the KDM drone, the

Oriole missile, and the Viking rocket. McDonnell had developed the KDD-1 drone and the abortive KUD-1. Republic had formed a missile division in 1952. Sperry was involved in the Sparrow missile program. A spreadsheet gave the companies' designations and design details: Convair 195311, Bell D163, McDonnell 97, Douglas SM 1834, Sperry (no designation), Chance Vought V387, Fairchild 424, Republic (no designation), Hughes H, and Martin 293. Of these, Convair had a single set of surfaces at the middle, Bell and McDonnell had them aft. Douglas and Sperry had wings of very low aspect ratio. Chance Vought, Fairchild, and Republic had wings amidships and a tail aft; Hughes and Martin used canard (small wings forward of the main wings) arrangements. Chance Vought's beam-riding missile was considered heavy (739 pounds) and too expensive ($7.93 million). The airplane would need a twelve-inch antenna boresighted (aligned) with an optical sight. Bell proposed that the airplane automatically track an infrared flare in the tail of the missile, correcting the missile course to the pilot's line of sight. The system, as BuAer's evaluator pointed out, was simple in theory but too complex in practice, as the missile had to be locked on after it was launched and needed a pulse command link. Also the sun and hot ground objects might cause interference. Bell's missile was the least expensive ($1.98 million). Republic's missile was the lightest (491 pounds) and the second-least expensive ($3.34 million) but had the least desirable type of guidance: automatic track and command, much the same as Bell's. Too, the missile was light, because it had a small (nine-inch) diameter, which precluded use of a standard bomb. McDonnell used an uncased warhead, which also would make it difficult to use a standard bomb. It was a radar-beam rider, with all the associated problems of boresighting and installation. The radar would use a ten- or twelve-inch dish, presumably the air-to-air type in some current jet fighters but not in the Skyraider. Since it had no wings, it would suffer from high angles of attack and large surface deflections. The surviving package shows no notes on the other designs. As a yardstick, BuAer's designers produced their own preliminary design, 126 inches long with

a 42-inch wingspan (no other data were given). Most entries were somewhat shorter, but Martin offered a 126-inch missile with 11.5-inch diameter and 33.8-inch wingspan, weighing 521 pounds at launch (417 with fuel exhausted). It just met the specification, with a slant range between a maximum of 15,000 yards and a minimum of 8,000. At a 20 January 1954 conference on the competitors, it was pointed out that most preferred an ammonium perchlorate solid-fuel rocket motor and that solid-fuel rockets had a longer storage life than liquid, about five years rather than one. A shaped charge was not required and would be less effective than a blast warhead if equipment was mounted ahead of it, as in the Martin proposal. As a warhead, one of the new streamlined (smaller-diameter) bombs would be just as effective but would add expense and might create logistical problems. They could probably be used in missiles designed for the larger standard warheads.

45. Sperry was superior in the inertial-only condition, except at extreme range, but that mattered only if ripple fire was important (it was rated desirable but not vital).

46. This account of MPQ-14 is based on Benjamin Huston, "The Flying Nightmares: An Operational History and Assessment of VMF(N)-513 at War in Korea 1950–1953" (master's thesis, Kansas State University, 1995), copy courtesy of the author. Dr. Benjamin Kristy, of the National Museum of the Marine Corps, sees airborne fire control, particularly continuously computed impact point (CCIP), as the successor to MPQ-14, but that does not apply to the Marines' need for what amounts to precision mobile artillery under the control of a ground commander.

Chapter 8. What Now?

1. Papers on the SCB 6A (CVA-58 *United States*) project, Naval Aviation History Division, NHHC. DCNO (Air) would to preside at the meetings. Most members were captains with wartime operational experience. According to an attached summary paper, by this time the *Essex* class was already inadequate, and it was foreseen the *Midways* too would be inadequate "if Naval Air was [*sic*] to exploit to the maximum the potentialities

of carrier warfare." Conversion of the *Essex*es was rejected as economically prohibitive, although construction of *Oriskany* was suspended for completion as a jet- and nuclear-capable carrier.

2. Op-06 was dissolved in November 1946, replaced by Op-36, which began to develop nuclear ordnance, and by Op-57, for missile development. Op-36 was headed by Rear Adm. William S. Parsons, who had been assigned to the wartime Manhattan Project as its ordnance expert and had headed the Atomic Defense Division in OpNav. Parsons' deputy was Cdr. Frederick L. Ashworth, who soon became secretary to the Atomic Energy Commission's Military Liaison Office, charged with developing requirements for future nuclear weapons. Op-57 was headed by Rear Adm. Daniel V. Gallery, who became a strong advocate of naval strategic nuclear weapons, as against those of the Air Force. Op-36 was part of Op-03, which was responsible for operations; Op-57 was part of DCNO (Air), Op-05.

3. VB Confidential file 1945, RG 72, NARA II. It was requested verbally. According to the cover sheet, this was one of several studies that, taken together, proved the feasibility of large carrier bombers. A 31 October memo in this file argued for further work, although "it is desired to emphasize that firm requirements . . . cannot be given due to the lack of information on form factors, weight, and operational characteristics of the payload" (presumably the nuclear weapon). Combat radius was to be increased as much as possible while maintaining a speed of at least 475 mph at 30,000 feet. Crew could be increased to three, since it would have to carry out the functions of radar operator, bombardier, navigator, observer (ordnanceman), and possibly copilot. The payload had to be accessible in flight: at this point nuclear weapons were manually armed en route. The cabin had to be pressurized and fixed guns eliminated.

4. An 11 December 1945 report to CNO is included in the NHHC package on the SCB 6A project.

5. On 28 December 1945, ser. 042P517 (Document 193018), SCB 6A project, NHHC. Also in the package is a 28 December 1945 letter from the chief of BuAer to CNO on future carrier design, referring to a recent letter to the General Board

and the bureaus directing a study of a carrier with the aircraft capacity of a *Midway* but displacing 35,000 tons. Studies by the informal board and the Bureau of Ships (BuShips) suggested a displacement of about 39,600 tons. It became Study C-1: "This latest study appears to incorporate the requirements for the immediate future as to carrying capacity of aircraft of present design, armor, armament, and gasoline tankage. However, looking further into a future of atomic warfare and radically different aircraft propelling units, a broader program appears indicated." The carrier should be able to make strategic attacks from within about a thousand miles of the target. Alternative aircraft complements would be: seventy VSB or VF (30,000 pounds, five-hundred-mile radius); reduced numbers of VSB to incorporate smaller interceptors; or twenty-seven VSB of 45,000 pounds (thousand-mile radius). The 39,600-ton carrier would be well adapted to general, all-round naval warfare, but an additional type should be developed to accommodate 100,000-pound aircraft with a two-thousand-nautical-mile radius. "This ship may be rather radical in design with, for example, no island and no hangar. A flight deck equivalent to the CVB 41 class would accommodate about fourteen planes of the above weight. Five hundred thousand gallons of gasoline would permit each plane about eight full-range flights. Such a carrier would be capable of long-range bombing as opportunity afforded; at other times it would be available for more conventional operations." This letter was signed by the bureau chief, Rear Adm. H. S. Sallada. It was the genesis of the *United States* program. On 8 January 1946 DCNO (Air), Admiral Mitscher, wrote CNO supporting the project and calling for sixteen to twenty-four large bombers and enough fuel for at least a hundred sorties. Assistant Secretary of the Navy for Air John L. Sullivan approved the project on 1 July 1946, pending a progress report. On 7 February DCNO (Air) was directed to proceed with a detailed design study of the heavy bomber, in coordination with the study of the new carrier. BuShips submitted initial characteristics of the big carrier (from its sketch design) to the Ship Characteristics Board in a 24 May 1946 memo; the short time involved demonstrates how urgent the project was.

6. Hydraulic catapults were inherently limited. They used wire rope to impart the motion of a hydraulic ram below decks to a shuttle that moved the airplane. The strength of their wires ultimately limited their power. In December 1944, the bureau chief ordered his catapult developers to begin work on an alternative concept, a slotted-cylinder catapult in which direct pressure moved the shuttle. The idea seems to have been inspired by the catapult the Germans had used to launch V-1 missiles (the British steam catapult had much the same origin). BuAer Confidential files, Catapult file (S83-2) vol. 4, RG 72, NARA II. BuAer catapult developers still hoped that there remained a bit more potential in hydraulics. Performance of a new hydraulic catapult would be limited by space available on a carrier. The new XH-8 was limited to launching a 15,000-pound load at 120 mph with a forty-second launch interval. It became operational on board the modernized carrier *Oriskany* in 1950. The H-8 was barely sufficient for early postwar jet fighters and far less than what the new heavy bombers needed. The BuAer catapult designers conceived a much more powerful H-9 for the new bombers but seem to have had little faith that it would work. To launch a heavy bomber, the catapult had to accelerate a 45,000-pound airplane to 120 mph—which meant three times the energy imparted by the H-8. The ultimate heavy bomber might weigh about 100,000 pounds. As for landing, initial preliminary designs of the 100,000-pound bombers indicated a landing speed of seventy-eight knots at 89,000 pounds. However, BuAer's Aircraft Design Research (ADR) office envisaged a 100,000-pound airplane that would stall at 105 knots and approach at 115; Director of Ships Installations Division to Director Design Research Division and Director of Piloted Aircraft Division, 10 November 1947, BuAer Confidential series, S83-2 file (Catapults), vol. 9, RG 72, NARA II. He needed data for catapult and arresting-gear design for the new carrier. By June 1946, BuAer was working on a slotted cylinder powered by hot gas; report of 13 June 1946 catapult conference, in BuAer Confidential file, Catapult file (S 83-2), vol. 8, RG 72, NARA II. By the beginning of 1947 BuAer favored high-energy solid propellant, such as that used in a gun. Alternatives were monofuel (nitromethane) and bifuel (liquid oxygen and gasoline). Steam seems not to have been considered for large catapults, although BuAer did try it for a small one. There was no question of simply propelling a piston carrying the shuttle launching the airplane, because it would be unacceptable to vent large volumes of hot gas through the slot in the deck (which would close behind the shuttle as it moved but would still release gas). Instead, the BuAer catapult designers envisaged a piston moving in a cylinder, which would be pierced at one end for a long piston rod carrying the shuttle. Alternatively, the piston could act like the hydraulic ram of earlier catapults, its motion multiplied by cable and pulleys. Through 1951 BuAer continued to promise DCNO (Air) that it would deliver a prototype internal-combustion catapult (XC-10) by June 1952, if it received sufficient priority and BuOrd (for the explosives) support.

7. The Aviation History Division file at NHHC on the AJ/A2J includes the 25 January 1946 letter asking for bids.

8. Alan David Rosenberg and Floyd D. Kennedy Jr., *Naval Strategy in a Period of Change: Interservice Rivalry, Strategic Interaction, and the Development of a Nuclear Attack Capability, 1945–1951*, study prepared for DCNO [Plans and Policy], October 1975, 1–155. This was an unclassified supporting study for the history of the strategic arms competition prepared for the Office of Net Assessment of the Office of the Secretary of Defense. When the civilian Atomic Energy Commission took over development of U.S. nuclear weapons, Ashworth became executive secretary of the Military Liaison Committee, which transmitted military requirements to the bomb designers. As plans and operations director of the Armed Forces Special Weapons Project, Hayward too was well placed to influence bomb development. His wartime experience in the Manhattan Project and his physics background led him to realize that it would be possible to build smaller bombs using less fissionable material. That perception was key to building carrier-capable strategic bombers.

9. Rosenberg and Kennedy, *Naval Strategy in a Period of Change*, describes (pp. 155–56) a decision taken

after the AJ-1 won the OS-106 competition, but BuAer records indicate that bidders were advised of the need for the larger bomb bay.

10. The NHHC Aviation History Division's AJ/A2J file includes the 3 June 1946 memo from the assistant chief of RD&E to the chief of BuAer evaluating the three proposals. North American submitted two basic designs, both with optional afterburners. One used opposite-rotation engines, which were not available, and hence was dropped. Weights reported in the comparison were for the afterburner, but the additional 190 pounds of thrust was not used in performance calculations. Afterburning did not seem to promise much higher performance. This airplane was powered by two R-2800s and one I-40 jet engine in the fuselage. Douglas' D-566 had the same engines and much the same layout. It had higher-aspect wings (North American's lower-aspect wings would be lighter). At extreme overload, the high aspect ratio would reduce fuel consumption and therefore gross weight, but at a radius of about six hundred nautical miles the lower aspect wing seemed satisfactory. There were no major deficiencies in the North American design, although the arresting hook had to be moved farther aft and the rudder needed slightly greater area. North American's was the only design that approached the desired landing weight of 29,000 pounds. It had the shortest takeoff distance, the best climb at altitude, and about the same maximum speed as the others. Combat radius was fourteen nautical miles short, because of a difference in opinion as to whether "auto-lean operation" (automatic adjustment of the fuel/air mixture with altitude and power to minimize fuel use) would be used during the cruise-out part of the flight. Added fuel would weigh less than the afterburner, which was not considered desirable—otherwise the airplane would grow—and the afterburner added little speed. Douglas recognized that its entry's radius was short and offered space for overload fuel. Its chief disadvantage was high cost. Consolidated offered one basic design with several alternative engine arrangements, all of which increased weight. Even the basic design exceeded the allowable landing weight, and only it was considered. It was powered by two R-2800s (turbo-supercharged) with two 24C jet engines, one in each nacelle. Takeoff distance and stall speed were greater than for the other, and combat radius was forty nautical miles shorter, because Consolidated had not allowed for a required 15 percent increase in piston-engine fuel consumption over the engine specification. The design was also judged to have poor calculated stability and control characteristics. The only outstanding feature was the fuel arrangement, the fuselage tanks permitting an unobstructed overhead in the bomb bay for hoisting gear. The Convair design is item 77 in Robert E. Bradley, *Convair Advanced Designs: Secret Projects from San Diego 1923–1962* (North Branch, Minn.: Specialty, 2010), 178–79. A sketch shows dimensions: span sixty-nine feet two inches, length sixty-four feet six inches (gross weight 44,372 pounds), wing area 640 square feet. Maximum speed was 485 mph at 35,000 feet. It had a crew of two. The engines had an augmented exhaust-thrust cooling system that would have made for smaller nacelles, and so less drag. Grumman assigned the designation G-76 to a planned OS-106 entry but never submitted it. According to Tony Buttler and Alan Griffith, *American Secret Projects 1945–1974*, 97, D-566 had an arrangement similar to that of G-76, with jet engines in each nacelle, D-566-9 having the fuselage jet engine arrangement. Unfortunately, no data for D-566 seem to have survived. Apart from Bradley and Buttler/Griffith, data are from the BuAer assessment of alternatives.

11. According to priority lists in BuAer Confidential VV file 1946–47, RG 72, NARA II. An evaluation dated 17 December 1947 is headed "Not to Be Compared to Other VA." It "must carry the highest priority until production requirements for the A2J are established."

12. These dates are in the AJ/A2J file in the Naval Aviation History Division, NHHC. The same file includes a 1957 history of the AJ/A2J program written in response to a congressional request; it gives the date of first carrier takeoff as 21 April 1950.

13. Data from DSIR 23/17889, a 1949 report of a British visit to BuAer and American firms. The report does not give the maximum speed of the AJ-1. According to the 25 July 1946 circular establishing the XAJ-1 designation, gross takeoff weight

was 42,000 pounds, including an 8,000-pound bomb, and service ceiling was 47,000 feet. When the AJ-1 designation was established on 21 November 1947, gross weight was given as 52,000 pounds. Maximum speed was given as 405 knots (466 mph) at 35,000 feet, compared to 500 mph at 35,000 feet in the 1946 memo. Combat radius was 695 nautical miles rather than 600–700 nautical miles, and service ceiling was given as 41,000 rather than about 47,000 feet. Power plants were two R-2800-44 piston engines and one J33-A-19 turbojet. The jet could provide nineteen minutes of maximum-speed flight.

14. Memo to Director Piloted Aircraft Division in Confidential VV series 1946–47, RG 72, NARA II, referring to a 2 December 1947 conference in the Fighter Design Branch: "Prior to proceeding with a new design it is recommended that a thorough study be made to determine definitely that the AJ type cannot be modified to meet the requirements for an escort." At this time between twelve and twenty-four AJs were planned, but production then shifted to the 1,200-nautical-mile A2J. The escort problem had arisen at the same time as the idea of the long-range carrier bomber. In November 1945 Military Requirements Branch had ordered a study of the practicality of an escort with a 500-mile combat radius, 300-knot cruising speed (at 30,000 feet), and a maximum speed of 480 knots. In April 1946 ADR submitted a report on a long-range escort, and in November 1946 the most urgent requirement in the program outline for FY47–48 was a 1,200-nautical-mile escort (for the follow-on A2J). However, for Military Requirements Division the most immediate need was an escort (for the AJ-1) with a radius of action of six to seven hundred nautical miles. In January 1947 ADR produced a report on the desired escort, and a conference proposed letting study contracts to Chance Vought and Curtiss (contracts were let that June). Several current fighters could be modified to meet this requirement. In May 1947 CNO made the fighter escort an interim objective, and in June the XF2H-1 (Banshee) contract was amended to include modification of one airplane as an AJ-1 escort, with six-hundred-nautical-mile radius. At this time the turboprop F2R-2 seems to have been considered the best initial 1,200-nautical-mile escort, but in March 1947 it was canceled owing to cutbacks in the turboprop program that killed its engine. As of December 1947, between twelve and twenty-four AJs were planned, all other aircraft of this class being A2Js. Escort performance requirements set at the December 1947 conference were a maximum speed of 485 knots at 35,000 feet at combat weight (a forty-knot advantage over the bomber, to allow for slight maneuvers or changes in formation) and the same combat radius as the A2J. Allowing for using maximum military power for the run in and back out, radius would actually be 1,370 nautical miles. Shipboard limits would be those of a *Midway*-class carrier, including its fifty-four-by-forty-six-foot elevator and its H4-1 catapult. Armament would be a quadruple 20-mm nose turret and a 20-mm rear turret, both with automatic gunlaying radar (AGL), the APG-25. Douglas won the design competition, but its airplane was canceled before any were built.

15. In the AJ-1, the third crew member was on the lower floor level; in the AJ-2 he was on the same level as the other two. Presumably the third crew member was the radioman, who also served as ordnanceman to arm the bomb in flight; a tunnel gave access to the bomb bay. Tommy H. Thomason, *Strike from the Sea: U.S. Navy Attack Aircraft from Skyraider to Super Hornet, 1948–Present* (North Branch, Minn.: Specialty), 54, notes that the patrol pilots who flew the airplane had the fighter-type stick replaced by a control wheel and the throttles moved to the center console. The wheel improved handling with an engine out or in a hydraulic emergency. Moving the throttles to the center made it possible for the bombardier/navigator in the right-hand seat to control the airplane in an emergency via the autopilot.

16. Letter to BuAer, 6 November 1952, AJ/A2J folder, Naval Aviation History Division, NHHC. The new airplane would use existing tooling but would be modified to fit the centerline elevator on an *Oriskany*-class carrier. Tail length would be reduced by four feet, and the wing-fold line would be moved four feet inboard on each side. The radome would fold manually. Power folding of wing and tail would reduce deck handling time. For better

performance, the airplane would use a more powerful (6,350 pounds of thrust) version of the J33 jet engine (alternatively, a J48) with a new retractable inlet, adding thirty knots over the target. North American also offered to incorporate many fleet suggestions, projecting that if production were approved late in 1952, the AJ-3 could be in production during the middle of 1954.

17. Proposal dated 25 November 1953. At this time the last AJ-2s were to be delivered in December 1953; they could have T34s installed when they received their initial overhauls (after eighteen months of service) in 1955 and 1956. North American claimed a shorter takeoff, improved rate of climb, and higher speed above 40,000 feet. A graph showed a speed of about 350 knots at 40,000 feet, compared to about 320 for a standard AJ. A mission-profile diagram showed the airplane climbing to initial cruise altitude, 35,000 feet, with the jet lit and cruising out at 355 knots for maximum range (jet off), then descending to sea level for a 100-nautical-mile run into the target, at a total distance of 615 nautical miles from the carrier. After attacking the airplane would run back at low altitude, then climb to 40,000 feet (jet on) before turning off the jet to cruise back at 410 knots, reaching a maximum of 46,000 feet before descending to land, with 5 percent fuel plus enough in the tanks for thirty minutes' loiter. Another mission profile was a "hi-hi-hi" attack, the run in and out of the target at 40,000 feet and 450 knots. Total range was 1,005 nautical miles.

18. Hugh L. Hansen, Structures Branch, Design Elements Division, memo to bureau chief, "Suggestions for Navy Assistance in Improving the Strategic Usefulness of the Air Force," 7 November 1947, BuAer Confidential VV series, RG 72, NARA II. Hansen argued against the Air Force approach to greater range ("Make it bigger"); quite likely the P2V Neptune could be modified economically (and rapidly) to offer equal or better range with a ten-thousand-pound bomb, as well as higher speed and better maneuverability. It would be much less vulnerable in the air and it would cost far less. It could, moreover, operate from ordinary four-to-five-thousand-foot runways. Modified Neptunes could be provided within a year. Because the Neptune was so much smaller than the B-36, it might surprise an enemy expecting much larger aircraft. Hansen did not speculate about carrier operation.

19. The undated directive for modification referred to the airplane as a "special long range ASW P2V." Its nuclear function was revealed by the requirement that a special bomb-bay shackle for a ten-thousand-pound bomb be installed (there was also special provision for access to the bomb in flight, another giveaway). BuAer used a 13 August 1948 curve showing a takeoff at a deck speed of ninety knots with thirty-five knots of wind over the deck after a five-hundred-foot roll. The airplane would have the most powerful available version of the R-3350 engine (-26W). Total tankage would be 4,050 gallons, including nose and waist tanks. Liquid oxygen might be used to supercharge the engines chemically (using nitrous oxide). A hydroflap (on the underside, in case of ditching) would be fitted. No tailhook was specified. On 1 December 1948 Hayward asked for some changes: two more aircrew and an ECM receiver the visual recorder of which would be visible to the special ordnance officer. The crew would include a second copilot (navigator), an ordnanceman (to arm the bomb), and a plane captain (the bomber/navigator was eliminated as such). The ECM equipment would be that of an AD-1Q or TBM-3Q, served by AS-124 blister antennas. It would cover 300–600 MHz; later aircraft would be given an APR-9 instead of the current APR-1.

20. Cruising speed would be 159 knots at 1,500 feet, the altitude to be maintained until within about seventy-five nautical miles of the target. Speed over the target would be 310 knots at 12,900 feet, or about 360 knots for sixteen minutes at 30,000 feet using chemical supercharging. Combat radius would be 2,235 nautical miles carrying a ten-thousand-pound bomb, flying one hundred miles at normal rated power at 20,000 feet plus five minutes over the target. According to an undated summary sheet, the airplane would be capable of 308 knots at 17,000 feet or 278 knots at 21,000 feet. This sheet gave the same combat radius; memo for Rear Adm. T. C. Lonnquest (Assistant Chief of BuAer for R&D), 3 August 1948. This figure was

revised downward in a 28 March 1949 memo for the director of the Piloted Aircraft Division. It was based on preliminary drag estimates and also on fuel specifics for the -26W engine, which was then just coming into service. Flight tests showed that it required 15 percent more fuel, although the formal characteristics sheet showed only 5 percent more. On the latter basis, radius would be 2,020 nautical miles at 155 knots, but better drag data drew it down to 1,825. Further corrections might be made on the basis of actual experience by VC-5—e.g., fuel capacity was 4,160 gallons rather than 4,050.

21. Vice Admiral Hayward clearly thought they could land on board. Thomason said (*Strike from the Sea*, 222) that Hayward told him in 1985 that he had made one "touch and go" on the carrier *Franklin D. Roosevelt*. He believed that he could have made a complete landing without a hook using reverse thrust from his propellers. Hayward said that the attempt to carrier-qualify the Neptune was abandoned both because the carrier-capable AJ Savage would soon be available and because getting a P2V safely down on a carrier would have required above-average pilot skill. The file on this airplane includes a 22 June 1948 memo on a proposed program to develop P2V landing techniques that refers to a report of carrier takeoff trials dated 21 May, presumably referring to the proof-of-concept flight. The Design Elements Division of BuAer recommended an initial study of carrier landing runs using light weights and minimum speeds, with reversible-pitch propellers and brakes. Structural Branch would establish a project (with Lockheed) to review the strength envelope of the airplane as related to carrier landings. That would be followed by simulated carrier landings at NAMC or NATC (the Naval Air Test Center at Patuxent River). A parallel memo referred to examination of the feasibility of operating these aircraft from *Essex*-class carriers. A 30 July 1948 conference decided that the program should begin with development of new techniques. It would evaluate the arresting power of brakes and of reversible-pitch propellers; landing weight would be about 42,000 pounds. Some field tests would be conducted at lower and higher weights. According to a 3 January 1949 report, trials had shown that Neptunes could take off from CVB-type (*Midway*-class) carriers, be handled on deck, and hoisted onto the deck using a floating crane. The deck runs indicated that the airplanes could operate at higher gross weights than 42,000 pounds and that they had adequate directional control. Although the landing gear of a Neptune could not withstand typical American-style, stall-out carrier landings, it seemed that British techniques involving gentler landings could succeed. As for the flight deck itself, *Midway*-class carrier with Improvement Program 1 (strengthened flight deck and nuclear-weapon provision) could withstand the shock of a P2V landing. If field landings were successful, BuAer proposed landings on board a *Midway*-class carrier. A BuAer memo dated 11 February 1949 cautioned that extreme care had to be exercised in operating Neptunes at overload gross weight, meaning above 60,000 pounds; specialized piloting technique would be essential. At 74,000 pounds, permissible acceleration was only 2 *gs*, and maximum smooth airspeed was 240 knots (150 to 200 in rough air). Above 68,000 pounds, the margin of excess power available immediately after takeoff was small until the landing gear was retracted and the flaps partly retracted, whereas flaps had to be retracted carefully and gradually at a safe altitude. The airplane could land at a gross weight between 54,000 and 57,000 pounds, but landings above 54,000 pounds were to be avoided if possible. Landing under "other than the best possible conditions" would cause structural failure; the pilot should minimize his sink rate and should brake gradually. According to a 21 April 1949 confidential telegram, Lockheed had just been asked for a design study to modify the airplane for arrested (i.e., hook) landings at a gross weight of 40,000 pounds. A 20 April 1949 memo refers to Hayward's request that BuNo 122969 (the eleventh P2V-3C) have a hook installed during other planned modifications. A 14 March 1950 memo described success at VC-5 with the P2V-3C, mentioned extensive landings by P2V-3s using reverse propeller thrust, and recommended carrier landings. Carrier suitability tests (ashore) were ordered, to include no fewer than fifty arrested landings. Subject to satisfactory completion, a Neptune would land on board a

Midway-class carrier. However, on 8 June the test airplane suffered wrinkling of its fuselage and engine nacelles and sheared some rivets while engaging a wire. The project died.

22. Tommy H. Thomason has pointed out to the author that although the P2V-3C could lift the weight associated with an atomic bomb, it could not accommodate the diameter of the Mk 4 bomb; it would have been limited to the Mk 1 "Little Boy." According to Hansen, "Suggestions for Navy Assistance," only five such weapons were built between August 1945 and February 1950, the weapon being retired in January 1951. Not surprisingly, by 1950 the Navy was intensely interested in a smaller-diameter bomb, which appeared as Mk 5.

23. Some of the P2V-3Cs seem to have survived as lead aircraft for groups of AJs on extended overwater flights, for which they were fitted with Loran receivers (APN-4s), as indicated in a 6 December 1952 Pacific Fleet speedletter to BuAer. As they became surplus, at least five P2V-3Cs became -3Bs, for heavy attack training. The chief of BuAer requested conversion of the remaining nine -3Cs to this configuration in a 30 November 1953 memo, referring to a requirement established in a 2 October 1953 DCNO (Air) letter. Two had been modified in response to a 2 November 1951 BuAer letter; ASB-1 bombing equipment and a CP-66 computer had been installed.

24. At this time the Atomic Energy Commission had custody of nuclear components, which would have to be flown to bomber units whether in the United States or at sea. When the *Midway*-class carriers deployed to the Mediterranean, they carried the nonnuclear components; the nuclear ones would have been flown to Port Lyautey by transport aircraft and then to the carriers on AJ-1 Savages. The bombs would have been assembled on board before being loaded onto the Savages for their attacks.

25. This account of the A2J program is taken largely from a 17 December 1957 BuAer report for the Preparedness Investigating Subcommittee of the Senate Committee on Armed Services.

26. According to the caption of a drawing of the initial version in the A2J file, Naval Aviation History Division, NHHC, span was seventy-one feet. The power plant was two XT40-A-6 turboprops and one J33-A-12 turbojet in the tail. Performance assumed a twelve-thousand-pound bomb load. Service ceiling was 36,200 feet (operational ceiling 42,500 feet), maximum speed was 450 knots at 20,000 feet, combat speed was 439 knots at 35,000 feet, combat radius was 1,220 nautical miles at 364 knots (8.2 hours), and combat range was 2,380 nautical miles at 345 knots (7.10 hours).

27. Op-55R-2/jk (SC) A16-1, ser. 0202P55R, 5 January 1948, BuAer Confidential VV file 1948, RG 72, NARA II.

28. Dated 12 March 1948, BuAer Confidential VV file 1948, RG 72, NARA II, referring a 10 March 1948 CNO memo changing the requirements. DCNO (Air) wrote that "limiting weight can well become the governing feature in the design of bombing aircraft, and can have an important influence on the role that the Navy will play in any future war."

29. Memo from Assistant Chief for R&D, 28 July 1948, A3D folder, Naval Aviation History Division, NHHC. It referred to a directive of 26 March 1948 and to an earlier ADR report dated 21 May 1948, presumably giving details of sketch designs of jet aircraft under ADR-64. Unfortunately, the enclosure to this memo giving design details is no longer attached. Two airplanes were sketched, one powered by the Allison T40A (then planned for the A2J and A2D), one by the alternative Allison T44A (which was never built). DR-65A would have an overload gross weight of only 85,000 pounds, but its radius would be only 1,400 nautical miles. DR-65B would have an overload gross weight of 130,000 pounds and a combat radius of 1,820 nautical miles; it might be cut back to between 115,000 and 120,000 pounds, in which case combat radius would be 1,700 nautical miles. It proved impossible to balance an airplane with the same wing sweep-back as the jet design, so ADR adopted a much more conservative approach, reducing the sweep to thirty-five degrees and also lessening the load.

30. Ed Heinemann, "The Whale . . . A-3 Skywarrior: Initial Concept, Research, Design, and Development of the A-3 Skywarrior (A3D) Manufactured by Douglas Aircraft," A3D folder, Naval Aviation

History Division, NHHC. Heinemann recalled becoming aware of the heavy-bomber project while attending Skyraider armament trials on the light carrier *Saipan* in 1948.

31. A sketch of this airplane dated April 1948 appears in Rene Francillon and Edward Heinemann, *Douglas A-3 Skywarrior*, Aerograph 5 (Arlington, Tex.: Aerofax, 1987). This drawing is marked XVA(H1) design study but is dated well before the XVA(H1) competition was announced. According to Francillon, Heinemann visited BuAer between 17 and 27 March 1948, showing staff information on both an 80,000-pound twin turboprop (T56 engines) and a 70,000-pound twin jet. Heinemann further refined the jet design on his return to El Segundo, bringing it to 78,450 pounds. Francillon and Heinemann also shows an earlier design study based on the Skyraider, with two underslung (but not pylon-mounted) jet engines and straight wings. No date is given, and this sort of design had been sketched as early as the late fall of 1943, before the Skyraider design was finalized. Francillon and Heinemann shows Study 1A (28 May 1948), powered by four podded engines (two per pod, side by side), and Model 593-2 (November 1948), which retained the semirecessed bomb. The latter had a tail-down undercarriage.

32. Formal evaluation of the bids, 21 February 1949, A3D file, Naval Aviation History Division, NHHC.

33. Martin offered the same Model 245 as it proposed for the special-attack competition described below. It was a seaplane/land plane with a detachable seaplane hull; without the hull it could land on a carrier, but its wings could not fold. There was no attempt to meet requirements for crew size, access to bomb bay, protected fuel stowage, wing folding, or a defensive turret. Had these items been added, weight would have been well over the 100,000-pound limit. Lockheed's four-engine design (J40s) was considered overpowered, with only half the desired radius. Republic used two J40s and two J46s and exceeded the weight limit even on its optimistic estimates. Its entry was marginal in size and had an unsatisfactory power-plant installation, with a complex fuel system (ten fuselage tanks, thirty-two wing cells). Fairchild's canard

design had six J46s and a jettisonable biplane wing for takeoff. It was judged the heaviest, most unconventional, and most expensive of all.

34. As formally ordered by DCNO (Air) letter Op-55106/jk (SC)A4-3/VV, ser. 0105P551, 29 April 1949, A3D file, Naval Aviation History Division, NHHC.

35. Summary of recommendations, 24 March 1949, BuAer Confidential VA file 1949, RG 72, NARA II. It refers to CNO letter Op-57-fcp (SC) F42-1/VV, ser. 025P54, 3 February 1948, and to BuAer confidential letter Aer-AC-3, 23 August 1948. The coding of the OpNav letter indicates that it was a research-and-development (R&D) rather than an operational-aircraft proposal. It was called XVA(H2). Proposals were dated October and November 1948. Given later experience, the supersonic performance predictions were wildly optimistic.

36. The planned introduction was announced 23 May 1956. There was a crew of seven: pilot, copilot/navigator, gunner (to control the tail turret), and an evaluator supervising three ECM operators. Aircraft served in twelve-plane VQ squadrons (VQ-1 in the Pacific and -2 in the Mediterranean). The mock-up was inspected in September 1954, and the first five were ordered at the same time as the first five A3D-2Ps. Twenty more were built. According to Francillon, *Douglas A-3 Skywarrior*, 34, the five A3D-1Q conversions were ordered as interim signals-intelligence aircraft while the A3D-2Qs were being developed. They had oblong "blisters" and short canoe-shaped fairings to house antennas. One A3D-2Q was converted into an executive transport (VA-3B). Later five TA-3Bs were modified into executive transports, but they retained their original designations.

37. According to a 29 September 1966 memo "as a result of the increased demand for tanker configured A-3Bs" a modification program had been established at Alameda to convert forty-eight aircraft; A3D folder, Naval Aviation History Division, NHHC. The KA-3B designation was established on 7 April 1967, but "EKA-3B" had been in place since 16 February. The A3D folder includes the memo establishing the EKA-3B designation; it referred

to the TACOS (Tanker Aircraft/Countermeasures or Strike) mission. Apparently aircraft had already been modified but had not been specially designated.

Chapter 9. The Era of "Nuclear Plenty"

1. Asked whether he was risking nuclear war over a trivial incident, Eisenhower replied that common sense would prevail; no one would blow up the world over something minor. He certainly exercised restraint himself. When the fortress of Dien Bien Phu was under siege in 1954, the French pleaded for a U.S. nuclear strike to relieve Viet Minh (North Vietnamese) pressure. A carrier strike was prepared—and vetoed by the president.

2. This and other plutonium bombs had cores ("pits") surrounded by explosives; the implosive force of their detonation compressed the plutonium (and created the critical mass that produced the nuclear explosion). The more explosive, the more massive the bomb. The alternative (as in the "Little Boy" used against Hiroshima) was a gun-type bomb, which could only be made of scarcer uranium. According to Chuck Hansen, *U.S. Nuclear Weapons: The Secret History* (Arlington, Tex.: Aerofax, 1988), 21, "Fat Man" was about 17.5 times as efficient as "Little Boy" in its use of fissionable material. Hansen wrote that two to three times as much fissionable material is needed in a gun-type bomb as in an implosion bomb. The only advantages of a gun-type weapon seem to have been smaller diameter and strength (penetrating weapons, according to Hansen, have been gun-type). The initial Mk 4 production weapon, derived from "Fat Man," was of about the same size; an AJ could barely accommodate it. The A3D could carry the next, sixty-inch-diameter bomb, the Mk 6. Like Mk 4, it had to be hand-armed, a process that took fourteen minutes. It was 128 inches long. It weighed 8,500 pounds (a later lighter-weight version weighed 7,600). It had better ballistics. Different cores offered yields of thirty to sixty kilotons (kt); Hansen, 132–33. Work on the Mk 6 seems to have begun at about the same time that Mk 4 was being prepared for production. Mk 6 was the first U.S. bomb to be produced in quantity, about a thousand being made. Hansen's account was based on declassified documents.

3. The Air Force had opposed development of lighter bombs, on the ground that they made less-efficient use of scarce fissionable material. Its attitude changed in about 1950, probably because NATO was being transformed into a military alliance that would depend on tactical nuclear weapons to overcome massed Soviet troops. F-84G fighter-bombers carried the Mk 5 externally, and their pilots could arm it remotely. This bomb was not streamlined and was not carried by naval fighter-bombers. Heavy naval bombers carried it internally (according to Hansen, manual arming in an A3D flying at 24,000 feet took eight minutes). Also according to Hansen, 128–31, in the summer of 1946 scientists at Los Alamos realized that they could develop much more compact bombs. On 31 October 1947 the AEC Military Liaison Committee pressed for a vigorous small-bomb program specifically for guided missiles (Regulus had a Mk 5 warhead). In August 1948 the AEC formally notified the Air Force that it was ready to begin work on smaller weapons. A September 1948 conference, including Navy representatives, advised the AEC that reducing diameter to between forty and forty-eight inches and weight to five or six thousand pounds would improve the performance of aircraft delivering such bombs. As the Mk 5 design was being firmed up, the Navy expressed willingness to accept a fifty-one-inch, six-thousand-pound Mk 5 immediately, because it would greatly improve AJ performance. Naval officers at China Lake were pressing for fighter-bomber bombs as early as 1949. New design concepts made Mk 5 far more efficient (and powerful) than its predecessors, to the point that it could be cut down to 3,175 pounds (43¾-inch diameter, length 132 inches). Quantity production of Mk 5s began in May 1952; about 140 were made (including missile warheads).

4. Design work was authorized in May 1950. Maximum weight as specified by military characteristics approved in May 1951 was 2,000 pounds, maximum diameter thirty and a half inches. A prototype warhead was tested in the fall of 1951. As developed, Mk 7 was 183 inches long and weighed between 1,645 and 1,700 pounds, depending on the type of pit, which determined its

yield: either between one and ten kilotons at the low end or from sixty to seventy at the high end (Hansen, 133). In all, 470 weapons were made. The follow-on Mk 12 was another implosion bomb, smaller than Mk 7, designed to be carried at speeds as great as Mach 1.4. Diameter was reduced to twenty-two inches and length to 155 inches (weight was 1,100 to 1,200 pounds). Smaller diameter cost yield; in two 1952 tests of the nuclear part of Mk 12, yields were twelve to fourteen kilotons; Hansen, 143. About 250 were made before the weapon was retired—between July 1958 and July 1962, before Mk 7 was. According to Hansen, Mk 12 was designed because the Mk 7 was too large for the F-86 to carry. The Navy did *not* consider Mk 7 too large for the related FJ-4; Hansen, 133–34, 137.

5. Hansen, 139–41. Initially both the Air Force and the AEC resisted the idea. However, in April 1948 the latter's Military Liaison Committee recommended development. The project was called "Elsie" (LC, for Light Case); it had a much thinner and lighter casing than the "Little Boy" on which it was based. It was much smaller than Mk 7: 14½ inches in diameter and 116 inches in length (it could be fitted with an additional penetrator nose), with a 25¼-inch tail ring; Hansen, 141. Carried internally, it was armed by removing a nose plug to insert the propellant and the projectile part of the gun. In an A3D, that took eighteen minutes at 22,000 feet. The external version had the necessary components in an extended nose section. Mk 11 was a redesigned and somewhat heavier (3,500 pounds) version. The Navy's Mk 91 bomb used the same internal mechanism in a much more streamlined shape with more conventional tail fins. It entered service in 1956.

6. In the first test of such a device, in November 1952, an entire atoll was wiped out.

7. Military characteristics for Mk 27 were proposed in August 1954 and approved in August 1955. Experiments in 1954–55 proved that H-bombs could be miniaturized. A prototype was tested in the spring of 1956, and about seven hundred were made. Diameter was 30.27 inches, length eleven feet ten inches, and weight 3,150 pounds. This weapon was not streamlined, since it was intended to be carried internally. It was intended only for naval use. Mk 27 was carried internally by A3Ds and was intended for the linear bomb bays of Vigilantes; Hansen, 148–51. The follow-on Mk 28 was streamlined for external carriage. Development was requested by the Joint Chiefs at the same time as for Mk 27, on 5 February 1954. Bomb designs were released for production in June 1957. Yield was comparable to that of Mk 27, reportedly a maximum of 1.45 megatons. A typical externally carried Mk 28 weighed about 2,040 pounds and was fourteen feet two inches long. Diameter was twenty inches. About 4,500, including missile warheads, were made. Mk 28 was the oldest U.S. nuclear weapon still in the inventory at the end of the Cold War; Hansen, 151.

8. Thomason, *Strike from the Sea*, 41, reproduces a drawing from the January 1953 *Naval Aviation Confidential Bulletin* showing such an attack, using a bomb set to burst at three thousand feet. The article does not give the appropriate release height but emphasizes that it took time for the blast and radiation to move out from the burst; the airplane should get away before they reach it. Jerry Miller, *Nuclear Weapons and Aircraft Carriers: How the Bomb Saved Naval Aviation* (Washington, D.C.: Smithsonian, 2001), 131, describes a variation used by Banshees in 1952–53: a dive- or glide bombing approach with speed brakes extended, release at 12,000 feet, then a 135-degree turn away and retirement at high speed and low altitude, the speed presumably keeping the airplane clear of the blast and radiation. Miller, 158, argues that it was soon obvious that a release altitude of 20,000 feet or more was needed for survival. That precluded acceptable accuracy. China Lake developed the BOAR (BuOrd Aircraft Rocket—see chapter 7, note 23), which would carry a Mk 7 bomb clear of a Skyraider, allowing it to turn away at a higher altitude. Burning for three seconds, the rocket accelerated the 2,070-pound BOAR to about Mach 1. Conceived as an interim weapon with a service lifetime of three years, BOAR served from 1956 through 1963, having undergone service tests in 1955.

9. Description of tactics (and attribution) from a presentation before the National War College, 29 January 1954, by Rear Adm. J. S. Russell, Naval Aviation History Division, NHHC. Some details are from Miller, *Nuclear Weapons*, 160–68, including a description of LABS. Naval delivery tactics were developed by VX-5 in 1951–53. Whatever the pedigree of "toss bombing," Ryan applied in 1955 for a patent (claiming no royalties) for a "Low Altitude Bombing System and Apparatus" and was granted it in 1959 (U.S. Patent 2898809A).

10. This figure is from Francillon, *McDonnell Douglas Aircraft*, 2:81, based on bureau numbers of known converted aircraft. However, in an article in *Naval Aviation News*, NHHC Naval Aviation History Division historian Hal Andrews gave a figure of fifty. According to Francillon, modified aircraft were in the aircraft-number range of 378 to 433, a total of fifty-six aircraft. Presumably Andrews based his figure on the OpNav notice establishing the -2B designation, which would have included the planned number. A twenty-five-airplane figure has also been given. These aircraft entered service by 1950, deploying on board *Midway*-class carriers. According to the Banshee history in the Naval Aviation History Division, trials were conducted on board USS *Wasp* during the second week of April 1952 to determine minimum wind over deck required for catapulting. The file in Naval Aviation History does not include the usual OpNav Notice establishing the -2B designation, possibly because no mention of nuclear weapons capability was releasable at the time. An F2H-2B successfully drop-tested a Mk 8 shape in the spring of 1951, and another drop-tested a Mk 7 on 10 July 1952.

11. Details of F2H-2B modifications from Steve Ginter, *Early Banshees: The McDonnell F2H-1, F2H-2/2B/2N/2P*, Naval Fighters Series (Warfield Center, Calif.: Ginter, 2008), 8). Presumably the refueling probe was a later addition, as in 1951–52 the fleet did not yet have aerial tankers.

12. Ginter, 12, includes a 1951 photo of one at Kirtland Air Force Base, near Albuquerque, N. Mex., with a Mk 8 shape underwing. VX-5 was established on 18 June 1951 specifically to develop procedures for nuclear attack. VC-3 was established

2 May 1949 and became the Pacific Fleet's night-fighter and special- (atomic) weapons unit. It is not clear when it received its first F2H-2Bs. VC-4 was established on 28 September 1948 as the AirLant night-fighter squadron. Ginter, 93, shows two of its F2H-2Bs flying over Greece, 28 March 1953, while assigned to Carrier Air Wing (CVG) 6. According to Ginter, 93–94, VC-4 F2H-2Bs deployed on board USS *Coral Sea* (as part of CVG-19) from 9 September 1950 through 1 February 1951 and again (with CVG-4) from 19 April through 12 October 1952. Others deployed to the Mediterranean on USS *Midway* (1 December 1952–19 May 1953) and to Korea on USS *Lake Champlain* (26 April–4 December 1953). The only other operator of the -2B seems to have been VF-101, which deployed them to the Mediterranean on board USS *Midway*, 27 December 1954–14 July 1955.

13. Tommy H. Thomason, *Scooter! The Douglas A-4 Skyhawk Story* (Manchester, U.K.: Crecy, 2011), 14. The special-weapons capacity does not show up in the 1951 Standard Aircraft Characteristics chart, but Thomason points to a special modification involving a beefed-up pylon under the port wing. He shows a VX-5 F2H-3 carrying a Mk 7 weapon (or shape).

14. A "B version" of the F9F-2 Panther had four bomb racks under each wing, the inner one for a 1,000-pound bomb and the outer three for 250-pound bombs or five-inch high-velocity aircraft rockets (HVARs). Later Panthers without special designation carried heavier bomb loads. The extra bomb racks were added to all production aircraft from BuNo 125083 on. Once most Panthers had been fitted, the *B* was dropped.

15. Chance Vought presented its A2U-1 proposal in June 1952. It was a minimum-time conversion of the F7U-3 Cutlass. Two of the four 20-mm guns were eliminated to save weight, although space for them was retained. The airplane would have more fuel, for increased combat radius. The A2U-3 would have two J65-W or two J47-GE-25 engines instead of the original pair of J46-WE-2 engines. It would carry a 3,570-pound load: a three-hundred-gallon drop tank plus a nuclear bomb. Both the A2U-2 and -3 proposals showed the

"Delta tip," a folding wing extension, and new "ailevators" on the wing trailing edge outboard of the vertical tails. The Navy proposal file includes a V-388 offered in October 1953. It was an A2U-1 with a new wing section between the side of the fuselage and the root of the A2U-1 wing. By that time it was pointless, since the F9F-8B, FJ-4B, and A4D projects were already well under way. V-381 (A3U-1) was the subject of an unsolicited 30 September 1952 proposal. It is described (and its model pictured) in Buttler, *American Secret Projects 1945 to 1974*, 2010), 117. It had a wide-chord swept wing with the new tip extensions (to increase aspect ratio, hence range) and a rooster tail. The engine was a J57-P-5; radius of action was over five hundred nautical miles (in its initial presentation, Chance Vought claimed six hundred nautical miles on a gross weight of about 22,000 pounds); sea-level rate of climb was given as 14,400 feet per minute. BuAer considered V-381 markedly inferior to the A4D and did not pursue it: lateral control would be poor, and the doubled effective dihedral with tips up at low speed would be unacceptable. Nor did BuAer like the absence of a centerline stores station. Buttler quotes a span of thirty-six feet four inches (twenty-four feet with tips up; gross wing area 283 square feet with tips down, 250 with them up), a length of forty-five feet five inches, and a gross weight of 18,455 pounds. Engine thrust was 10,500 pounds. Maximum speed was 698 mph at sea level or 662 mph at 35,000 feet. Armament was four 20-mm cannon and at least 3,000 pounds of stores.

16. Confidential Correspondence 1952, VV file, RG 72, NARA II. According to Thomason, *Scooter*, 48, at least one Skyknight (F3D-2) night fighter was evaluated by VX-5 as a light nuclear bomber but had been dropped before the Cougar and Fury were selected.

17. Copy in Grumman history file; it gave BuNos of aircraft affected: 131063 through 131258, 134336 through 134386, 133831 through 133898, and 14030 through 141173, a total of 459 aircraft, far beyond what was actually done. Steve Ginter, *Navy and Marine Fleet Single-Seat F9F Cougar Squadrons*, Naval Fighters Series (Warfield Center, Calif.: Ginter, 2006), 69, gives squadron assignments.

They seem to have included VX-5 (nuclear weapon tactical development), VA-54, VA-56, VA-63, VA-76, VF-81, VA-86 (probably), VF/VA-94, VF-103, VF-106, VF-112, VF/VA-113, VF-144, VA-146, VF/VA-153, VA-156, VA-212 (briefly), and VA-192. Whatever their designation, many of these units operated a mix of fighters and bombers. In contrast to FJ-4B units, many of them operated in the Atlantic/Mediterranean.

18. The statement of planned introduction of the FJ-4B dated 12 June 1956 mentioned deployment in both the Atlantic and the Pacific Fleets, in fourteen-plane VA squadrons. According to the FJ file, Naval Aviation History Division, NHHC, in VA-23 the FJ-4B replaced F9F-8s. They replaced F9F-8Bs in VA-56, -126, -146, and -212; they replaced Skyraiders in VA-55; and A4D-2s in VA-216. In each case the FJ-4B was replaced by the Skyhawk. Three Marine attack squadrons also received FJ-4Bs: VMA-212, -214, and -223.

19. Grumman's Design 98S proposal was dated 15 June 1956. Compared to the F11F, the proposed airplane had a much larger wing and a somewhat modified cockpit; it was also somewhat longer. Grumman argued that the afterburning J65 engine offered vital performance growth over the 1958–62 period. The emphasis was on delivering high-yield weapons at ranges beyond 600 nautical miles. Combat radius would be up to 780 nautical miles with a 1,736-pound bomb, with a "cycle time" (launch to recovery) of three hours and a speed of 650 knots (Mach 0.99) at sea level. The airplane would escape at supersonic speed. Maximum load was over 6,200 pounds. Alternative engines were the J52-P-2 (best for radius and cycle time, lightest airplane), the J65-W-18 (best in availability), and the J79-GE-3 (best in speed and takeoff power). Avionics would include the APG-46 air-to-ground ranging radar, Aero 26C LABS gear, Aero 1A dive-bomb computer, and a rho-theta (calculating with reference to a beacon) dead-reckoning navigation computer for low-altitude approach. The EX-16 fire-control system of the F11F would be retained but modified for dive-bombing; Grumman History Archive. Thomason, *Scooter*, 80, sees Navy interest in the A2F as a means of forcing Douglas to take criticisms of

its A4D seriously. He views the FJ-4B in that light as well. However, BuAer files show that the FJ-4B was conceived well before the A4D flew. The chronology of entry into service obscures the earlier origin of the FJ-4B and probably explains why BuAer chose to buy the A4D on an accelerated basis.

20. According to Francillon, *McDonnell Douglas Aircraft*, 1:477, Heinemann's initial goal was a thrust-to-weight ratio of unity, compared to 0.36 for the Sabre and 0.39 for the Panther. Thomason, *Scooter*, 18, shows a sketch of the fighter, which would have been powered by two small turbojets and weighed eight thousand pounds.

21. According to Francillon, *McDonnell Douglas Aircraft*, 1:477–78, BuAer asked for a maximum speed of 500 mph, a combat radius of 345 miles, and a 2,000-pound payload, at a maximum gross weight of 30,000 pounds; two weeks later Heinemann was back with a design that more than met all the requirements on less than half the weight. Navy files do not reflect these requirements; presumably they were in Douglas files that Francillon saw. According to Thomason, *Scooter*, 19, the acting BuAer chief, Rear Admiral Soucek, told Heinemann to apply his weight-saving ideas to an attack bomber. Heinemann then offered to design a 12,000-pound airplane. The 30,000-pound figure seems to refer to a BuAer Aircraft Design Research study.

22. Assistant Chief of BuAer for R&D, memo, 16 July 1952, Confidential 1952 VA file, RG 72, NARA II, justifying a direct contract award to Douglas. Approval to proceed was granted by a personal note from Rear Admiral William H. Coombs, hand carried to a conference with Mr. John F. Floberg (Assistant Secretary of the Navy for Air) on 26 July; memo on A4D, recommended procurement procedure, 24 July 1952.

23. Thomason, *Scooter*, 39, explains that Heinemann adopted the short fuselage (with rooster tail) to save weight. Wind-tunnel tests showed that extending the fuselage aft about six feet offered both significant drag reduction and less possibility of "snaking" (yaw). The extended fuselage also allowed larger and better-positioned speed brakes and structurally more efficient support for the tail.

The vertical fin, horizontal tail, and the thrust line of the engine were all raised about three and a half feet. The rudder was enlarged. At about this time NASA published the "Area Rule": transonic shocks could be avoided by causing the total frontal area of an airplane to change more smoothly from forward to aft. Engine inlets were changed to suit. However, a proposed fuselage fillet (to smooth airflow) abaft the wing was abandoned as not worthwhile.

24. This was the folded width of current aircraft such as the FJ-4. The origin of this twenty-seven-foot six-inch figure is not clear. In mid-1949 a twenty-five-foot four-inch limit was set by the requirement that two airplanes be able to pass each other anywhere on an *Essex*-class hangar deck. It was increased to twenty-seven feet six inches in the early 1950s. Thomason, who tried and failed to find an official rationale, suggests that the key was a requirement that two aircraft fit side by side on an *Essex*-class centerline elevator. His drawing (*Scooter*, 23) shows that this was true for the A4D.

25. The F2H, for example, had an aspect ratio of 6.8:1, about twice that of the A4D, which was 2.9:1. For a rectangular wing, aspect ratio would be span divided by chord. In a delta, it is span divided by average chord.

26. Thomason, *Scooter*, 26, explains that pure deltas require greater span (larger wings) than conventional aircraft, because they cannot have flaps to increase their wing area when needed. Extending flaps in a delta wing causes the nose to pitch down, and there is no conventional tail to counteract that; however, extra area is needed to provide enough lift near the stall for a carrier approach. In the Skyhawk, the wing does have the requisite flaps; the conventional tail solves the pitch-down problem.

27. Thomason, *Scooter*, 31, describes some of the issues. Direct increases were 107 pounds for the all-moving tail and 144 pounds for the irreversible boost activator. Thomason considers the BuAer Airframe Design Division right to have disagreed with Heinemann.

28. Thomason, *Scooter*, 20–21. BuAer also accepted a lower turnover angle, a measure of the airplane's stability on its wheels.

29. Entry P7, RG 72 (formerly RG 402), NARA II. A Standard Aircraft Characteristics chart dated 12 September 1952 showed a span of 27.5 feet, a length of 38.4 feet (39.0 feet parallel to the ground, with wheels down), and a height of 15.1 feet. The main wheels were 8.1 feet apart. The Characteristics chart suggests that Heinemann was unaware of what weapons BuAer had in mind. He calculated performance for a 1,050-pound bomb, whereas the one that mattered weighed about 3,250 pounds. Bizarrely, he showed a torpedo as a possible load. He offered a fighter version, which could not possibly have competed with the supersonic aircraft BuAer was then buying. The engine was a J65-W-2 rated at 6,400 pounds thrust (7,220 pounds military rating). With the 1,050-pound bomb, combat range would be 870 nautical miles (gross weight 14,000 pounds).

30. Thomason, *Scooter*, 41, gives the four performance guarantees in the specification BuAer wrote: maximum speed at combat weight at sea level with military thrust, 595 knots; combat ceiling at combat weight with military thrust, 49,500 feet; stall speed with engine out, 106.1 knots; and deck run with twenty-five knots of wind over deck, 725 feet. Combat weight was 14,400 pounds, including a 1,050-pound weapon (without the bomb and with 40 percent fuel it was 11,454 pounds). Empty weight did not include armor around the pilot (sixty-eight pounds in front and eighty-seven below). These figures were later changed somewhat, so that in March 1957 the guaranteed maximum speed at sea level was 593 knots, guaranteed power-off stall speed 108.6 knots, and guaranteed takeoff run 780 feet.

31. The wings were fitted with two rows of "vortex generators," near the tips of the leading-edge slats and in front of the ailerons, to correct buffeting at high speed and high attitude. A "sugar scoop" fairing was added above the exhaust to smooth the airflow. A further rudder modification was introduced in the A4D-2 to cure rudder buzzing.

32. According to Thomason, *Scooter*, 68, 28 percent of the structure was changed and both elevators and ailerons now had irreversible actuators (in the -1 only the ailerons had them). A dead-reckoning computer was added, but it was never satisfactory; Skyhawk pilots navigated visually at low altitude.

33. APG-53 had a ground-mapping mode (search), an air-to-ground slant-range mode, and two terrain-clearance modes. The terrain-clearance modes could show the pilot either features projecting above a selected altitude or a profile of terrain directly ahead. To Op-55, the key improvement was that in bad weather an A4D-2N could operate effectively at a thousand feet. Neither mode could compare to the terrain-avoidance built into the A-6 Intruder; Thomason, *Scooter*, 97. The facing page is a reproduction of the radar modes shown in the pilot's handbook.

34. Thomason, 91.

35. Thomason, 113. This proposal seems not to have survived in the sets of BuAer proposals in NARA II. It is not clear from the record why the designation was not reused for the next variant, but it is possible that what became the A4D-5 was proposed alongside it. Thomason dates the A4D-5 proposal to February 1959.

36. A-4G was for the Royal Australian Navy; A-4H for Israel; A-4K for New Zealand; A-4KU for Kuwait. A-4N was an A-4M drastically modified for Israel. A-4PTM was a Grumman-modified A-4C or -4L for Malaysia.

37. Michael Grove and Jay Miller, *North American Rockwell A3J/A-5 Vigilante*, Aerofax Minigraph 9 (Arlington, Tex.: Aerofax, 1989). Inertial navigation was likely to be more accurate than the dead reckoning then planned for low-altitude flights and also better than following strip maps, which often had to be based on obsolete information. In an article in the Winter 1980 *Hook*, "Vigilante: Beauty, Beast, or Both?," Grove ascribes the inertial-guidance concept to Frank G. Compton, who was then chief designer for North American Aviation. At the time (1953) the Navy liked turboprops, so Compton began with an F-82 powered by the largest possible turboprops, with a rocket engine in the tail of one of the two fuselages and a linear bomb bay in the rear of the other. It was soon clear that the airplane had to be jet propelled.

38. Grove's and Miller's book includes a redrawn version of the patent drawing.

39. From a summary of Vigilante documents in the Vigilante file, Naval Aviation History Division NHHC. Unfortunately, this summary includes no data on the design, and neither the NAGPAW I nor

the NAGPAW II proposal is in the BuAer proposal file. On 16 November 1954 North American submitted a NAGPAW study that presented a concept for adding high-altitude penetration to the earlier low-altitude technique. Although the preliminary study dealt only with the linear bomb bay, it was soon extended to a full design study of NAGPAW II, including alternative versions with conventional bomb bay, escape capsule in place of ejection seats, and an alternate bombing-navigation system. An R&D Project Card (file) in Naval Aviation History Division, NHHC, describes CA-01501 (Revised) as an 8 April 1955 outline requirement for a high-performance nuclear-attack airplane capable of attacking unescorted without regard to weather; it is not clear, however, whether it was simply written around the North American proposal. CA-01501 (Revised) was cited as partial justification for the NAGPAW contract. On the basis of North American documents, Tommy Thomason gives data for NAGPAW I (span thirty-five feet, length 47.25 feet, takeoff weight 26,050 pounds, combat radius five hundred miles, maximum speed at 35,000 feet, Mach 0.777). Engines were two J46s (6,075 pounds thrust each). NAGPAW II as proposed had a wing area of seven hundred square feet (aspect ratio 4, span fifty-two feet eleven inches). Data courtesy Tommy H. Thomason.

40. Thomason, *Strike from the Sea*, 110.

41. According to the chronology in the Naval Aviation History Division file, NHHC, North American submitted its NAGPAW II paper on 16 November 1954, adding as virtues high-altitude attack capability and a very high ceiling. Groves' and Miller's date of January 1955, assuming it is correct, refers to OpNav's reaction to this new proposal. It would seem that North American initially offered higher altitude and only slightly later became interested in much higher speed.

42. This big (769 square feet) wing had nearly full-span blown flaps for good low-speed performance. Because conventional ailerons could twist a wing at high speed, North American used a combination of spoilers and deflectors. With both open, air could flow *through* the wing. The combination could add lift on one side and reduce it on the other, causing the airplane to rotate about its axis.

43. North American described the A3J weapon system in detail in its confidential brochure "A3J-1 Weapons System Capabilities," NA57H-480,16 September 1957, RG 343: Test and Evaluation series, NARA II. The brochure mentions a reconnaissance variant as one of several possible special versions. VERDAN was the first general-purpose digital computer in a U.S. combat airplane. It was incorporated in many other systems, such as the prototype CAINS (see below), early Ship Inertial Navigation Systems (SINS), and the A-6. For example, VERDAN and other Autonetics components were in the inertial system that navigated the submarine *Nautilus* to the North Pole in 1958. About 1,700 VERDAN computers were made. The supplemental radar was North American's NASARR (North American Search and Ranging Radar), which was also used in many F-104 light nuclear bombers.

44. The display was similar to that in the A4D-2N. It indicated whatever projected above a set altitude (in the A4D-2N that was fixed at a thousand feet). The display could show the radar picture or computer-generated attack data. North American offered a new Doppler system as an alternative (or supplement) to inertial; it too could measure movement over the sea.

45. Vigilante folder, Naval Aviation History Division, NHHC.

46. OpNav (Op-55), memo to Chief, BuAer, 4 November 1957, Vigilante folder, Naval Aviation History Division, NHHC.

47. Thomason, *Strike from the Sea*, 114–16. Thomason's account is clearly based on contemporary official reports. He found no evidence of the frequently claimed problem that ejected bombs sometimes kept formation with the bomber. There were few tests of the linear bomb bay. In the initial tests, both high loft (sixty-five-degree angle) and over-the-shoulder tests were made. Guaranteed CEPs were 1,000 to 1,500 feet. In loft shots, the airplane missed this standard five of six times. Performance in over-the-shoulder drops was the same. Misses were at twice the guaranteed CEP. In level attacks, the airplane missed six times out of eight. In high-altitude, radar, level attacks, it made two out of five, owing partly to poor

radar performance at 40,000 feet and above. Tests using a simulated Mk 43, the Mk 28's successor, were canceled in October 1962; by that time the Vigilante was being turned into a fast reconnaissance airplane. Performance was far better when bombs were carried externally; for example, in loft attacks the airplane made six hits in six tries.

48. Only the carrier *Enterprise* had the required inertial system to initialize the airplane. Operations on board *Enterprise* began on 26 January 1962, and the first full-squadron deployment, on the same ship, began on 2 August 1962 as part of NATO Exercise Riptide III. The squadron subsequently operated from the carrier with the Sixth Fleet.

49. McDonnell Report 6742, 27 April 1959, in the same RG 343 series as the A3J brochure, NARA II.

50. This report, dated 25 July 1958, is in the same box as the 1957 North American brochure and the report of the F-4 attack bomber, in the RG 343: Test and Evaluation series, NARA II. The maximum rating of the J58 was 30,600 pounds of thrust, compared to 16,500 for the J79-GE-8; sustained military rating was 21,600 pounds. North American's expertise in such aircraft was reflected in Air Force contracts for the Mach-3 XB-70 bomber (contract 23 December 1959) and for the Mach-3 F-108 fighter.

51. The cover sheet of the Temco brochure is dated 12 September 1958. The undated brochure is in entry P7, box 543, RG 402: "Proposal Files" for Airplanes, Helicopters, and Missiles 1934–1961, NARA II.

52. Corvus missile proposal file. Unfortunately, this brochure is undated. Required CEPs were 1,000 feet against a point target and 5,000 against an area target. When launched at Mach 1.2 (i.e., from a Vigilante) at 40,000 feet, Corvus II could reach 214 nautical miles (CEP 3,000 ft). Launched subsonically (Mach 0.8, meaning from an A3D or A4D) at 40,000 feet, range would be up to 170 nautical miles, and CEP would be 1,500 feet. CEP would be much smaller at shorter ranges: 500 feet at 30 nautical miles, 1,000 feet at 75. In its brochure Temco pointed to limitations imposed by navigational accuracy. Corvus was powered by a rocket engine powered by storable liquid fuel.

53. Vought's Report E8R-11625 dated 13 June 1958 in same series as the Corvus brochures. No BuAer evaluation is attached. The proposal reflected the new reality that with the advent of very-light-weight nuclear warheads, a large air-to-air missile could become an effective long-range air-to-surface weapon.

54. This was Project Rainbow. Thomason, *Scooter*, 151.

55. AGM-78A (in service February 1968) had a modified Shrike seeker operating only in S-band. A-2 had a phosphorous smoke marker. A-4 had an improved seeker to counter specific early-warning and ground-control-intercept radars. AGM-78B had a Maxson Electronics Corporation seeker with three frequency bands (S, C, and X); it entered service in 1969. B-2 had a unified seeker covering all bands. AGM-78C (1971) had a three-band Bendix Radio Division seeker and other improvements. C-2 had a more sensitive seeker, a modified fuze, and a better rocket motor. AGM-78D (December 1972) had a new General Dynamics seeker with wider coverage over multiple bands between low S (2 GHz) and X (10 GHz). D-2 was the final production version. AGM-68 was first used in combat on 6 March 1968; in all, 708 Standard antiradiation missiles (ARMs) were fired during the Vietnam War, far below the expected rate (in mid-1967) of 700 each month. The Navy fired 299, of which 225 were expended in 1972 as part of Linebacker. Production of the original AGM-78A ended in FY70; of AGM-78B in FY71; of AGM-78C (for the Air Force) in FY71–72; and of AGM 78D (developed FY70–71) in FY74. Deliveries of the final AGM-78D-2 version were completed in August 1978. Total production amounted to three thousand.

56. HARM originated in a 1969 Navy study, and Texas Instruments (now part of Raytheon) was selected on 24 May 1974. Performance objectives were broad frequency range, increased sensitivity, countermeasures resistance, and a smokeless motor (for low detectability). The missile was first used in combat during the 1986 raid on Libya. More than two thousand were fired during the Gulf War, and HARM was later used in Bosnia (Allied Force) and over Iraq (Desert Fox and Northern and Southern Watch). Like Standard ARM, HARM has a strapdown (without gimbals)

inertial guidance system. It can be fired in pre-briefed, self-protection, or target-of-opportunity (missile as sensor) modes. In the long-range pre-briefed mode, the missile is set to acquire the target well after launch, the target signature having been input into its seeker. The seeker is activated at a point from seven to ten miles from the target (before that the missile flies under inertial control) so that the missile can follow an efficient trajectory. Depending on programming, if the target radar is turned off the missile may either fly to the assigned coordinates or search for an alternative target. Typically (reportedly) the missile is programmed with target range, bearing, and frequency before being fired. Given a preset range, the missile seeker can be set to activate seven to ten nautical miles from the expected target position, to protect the missile from distraction by other emitters resembling the target. HARMs must be pointed to within fifteen degrees of the direction of the target (i.e., the airplane must be pointed within 7.5 degrees of the target bearing at the moment of firing). In 1974, when HARM was being developed, desired maximum range was given as a hundred nautical miles. In 1988 it was reported that maximum range was eighty nautical miles. Missiles are usually fired at between thirty-five and fifty-five nautical miles, to ensure that their seekers acquire the targets. The thousandth missile was delivered on 26 August 1986, and the five-thousandth in September 1988. In all, 1,961 HARM were fired during Desert Storm. It is not clear how many were fired during the Bosnian and Kosovo wars or in enforcement of the no-fly zones in Iraq. Another 408 were fired during the 2003 Gulf War. By about 2005, HARM was considered so reliable that the airplane carrying it could use its seeker for electronic intelligence, its guidance system detecting emissions from shipboard radars and on the basis their signatures identifying the ships. A follow-on Advanced Anti-Radiation Guided Missile (AARGM) is currently under development.

Chapter 10. The Return of Limited War

1. This group was created in 1954 by the CNO, Adm. Robert Carney. In effect it resumed the future study begun by the General Board in 1949–50; the Korean War had effectively suspended such long-range efforts. Burke had been secretary of the General Board during its future study. Remarkably, the Ad Hoc Committee mentioned "the inherent threat to Russia caused by a growing Chinese power on her Eastern flank," which would tend to reduce the Soviet manpower advantage in Europe. I am grateful to Dr. Alan David Rosenberg for a declassified copy of the December 1955 report. It was signed by Vice Adm. Thomas S. Combs, who was senior member, relieving Vice Adm. R. A. Ofstie. Members included Rear Adm. David L. McDonald, a future CNO. At about the same time the Royal Navy, led by the First Sea Lord, Adm. (later admiral of the fleet) Louis Mountbatten, used this logic to justify its carrier force, which would deal with what it called "warm war," as opposed to a "hot" global thermonuclear war or the Cold War of political pressure and subversion. There was a close working relationship between Mountbatten and Adm. Arleigh Burke. By the time Admiral Burke received the Ad Hoc report as CNO, he had created what he hoped would be its permanent successor, the Long Range Objectives Group (LROG, Op-93). I am grateful to Dr. Rosenberg also for a declassified copy of the initial June 1956 LROG report.

2. The report hedged: high-altitude capacity should be retained to force the enemy to invest in a wider range of defenses and also to provide commanders with greater flexibility to meet "special situations." In a sustained campaign, intermediate-level attacks might become possible once enemy air defenses had been reduced by antiradar missiles.

3. Eisenhower administration policy was to consider nuclear weapons no different from others, so they would be usable in the expected small wars.

4. In October 1956, according to Thomason, *Strike from the Sea,* 125. He quotes George Spangenberg, the longtime BuAer senior engineer, as claiming that the program was begun for the Marines and that OSD (the Office of the Secretary of Defense) would not approve it until the Navy added its own mission. However, the chronology of the long-range studies suggests the opposite, that the Navy did not become interested in the Marine close-support mission until after the carrier attack mission had been set. Unfortunately it is not clear from available sources when the Marines began to

demand a Skyraider replacement. Marine aviation as such is not mentioned in either the 1955 or the June 1956 long-range paper. CA-1504 mentions both missions.

5. The text of a 19 February 1957 copy of TS-149 is in the RG 72 proposal file, entry 1048, NARA II.

6. Raywinder would home on reflected X-band radiation from the attack radar. It materialized as Sidewinder 1C (which apparently was never carried by Intruders).

7. The search radar would have long-range (120 nautical miles) and short-range precision (30 nautical miles) modes, with aided tracking of selected aim points or targets. Effective resolution would be 0.8 degrees or better using monopulse technique, 1.4 degrees without. Accuracy of passive bearings would be 0.3 degrees or better using monopulse technique. Prominent fixed targets would be detected beyond 100 nautical miles (line of sight), with solid-ground mapping over average terrain at 30 nautical miles or more. The search radar would have an airborne moving-target indicator mode over at least a thirty-degree area at an aircraft speed of four hundred knots (sixty degrees at two hundred knots). It would detect a "six-by-six" truck moving at five knots at a distance of 5 nautical miles (at seven knots, 10 nautical miles, or at twenty knots, 15 nautical miles) and a twenty-five-knot train at 25 nautical miles. (Speeds were all the components of actual target speed along the line pointing at the airplane.) The search radar would be able to lock onto air targets (for self-defense) and to track enemy radar emissions. In addition to radars, there would be an optical sight. Bombing accuracy requirements were fifteen bombing mils for level bombing, ten "Navy mils" for glide or dive-bombing, ten Navy mils for toss bombing. (A "bombing mil" was bomb error measured on the ground divided by one one-thousandth of altitude. A "Navy mil" was an angle whose tangent was 0.001.) For special weapons they were a four-hundred-foot CEP for lofting and two hundred feet for laydown (i.e., dropped at low altitude, detonation delayed—perhaps by a drogue [see chapter 9]—to give the aircraft time to get clear). Rocketry and gunnery were to be within five mils.

8. The V-416A brochure showed a span of 36 feet 2.5 inches (18 feet 0.5 inches folded), a length of 49 feet 6 inches, and a height of 16 feet 4.8 inches. The engines were 9,800-pound-thrust J52-P-(J8B-1)s. Three of the weapon stations were clustered under the fuselage; there were two more under the outer wings. Vought claimed that maximum sea-level speed with all stores for the 300-nautical-mile mission would be 601 knots (Mach 0.91); on the long-range mission it would be 605 knots (Mach 0.915). Chance Vought had previously designed a V-406 turboprop attack airplane, which is described by Buttler, *American Secret Projects 1945–1974*, 122–23. It is not in the RG 72 proposal file. V-406A (9 May 1956) was powered by a single T56 and had two weapon stations under its wing roots and four more under its wings. The four inner ones could carry Bullpups. A search radar in a "bathtub" could be lowered from the rear fuselage. Two J85 boost engines could retract into the rear fuselage. V-406B (7 May 1956) used twin Dart turboprops plus a retractable J85 in each nacelle. The search radar was under its nose. Two heavy pylons were mounted between body and engines and four others (for Bullpups) under the wings. One wingtip carried a pod for a tracking radar, the other one for a radio direction finder. Light pylons for Sidewinders were mounted under the outer wings. V-406C (8 May 1956) used two T56s (no jets) and had its two heavy pylons under its engines, with Bullpup pylons under its body and under its wings, plus two Sidewinders farther outboard and the wingtip pods of the earlier design. V-406C span was fifty-one feet eight inches (area 330 square feet), length was forty-nine feet ten inches, and gross weight was 31,500 pounds. Combat radius was 330 nautical miles. Sea-level speed was 478 mph, and normal bomb load was 2,000 pounds. The other V-406 versions were similar; none approached the TS-149 requirement.

9. Proposal files, entry 1048, accession 61A2377, box 515, RG 72, NARA II. D-715 span was 46 feet (folded span 27.5 feet), length 54 feet, and height 16.4 feet. Basic catapulting design weight was 49,888 pounds. D-725 data from Buttler, *American Secret Projects 1945–1974*, 125–26.

10. There are no details in the BuAer proposals file. Buttler, 126–27, gives gross takeoff weight as 33,275 pounds; claimed sea-level rate of climb was 6,700 feet per minute, combat radius 330 nautical miles, and service ceiling 44,000 feet. Span including tip tanks was forty-six feet (wing area 340 square feet), and length was forty-five feet nine inches; maximum speed was 637 mph at the full gross weight.

11. Proposal files, entry 1048, accession 16786, box 507, RG 72, NARA II. Martin's brochure was titled "XVA-AW." For Model 346, span was sixty-five feet (thirty-five feet two inches folded), length was fifty-five feet, and height was fifteen feet eight inches. Maximum sea-level speed was given as 518 knots at a combat weight of 19,374 pounds. Martin claimed an average cruising speed on the short-range mission of 405 knots. For Model 345, maximum sea-level speed was 532 knots. Span was fifty-five feet (thirty-four feet one inch folded), length was fifty-five feet, and height was fifteen feet eleven inches.

12. Proposal files, entry 1048, accession 61A2377, box 544, RG 72, NARA II. This proposal was dated 19 August 1957. Span was thirty-seven feet (folded span twenty-four feet); length was fifty-four feet eleven inches, and height was sixteen feet three inches. For the short mission, takeoff weight was 27,400 pounds, 37,086 pounds for the long carrier mission. Maximum sea-level speed was 627 knots. Stall speed with approach power was 79 knots in the close-air-support mission and 83 knots in the long-range carrier mission.

13. Proposal files 1934–1961, accession 61A3012, box 567, RG 72, NARA II. The engines would have been Allison T56-A-7s. Span would have been forty-one feet 7.4 inches (twenty-seven feet 7.1 inches folded), length fifty-two feet ten inches, and height thirteen feet two inches (basic gross weight 26,317 pounds). Maximum sea-level speed would have been 518 knots. Catapulting weight (for the long-range mission) would have been 38,053 pounds. Boeing claimed it could meet the TS-149 combat-radius requirement. This design seems comparable in size and power to Vought's V-406C, but it offered much higher speed, greater carrying capacity, and longer range.

14. The airplane would have a gross weight of 23,000 pounds in the vertical-takeoff-and-landing (VTOL) configuration. It would have a radius of nearly 508 nautical miles and be able to take off in 1,400 feet over the fifty-foot obstacle. In overload condition (32,380 pounds) it would have a combat radius of 856 nautical miles, including a 200-nautical-mile run in and out at sea level. This overload needed catapult launch, but the airplane would land vertically. The design had straight shoulder wings. Span was 34 feet 10.4 inches (the wings did not fold), length was 49 feet 5 inches, and height was 14 feet 7 inches. Bell claimed a maximum speed of 550 knots at sea level.

15. At this stage, span was 51 feet, length 53 feet 2½ inches, height 16 feet 1 inch. For the short-range mission, takeoff weight was 32,509 pounds, but for the long-range mission it was 43,265; maximum catapulting weight (with a C-11 catapult) was 52,000 pounds. With two Mk 83 bombs, sea-level speed would be 576 knots. Takeoff distance with military power was 1,460 feet, which was within the requirement.

16. Thomason, *Strike from the Sea*, 125. Unfortunately, the usual BuAer summary comparing the alternatives has not been released. The engineering recommendation went out on 24 December 1957.

17. This account of Grumman reasoning is based largely on Lawrence M. Mead, "A2F-1 (A-6A) Development 1958–1963," a talk given to the AIAA Long Island Section, 15 May 1985, text courtesy of Grumman Aerospace. An early decision was to use two jet engines (rather than turboprops), for reliability and performance, and to place the two aircrew side by side. Mead showed slides of some rejected configurations. Design studies were numbered G-128A through G-128Q. G-128E and -128F were twin turboprops with straight wings and twin tails. G-128M and G-128N had a low wing, with engines in its roots. G-128M4 (16 March 1957) had podded engines on a "W-wing" (inner wings swept back, outer wings forward). G-128P was much the configuration adopted, except that its wings had "compound sweep" (inner wings swept back more sharply than the outer wings). Thomason, *Strike from the Sea*, points out that Grumman considered twin-engine reliability

mandatory for any long-range airplane, particularly one carrying such elaborate avionics. Turboprops offered greater cruising efficiency but would have been so far apart that the airplane would be difficult to control if one were stopped, and their efficiency came at an unacceptable cost with respect to the speed needed to deliver nuclear weapons. The airplane would not operate solely at sea level (where they were best), and fuel economy (which they offered) was not the key consideration The lightweight J85 was rejected because of a mismatch between the short-range mission (which required four) and the long-range one (six). That left the J52 and the J79; the latter was more powerful than was needed.

18. Aspect ratio was kept to 5 to avoid high-lift pitch up. That meant limiting span, which simplified wing folding and made it easier to keep all the store stations inboard of the fold. Shortly after the September 1958 mock-up meeting, Grumman engineers found an error in their cruise calculations (the wrong reference wing area had been used in processing wind-tunnel data). They immediately added two feet to the wingspan. That increased aspect ratio to 5.31. They also found space for another thousand pounds of fuel.

19. The computer initially used had a magnetic-drum memory, the drum revolving and the computer waiting for the appropriate data to come up in order to use it. As the software developed, computer memory had to be doubled, filling the space in the cockpit between the bomb/nav's legs. Computer and air-conditioning capacity were first increased in June 1959.

20. According to the Mead talk, on its first flight Grumman found that the system hardly worked. Grumman had not assembled the system in a ground lab before flying it. Contemporary digital warships encountered similar problems, the lesson being that a test site for the entire integrated system was essential. The first avionics test flight was in December 1960, and the first attempt to use the terrain-clearance mode and display came at the end of February 1961. There were serious reliability problems. Navy tests in November 1961 showed that the displays were not bright enough and had insufficient resolution; a change approved

in April 1962 replaced the five-inch displays with a seven-inch CRT for the bomb/nav and a direct-view storage tube for the pilot. This change delayed avionics Board of Inspection and Survey trials and the start of fleet training by nearly a year. According to Mead, this effort finally yielded CEPs of seven to eight mils.

21. This information was provided by using the antenna as a phase interferometer, exploiting the fact that returns from objects at different heights arrive at slightly different times.

22. Thomason, *Strike from the Sea*, 130, partly quoting the Board of Inspection and Survey report. Mean time between failure was 7.4 rather than the required fifty-five hours for the search radar, 17.2 rather than seventy-five for the track radar, 14.3 rather than twenty for the ballistic computer, 12.8 rather than 150 for navigation, and 333 rather than 610 for the optical sight. Only 60 percent of scheduled flights were launched during 1963 service acceptance trials.

23. According to Thomason, *Strike from the Sea*, 132, VA-75, the first squadron to deploy, lost two and perhaps three aircraft when electrical fuzes accidentally detonated bombs immediately after they were released. VA-85 lost ten of its twelve aircraft. The program was almost canceled in 1966, when a third squadron, VA-65, was preparing to deploy. Its aircraft were sent back to Grumman to have their systems fixed. In response to unexplained losses in low-level attacks, a new tactic was developed in which the airplane turned sixty degrees immediately after dropping at eight hundred feet, the bomb continuing about a mile before exploding. It took time for the fleet to appreciate that the A-6 should be used at night rather than as a daylight heavy bomber.

24. Mods 0 and 1 had their APQ-112 tracking radars and ASQ-61 ballistic computers removed; they retained their APQ-92 navigation radars. In the ten Mod 0 aircraft, the primary sensor was the APS-107 passive radar. They could be distinguished by small diamond-shaped antennas, two on the upper part of the radome and two on the lower parts of the air intakes. Mod 0 deployed with VA-75 late in 1967 with Shrikes. Mod 0 Update (1969) carried Standard ARMs, the sensor being upgraded to APS-107B (replaced by ALR-55/57 beginning in

1972). As modified, they had small bomb-damage-assessment antennas on the trailing tips of their horizontal stabilizers. Mod 1 (five aircraft) had Target Identification Acquisition System (TIAS, APS-118) sensors, with multiple "button" antennas spread irregularly over their noses; the base of the tail was also modified, with numerous antennas. The final version was passive angle tracking/ARM (PAT/ARM, fitted on three aircraft) using the APR-25 radar receiver, whose antennas were embedded in the existing tracking antennas and thus were not visible externally. This version retained full A-6A capability. Another fifty-four bureau numbers were assigned to planned new-build aircraft that were never built.

25. "TRAM" meant Target Recognition Attack Multi-Sensor. The turret carried a forward-looking infrared radar (FLIR) for target detection, a laser designator, and a laser-spot detector, the latter allowing a bomb/nav to see a target designated by another airplane. The first TRAM airplane flew in March 1974, but the first full-system airplane was not delivered until December 1978.

26. Francillon, *Grumman*, and Thomason, *Strike from the Sea*, disagree as to total A-6E production. According to Thomason, 137, about half the A-6As (240) were converted into A-6Es; another ninety-five A-6Es were built before TRAM was adopted, followed by seventy-one with TRAM, and forty-one with the Systems/Weapons Improvement Program (SWIP), with new wings; ultimately 188 A-6Es received the SWIP upgrade, but many were not rewinged. The Navy retired the airplane in 1993, well before it ran out of airframe life.

27. When the Army began buying its own small airplanes in the 1950s, BuAer was its design agent.

28. Gross weight was 14,500 pounds, span thirty-one feet five inches (twenty-two feet six inches folded), length thirty-two feet, and height twelve feet eleven inches.

29. Performance based on the current guaranteed engine rating of 2,570 equivalent shaft horsepower (eshp), which was expected to increase to 3,800.

30. Director of Production Division to Director of Aircraft Division, BuAer, 8 May 1959, in "New Light Attack" folder, NARA II, responding to a request for the cost of a stripped-down A2F. It includes a table of costs with V-433 data penciled in. The North American Lightweight Versatile Attack Aircraft (LVA) had not yet been proposed, but it seems to fit the same set of requirements. The file includes a 2 July 1959 memo from the director of the Production Division to the director of the Evaluation Division pointing out that the V-433 proposal included producibility features not normally found on high-performance aircraft. This was an extension of the May 1959 cost memo. An earlier paper on a stripped-down A2F in the same file is dated 22 April 1959. The Avionics Division commented that the V-433 might be unsuitable, because so much of its delicate avionics had to be podded underwing. The sheer amount of analysis done by different BuAer divisions suggests how seriously V-433 was taken.

31. This proposal is in entry P7, RG 72 (formerly RG 402), NARA II. Span was 39.58 feet (24.5 feet folded), length 37.02 feet, and height 14 feet. North American claimed that it would not only satisfy the Navy's requirement for large numbers of aircraft for conventional war missions but also supplement all-weather-attack aircraft in all-out nuclear war. It would be capable of high subsonic speed for quick turnaround. Its inertial navigation system eliminated any need for a sophisticated long-range target-acquisition radar. That left a requirement for a radar to confirm navigational checkpoints, aid safe landings in weather, and target ranging. The long-range mission was probably derived from that demanded of the VAX, since it entailed a three-hundred-mile run in and out at sea level; full range on internal fuel would be 940 nautical miles. This mission was to deliver a Mk 43 bomb, successor to the Mk 28 of the TS-149 mission. The short-range-support mission entailed delivery of six thousand rather than two thousand pounds of bombs. Maximum sea-level speed was 470 knots (500 at altitude). Avionics would include a Mk 10 bomb director (for LABS).

32. Douglas offered details of 774-D in a 7 January 1960 technical proposal. It was powered by a single ten-thousand-pound-thrust XTF30-P-2 engine. Span was 37 feet 4 inches (27.5 feet folded), length 45 feet, and height 15 feet 9 inches. In the close-support mission, with two thousand pounds

of bombs, 774-D would take off in 1,690 feet at 18,832 pounds. At a radius of three hundred nautical miles, with two 1,000-pound bombs, it could loiter over the target area for an hour (as in TS-149). Douglas envisaged a Doppler ground sensor for accurate aircraft-velocity measurement, feeding a digital computer that would provide aircraft position and true heading. The airplane would also have a Ka-band search and terrain-avoidance radar (also terrain position fixing). The airplane would be compatible with the Marines' TPQ-10 system. Douglas estimated that the entire avionics system would weigh about five hundred pounds, using system integration to hold down weight and volume.

33. V-433 was revived about March 1965. It was somewhere between the new A-7A (Corsair II) and the OV-10A. It seems to have attracted considerable Navy interest. Estimated flyaway price for five hundred aircraft was now $650,000, and if a contract were signed in July 1966 the first production airplane could be delivered in January 1969. CNO was interested, and data were provided to him in late October or early November 1966. However, the project died.

34. Richard Glenn Head, "Decision-Making on the A-7 Attack Aircraft Program" (PhD dissertation, Syracuse University, 1971), 171. Head had been an Air Force major at the time, and he was able to interview many of the participants. The thesis was approved on 18 December 1970. Head cites a 9 March 1961 memo from the Assistant Secretary of the Navy for Research and Development. The Navy went further, arguing that for future nonnuclear wars the Defense Department should buy existing Navy aircraft: Skyhawks for day close support and Intruders for all-weather close air support and interdiction. Phantoms would meet the air-superiority requirement and could and should be followed by (proposed but soon canceled) Missileers. It seems unlikely that anyone in the Navy expected the Air Force to accept this argument, but soon the Air Force did find itself buying Phantoms instead of F-105s.

35. Head, 173, quoting Cdr. Robert F. Doss, USN.

36. Letter Op-506/pep ser. 06047P50 of 18 April 1961, Bureau of Weapons (BuWeps) proposals file, folder: New VA(X) Light April 1961, entry P7, RG 72 (formerly RG 402), NARA II.

37. Head, "Decision-Making," 181, quoting George A. Spangenberg, who was then Director Evaluation Division, NavAir.

38. The Boeing proposal was often described as a half-size version of its failed TFX offering, to the extent that it was sometimes called the "F-55.5."

39. The Boeing Wichita VAX studies slides are dated 24 June 1961.

40. Boeing listed current proposals in a 26 March 1963 sheet in the BuWeps proposal file. Model 837-665 was a twin-engine airplane with a scalable turbofan with afterburners. It met the current VAX development characteristics. 837-657A had a variable-sweep wing and a single TF30 with afterburner. It satisfied the development characteristics except that it had one rather than two engines. 837-678 was a subsonic airplane with fixed wings; it met the characteristics except for supersonic capability and the number of engines. 837-679A had a single TF30 without afterburner. It met the characteristics except for number of engines and supersonic speed. 837-680 was a twin TF30 airplane with afterburner; it was supersonic and had variable-sweep wings. It met the characteristics and offered growth to vertical/short takeoff and landing (VSTOL). The contemporary Boeing brochure showed VSTOL VAX configurations that could be operational in 1973, plus a simpler STOL (short takeoff and landing) design with variable-sweep wings. The VSTOL alternatives were a vectored-thrust airplane that looked like a big Harrier, an airplane with wing lift fans, and one with a combination of vectored-thrust and lift engines, which would lift something heavier. The rating of the envisaged vectored-thrust engine was 40,000 pounds, which was far beyond what Rolls-Royce would achieve with its Pegasus engine. The combination airplane used a 16,200-pound-thrust vectored-thrust engine (which was far more realistic) and four lift engines (6,200 pounds of thrust each). Boeing proposed further technology studies, such as improvements for short takeoff (vectored thrust, lift engines, boundary-layer control, and advanced mechanical flaps), vertical takeoff, passive

defense (reduced radar cross section, IR suppression), and improved cockpit displays. Examples of its research included a stealthy Model 853 with a low wings, a V-tail, and a chine at its bottom. Although its radar cross section peaked at a ninety-degree aspect angle, it was substantially smaller than for a conventional airplane.

41. This "Parametric Variable-Sweep Wing Design Study" was performed under BuWeps Contract NoW 60-0443-c, the number suggesting that it was awarded under the FY60 program.

42. This paper is undated. The subsonic airplane would have a span of 42.7/59.8 feet (swept/extended), a length of 59.1 feet (63.6 feet overall), and a height of 16.9/18.2 feet (folded/unfolded), and it would have seven store stations and a total capacity of 13,500 pounds. The supersonic airplane would have a span of 37.3/60.9 feet, a length of 59.7 feet (70 overall), and a height of 16.9/22.7 feet. Its six external store stations would give it a total load of 12,400 pounds of ordnance. In its bomb bay it could carry one Mk 28 thermonuclear bomb or three GAR-9 air-to-air missiles (an Air Force, not Navy, weapon).

43. Requirements quoted in Boeing 837 VAX brochure, 28 September 1962. The file in NARA II includes a 4 April 1962 brochure for earlier versions of Model 837. This brochure shows a subsonic fixed-wing Model 837-430. Another brochure, dated 9 March 1962, compared the fixed-wing version (in this case, 847-434) with two variable-sweep airplanes, one subsonic and the other supersonic. A 29 March report described a delta-wing 837-333. Light attack studies are in "Light Attack Studies 1961, -62, -63," box 3, entry UD-WW 441, RG 343, NARA II.

44. Grumman memo, 9 April 1962, referring to the 9 January 1962 development characteristics as modified that February. Grumman had been sketching subsonic aircraft, on the reasonable theory that the program would go in that direction (Grumman had the original Development Characteristic, dated May 1961). Grumman's memo writer considered the demand for two engines the key change since 1961, since this did more than force an engine arrangement: it required an engine in a nonexistent size category. Since General Electric

was still designing the MF 295 turbofan, however, Grumman could scale the engine as required (basic design thrust was ten thousand pounds).

45. According to a Grumman memo dated 30 August 1962 (in the Navy VAX file), the company received a revised Development Characteristic dated 22 May 1962 (an earlier Development Characteristic had been received on 5 January). The company's engineers had overrun the 30,000-pound target by about 6,000 pounds, and they expected to miss the new goal by a similar amount. In the current G-303-25, with fixed wings, they had needed two 300-gallon and one 230-gallon drop tank, but now they had to increase internal tankage, which would add over a thousand pounds to the empty weight of a new airplane. The earlier strength criterion of 6.0 gs at full internal fuel had been based on the May 1961 Development Characteristic. Grumman estimated that the change would cost a thousand pounds.

46. Fighters of this era could fly at supersonic speed for no more than a few minutes, if that. At Mach 2, which might be about 1,300 mph at altitude, five minutes equaled about 108 miles (about ninety-four nautical miles). Boeing's VAX brochure included a graph of supersonic-dash radius, varying from fifty to about 240 nautical miles depending on the total radius demanded, given the high fuel consumption involved. Supersonic flight heated aircraft structure and could drastically weaken the usual lightweight materials, such as aluminum alloys. The main issue was how to stretch the operating envelope of the airplane so that it could fly supersonically at altitude and nearly supersonically at low altitude. That led most companies toward variable geometry—"swing wings"—which NASA was then promoting. The A-5 proposal file includes a set of undated slides (presumably from North American, marked "ESA 4214") describing a NASA project to use a Vigilante for swing-wing tests. NASA claimed that on internal fuel the resulting airplane would have a supersonic radius of 1,000 nautical miles at Mach 2.5, compared to 260 nautical miles for a clean A3J-1. A drawing showed that it would have a maximum sweep of sixty to seventy-two degrees and a minimum of twenty-five or thirty, depending

on how the wing was arranged. The swing-wing offered improved carrier suitability for high-speed aircraft. It might minimize trim drag and reduce supersonic interference.

47. As described in Douglas Report 43613A, a brochure revised 2 October 1962, in the BuWeps proposals file, NARA II. The earlier version is not in this file. The single engine was a JTF10A-20 (18,500-pound thrust); the twin engines would be Pratt & Whitney STF 182As (scaled to 0.595) producing 11,600 pounds of thrust each with afterburners lit. There were also the fixed-wing, single-engine D-819A and twin-engine D-825A. D-819A dimensions were span thirty-five feet six inches (folded twenty-six feet two inches), length forty-three feet four inches, and height fifteen feet. This airplane had a maximum weapons load of 9,000 pounds, compared to the 12,000 desired for VAX. The engine was a single JTF 10-A-8. Spotting factor was 1.1–1.2. The Douglas brochure did not show dimensions of the swing-wing airplanes.

48. V-456 was called the "Attack Crusader." The June 1962 brochure was addressed to Rear Adm. Frederick Ashworth, now assistant chief of R&D at BuWeps (a post he held between February 1961 and May 1963); a hand-written Vought memo attached to the brochure is dated 16 July 1962.

49. The Grumman brochure compared its VAX to the requirement, to a modified A4D-5, and to a modified F4H Phantom. VAX demanded seven store stations with a total capacity of 15,500 pounds; the A2F-VAX would normally have seven (22,000 pounds) but could have nine (24,000 pounds). This dwarfed the 7,000 pounds of the A4D-5 and the 13,000 of the Phantom. However, the Intruder far exceeded the required empty-weight limit of 13,000 pounds. On its much greater weight the modified A2F carried 16,020 pounds of internal fuel, compared to the 6,800 envisaged for VAX, and it could also carry more external fuel. Grumman also pointed to the possibility that Intruders carrying four long-range Eagle air-to-air missiles could work with Missileer fighters, whose radars could be used to guide these missiles as well as the Missileers' own. Instead of the usual six Eagles, each Missileer could carry two and guide four more from an A2F.

50. This was in a folder of "New Light Attack: 1961 Cost, Etc" that was also marked "Interim VAL." The folder included a 6 December 1960 brochure for what North American then called an "FJ-4BF." It also includes a 29 November 1960 set of BuWeps cost estimates for FJ-4BF production that describes North American projections as very realistic. In cost, it was considered competitive with the A4D-5 as then specified. This airplane differed from the later one, having been modified to use two CF-700-2B turbofan engines (4,300 pounds of thrust each) and an additional store station (seven total), plus terrain-clearance and search radar (APG-53A). The same wing was used as in the later proposal but with increased span. For the two engines, the airplane was given intakes in the wing roots. The new attack radar was in the nose. According to a 6 February 1961 memo, the initial North American proposal was hand-carried into the bureau about 19 December 1960. The assistant chief of BuWeps for program management wrote that his chief appreciated the idea and the effort and that the improved FJ-4 (which BuWeps called the FJ-4BF) would certainly "provide many attractive and desirable performance features not previously available in light attack aircraft" but not enough to justify changing the existing program. North American later offered a less heavily modified FJ-4, in a brochure dated 7 March 1962. North American called it "an interim light attack aircraft for limited war" with an eight-thousand-pound payload. It would be a low-risk program, offering fleet delivery in twenty-eight months.

51. Thomason, *Scooter*, 159, shows a sketch of CL-706-15-4, a twin-engine, swing-wing airplane, with the engines above the wings in pods attached to the fuselage. Span would have been forty feet. All versions of CL-706 were less than fifty feet long. There were at least fifty-four studies and sub-studies. CL-706-13 had twin jet engines plus six J85 lift engines in two wingtip pods. It had four main weapon pylons under its wings between pod and landing gear, plus two tandem centerline triple ejector racks. Span was twenty-seven feet six inches and length fifty feet. CL-706-15-4 was a conventional twin-engine design, with a forty-foot wingspan and forty-nine-foot length. The

variable-sweep wings could fold into a span of twenty-five feet eight inches. CL-706-16 had a single TF30-P-2 engine (without afterburner) fed by an intake abaft the canopy. It had straight wings (span forty-one feet 2.7 inches) and was forty-five feet nine inches long. CL-706-23 had two podded "advanced turbofan" engines with afterburners (10,600-pound thrust each) near its fuselage; span was forty-five feet and length forty-three feet ten inches. CL-706-49 had two podded Lycoming T-55 turbofan engines (4,650-pound thrust each) carried above its fuselage, as in an A-10, plus paired (side-by-side) Continental 365-5 lift engines (5,220 pounds of thrust at sea level) forward of and abaft its tapered straight wing (span thirty feet); length was forty-one feet seven inches. The sketch shows GE 7.62-mm miniguns in the wing roots, and the canopy appears to have been designed for side-by-side seating. CL-706-54 showed a centerline TF30-P-1 engine fed by an intake above and abaft the cockpit plus underwing pods each containing four J85 lift engines. The wing could swing back, outboard of the pods. Each of four underwing multiple ejector racks could carry ten 250-pound Mk 81 bombs, and two more ejector racks were on the centerline. Span (fully extended) was thirty-five feet four inches, and length was forty-six feet five inches. No performance data were given. Data courtesy of Tommy S. Thomason.

52. According to Head, "Decision-Making," 182–83, OSD System Analysis thought that the Navy could replace the A-4 with the F-4, since the latter could also carry bombs. They did not understand that a fully bombed-up F-4 could not always be catapulted.

53. Head, 184. The main System Analysis representatives were Russell Murray (later Secretary of the Navy) and Dr. Dieter Schwebs, who had been instrumental in killing the Air Force A-4. Murray was an aeronautical engineer who had worked at Grumman, ultimately for eight years as assistant chief of its Operations Analysis Group.

54. Head, 185–86, based partly on contemporary published accounts. The study has not been released, but it is possible to understand how the system worked.

55. Its references to all possible engines, including turboprops, suggest that the study included the

V-433 and probably the turboprops Douglas had suggested in the run-up to VAX. Probably it also included North American's LVA. Presumably the existing A4D-2N and the Intruder were also taken into account. Boeing's 29 April 1963 brochure on VAX studies included some STOL and VSTOL aircraft, which may have been considered a separate category. VAX studies were probably used to measure the effect of demanding supersonic speed. These aircraft were intended to reach such speeds only at high altitude, but some accounts of the lessons of the study were written as though the question were the effect of a low-level supersonic dash.

56. Head, "Decision-Making," 190, quotes Murray that pound for pound a supersonic airplane would cost about twice as much as a subsonic one, not counting avionics, and it would probably also be larger. An estimate of survivability at Mach 0.9 and Mach 1.2 showed little difference. The supersonic airplane might even do worse, because it would have to fly higher. Note, however, that the Navy had never been particularly interested in an airplane capable of sustained low-level flight beyond Mach 0.9.

57. Because the supersonic airplane would have to fly higher to avoid terrain, it and a subsonic airplane would be visible to the enemy for about the same length of time. The supersonic airplane would cost about three or more times as much, so if the enemy shot down both types at the same rate, more subsonic airplanes would survive to hit their targets for the same price.

58. Span increased to thirty feet, length to 46.5, and height to 15.8. Dimensions from a "Standard Aircraft Characteristics" chart dated 1 August 1962 enclosed in the Douglas brochure; it was posted on the Internet by Tommy Thomason. The "A-4F" designation was later used for a single-seat version of the TA-4F trainer.

59. Head, "Decision-Making," 194. A synopsis of the request for proposals was issued on 24 May 1963 and a formal RFP on 29 June.

60. The 9 August 1963 summary report is in entry UD-WW 441, accession 71A-3900, box 2, RG 343, NARA II, a set of NavAir Evaluation reports that includes material on the RA-5C. This box also includes parts of the Vought A-7C proposal. To

maintain fore-and-aft balance without the afterburner, the forward fuselage had to be made substantially shorter than that of the F-8 fighter. According to Head, 200, the airframe structure was only 13 percent common by weight and major tooling only 7.6 percent. None of the wing was identical to that of the F-8E, but 96 percent of wing parts were similar. According to Head, Navy decision makers had already acknowledged that there would be little real commonality between the proposed aircraft and earlier production types. However, the issue was brought up in congressional hearings in January 1964 and by some individuals in OSD Systems Analysis.

61. Head, 205, quoting an interview with George Spangenberg, who supervised the BuWeps evaluation. LTV quoted a price of about $1 million per airplane, compared to $800,000 for the Douglas proposal.

62. According to Erwin J. Bulban, "A-7A Contract Defines Scale of Penalties," *Aviation Week*, 15 June 1964, 112, one V-463 was equal in effectiveness to two A-4Es at short range or to four of them at long range. This difference was exactly the dramatic improvement OSD wanted.

63. Congress had two weeks to demand hearings into this reprogramming. The relevant committees were already unhappy with massive reprogramming over the previous two years. The hearings were held in January 1964.

64. Head, "Decision-Making," 381–82, interview with Commander Doss. Doss pointed out that the analog computer alone cost half as much as the Sparrow missiles Navy fighters fired. Surely an attack bomber needed something better.

65. Avionics description from Thomason, *Scooter*, 162; Board of Inspection and Survey data on page 164. Oddly, the quoted CEP was 150 to 250 feet for the A-4E versus 150 feet for the A-7A.

66. Without catapults to boost its underpowered A-7, the Air Force wanted more powerful propulsion. Its Aeronautical Systems Division recommended against an afterburner and in favor of a more powerful engine, if a suitable one could be found. The only candidate seems to have been the Spey. The Air Force had bought A-7s to head off an Army attempt, citing a lack of Air Force interest in close air support, to develop its own. In November 1961 it was rumored that the Army would receive as many as eleven squadrons of A-4s. The Air Force bought its A-7s through a joint project office it shared with the Navy.

67. Head, "Decision-Making," 324, describes the three alternatives the Air Force considered.

68. Head, 380.

69. According to the pilot's manual, there were ten modes: beacon, terrain avoidance (TA), ground mapping with pencil beam (GMP), ground mapping with a shaped beam (GMS), air-to-ground ranging (AGR), combined terrain avoidance and terrain following (CSTA), combined terrain following and ground mapping (CSGMP), terrain following (TF), Shrike improved display (SIDS), and television (TV), the latter for use with the Walleye missile. There was also a cross-scan pencil mode, in which the radar covered a narrower (twenty degrees rather than forty) area ahead of the airplane. One source describes APQ-126 as an improved version of the APQ-116 with an added maritime mode. Reported maximum ranges were: 150 nautical miles for high-altitude maritime search, 106 from low altitude, fifty-one against a small contact, and twenty for air-to-air search.

70. Thomason, *Strike from the Sea*, 153–54, credits the A-7A/B with a CEP of 100 to 150 feet in training and twice that in combat. A-7D/E achieved 50 feet in training and, anecdotally, 100 in combat. Moreover, many fewer training flights were needed. However, a five-hundred-pound bomb had to be delivered within 25 feet to destroy a truck, so these figures were still not good enough. Claimed improvements were an order of magnitude in navigational accuracy and a threefold improvement in weapon-delivery accuracy.

71. According to page 5 of the May 1967 Technical Development Plan (TDP), early efforts concentrated on a low-PRF (pulse repetition frequency) Doppler radar combined with an integrated avionics system emphasizing air-to-air and air-to-ground combat equally. This must have been an unhappy effort, because low-PRF Doppler is nearly a contradiction in terms (the lower the PRF, the lower the speed at which the radar is "blind"). As

a result of CNO emphasis on air-to-air (SOR 11-06R1 of 24 April 1967) and delays in obtaining a go-ahead for VFAX, the ongoing advanced development program was reviewed; it became clear that a high/low PRF system could be developed in time for VFAX. It would provide the desired performance against enemy aircraft at all altitudes of both target and interceptor and at both high and low closing velocities. The existing ILAAS system was studied because it was an integrated approach satisfying much of the VFAX attack requirement. Also studied was the possibility of using a data bus (coaxial cable) instead of the usual point-to-point, hardwired bundles. Hughes received a contract to modify an APQ-72 fighter radar for multimode operation. Major radar manufacturers were asked how they would design a truly multimode radar. A "vu-graph" (text on a sheet of transparent plastic about ten inches square, magnified and displayed on a wall screen by an "overhead projector") describing the TDP summarized required air-to-ground capabilities: terrain avoidance; automatic terrain following (two hundred to five hundred feet); "squinted" high-resolution ground mapping (resolution sixty-five to ninety-five feet at fifty miles [ten-mile swath], thirty-five feet at two to ten miles [2.5-mile swath]); beam-sharpened forward ground mapping (seven mils); air-to-ground ranging (0.8 percent of slant range for twenty degrees grazing angle); air-to-ground tracking of prominent targets (0.75 mils for a ship in calm water, two to ten mils for a target ten decibels above clutter [e.g., a bridge, oil/petroleum facility, docking area, building, ship, possibly a SAM site]); ground moving-target detection and ranging (radial velocity 5 mph for targets within fifteen degrees of ground track, fourteen for targets fifteen to thirty degrees off).

72. In the strike model, alternative air groups were twelve F-111Bs, twenty-four VFAXes, and thirty-four A-7s; eighty-one VFAXes; thirty-five VFAXes, and forty-six A-7s; and twelve F-111Bs and fifty-eight A-7s (currently planned). The number of strike sorties (or tons of ordnance delivered) was plotted against initial numbers of enemy attack aircraft the carriers would have to defeat. Numbers of strike sorties fell as the number of enemy attackers rose. The worst air group was the current one; the best was a mix of F-111Bs, VFAXes, and A-7s. The A-7s had Sidewinder missiles. The amphibious assault faced ten fighter-bomber regiments (thirty-seven aircraft each), one to an airfield, plus a reserve of fifty bombers in either a neighboring country (Russia or China) or in a sanctuary (e.g., Hanoi). Thirty percent of the enemy fighter-bombers had Sparrow-equivalent missiles and fire-control systems capable of detecting low fliers. All the fighter-bombers were assumed to be within range of the carrier force once it was two hundred nautical miles offshore. All aircraft mixes included three F-111Bs. The other aircraft were either twelve F4Js and thirty-nine A7Bs, twenty-four VFAXes and thirty-four A7Xes, or fifty-eight A7Xes and no VFAXes (A7X was armed with Sidewinder). Phase I of the amphibious operation was the fight to gain air superiority by killing at least 85 percent of the enemy aircraft. Once that was done, the Marines could land (Phase II), protected by three combat air patrol (CAP) stations over the beachhead. Phase III was close air support (70 percent) and interdiction (30 percent). The presence of bombers in sanctuaries required that F-111Bs remain on CAP stations and on "deck alert." VFAX and F-4 aircraft loaded with Sparrows and air-to-surface weapons would accompany A-7X bombers. Not surprisingly, an "all A-7B" force was rated unsatisfactory in CAP performance over the beachhead.

73. A full copy of the confidential part of the TDP survives in VFAX files in RG 343, NARA II, with an 18 March 1966 endorsement by Deputy Chief of Naval Materiel (Development). Unfortunately there is no copy of the 1967 or later TDP. A separate secret discussion of the threat is not included in the document.

74. Op-506, confidential memo, 4 November 1966, no. 0301-506, VFAX correspondence folder, proposal file, RG 343, NARA II. Given the Op-501 comments, it seems odd that the file includes a report of the FAX (i.e., VFAX-FX) Joint Study Review Team, established by a secretary of defense memo dated 3 May 1966 (this report is dated 14 June 1966). A comparison between Air Force and Navy fighter specifications was dated 15 March

1966. This file includes an undated table of data describing FX version 7, as well as a summary of details of Air Force WS-328A, which was its FX.

75. The VFAX correspondence folder includes vu-graphs titled "VFAX Candidate Aircraft for Scope," which was presumably NavAir's estimate of what was needed. It is dated, in pencil, 28 December 1966, but it gives details of the "specific operational requirement" (SOR), which was probably issued early in 1967. They include a hundred-knot approach speed for landing, zero wind for catapulting and arresting; STOL takeoff and landing (takeoff over a fifty-foot obstacle in 2,500 feet on a ninety-degree Fahrenheit day, landing over five feet with a two-thousand-pound bomb in the same distance); and the maneuverability requirements DCNO (Air) disliked. In these studies, the assumed engine was a GE 1/10F1B with J79-type nozzle, producing 8,685 pounds of military thrust as specified, and 15,330 pounds maximum thrust with afterburner (in each case the engine would lose 7 to 10 percent of its thrust as installed). As a measure of what sort of computers then existed, these sketch designs incorporated a 30-bit computer with memory capacity of 12,800 words (30-bit words) and a clock rate of 750 kHz (0.75 MHz). In one case memory was 9,000 words. Modern computers are rated in GHz (thousands of MHz).

76. For example, when two members of the Military Aircraft Panel of the President's Scientific Advisory Committee (PSAC) saw a presentation on VFAX in March 1967, they showed considerable interest in which Navy requirements would support VTOL systems.

77. From the March 1967 PSAC presentation, which also traces the multimission concept back to the 1963 Sea-Based Air Striking Forces study. VFAX would replace the F4J (or other follow-on F-4), some A-7s, and possibly some A-6s. Studies suggested that the carrier should operate two squadrons (twenty-four aircraft) of VFAXes, probably replacing twelve F-4s and ten A-7s (the "ten" may be an error). The PSAC presentation explains why VSTOL would be inferior despite its ability to operate under some conditions of carrier damage (to the flight deck) that would otherwise preclude operation. To achieve the same performance

a VSTOL had to be considerably larger. It would also demand more maintenance.

78. The contractor designs were Boeing's Model 959-007, McDonnell's 199-302, Grumman's April 1967 version of its basic VFAX, North American's point design taken from a study of the effects of varying engine-bypass ratio, and Douglas' D-879-3. An LTV study (V-483) seems not to have figured in the 1967 estimate. The TDP envisaged a fifty-nine-foot-long airplane with 410-square-foot swing wings (maximum sweep seventy-five degrees, span thirty-two feet six inches; minimum sweep twenty-five degrees, span fifty-five feet). The airplane would weigh 45,000 pounds at takeoff in fighter configuration, with a spotting factor of 1.25 and internal fuel capacity of 13,300 pounds. Empty weight would be 27,900 pounds. Height (unfolded) would be fifteen feet six inches. Internal armament would be a single 20-mm Mk 11 gun. Wing aspect ratio would be 7.38. Avionics (black box) weight would be 2,084 pounds, compared to 670 pounds for gun and ammunition. The airplane would be powered by two turbofan engines (bypass ratio 0.9) with a normal military rating of 10,443 pounds of thrust and a maximum rating of 19,950 pounds (fighter thrust-to-weight ratio would be 0.887). Maximum speed would be Mach 2.4 (McDonnell was offering Mach 2.5 and Boeing 2.4). Combat-air-patrol radius would be 150 nautical miles, with a loiter time of 2.7 hours, allowing for two 1.5-minute periods of combat and acceleration. Deck-launched-interceptor radius would be 190 nautical miles. Escort radius was 600 nautical miles, including two 1-minute periods of combat. For interdiction the airplane would fly a high/low profile (500 nautical miles high in and out, 100 nautical miles low, or 400/200 nautical miles). Close-support radius was 300 nautical miles, including a half-hour loiter at five thousand feet. For long-range strike, the profile was 600 nautical miles high and 400 low, in and out, for a total radius of 1,000 nautical miles, carrying two 450-gallon drop tanks. The airplane would need 2,500 feet of runway to take off over a fifty-foot obstacle at a close-support weight of 42,000 pounds. Another set of VFAX vu-graphs showed a CAP mission in which the airplane would loiter for three hours at 35,000 feet 150 nautical miles from the carrier, with

sufficient fuel for four minutes of combat, including acceleration to and maneuvering at Mach 1.5. A spotting-factor comparison showed the F-111B at 2.00, the A6A at 1.85, the F4B Phantom at 1.66, the F8E Crusader at 1.41, and VFAX at 1.25 (with its wings at eighty-four-degree "oversweep," i.e., for deck stowage). An early NavAir vu-graph summarizing the VFAX program listed both the five contractor studies and work on VAX, modifications to the F-111, and ADO-12 (Air Development Objective 12, unfortunately not described). A table of fighter characteristics included the proposed swing-wing version of the F-4, which would have weighed far more than VFAX.

Chapter 11. Vietnam

1. This account of the political background is based on my Cold War history, *The Fifty-Year War* (Annapolis, Md.: Naval Institute Press, 2000). A North Vietnamese journalist told his Soviet counterpart that the Soviet Union was providing 80 percent of what was needed to fight the war but that compared to the Chinese it was exercising perhaps 4 or 5 percent control. The Russian thought the Soviet Union had 15 to 20 percent control over the North Vietnamese. The only wartime improvement to North Vietnamese Fan Song B radars was an optical sight (Fan Song F). The Soviets were perfectly willing to risk compromise by exporting systems to Egypt (the Israelis captured a Fan Song in 1967 and later traded it to the United States) and to other quasi allies. For example, Soviet equipment including the SA-3 system was given to Indonesia in competition with the Chinese; it was compromised after the anticommunist coup in that country. However, according to Steven J. Zaloga, *Red SAM: The SA-2 Guideline Anti-Aircraft Missile* (Oxford, U.K.: Osprey, 2007), 22, in 1970 the Soviets agreed to sell Vietnam the S-125 (SA-3). Two Vietnamese regiments were training on the missile when the United States renewed strategic air attacks in 1972; it is not clear to what extent the Soviets were stalling delivery of this weapon. Later it was estimated that 85 percent of Soviet aid to North Vietnam went into air defense. For limits to SA-2 improvements, see Price, *History of U.S. Electronic Warfare*, 3:231.

2. Carriers also provided support inside South Vietnam while air bases were built up. Their short-lived Dixie Station in the southern Tonkin Gulf was established after aircraft from three carriers struck a mountain northwest of Saigon on 15 April 1965. Gen. William Westmoreland, commanding U.S. forces in South Vietnam, was sufficiently impressed that he asked for a permanent carrier station off South Vietnam. Initially one carrier was typically on Dixie Station and three on Yankee Station, but a fifth carrier was authorized in mid-1965. Dixie Station carriers also supported a major amphibious operation (Operation Jackstay, March 1966). In August 1966, when more airfields had been completed, Dixie Station was disestablished.

3. Air Force aircraft spent much longer in the view of North Vietnamese radars, even if they were flying outside North Vietnam. The Air Force preferred larger numbers drawn from multiple airfields and units. In a typical operation, of 120 aircraft launched from bases in Thailand, eighty-eight penetrated North Vietnam and thirty-two actually attacked—the difference was a measure of the extent to which North Vietnamese defenses diluted any attack. Operations from multiple bases required far more radio chatter. It did not help that the Air Force rotated all its pilots through Vietnam, precluding creation of any experience base.

4. According to the 1974 edition of *Striking Force Operations*, NWP 20(D) (declassified 1982), which reflected "Vietnam War experience," "a strike support ship (SSS) and CVA [attack aircraft carrier] AEW [airborne early warning] aircraft now act as a well-balanced team and provide flexible, complete, support services to strike aircraft." The Red Crown ship was a missile cruiser or guided-missile "destroyer leader" (later called cruisers) equipped with the computer Naval Tactical Data System (NTDS). Only a ship so equipped could keep track of the numerous aircraft involved. Although strike aircraft had used radar-picket ships as navigational references late in World War II, "the full value of strike support services was not realized until the limited warfare in Vietnam [when] the naval task force operated for protracted periods in proximity to the target area and strike aircraft flew repeated

sorties over enemy territory." AEW aircraft might be assigned to strike-support missions, even at the expense of their air-defense ones. Their crews had the advantage of being briefed with the strike pilots on board the carrier and thus had detailed knowledge of strike group's composition and planned tactics.

5. Strike controllers were typically on board AEW aircraft. The strike controller identified the strike aircraft on his screen, determining which were in the strike group by means of either secure UHF voice contact with the carrier or through readout of IFF codes. The controller might be assigned close control of TARCAP (Target CAP) should there be an enemy fighter threat over the target area. The strike controller would issue MiG and SAM warnings as the strike group neared the target; NWP 20(D), 1974. References to communication security contrast with extensive Air Force chatter. The National Security Agency (NSA) found that extremely poor radio discipline by the Air Force gave the North Vietnamese very early warning of air raids and of their objectives. Apparently the Air Force repeatedly refused to encrypt its communications. It does not appear that the Navy, which placed a high premium on communications security, suffered similarly.

6. Lt. Cdr. Douglas M. White, USN, "Rolling Thunder to Linebacker: U.S. Fixed-Wing Survivability over North Vietnam" (master's thesis, Army Command and General Staff College, 2014, National Technical Reference Library ADA615296), 50. Notwithstanding its title, this paper gives a very detailed account of the bombing campaigns against North Vietnam, including their tactical and strategic rationales. Much of White's account of strategy is based on the Air Force CHECO papers (see note 51 below).

7. The Chinese provided seventeen MiG-15s and -17s immediately after the August 1964 U.S. air strikes and thirty-six more that December. The Soviets provided the bulk of North Vietnam's aircraft, including supersonic MiG-21s, but in 1966 the North Koreans agreed to provide pilots for a fighter regiment (two companies of MiG-17s and one of MiG-21s).

8. According to an NSA history, Robert J. Hanyok, *Spartans in Darkness: American SIGINT* [signals intelligence] *and the Indochina War, 1945–1975* (Fort George G. Meade, Md.: NSA, Center for Cryptologic History, 2002), 250–61 (235 in the redacted version, PB209106900), initially the North Vietnamese system employed about forty visual observation posts that reported to a filter center in Hanoi, which in turn sent tracking information to sector headquarters controlling antiaircraft guns. Messages were sent by high-frequency (HF) Morse radio, which could easily be intercepted, in a very simple code that was immediately broken. The system also at first passed information extremely slowly, a tracking message sometimes taking as long as thirty minutes to get through. The basic structure remained in place through the war, and American cryptologists exploited it. The question was always to what extent that exploitation would be used tactically. As the North Vietnamese acquired air-defense radars, they set up additional filter centers. At the same time the North Vietnamese integrated their air-defense system with that of the Chinese, with links to both Kuangchou and Kunming. By early 1966 the North Vietnamese had an Air Defense Headquarters at Bac Mai airfield, working closely with their air force headquarters there. It contained an Air Situation Center and an Air Weapons Control Staff, the former using both Vietnamese and Chinese data. It issued advisories to the Air Weapons Control Staff and other elements of the system, and also to the Chinese. The Air Weapons Control Staff plotted threat tracks and assigned targets to air force and antiaircraft units. Track information was generally sent by radio, which could be and was intercepted. According to Hanyok, 237, messages were generally either in the clear or in low-grade codes, as it was more important to get them through than to protect them. By 1965 the system cycle time, from radar tracking to advisory (alert) to a defense unit, was about five minutes. Effective range was up to 150 miles beyond the border of North Vietnam, presumably depending on the altitude of the approaching airplane. The senior air controller at Bai Mac assigned targets to subordinate controllers near the major MiG bases, who actually scrambled and controlled the interceptors. Separate controllers handled air navigation and directed fighters; they

generally tried to place their aircraft above and behind the U.S. aircraft and also to warn their MiGs that they were about to be attacked. Late in the war senior Vietnamese pilots were sometimes used as controllers. The U.S. Air Force's Security Service (its signal intercept branch) set up an intercept station on Monkey Mountain overlooking Da Nang in 1962 but obtained almost no signals until the North Vietnamese began operating jet fighters.

9. Zaloga, *Red SAM*, 19.

10. Price, *History of U.S. Electronic Warfare*, 2:136. Soviet control systems used separate vertical and horizontal antennas, each of which scanned. In contrast, Western trackers kept themselves pointed at a single target by measuring deviation as the tracking beam spun. That associated deviation in angle with variation of target strength as the beam kept spinning. If the radar received a signal much stronger than the echo, the radar would indicate the wrong target direction, because it depended on timing to measure direction. Each of the two Soviet tracking radars also had a cycle, so a jammer that broke the association between timing and angle worked against them too. Jammers detected radar pulses and returned stronger pulses far enough from the returning echoes that the radar and operator would mistake them for real ones. Deception became possible with the advent of broadband amplifiers (TWTs); they were later the basis of software-controlled radars.

11. Fan Song employed two scanning trough antennas linked to form a *T* that could turn and elevate. It showed target position in the two directions. A secondary pair of antennas tracked a beacon, whose position was also shown on the displays. Each beam was quite narrow (ten degrees), so the system was typically cued by a broad-beam search radar, typically Spoon Rest. As the operator tracked a target, feeding apparent rate of motion into a fire-control computer, which calculated an impact point, to which the missile was commanded. To obtain a sharply enough defined radar image of their targets, operators typically doubled their pulse rates thirty to forty seconds before firing a missile, which provided launch warning. The guidance signal was turned on no more than four seconds after launch; the missile transponder beacon

was uncovered when the booster fell away. SA-2 operators learned to use other radars for initial target warning, never transmitting in the long-range, low-PRF mode and thus denying American pilots the doubled-PRF launch warning. The missile had a proximity fuze. A dish transmitted command signals to the missile. After an optical adjunct was added in 1968, operators could track low-flying aircraft (down to a thousand feet), and there were no longer tracking signals to warn pilots. The only remaining signals were the guidance command and the missile downlink. According to Bernard C. Nalty, *Tactics and Techniques of Electronic Warfare: Electronic Countermeasures in the Air War against North Vietnam 1965–1973* (16 August 1977, National Technical Reference Library AFD-110323-034) [hereafter Nalty, *ECM in Vietnam*], the optical version made it possible for the North Vietnamese to set up missile ambushes, using Fan Song transmissions to attract attention and then attacking silently (optically) with a second missile battery. By 1972 Fan Song had been modified to track "low fliers" (possibly as low as five hundred feet, but the missile could not arm in time to engage a target flying below a thousand feet). In 1972 some Fan Songs were apparently modified to operate in the X- rather than S-band.

12. Price, *History of US Electronic Warfare*, 3:55. That possibility placed a premium on knowledge that an SA-2 had been launched. According to Price, 3:58–59, the key was the weak command uplink. The CIA launched drones specifically to cause missile fire, the drone intercepting the link and transmitting it to an RB-47H waiting offshore. A successful 13 February 1966 mission intercepted not only the uplink but also the proximity fuze signal.

13. Zaloga, *Red SAM*, 24, tabulates missiles per aircraft kill, presumably based on a combination of published Russian, Vietnamese, and U.S. data.

14. The North Vietnamese generally reduced batteries to three rather than six launchers to make them easier to move. Normally it took three hours to shut down a site before moving and then four to six hours to set up at a new site.

15. Lt. Col. James R. Brungess, USAF, *Setting the Context: Suppression of Enemy Air Defenses and Joint War Fighting in an Uncertain World* (Maxwell

AFB, Ala.: Air University Press, June 1994), 96–97, credits Navy defense-suppression tactics with the success of the Benghazi raid in 1986, when the Navy was able to blanket surprised Libyan defenses. He argues that the same tactics were not so successful in Iraq in 1991 because the conditions there did not match the usual Navy criteria of specific objectives, short duration, and limited force. He holds that "single-minded piecemeal" tactics were inefficient in an extended campaign.

16. According to what is known as the "Night Song report" on the air war, prepared by an interservice study group headed by Maj. Gen. B. McPherson USAF, and declassified with some redactions: *An Examination of U.S. Air Operations against the NVN Air Defense System*, PB 2010113655 (Washington, D.C.: Joint Staff, 30 March 1967), redacted copy in the National Technical Research Library, available online. (Redactions appear to concern communications intelligence.) Most of the airplanes involved were Air Force and Navy EC-121s, which were too slow: the observed bearings of the radars did not change enough while the radar was active.

17. Zaloga, *Red SAM*, 20.

18. Emerson's lower-frequency ALQ-35, intended to jam search radars by generating false targets, was canceled because it could not jam enough of them simultaneously. Melpar's ALQ-38, intended to jam fighter-control links, failed flight testing.

19. According to Price, *History of US Electronic Warfare*, 2:250, at least at the outset they proved quite unreliable because the shock of being catapulted could push the TWT out of the necessary precise position. The standard outfit was two of each, facing fore and aft for full coverage. Aircraft were also to have carried one each of the abortive ALQ-35 and -38.

20. These were track breakers. Others were ALQ-49 for C-band and ALQ-69 for Ku-band (J-band). By 1962 two other countermeasures were assigned to attack aircraft: ALQ-35 to generate false targets for early-warning radars and Sanders' ALQ-55 to disrupt the communications used to control interceptors. In 1962, according to an official film on naval air ECM (courtesy of Tommy Thomason), the Intruder was expected to carry only the S- and X-band track breakers. It had antennas inside and outside its air intakes and in its tail. The A-3 was expected to carry all three jammers. It had antennas forward, in its waist, and in its tail (including at the top of the tail). The movie makes it clear that at the time no radar homing/warning sets were planned, because there was no intent for aircraft to survive by maneuvering radically when they detected surface-to-air missile batteries. Instead, they would turn on jammers when their systems received warning (the jammers all had receive-only modes). The 1962 film did not mention the ALQ-38, which had presumably already been superseded by ALQ-55.

21. Some AD-5Qs were electronic-intelligence (ELINT) reconnaissance aircraft, not jammers.

22. Their nose radars were replaced by an APA-69 direction finder, an ALR-8 receiver, and an ALA-3 pulse analyzer. Underwing pylons could carry jammers (ALT-2 or -6B) and chaff-dispenser pods.

23. APR-24 was a related device used in F-4s, employing a pair of antennas atop the tail fin to give 360-degree coverage. Some A-4s had the APR-28, which had coverage of the L- and P-bands (long-range radars) in addition to the other frequencies; the Navy bought fifty, and they entered combat in 1965. Total purchases amounted to 150 APR-23s plus fifty APR-28s. The APR-24 in particular was much disliked: they produced warnings even for relatively distant radars, to the point that pilots preferred to turn them off. A few aircraft had the follow-on APR-30.

24. There was no Air Force equivalent to APR-23. Soon after the first losses to surface-to-air missiles, both the Air Force and OSD laid down urgent requirements for a RHAW (radar homing and warning) system. They chose Applied Technologies' APR-25 (Vector IV), which was initially installed on board some F4Bs. Apparently A-4s began to be fitted with it in 1968. According to Jim Winchester, *Douglas A-4 Skyhawk: Attack and Close-Support Fighter-Bomber* (Barnsley, U.K.: Pen & Sword, 2005), 97, APR-25 was associated with the "hump" in the A-4F. He claims that the A-4Fs that deployed on board *Ticonderoga* in 1968 (VA-23 and -195) had the necessary wiring but that some lacked the display, offering only a flashing "SAM" light and sound. These devices may actually have been

APR-27s. A prototype of Melpar's APR-30, under test early in 1967, had separate homing and warning bands, the latter being broader, in the S-, C-, and X-bands. It was described as a radar warning receiver. The A7A had APR-25/27, and Navy F4Bs had APR-27 or APR-30. A6As had the APR-27 warner and also ALR-15. Aircraft also had a variety of chaff dispensers (ALE series).

25. The ALR-45 was fitted in the A4M, A7E, AV-8C, KA-6D, and F4S. Versions through ALR-45E were hardwired. An "engineering development model" of the digital ALR-45F was completed in May 1978. It used the same processor as the ALR-67 in the F/A-18 and could interface with, for example, the HARM missile and various jammers. Production began in 1981, and the first three of a 108-system contract were delivered in FY83.

26. According to Price, *History of US Electronic Warfare*, 3:57, the deception pulses produced by ALQ-51A were designed to confuse a radar as to both direction and range. Range deception in practice was ineffective, but the Fan Song angle tracker wandered when jammed, causing the system to order constant corrections that led the missile either to jam or to run out of energy. If the operators concentrated on range jamming, that slowed system reactions so much that even a shallow turn and a shallow dive could save an airplane. Often a violent maneuver (with bombs jettisoned) would do so. A jammer must react to every pulse; ALQ-51A had a capacity of 20,000 pulses/sec. To see whether that was enough, in the summer of 1967 six special recorders (Sidesaddles) were fitted to ALQ-51As. They picked up no more than 8,000 pulses/sec.; Price, 3:108. The ALQ-100 in an A-7A combined the functions of the S-band ALQ-51A and the C-band ALQ-49 in the same volume that each had required alone. According to Night Song, ALQ-100 operated in the 1.8–8.0 GHz band against pulse radars and in the 4–8 GHz band against CW radars; it would simultaneously deceive radars in more than one band or mode of operation. It would also prematurely detonate or "dud" (disable) proximity fuzes in missiles and antiaircraft shells. The novel techniques that squeezed so much into this volume were sensitive: the ALQ-100 was set to self-destruct if the airplane crashed. Unfortunately it could also self-destruct after heavy deck landings; seventy-four jammers were destroyed this way before the self-destruct was neutralized; Price, 3:146.

27. Price, 3:30, 49–50.

28. Price, 3:51, 147. Price also claims that ALQ-92 was intended to jam long-wave early-warning radars (a capability that was rarely used).

29. According to Grumman's confidential brochure for the A2F-1H (July 1962), which shows a mock-up in Navy markings. The Grumman file includes the original (unfortunately undated) G-128D proposal for the A2F-1Q, which it describes as a carrier-based tactical ECM aircraft. It differs from the brochure in that the jamming frequency range is given as 40 MHz to 11 GHz; equipment that would expand the range to 20 GHz was under development. Perhaps more significantly, the 29 September 1961 BuWeps memo establishing the A2F-1Q designation makes no reference to the Marines. The A-6 file in the Naval Aviation History Division, NHHC, included a 10 October 1961 DCNO (Air) memo restoring procurement of the twelve A2F-1Qs (EA-6As) to the FY63 budget without mentioning their Marine role. Work accelerated during 1961, and the Marines placed an initial development contract in March 1962. The 16 October 1964 OpNav memo establishing the EA-6A designation associates the airplane only with the Marines, claiming that it had been developed to meet their requirements. At that point the program amounted to twelve aircraft.

30. Declassified EA-6A tactical manual, NAVAIR 01-85AD8-1T(A), 15 June 1969, NARA II. The ALQ-55 blade was moved forward (another was aft), and another blade, somewhat farther aft, served an APR-27 missile warner.

31. According to Night Song, EA-12, the Marines were authorized in 1963 to procure twelve EA-6As, which were delivered in 1965; they were one-for-one replacements for EF-10Bs. Five were deployed immediately to Southeast Asia; the others were retained in the United States for training and to meet depot-repair requirements. To meet the increasing electronic-warfare requirements in the Demilitarized Zone (DMZ) and in Route Package I (just north of the DMZ), where the Marine strike aircraft were operating, the Marines needed at least

fifteen more on an urgent basis, of which twelve would go to Southeast Asia. That is what they got. The Navy wanted at least four EW aircraft per carrier air wing. It decided to buy the EA-6B instead of the existing EA-6A, but by early 1967 the Navy was trying to increase planned EA-6B strength.

32. The major series are XCAP (Expanded Capability), ICAP-1 (Increased Capability), ICAP-2, "Block 86," and ADVCAP (Block 91). XCAP (1973) added four higher-frequency bands, which were important outside Vietnam. It doubled computer memory and added a digital recorder. New computer-operated modes allowed simultaneous automatic coverage of several enemy radars. There was also a new exciter. In this version, the front-seat electronic countermeasures officer and one of the back-seaters operated the main jammer, the other back-seater being assigned to the ALQ-92 communications jammer. The front-seater was also the navigator, giving him a disproportionate workload. ICAP-1 (1977) added digital receivers that fed a computer display system. The entire track/jam function was moved aft to the back seats; the front-seater became the ALQ-92 operator, navigator, and radar operator. Problems with operator speed were noticed in ICAP-1, so ICAP-2 was fully automated, using a faster AYK-14 computer. It introduced a universal exciter. It had an automatic signal processor, a mission planner, and improved band coverage. The universal exciter substituted a computer-generated signal for the previous analog-processed threat signal. The antennas were electronically steered, and the jamming pods were much more flexible. A more basic change was from semi-automatic operation to computer-initiated reaction with human monitoring. "Block 86" was the final version, using the ALQ-99. ADVCAP (1991) introduced an entirely new "receiver-processor group" system. It began full-scale development in 1983.

33. Because there were many more American than enemy aircraft in the air over North Vietnam, the use of long-range air-to-air missiles was prohibited, and pilots had to confirm visually that their targets were hostile. However, a missile had to reach a minimum speed (i.e., had to fly a minimum distance) before its control surfaces were effective.

Pilots also often fired from outside effective firing zones. Solutions to the latter included much better training, such as the Navy's Top Gun program.

34. Price, *History of US Electronic Warfare*, 3:195.

35. According to Night Song, the most promising current program was TRISAT (Target Recognition Through Integrated Spectral Analysis Technique), which applied spectral analysis to modulation—caused by machinery vibrations, airframe vibrations, and rotating engine blades of a detected aircraft—in a returning radar echo. TRISAT could analyze radar returns from up to ten selected airborne targets; its computer sorted the returns and compared them digitally with stored modulations from known aircraft. The Navy eventually adopted telescopes, which made identification possible. In the Libyan combat of 1981, the indication of hostile intent was the fact that the Libyan fighters were carrying air-to-air missiles, which normally they did not. That was obvious in the telescope picture.

36. Night Song lists communications jammers but states that they had been "secured," i.e., kept in reserve but not used. (Price, *History of US Electronic Warfare*, 3:143, mentions NSA vetoes on the ground that the intercepted communications were too valuable.) According to Price, ALQ-59 was developed under an Air Force project, Combat Martin, specifically to jam VHF fighter-control links. It rebroadcasted the VHF fighter-control messages with a slight delay, so that they interfered with the original messages. The 1977 Air Force history of EW in Vietnam mentions cases in which Navy aircraft successfully jammed MiG communication links using ALQ-55 jammers; in at least one instance an Air Force RB-66E was also involved. The ALQ-92 was also sometimes used. Jamming may have been allowed in 1968, after bombing was restricted to southern North Vietnam. There was an indication that when the missile cruiser *Long Beach* shot down a North Vietnamese interceptor in 1968, jamming prevented a missile warning from getting through to the MiG. There was no further use of such jamming until 10 May 1972, when it was involved in the destruction of one MiG-21 and seven MiG-17s; Nalty, *ECM in Vietnam*, 65.

37. In 1967 new time-of-arrival techniques (Compass Strike/Eels) offered the potential to reduce the area of uncertainty to a nine-hundred-foot circle, which a pilot could search visually.

38. According to the NSA history *American Cryptology during the Cold War, 1945–1989*, book 2, *Centralization Wins, 1960–1972* (Fort George G. Meade, Md.: NSA, 1995, redacted PB2009106500), 549, Iron Horse linked the electronic output of a standard intercept position (a grid plot) to a radarscope by passing it through a computer. An Air Force SIGINT analyst selected plots to be sent to Seventh Air Force; there they were integrated with the usual radar plots in the Air Force BUIC II command system. Plots sent to Task Force (TF) 77 and to the Marines were given signatures indicating that they had not come from U.S. radars. Overall, Iron Horse brought throughput time from between twelve and thirty minutes down to between eight seconds and three minutes. Charger Horse built on existing Navy practice, in which each ship with a SIGINT detachment on board copied North Vietnamese air-defense-net signals, including radar tracking and VHF air/ground voice, to support TF 77 operations. Charger Horse linked the detachments so that they could divide up the nets they were monitoring so as to intercept lower-level communications and reduce the lag time in reporting to the TF 77 command (page 571). The great lesson of Iron Horse and Charger Horse was that combining many different sources of information protected the SIGINT source, because the enemy could not be sure whether, or if so how, his communications had been compromised. Previous practice had been to avoid using SIGINT for fear of compromising it. Description of integrated systems are taken in part from the Night Song report. This report called for acceleration of the NSA Yogi Bear project, which was intended to make special intelligence immediately available to operating forces; presumably it was connected to Iron Horse. The document also describes the Navy's Charger Horse project.

39. This account is based on Hanyok, *Spartans in Darkness*, 250–61. A comment on page 276 is telling: the Navy did far better than the Air Force in air-to-air combat, with a 5.5:1 ratio before Teaball became operational, compared to the Air Force's 1.3:1. Through the end of Rolling Thunder, MiG-21s shot down U.S. aircraft at a rate of better than 3:1 (page 244). Of 109 North Vietnamese aircraft shot down through December 1967, twenty-five were MiG-21s, a type that shot down twenty-six American aircraft during that period. According to Hanyok, 248, the only advantage enjoyed by the MiG-17 was that it could turn inside U.S. aircraft, an advantage that unaggressive North Vietnamese pilots often wasted. These aircraft were ineffective in general air combat and were used mainly for low-level point defense. MiG-21s were much more effective, because they could outmaneuver and outaccelerate their U.S. counterparts. After the 1968 disappointments, the Navy created the Navy Strike Fighter Tactics Instructor program, better known as "Top Gun." Hanyok, 270, adds that the Air Force failure was critical in that it endangered F-4s using Pave Knife laser-designator pods.

40. "What made the Navy so effective during Linebacker was the integration of all naval air intelligence and command-and-control functions with the Red Crown [PIRAZ, or Positive Identification Radar Advisory Zone] controller. In August, Red Crown could claim to have downed twelve MiGs, while USAF pilots under Teaball control could claim only one. So effective was Red Crown, that Air Force pilots preferred to use it instead of DISCO or Teaball." (Hanyok, 276; Teaball and Disco were Air Force controllers.) "More importantly, and often overlooked, was that Navy GCI [ground intercept control] operators trained with their pilots. The critical factor for the Navy's later high performance against the North Vietnamese was the working familiarity between the carrier pilots and their GCI operators." That is, the Navy emphasized the integrated-system aspect of air combat. At the beginning of Linebacker, the Navy was shooting down MiGs at a rate of 5:1; the Air Force was barely holding its own at 1.5:1, and during June and July it did far worse (eleven lost for three MiGs in June).

41. Even when fully integrated, the Teaball picture was somewhat "time-late" (usually by about two minutes) because of delays in passing intelligence information, making it most useful in providing early information of enemy activity.

42. This account is based on material posted online by Jared A. Zichek, particularly in his e-zine *Retro-Mechanix*, retromechanix.com. He cites material

from the Greater St. Louis Air and Space Museum, which apparently has files from the former McDonnell Douglas Corporation.

43. CNO, "The Navy of the 1970 Era," January 1958; my copy is courtesy of Dr. Alan David Rosenberg. This pamphlet summarized the conclusions of the then-recent Long Range Objectives study, LRO-57.

44. *History of the Navy at China Lake, California*, vol. 4, *The Station Comes of Age: Satellites, Submarines and Special Operations in the Final Years of the Naval Ordnance Test Station, 1959–1967* (China Lake, Calif.: Naval Air Warfare Center Weapons Division China Lake, 2017, National Technical Reports Library AD 1056938), 125.

45. The names were chosen by George Spangenberg, the longtime senior aeronautical engineer in BuAer. BuWeps' *Free-Fall Nonnuclear Ordnance RDT&E* [research, development, test, and evaluation] *Program* (September 1960) was a revised version of the 1959 China Lake document; *Station Comes of Age*, 128. Existing weapons were "inefficient in terminal effectiveness, incompatible with jet aircraft, of doubtful ability to withstand ejection forces, dimensionally incompatible for efficient magazine stowage and handling[;] . . . explosive fillers are TNT or one of the older compositions. The fuzes are air arming and utilize distances of air travel to provide arming. Since they are designed for aircraft in the 200–300 knot category, they do not provide sufficient separation from aircraft on completion of arming at release speeds of modern jet attack aircraft. . . . Mechanical design of the fuzes is such that some will not withstand the torque loads imposed by arming at high release speeds."

46. The full series, including weapons not ultimately adopted, was Bigeye, Briteye, Bugeye, Chaffeye, Deadeye, Deneye, Evileye, Fakeye, Fireye, Gladeye, Hawkeye (later named Rockeye), Marceye, Misteye, Padeye, Rockeye, Sadeye, Smokeye, Snakeye, Walleye, and Weteye; Walleye was a guided missile, not an unguided munition. Weapons not produced included the Briteye, a flare suspended under a hot-air balloon; the Deneye, an air-launched mine dispenser; the Fireye, gelled-fuel flame weapon with a more controllable area of effect than napalm; the Gladeye, gas bomb version; the Padeye Mk 23 cluster bomb; and the Weteye chemical dispenser

(spray). Unfortunately, the China Lake history does not describe others not produced.

47. A postwar development used a "ballute" (balloon-parachute) to retard the bomb. Ninety percent of bombs dropped by Navy attack aircraft in Vietnam were Mk 82s, nearly all the rest Mk 83s.

48. Cluster bombs were valued because their area coverage could make up for inaccuracy in delivery. Existing ones could not be carried at high speed, because they were not streamlined. Rockeye I (Cluster Bomb Mk 12) was an antiarmor weapon carrying ninety-six bomblets in a 750-pound case. It was not produced, but Rockeye II (Mk 20, with 247 antiarmor bomblets) entered development in 1961 and production in 1967. It was designed to burst at a preset altitude, using a linear shaped charge to split its casing. Rockeye began as "Hawkeye" in 1959 and was presumably redesignated when the E-2 airborne radar plane became the "Hawkeye"; it was released for pilot production in 1967. Cluster Bomb Mk 15 (Sadeye) was a "universal weapons dispenser" carrying 2,100 bomblets, suitable for dropping at high speed at moderately low altitude, bursting away from the bomber. Gladeye (Mk 17) had two separate time-delay cartridges so that its canisters could be dropped singly, in sequence, or in salvo. Its canister was carried on a standard multiple bomb rack; each of seven on such a rack carried 330 antipersonnel bombs.

49. *History of the Navy at China Lake, California*, vol. 3, *Magnificent Mavericks: Transition of the Naval Ordnance Test Station from Rocket Station to Research, Development, Test, and Evaluation Center, 1948–58* (Washington, D.C.: Naval Historical Center and Naval Air Systems Command, 2008), 456.

50. Peter deLeon, *The Laser-Guided Bomb: Case History of a Development*," Report R-1312-1-PR (Santa Monica, Calif.: RAND, June 1974, National Technical Reports Library ADA017126). In bang-bang guidance, the fins are deflected all the way each time they are deflected at all; the missile follows an undulating path toward the target. That is inelegant but also inexpensive—the cost of the Texas Instruments approach was a third of that of its competitor, North American Autonetics. The

service version had a nose seeker integrated with canard control fins; the bomb had fixed tail fins instead of the rear control fins of the prototypes.

51. Maj. Calvin R. Johnson, USAF, "Project CHECO Report on Linebacker Operations, September–December1972," 31 December 1978, Office of History, HQ PACAF, Joint Base Pearl Harbor–Hickam, Hawaii, ADA487182), 37–39. CHECO was Contemporary Historical Examination of Current Operations, and the released reports are analyses of Vietnam air operations.

Chapter 12. War at Sea

1. In the eastern Mediterranean, radar-trapping ducts formed above the surface nearly all the time. The ducts trapped shipboard radar signals, so that strike aircraft could often approach Soviet ships undetected until they became visible. The carriers, however, could detect ships through the ducts by using airborne-early-warning aircraft such as the E-2 Hawkeye. Sixth Fleet pilots practiced "time-on-target" operations, in which strike aircraft appeared simultaneously above Soviet ships. Ironically, the Soviets also benefited from these ducts, which made it possible to project radar signals well beyond the horizon. In support of their own long-range missiles, the Soviets developed a series of over-the-horizon naval radars exploiting both ducting and troposcatter (reflection from the troposphere). The relevant radars were contained in large radomes that NATO nicknamed "Band Stand." These devices made small Soviet Nanuchka–type (Project 1234) missile boats effective anticarrier weapons—if they could survive long enough to shoot. It is not clear whether the presence and use of the ducts was widely understood in the West before the 1980s. Band Stand certainly was not.

2. The author remembers hearing Gen. Johnny Vogt, who had just retired as NATO's Allied Forces Southern Europe (AFSOUTH) commander, saying that he valued the carriers as his nuclear strike force. The ideas that, on one hand, the carriers were vital to AFSOUTH but, on the other, that they might have to withdraw at the outset could be reconciled if it was assumed that the war would begin with a lengthy nonnuclear phase, during which Soviet surface ships in the Mediterranean would be attacked. The question of carrier survivability (and value) was raised particularly by the CNO of those years, Adm. Elmo Zumwalt Jr.

3. In the late 1970s it was suggested that the four *Iowa*-class battleships be reactivated to replace the carriers in the initial-presence role. The Soviets would feel compelled to attack them at the outset of a war. They would be far more difficult than carriers to sink, and they could help deal with the attackers.

4. Later it became apparent that the multiplicity of systems was really evidence that the Soviets had little confidence in any one system.

5. Soviet missiles, even those carried by small missile boats, had much larger warheads. However, in the 1980s the Soviets decided to replace them with Kh-25 (SS-N-25) missiles, analogous to Harpoon.

6. Tactics described by Capt. Don Watkins, an A-6 pilot and squadron commander, as quoted in Lon O. Nordeen, "USN Airborne ASUW [antisurface warfare] in the 1980–Gulf War Era," *The Hook* (March 2014), manuscript courtesy of Mr. Nordeen. Watkins wrote that he was responsible for the VA-34 standoff tactics discussed here, which he thinks other squadrons developed at about the same time. The article describes the attack on the Iranian frigate *Sabhalan* in April 1988 by what seems to have been a standard antiship package—two A-6Es, six A-7Es, and an F-14 CAP. Weapons were Harpoons, laser-guided bombs, and Rockeyes for the Intruders and Walleye IIs and unguided bombs for the A-7Es. There were also an EA-6B and an E-2C Hawkeye in support. Because there were somewhat similar Royal Navy frigates in the area, to confirm its identity the lead A-6E flew down to the ship, which fired antiaircraft missiles at it. The A-6E pulled back, fired a Harpoon, and then followed up with two AGM-123 Skipper laser-guided short-range missiles. Then it dropped a 500-pound laser-guided bomb. More ASUW aircraft soon arrived, called up by the first attacker, which acted as strike coordinator. The second A-6E fired another Harpoon and then a Skipper, which had a much larger (5,000- rather than 250-pound) warhead.

7. American analysts seem not to have realized the extent to which the Soviets redirected naval investment to counter the Polaris missile system—which

made sense, since Polaris displaced the carriers in the strategic role. Thus Kresta I (in Soviet nomenclature Project 1134A) was designated an ASW ship despite her antiship armament, and later ships of this type (Kresta II [Project 1134B]) were armed with an antisubmarine missile (SS-N-14). In the 1970s the major surface ship anticarrier threat was the SS-N-3 Shaddock, which could be launched by only four Kynda-class (Project 58) ships and four Kresta I's. Submarines and bombers were a much more serious threat, although that apparently was not obvious at the time. Zumwalt also does not appear to have realized how limited the missiles were by the need to identify and locate targets for them well beyond the horizon. Later there were considerably more Soviet surface warships with substantial anticarrier capacity, and better anticarrier missiles.

8. A letter of interest calling for a design for a VSTOL fighter was published in the *Commerce Business Daily* in the fall of 1971. Eleven designs were submitted. NavAir preferred the Convair Model 200, with Republic's FR-150 and LTV's V-517 as runners-up, in that order. North American Rockwell offered a very different type of VSTOL, using air drawn in by its engine and released over its wings to entrain further air circulation. Admiral Zumwalt's materiel chief, Adm. Tom Davies, felt that he had a mandate to pursue the most revolutionary possible technology. In March 1972 he chose North American's radical NR-356 (XFV-12A) even though NavAir refused to recommend it. Two prototypes were ordered on 18 October 1972. At that time the first conventional flight was planned for November 1974 and the first vertical flight for February 1975. Once Zumwalt and Davies retired, the program lost momentum; the prototype was rolled out only on 26 August 1977. Rockwell had calculated that hot air from the engine flowing through canards and wings would entrain 50 percent more air, generating lift. When tested, the airplane achieved 6 percent augmentation through its canards and 19 percent through its wings; it never lifted itself. The program was closed down in 1981. Some of those involved thought that there was still potential in the thrust-augmented wing but that the FV-12A had been premature. The more conventional VSTOL aircraft in the competition would probably have been successful.

9. As described in J. T. Tyler (OpNav), C. E. Chambers (CAN), and R. G. Perkins Jr. (NavAir). "An Assessment of Sea-Based Air Master Study," AIAA-80-1820, paper delivered at the American Institute of Aeronautics and Astronautics (AIAA) Aircraft Systems Meeting, 4–6 August 1980, at Anaheim, Calif. The study was commissioned in 1977 to establish a database for program decisions, determine budget requirements, develop a long-range master plan for naval aviation, and analyze future sea-based aircraft requirements. It looked at numerous alternative sea-based aircraft. It did not consider a separate attack mission but instead a variety of fighter/attack aircraft: supersonic VSTOL or CTOL (conventional), or STOL (short) or STOVL, in both high and low categories. STOVL meant short takeoff and vertical landing; VSTOLs could take off and land vertically. An additional "V/STOL" category could take off vertically but would take off conventionally for more demanding missions. Alternative carriers were the *Nimitz*-class CVN, a *Kennedy*-class conventional carrier, a 60,000-ton "CVV," a 35,000-ton V/STOL support ship (VSS), a 50,000-ton STOL support ship (SSS), and 22,000- and 11,000-ton VTOL ships, the latter an enlarged destroyer and the former close to Zumwalt's Sea Control Ship (SCS). Scenarios were strike warfare (as in Vietnam or Korea), AAW defense against a large bomber force, AAW against enemy surface-to-surface missiles, sea control (convoy), and amphibious warfare. The question was how to allocate a fixed amount of money over an expected fifteen-year service life. The study showed that fighter and attack-aircraft requirements would dominate planning for sea-based air forces. Large carriers were more efficient and offered the greatest aggregate offensive capability. Generally V/STOL aircraft cost about a quarter more than conventional ones; a force equipped with conventional aircraft generally had about a quarter more ships and aircraft than a comparable V/STOL force. The V/STOL force was most effective at shorter ranges. A V/STOL air wing operating from a carrier of given size was more effective than the equivalent CTOL air wing—*if* all aircraft

were assumed to have equal performance. The key point was probably that V/STOL meant less stand-off from a target at a higher cost per airplane.

10. This HIPASS (High Performance Aircraft System Study) is mentioned by George Spangenberg, the veteran BuAer/NavAir design analyst, in a 22 October 1975 memo for William P. Clements, at that time deputy secretary of defense, opposing Clements' advocacy of the F/A-18; exhibit VF-17 in a series of exhibits to complement Spangenberg's oral history. I have not found a copy of the HIPASS study.

11. In Spangenberg's view Grumman had blundered in signing a contract fixing the unit cost of airplanes in each lot of F-14s whatever the number ordered. The Navy ordered fewer than expected, so the fixed part of the cost of each lot made airplanes much more expensive than projected. Spangenberg comments drily that his NavAir negotiators were aware that Grumman was erring but did not let it in on the problem.

12. Memo from Deputy Secretary of Defense to Secretary of the Navy, 7 June 1973, exhibit VF-8, Spangenberg oral history. In another memo, Spangenberg noted that in 1971 Deputy Secretary of Defense Packard had ordered a study of a carrier version of the Air Force F-15. McDonnell submitted the F-15N proposal in September; this airplane had the Air Force fire-control system and a one-man cockpit (when choosing the Phantom in 1958 the Navy had decided that a two-man cockpit was essential). Navy analysis showed that the F-15N would cost about 10 percent more than continuing to buy F-14s, even though it was considerably less capable. Presumably the added cost represented redesign to strengthen the airplane and reduce its landing speed, both for carrier operation. The Air Force was asked about redesigning the F-15 to carry Phoenix missiles and to add a second place in its cockpit. It replied that would be impractical, given current resources. McDonnell claimed that it had a modestly priced Phoenix version of the F-15 design, but NavAir evaluation of the limited available data suggested that it could not be ready for at least three years and would be more expensive. Even so, Deputy Secretary of Defense Clements asked McDonnell for further information on three single-seat designs: the original F-15N, an F-15

carrying four Phoenix missiles (tracking twelve targets, displaying six, and engaging four, with an overload of six missiles), and an alternative armed only with Sparrow missiles. OSD ordered a study of alternatives, both by its own expert and by the Navy (the latter became the Naval Fighter Study). The Navy rejected all versions of the F-15. OSD then ordered the Navy to submit plans to implement a non-Phoenix program using versions of the F-15, F-4, and F-14, which would compete in a fly-off, the choice of a production fighter to be made in mid-1976. The F-4 was dropped before the project was presented to Congress. The Navy was able to show that the F-15 alternative was far more expensive than continuing F-14 production in reasonably economical numbers.

13. According to James P. Stevenson, *The Pentagon Paradox: The Development of the F-18 Hornet* (Annapolis, Md.: Naval Institute Press, 1993), 86, in January 1969 the Navy Aviation Plan envisaged a next-generation fighter capable of Mach 4.5, with a dash speed of Mach 6, armed with long-range missiles, as an F-14 successor. What Stevenson calls a "navy fighter mafia" considered this absurd, given the reality of air warfare in Vietnam. The "mafiosi" included a small group at the David Taylor Naval Ship Research and Development Center in Maryland, which became interested in a close-coupled canard wing for higher lift. Convair submitted a proposal to study such an airplane and received a small ($70,000) contract. At a March 1970 AIAA meeting a Navy lieutenant commander who had flown 172 F-8 missions presented a paper on the "Southeast Asian Aerial Combat Environment." Among his points was the need for specialized air-to-air training—what became Top Gun.

14. This account of Navy requirements is based on the Grumman G-623 proposal of 15 July 1974, in the Grumman History Center. The Grumman file includes the operational requirement dated late August 1974 on which the figures here are based. Prior to VFAX, the Navy convened Navy Fighter Study Group IV, which (Stevenson, 160) recommended a lower-cost complement to the F-14 to be bought once the fleet's need for F-14s had been met (one squadron per carrier and four for the Marines). That could be read either as support for

a low-cost fighter or as yet another way of beating off OSD. Nine lower-cost fighters were considered in the group's 15 November 1973 interim report. One was a navalized F-15. There were four versions of the F-14: AF-14X (four Phoenix), AF-14X (A-6 type air-to-ground radar), F-14X (F-15 type APG-63 radar and Sparrows), and F-14X with the Westinghouse WX-250G radar ultimately used in the F-16. General Dynamics offered a navalized version of its F-16; Stevenson does not describe the other three proposals. At the time the light fighter was called "VFX," but the Fighter Study Group recommended a strike fighter with all-weather capability to replace the A-7. That was VFAX.

15. According to Stevenson, *Pentagon Paradox*, 174, the interdiction mission was self-escorted, with a threshold radius of 550 nautical miles and a goal of 600. Fighter-escort radius was 400 nautical miles (goal 550). For close air support the desired time on station was 1.5 hours, with two Sidewinders, two Sparrows, eight Mk 82 high-drag or twelve Mk 82 low-drag bombs. Radar look-down detection range on a five-square-meter target was 30 nautical miles (45 desired). Desired operational readiness rate was 0.75 (0.85 goal). Maintenance man-hours per flight hour was eleven (goal six). Stevenson does not describe any of the VFAX competitors.

16. Stevenson, 150, attributes the supercruise remark to Capt. Jerry O'Rourke, who was involved in the competition for the fighter for the SCS. Stevenson recounts that two of the proposals offered "supercruise," which meant supersonic flight without an afterburner. The San Diego Aircraft Engineering (Sandaire) consulting company offered the lightest of the lot, 17,377 pounds. O'Rourke asked Sandaire to design a lightweight conventional fighter. It would be powered by a single J101 engine and armed with two short-range Agile (Sidewinder successor) missiles plus one or two 20-mm cannon. The only sensor would be a FLIR. Thrust-to-weight ratio in combat would be at least 1:1 (1.3:1 desired). Maximum speed with afterburner should be Mach 2.16 (2.4 desired), and maximum speed on military power Mach 1.02 (1.6 desired)—in supercruise. Wing loading at takeoff (a measure of maneuverability) should be sixty pounds per

square foot. The airplane should be 20 percent composites, and its spotting factor should be less than that of an A-4. It is not clear what impact Sandaire's feasibility study had.

17. Convair (a branch of General Dynamics) offered its Model 218, which was based on its Model 200 naval VSTOL. General Dynamics Fort Worth offered a stretched version (with greater wing area) of the YF-16 that it was offering the Air Force; it had wingtip Sparrow launchers. Grumman offered a fresh design that could be built in either conventional or VSTOL form. Northrop offered its P 630, a stretched version of its Lightweight Fighter contender (YF-17) with a nose sufficiently enlarged to accommodate a twenty-eight-inch radar dish. McDonnell offered its Model 263, which had canards abaft its air intakes; it had been offering this airplane since about October 1972. Vought had been promoting a lightweight fighter (V-523) since at least August 1972. It had a ventral air intake somewhat reminiscent of that of the Crusader III of the late 1950s, but it had a longer nose. For VFAX, Vought enlarged it as V-526, which better met the requirement.

18. Stevenson, *Pentagon Paradox*, is essentially the story of the "light fighter mafia," whose heroes were Col. John Boyd. USAF, Dr. Pierre Sprey of OSD, and Charles E. "Chuck" Meyers Jr. The "paradox" of the book's title is that the Navy was able to resist this tide when it bought the F/A-18. The title avoids the *A* (attack) side of the airplane.

19. Under its agreement with McDonnell Douglas, Northrop would lead on any land-based derivative. It had great hopes for this airplane, and in the late 1970s the Shah of Iran was likely to be lead customer. This program died when the shah was overthrown in 1979. Users of the F/A-18 ashore (Canada, Finland, Kuwait, and Switzerland) have had to buy the naval version, with a penalty for carrier capability.

20. No one would have guessed at the time, but some years later the F-16 suffered a number of crashes caused by problems with its wiring harness that disabled its fly-by-wire system.

21. Thomason, *Strike from the Sea*, 165, tabulates Navy requirements: a fighter radius of 400–450 nautical miles, a strike radius of 550 nautical

miles, maximum speed (intermediate thrust at 10,000 feet) of Mach 0.98–1.0, combat ceiling of 45,000 to 50,000 feet, specific excess power of 750–850 feet/second, acceleration time from Mach 0.8 to Mach 1.6 of eighty to 110 seconds, sustained buffet-free load factor of 5.5 to 7.0, minimum carrier approach speed of 115–125 knots, and single-engine rate of climb of 500 feet per second. Vought's 1602 was unacceptable in speed with intermediate thrust (Mach 0.94), in combat ceiling (42,650 feet), in specific excess power (723 feet/second), and in sustained buffet-free load factor (5.26). McDonnell's design was unacceptable only in carrier-approach speed (130 knots), which could be fixed. Both airplanes more than met the range requirement, and McDonnell's offered very much better acceleration (eighty-eight seconds) and a better buffet-free load factor (6.6). The better acceleration reflected greater specific excess power.

22. Supporters of the F/A-18 argued that the A-7 was underpowered. In May 1977, however, LTV offered a twin-engine version of the A-7 (Model 529D) using two non-afterburning turbofans (F404s, the engines used in the F-18). Once Hornet production was under way, unit price rose, bringing the future of the program into question. In 1981 LTV again promoted an updated A-7 as an alternative. It offered two versions of its A-7X, the "Next Generation Corsair." One was the twin-engine airplane previously promoted; the other was a single-engine airplane powered by an F101 engine that resembled an updated Crusader. Maximum speed would have been Mach 1.4. At this time the F101 was the candidate to replace the TF30 engine in the F-14. Avionics upgrades would have included the same multifunction displays planned for the F/A-18 and a nose FLIR in place of the pod that A-7Es carried.

23. The F/A-18 had three multifunction displays and a head-up display. All controls were concentrated in one stick, an arrangement called HOTAS (Hands on Throttle-and-Stick).

24. Thomason, *Strike from the Sea*, 171, points out how tricky the range issue could be. On internal fuel, the combat radius of the F/A-18 with two 1,000-pound bombs was 220 nautical miles, compared to 400 for the A-7. However, with two 315-gallon drop tanks, the F/A-18 had pylon space for six 1,000-pound bombs, two Sidewinders, and two Sparrows and could carry that load as far as the A-7 could carry the bombs and two Sidewinders.

25. See my *Fighters over the Fleet: Naval Air Defence from Biplanes to the Cold War* (Annapolis, Md.: Naval Institute Press, 2016; Barnsley, U.K.: Seaforth, 2016) for a discussion of the fighter end of this issue, the "Outer Air Battle."

26. According to Thomason, *Strike from the Sea*, 180, McDonnell proposed to increase wing area by thirty-two square feet and to modify the flaps to allow for 13 percent greater gross weight (to 55,270 pounds), including three thousand more pounds of internal fuel and two 460-gallon drop tanks. F404 thrust would have been increased by 15 percent. The airplane would have had automatic terrain-following and offset-bombing capacity, and the rear cockpit displays would have been enlarged.

27. Thomason, *Strike from the Sea*, 180, suggests that the A-6F was virtually an all-new airplane in the envelope of the A-6E.

28. There were other economic attacks, perhaps most notably the successful attempt to hold down oil prices (oil being the main Soviet export, the prop of the Soviet Union's economy).

29. For example, in July 1971 the Air Force produced a series of confidential vu-graphs describing its prototyping program, including a "very low RCS [radar cross section] vehicle," in FY72, the first year of the program (another prototype that year was the lightweight fighter that became the F-16). The separate sheet on the vehicle described it as an unmanned subsonic airplane (about three thousand pounds gross weight) intended to verify the sort of signature reduction (to 0.002 square meters) that had been demonstrated by nonflying models. It "could have significant application to strategic and tactical systems." First flight was expected in twenty-four months, test program completion in thirty. Three vehicles were planned. These vu-graphs are in a package of declassified papers (RG 343, NARA II) justifying the F/A-18 program. The Air Force vehicle was the basis of the F-117. The cited NARA file includes a photocopy of the unclassified transcript of a 9 September 1971 hearing before the Senate Committee on Armed Services

in which Deputy Secretary of Defense Packard explained prototyping. The file includes a crude sketch of the low-RCS test vehicle, which clearly had a faceted shape (as in the F-117) and seemed to have a V-tail. Objectives listed were radar invisibility, planform design, and materials exploitation, the latter implying the use of radar-absorbing material. Wingspan was deleted from the published transcript, but length remained, about nineteen feet. "We have demonstrated new material on a laboratory mock-up and we think the objective [radar invisibility] ought to be flight-tested." It is not clear to what, if any, extent the aeronautical press took notice. Accounts of the beginning of the U.S. stealth program generally date it several years later and attribute it to DDR&E and DARPA. It does seem clear that DARPA was responsible for the Project Harvey competition (August 1975) that led to the F-117. The competitors for the Experimental Stealth Testbed (XST) were Lockheed, Northrop, and McDonnell Douglas. Lockheed won, in April 1976. This airplane was highly classified, but on 21 August 1980 Secretary of Defense Harold S. Brown announced that the United States was developing an Advanced Technology Bomber (which became the B-2) whose stealthiness would "alter the military balance significantly." He did not say how, and his announcement was dismissed as an excuse for the administration's decision to cancel the more conventional B-1. According to James P. Stevenson, *$5 Billion Misunderstanding: The Collapse of the Navy's A-12 Stealth Bomber Program* (Annapolis, Md.: Naval Institute Press, 2001) 54, soon after being inaugurated in 1977, President Carter was told that experiments had shown that aircraft could be built that would be nearly impossible to detect by radar, hence could not be engaged by any existing air-defense system. Investment in the new technology was increased tenfold, and the existence of the program was classified. Stevenson's source was William J. Perry, who became DDR&E on 11 April 1977. Stevenson cites numerous knowledgeable insiders who doubted Perry's confidence in the technology and points to the demand, in the attack on Baghdad in 1991, that stealthy F-117s always be accompanied by jammers and always fly at night

to frustrate visual antiaircraft systems. I have relied heavily on Stevenson because he in turn based his book mainly on material that emerged in the lawsuits surrounding cancellation of the A-12. Without those suits, classification would have made this degree of detail entirely impossible.

30. At 16 GHz (high X-band), Lockheed's approach offered a relative radar reflectivity of 1.3, compared to 4.0 for Northrop's flying wing; Stevenson, *$5 Billion Misunderstanding*, 23, quoting the Have Blue briefing. At 2.3 GHz (low S-band, used by a missile fire-control radar) the figures were 1.5 and 1.0. At low frequency (0.175 GHz, as in Soviet search radars), the figures reversed: 13 for the Northrop design and 1,000 for Lockheed. Northrop's flying wing used curved edges to reduce radar cross section. In Northrop's view, a stealthy airplane had to be stealthy across the spectrum.

31. Stevenson, *$5 Billion Misunderstanding*, 24, shows the General Dynamics Cold Pigeon (Model 100), which was not too far from the A-12. Visible are two exhausts above its wing, to limit IR visibility from underneath. Leading edges and wingtips were all curved, and there was no fuselage beyond the point of the flying wing. General Dynamics tried and failed to join the Advanced Technology Bomber program.

32. Stevenson, *$5 Billion Misunderstanding*, 28

33. Lehman said that the Navy should buy perhaps two squadrons of "silver bullet" aircraft. They would be kept in the desert, practicing carrier landings at night so that they would never be seen. Stevenson, *$5 Billion Misunderstanding*, 49.

34. Stevenson, *$5 Billion Misunderstanding*, 41, quotes Assistant Secretary of Defense William H. Taft IV to the effect that the need for this airplane had been validated by the Blue Ribbon Panel, which, Stevenson argues (pp. 37–38) was anything but the case. This difference may be explicable by the administration's use of black programs specifically to drive Soviet spending, a point Stevenson never addresses. Lehman told Stevenson that Taft had engineered a compromise between the OSD position (that the F-14 and A-6 should be killed in favor of the stealth airplane) and his own intent to go ahead with the upgrades for the existing aircraft and also bring the stealthy airplane into service in 1997. He thought that he had won on the

in-service date, but that was later pushed back to 1995. According to Stevenson, 42, OSD pushed the in-service date back to prevent the Navy from buying A-6Fs instead of investing in stealth. Lehman did not believe that it was possible to combine the range and payload of an A-6 with stealth. Nor could it have the same basic air-to-air capability, the same *q* (energy) and ability to keep energy up over the target, and so on. He doubted that materials technology was ready.

35. Navalizing Lockheed's F-117 was not an option; it would have required such drastic changes that an entirely new design would have been less expensive. Nevertheless, in 1993 Lockheed offered a fully navalized F-117, which attracted some interest. The most obvious changes were a new less-swept foldable wing with trailing-edge flaps and leading-edge slats, plus a horizontal tail for better control during landing. Spoilers offered better roll control at low speed, and fuselage speed brakes improved maneuvering. The engines were afterburning F414s, as in the F/A-18E/F. The bomb bay was enlarged to 10,000-pound capacity; it could carry air-to-air missiles on the insides of its doors. Each wing could carry two removable stores pylons. With a ten-knot wind over the deck the airplane could take off at 68,750 pounds (the original F-117A had a gross weight of 52,500 pounds). Mission radius with 6,600 pounds of ordnance was seven hundred nautical miles, and the airplane could land on a carrier with 11,600 pounds of fuel and ordnance. Radar cross section was an order of magnitude less than that of the F/A-18E/F. The airplane was marketed as the A/F-117X or F-117N Seahawk. The Senate Armed Forces Committee wanted the FY96 defense bill to include $175 million for the beginning of a program-definition phase. The final defense authorization included $25 million for a six-month program-definition effort for an A/F-117X. Lockheed tried to make the airplane affordable by convincing the Air Force to buy a similar F-117B, but the Air Force refused, because it was concentrating on the F-22.

36. Stevenson, *$5 Billion Misunderstanding*, 50, gives some of the 1984 development estimates made by the OSD Cost Analysis Improvement Group: $2.53 to $3.08 billion in FY85 dollars, leaving only $800 million for avionics and engines (hence the

importance of using existing systems). Lehman did not want to earmark any money until 1990, so that the Navy would obtain a much better (if more expensive) airplane. OSD badly wanted airplanes to enter service earlier; it pressed unsuccessfully for first procurement in FY91 (twelve aircraft), which meant a less sophisticated design using A-6F avionics.

37. According to Stevenson, *$5 Billion Misunderstanding*, 61, NavAir forced the companies to adopt these arrangements.

38. Stevenson, *$5 Billion Misunderstanding*, 70, tabulates some characteristics. He describes the three as strike (flying wing), multimission (flying wing), and intermediate (supersonic). For a deep-strike mission carrying two air-to-surface missiles, two Sidewinders, and two AMRAAMs (Advanced Medium-Range Air-to-Air Missiles), radii would be, respectively, 1,028, 977, and 977 nautical miles, compared to 560 nautical miles for an A-6E carrying six Mk 83 laser-guided bombs, two Sidewinders, and a four-hundred-gallon belly tank. Takeoff gross weight would be, respectively, 63,938; 66,652; 73,202; and 53,920 pounds (the last for the A-6E). For an interdiction mission (maximum internal load of Mk 83 laser bombs), radius would be, respectively, 1,009 nautical miles (six bombs), 960 nautical miles (six bombs), and 1,006 nautical miles (three bombs). The maximum number of Mk 83 bombs, internal and external, which could be taken aloft with a twenty-knot wind over the deck was, respectively, fourteen, twelve, and thirteen. The A-6E could carry thirteen such bombs (359-nautical-mile radius). Combat-air-patrol time at 400 nautical miles from the carrier was also tabulated. Stevenson, *$5 Billion Misunderstanding*, 94, provides a detailed comparison. Dimensions were a seventy-foot (later seventy-four-foot) span and thirty-five-foot length for General Dynamics–McDonnell versus eighty and forty-six feet for the Northrop team, with wing areas of 1,250 and 1,720 square feet. The General Dynamics design used leading-edge devices for lift and set its crew in tandem; Northrop raised its wingtips for carrier landing and takeoff, and its crew sat side by side. Both alternatives used radar arrays in their wing leading edges. General Dynamics planned to use active arrays (i.e., with

distributed transmitters and receivers); Northrop used more conventional passive arrays. Takeoff weights were, respectively, 69,316 and 69,713 pounds, with weapon (and, presumably, personnel) weights of 5,550 and 5,160 pounds; fuel loads were 24,538 and 21,322 pounds. As a measure of maneuverability, wing loadings were 55 and 41 pounds per square foot, and spotting factors were 1.32 and 1.44 (later 1.37 and 1.59). The General Dynamics airplane needed nineteen knots of wind over deck to launch (presumably using a catapult), but Northrop's airplane could launch with a negative wind. In a comparison of strengths and weaknesses of the two proposals, the General Dynamics–McDonnell airplane was rated as having achieved maximum practical compliance with the most important requirements, whereas Northrop's reflected major deviations in survivability, structures, avionics, and subsystems. The General Dynamics–McDonnell airplane was rated a good technical proposal overall, whereas Northrop's was weak or incomplete in many areas. General Dynamics–McDonnell did not meet the requirements for radar cross section or air-to-air detection range, and the design of its flight-control system was considered weak. (Stevenson, *$5 Billion Misunderstanding*, 101, summarizes comparative strengths and weaknesses.) Stevenson reports (page 99) that the key decision memorandum that authorized the Demonstration/Validation phase of the project stated that "ATA [Advanced Technology Airplane] size, weight, and range will be bounded by designs that utilize the GE F404 engine or a re-fanned derivative thereof."

39. Stevenson, *$5 Billion Misunderstanding*, 74–75, quotes Air Force colonel Paul Kaminski, who managed the OSD stealth advisory committee between 1977 and 1984, to the effect that a straight trailing edge (as in a pure delta) can cause radar signals striking it to reradiate out the front of the airplane. Hence Northrop chose a sawtooth trailing edge for the B-2. General Dynamics considered it unlikely that its airplane would be heading directly toward a radar, so the effect would not be important. Moreover, it planned to cover the wing with radar-absorbing material.

40. The General Dynamics–McDonnell team was slightly over the terms of the request for proposals,

offering eight aircraft for $3.981 billion (average "flyaway cost"—that of production and production tools—$31.2 million, less engines, which the firm would not manufacture itself. Presumably the difference represented design and development). Northrop wanted a minimum of $5.515 billion (ceiling $6.6 billion). Average flyaway without engines would have been $50.4 million. Prior to the lawsuit, no one would disclose the gap between the two bids or Northrop's refusal to accept a fixed-price contract.

41. Summarizing the proposals in March 1987, Nav-Air found General Dynamics–McDonnell Douglas particularly attractive because it offered a much longer mission radius, 1,100 versus 850 nautical miles. Stevenson (*$5 Billion Misunderstanding*, 111) considers Northrop's choice to scale down the B-2 a serious mistake, because the Navy did not plan to use its new attack bomber the same way. The B-2 would fly high, so its design countered look-up radar. Its upper surface did not have to be stealthy. The Navy did not have high-altitude attack weapons and planned to fly at lower altitudes, where enemy radar would see air intakes atop the wing as well as other upper-wing-surface details. Northrop fought to keep the Navy from testing its model in an inverted position against radars, as that would have revealed its low-altitude deficiencies.

42. Requirements from Stevenson, *$5 Billion Misunderstanding*, 145. Span was 70.27 feet (36.27 folded, area 1,317 square feet, and aspect ratio 3.796), length 37.26 feet, and height 11.55 feet. There were four internal weapon bays, to allow the airplane to carry weapons without sacrificing stealth: two inner ones for air-to-ground weapons (two 2,200-pound TSSAMs [Tri-Service Standoff Attack Missile], eight Mk 84 bombs [2,000 pounds each], or ten Mk 83 bombs [1,000 pounds each]), and two outer ones for either a Sidewinder or an AMRAAM. The TSSAM was canceled before the A-12 was.

43. According to Stevenson, *$5 Billion Misunderstanding*, 161, black clearances applied to specific programs, not (as with normal security) to classes of information. For a long time no one was cross-cleared for access to both the B-2 and the A-12

programs, so there could be no sharing of information. By the time information was shared, early in 1990, the A-12 design was complete. However, some government officials told Stevenson that the cross-clearing argument was a sham, in that engineers involved in the A-12 had previously been involved in other stealth programs that should have provided the necessary insight. It seems fair to say that the General Dynamics–McDonnell team understood stealth shaping issues and also that McDonnell Douglas had some insight from its involvement in the unsuccessful F-23 stealth fighter program. By the time the A-12 demonstration contract had been signed, Dr. Lehman had resigned. He had considered access to the B-2 program a prerequisite for success in the A-12 program.

44. Stevenson believes (*$5 Billion Misunderstanding*, 3) Cheney may already have been told that the A-12 was at least $500 million over budget and one to two years behind schedule but that the actual situation was bad enough to have justified Cheney's anger. It is not altogether clear from Stevenson's account that Cheney himself was in fact aware of how serious the problem was. The A-12 program manager claimed that the program was under control. The $500 million had to be considered in the context of a contract allowing for a fixed ceiling price of $4.8 billion and a fixed target price of $4.4 billion. Fixed pricing meant that the contractors would have to absorb the overrun; the Navy was protected. Defense procurement regulations triggered publication of an overrun beyond a fixed percentage. In January 1990 the Navy evaded the problem by increasing the number of A-12s to be bought, including Marine aircraft even though nine months earlier the Marines had decided not to buy the A-12. Initially Cheney was told that the A-12 would be a few hundred million dollars over budget, but late in March 1990 he was told the true figure was at least $1 billion, which shocked him. He was also told that the A-12 was the only one of the aircraft being reviewed with serious cost and schedule concerns. Cheney still briefed Congress that the A-12 was the best of the aircraft being reviewed.

45. Thomason, *Strike from the Sea*, 189.

46. NavAir (RG 343) records in NARA II include a 4–5 June 1975 trade-off study (design review) on which this account is partly based. The AV-16A program was a joint effort sponsored by the British Ministry of Defence and the U.S. Marines, formally begun with a 12 April 1973 memorandum of understanding. It was to have had performance "normally associated with a full second generation development, but at a much lower risk and cost." The engine would have been the Pegasus 15, fitted with a further zero-stage fan and compressor to increase thrust to 31,700 pounds and reduce specific fuel consumption to 0.63. The fan was the turbofan, driven by the low-pressure (LP) turbine. It and the LP turbine would have been replaced. The existing high-pressure turbine would have driven a new compressor. Improved cooling would have enabled it to withstand higher temperature. At this time it seemed that the Pegasus 16, which might be available in 1978, would increase thrust to 35,500 pounds but that fuel economy would have been reduced. By way of comparison, the engine in the AV-8A was rated at 20,930 pounds of thrust for a short lift, using water injection (i.e., "wet"). The original engine in the AV-8B had a similar rating (21,550 pounds thrust for a short lift using water injection). In each case, sustained thrust was less. Specific fuel consumption was 0.735. These figures show how much development was needed to attain AV-16A goals. The main changes arising from the 1975 design review were elimination of the Aden gun pod under the fuselage in favor of provision for internal guns and provision for an ECM pod and a chaff dispenser pod. As built the AV-8B had provision for two underbody gun packs, one containing a 25-mm GAU-12 gun and the other its ammunition.

47. The lift-improvement devices were underfuselage strakes and a retractable fence forward that would capture hot air reflected from the ground and thus generate additional lift. The wing was larger (230 versus 201 square feet) but lighter because of the use of carbon-epoxy material.

48. Of the armament stations, the two outboard would normally carry Sidewinders. The other four underwing stations could carry multiple bomb-ejector racks. The two inboard wing pylons were

stressed for 2,000 pounds, compared to 1,135 for the AV-8A. The next outboard pylons were stressed for 1,000 pounds, as was the fuselage station. Payload capacity was about twice that of the AV-8A. Each of the four main wing stations could carry a drop tank. In the AV-8A only the two inboard stations could carry drop tanks.

49. The detailed summary dated 28 February 1979 is in the same RG 343 file as the 1975 AV-8+ trade-off summary.

Chapter 13. After the Cold War

1. The F/A-18 was expected to be 40 percent more available than the A-7E, reducing manning (for maintenance) by about 35 percent. The F/A-18 never had anything like the weapon capacity of the A-6 or A-7. Proponents pointed to improved survivability (due largely to much higher performance) and higher sortie rates. It was also argued that no airplane could make more than two passes over a heavily defended area or thereby have more than two shots with laser-guided bombs, which were the real forte of the airplane. Only with the advent of GPS-guided bombs could multiple weapons be dropped precisely on the same target.

2. The air-to-ground possibility was inherent in a radar such as this, which had software-generated waveforms. In this case, according to Thomason, *Strike from the Sea*, 192, Grumman took the necessary software directly from the APG-70 of the F-15E Strike Eagle. It added inverse synthetic aperture radar (ISAR), capability valued for its ability to support identification of ships. Sea-search and terrain-avoidance modes were also added.

3. According to Thomason, *Strike from the Sea*, 193, this responded to the creation of the NATF office, which considered a Navy version of the Air Force ATF (F-22) fighter. Gross weight would have increased 3,000 pounds to 76,000. After the A/F-X program was canceled in 1994, Grumman proposed a further Advanced Strike F-14 (ASF-14) incorporating new technology, such as 3-D thrust vectoring for its engines, an increased degree of stealth, and new avionics. A conformal radar would have been installed on the wing leading edge.

4. Frederick H. Lund (Naval Aviation Depot Jacksonville), "Evolution of the F-14 Bombcat," AIAA 2001-0314, paper delivered at the 39th AIAA Aerospace Sciences Meeting and Exhibit, 8–11 January 2001 at Reno, Nev. His notes on the Bombcat come only at the end of the paper.

5. Dates of tests from Thomason, *Strike from the Sea*, 192.

6. This precision-strike system was proposed in 1993, and a demonstration system was flown at Patuxent River in 1995. Lockheed Martin was chosen as sole supplier in November 1995. Other improvements were a digital flight-control system designed for the Eurofighter, an embedded GPS navigation system, a low-altitude warning system, improved ECM, and night-vision systems.

7. Thomason, *Strike from the Sea*, 196.

8. Thomason, 196. Thomason points out that radar cross section was still considerably larger than that of the Air Force F-22.

9. Reportedly the rotating wing joint of its swing-wing was a particular maintenance problem.

10. This discussion is based on Thomason, *Strike from the Sea*, 196–97.

11. This account of the airframe part of the JSF program is based on Paul M. Bevilaqua, "Genesis of the F-35 Joint Strike Fighter," *Journal of Aircraft* 46, no. 6 (November–December 2009). Bevilaqua was lead engineer on JSF at Lockheed Martin.

12. Bevilaqua. Gerard Keijsper, *Joint Strike Fighter: Design and Development of the International Aircraft* (Barnsley, U.K.: Pen & Sword, 2007), lists and describes the NASA alternatives tried in 1987: General Dynamics E-7, McDonnell Douglas 279–3, Rockwell Baseline, Rockwell Alternative, and Vought TF120. The E-7 was an ejector-augmentation airplane based on the company's F-16, with swept wings and swept canards. McDonnell Douglas used a Harrier-type engine. Vought used tandem fans with a dorsal intake; its design featured close-coupled delta wings and swept canards, with twin tails.

13. Bevilaqua calls this engine operation "dual-cycle": the engine operates in different lift and cruise modes. A 1986 review of the NASA work found the tandem fan among concepts worth pursuing. Alternatives were advanced vectored thrust (a bypass nozzle downstream from the compressor, as

in the Harrier), ejector lift (using a separate ejector nozzle, compressor air passing through it entraining other air), and remote augmented lift (air bled from the compressor ducted forward and down). They are illustrated in Ian A. Maddock, "The Quest for Stable Jet Borne Vertical Lift: ASTOVL to F-35 STOVL," delivered at the AIAA Centennial of Naval Aviation Forum, 21–22 September 2011, Virginia Beach, Va. At this time NASA was working with Lockheed's Skunk Works on a stealthy STOVL derived from the F-117. Maddock credits the DARPA work on supersonic STOVLs to the 1988 Nunn-Quayle Research and Development Initiative to fund cooperative projects with NATO allies, in this case the United Kingdom (the Royal Navy). Study results led to concentration on fan concepts.

14. Bevilaqua points out that lift required sufficient thrust, which is the product of mass flow and velocity. Boeing used limited mass flow at relatively high velocity; Lockheed Martin used the large mass flow of its fan (about 2.5 times Boeing's) moving at relatively low velocity. The difference between Boeing's arrangement and that in the Harrier is that Boeing used two rather than four streams of gas.

15. Maddock, "Quest for Stable Jet Borne Vertical Lift," illustrates the proposals. Boeing offered a modified delta wing with twin fins (later moved inboard to become twin tails), with a chin inlet under the cockpit. Directly abaft the engine were two lift nozzles, into which the exhaust was diverted for vertical flight. Some of the engine fan flow was diverted into a "jet screen" forward, not for lift but to protect the engine from ingesting hot exhaust gas. Some of the exhaust was diverted into two roll nozzles in the wings and also into a pitch nozzle near the exhaust; there were also two yaw nozzles. The Pratt & Whitney SE-614 (derived from the F119) produced 34,000 pounds of thrust (42,000 with afterburning). The key design feature was to place the engine forward, near the center of gravity (and lift), so that the weight of the airplane could be balanced on just one pair of lift nozzles, supplemented by small yaw nozzles. Boeing's X-32B flew on 29 March 2001.

16. The primary requirements of the Navy variant were to take off and land on a carrier in less than three hundred feet in twenty knots of wind over deck; Bevilaqua, "Genesis of the F-35." Lockheed Martin considered the STOVL version (which sacrificed payload and range), a version in which the roll-control jets would blow the wing flaps, and a version with a larger wing (enlarged flaps and slats and a wingtip extension). The larger wing offered greater range both by providing more fuel volume and by reducing induced drag. It appears that the short takeoff requirement has since been abandoned in favor of catapulting and arrested landing.

17. This designation simply substituted an *F* for the *X* in "X-35." No fighter numbers had been assigned beyond the F-23, the unsuccessful competitor to the F-22.

Bibliography

Andrews, Harold. *The Curtiss SB2C-1 Helldiver.* Aircraft Profile 124. Windsor, U.K.: Profile, 1971.

Bowers, Peter M. *Curtiss Aircraft 1907–1947* London: Putnam, 1979.

Bradley, Robert E. *Convair Advanced Designs: Secret Projects from San Diego 1923–1962.* North Branch, Minn.: Specialty, 2010.

Brungess, Lt. Col. James R. USAF. *Setting the Context: Suppression of Enemy Air Defenses and Joint War Fighting in an Uncertain World.* Maxwell AFB, Ala.: Air University Press, June 1994.

Buttler, Tony. *American Secret Projects: Bombers, Attack, and Anti-Submarine Aircraft 1945 to 1974.* Hersham, U.K.: Ian Allen [Midland], 2010.

Buttler, Tony, and Alan Griffith. *American Secret Projects: Fighters, Bombers, and Attack Aircraft 1937–1945.* Manchester, U.K.: Crecy, 2015.

Campbell, N. J. M. *Naval Weapons of World War Two.* London: Conway Maritime, 1985.

Celander, Lars. *How Carriers Fought: Carrier Operations in World War II.* Philadelphia: Casemate, 2018.

Cunningham, Bruce. *Douglas A3D Skywarrior.* Part 1, Design/Structures. Simi Valley, Calif.: Ginter, 1998.

Cunningham, Bruce, and Steve Ginter. *"Fleet Whales": A-3 in Fleet Service.* Simi Valley, Calif.: Ginter, 2004.

Dorr, Robert F. *Douglas A-1 Skyraider.* London: Osprey, 1989.

———. *Grumman A-6 Intruder.* London: Osprey, 1987.

Doyle, David. *TBF/TBM Avenger: Grumman's First Torpedo Bomber in World War II.* Atglen, Pa.: Schiffer, 2020.

Elliott, John M. *The Official Monogram U.S. Navy and Marine Corps Aircraft Color Guide* [includes markings]. 4 vols. Sturbridge, Mass.: Monogram Aircraft, 1987–1993.

Elward, Brad. *McDonnell Douglas A-4 Skyhawk.* Ramsbury, U.K.: Crowood, 2000.

Francillon, René J. *Grumman Aircraft since 1929.* Annapolis, Md.: Naval Institute Press, 1989.

———. *McDonnell Douglas Aircraft since 1920.* 2nd ed. 2 vols. Annapolis, Md.: Naval Institute Press, 1988.

Francillon, René J., and Edward Heinemann. *Douglas A-3 Skywarrior.* Aerograph 5. Arlington, Tex.: Aerofax, 1987.

Friedman, Norman. *The Fifty-Year War.* Annapolis, Md.: Naval Institute Press, 2000.

———. *Fighters over the Fleet: Naval Air Defence from Biplanes to the Cold War.* Annapolis, Md.: Naval Institute Press, 2016; Barnsley, U.K.: Seaforth, 2016.

Ginter, Steve. *Douglas A-4A/B Skyhawk in Navy Service.* Simi Valley, Calif.: Ginter, 2001.

———. *Douglas A-4C/L Skyhawk in Navy Service.* Simi Valley, Calif.: Ginter, 2019.

———. *Douglas A-4E/F in Navy Service.* Simi Valley, Calif.: Ginter, 2001.

———. *Douglas AD/A-1 Skyraider.* 2 vols. Simi Valley, Calif.: Ginter, 2014.

———. *Early Banshees: The McDonnell F2H-1, F2H-2/2B/2N/2P.* Naval Fighters Series. Warfield Center, Calif.: Ginter, 2008.

———. *McDonnell Douglas A-4M Skyhawk.* Simi Valley, Calif.: Ginter, 2002.

———. *Navy and Marine Fleet Single-Seat F9F Cougar Squadrons*. Naval Fighters Series. Warfield Center, Calif.: Ginter, 2006.

Grove, Michael, and Jay Miller. *North American Rockwell A3J/A-5 Vigilante*. Aerofax Minigraph 9. Arlington, Tex.: Aerofax, 1989.

Hansen, Chuck. *U.S. Nuclear Weapons: The Secret History*. Arlington, Tex.: Aerofax, 1988.

Head, Richard Glenn. "Decision-Making on the A-7 Attack Aircraft Program." PhD dissertation, Syracuse University, 1971.

Heinemann, Edward H., and Rosario Rausa. *Ed Heinemann: Combat Aircraft Designer*. Annapolis, Md.: Naval Institute Press, 1980.

Jackson, B. R., and T. E. Doll. *Grumman TBM/TBF Avenger*. Fallbrook, Calif.: Aero, 1984.

Jenkins, Dennis R. *Grumman EA-6A Intruder, EA-6B Prowler*. Aerofax Minigraph 7. Arlington, Tex.: Aerofax, n.d.

Johnson, E. R. *United States Marine Corps Aircraft since 1913*. Jefferson, N.C.: McFarland, 2018.

———. *United States Naval Aviation 1919–1941*. Jefferson, N.C.: McFarland, 2011.

Keijsper, Gerard. *Joint Strike Fighter: Design and Development of the International Aircraft*. Barnsley, U.K.: Pen & Sword, 2007.

Kilgrain, Bill C. *Color Schemes and Markings U.S. Navy Aircraft 1911–1950*. Self-published, 1981.

Kinsey, Bert. *A-7 Corsair II in Detail and Scale*. Blue Ridge Summit, Pa.: Aero, 1986.

———. *SB2C Helldiver in Detail and Scale*. Carrollton, Tex.: Squadron/Signal, 1997.

———. *TBF & TBM Avenger in Detail and Scale*. Carrollton, Tex.: Squadron/Signal, 1997.

Kowalski, Bob. *Martin AM-1/1-Q Mauler*. Simi Valley, Calif.: Ginter, 1994.

Larkin, William T. *U.S. Navy Aircraft 1921–1941 and U.S. Marine Corps Aircraft 1914–1959*. New York: Orion, 1961.

Levinson, Jeffrey L. *Alpha Strike Vietnam: The Navy's Air War, 1964 to 1973*. Novato, Calif.: Presidio, 1989.

Miller, Jerry. *Nuclear Weapons and Aircraft Carriers: How the Bomb Saved Naval Aviation* Washington, D.C.: Smithsonian, 2001.

Millot, Bernard. *Les Avions Vought*. Paris: Docavia, 1982.

Morgan, Mark, and Rick Morgan. *Intruder: The Operational History of Grumman's A-6*. Atglen, Pa.: Schiffer, 2004.

Morgan, Rick. *A-6 Intruder Units 1974–96*. Oxford, U.K.: Osprey, 2017.

Nordeen, Lon O. *AV-8B Harrier II Units of Operation Enduring Freedom*. Oxford, U.K.: Osprey, 2014.

Pardini, Albert L. *The Legendary Norden Bombsight*. Atglen, Pa.: Schiffer, 1999.

Powell, Cdr. Robert R. *Vigilante! 1200 Hours Flying the Ultimate U.S. Navy Reconnaissance Aircraft*. Forrest Lake, Minn.: Specialty, 2019.

Price, Alfred. *The History of U.S. Electronic Warfare*. 3 vols. [Alexandria, Va.]: Association of Old Crows, 1984–2000.

Smith, Peter C. *Curtiss SB2C Helldiver*. Ramsbury, U.K.: Cro-wood, 1998.

———. *Dive Bomber!* Annapolis, Md.: Naval Institute Press, 1982.

———. *Dive Bombers in Action*. London: Blandford, 1988.

———. *Douglas AD Skyraider*. Ramsbury, U.K.: Cro-wood, 1999.

———. *Douglas SBD Dauntless*. Ramsbury, U.K.: Cro-wood, 1997.

Snaples, N. L., Jr. "Institutionalizing Aircraft Procurement in the U.S. Navy 1919–1925." PhD dissertation, Texas A&M University, August 1999.

Stevenson, James P. *The $5 Billion Misunderstanding: The Collapse of the Navy's A-12 Stealth Bomber Program*. Annapolis, Md.: Naval Institute, 2001.

———. *The Pentagon Paradox: The Development of the F-18 Hornet*. Annapolis, Md.: Naval Institute, 1993.

Swanborough, Gordon, and Peter M. Bowers. *United States Navy Aircraft since 1911*. Annapolis, Md.: Naval Institute Press, 1990.

Thomason, Tommy H. *Scooter! The Douglas A-4 Skyhawk Story*. North Branch, Minn.: Specialty, 2011.

———. *Strike from the Sea: U.S. Navy Attack Aircraft from Skyraider to Super Hornet, 1948–Present*. North Branch, Minn.: Specialty, 2009.

Thornborough, Anthony M., and Peter E. Davies. *Grumman A-6 Intruder Prowler*. London: Ian Allan, 1987.

Tillman, Barrett. *Avenger at War*. London: Ian Allan, 1979.

———. *TBD Devastator Units of the U.S. Navy.* Oxford, U.K.: Osprey, 2000.

———. *TBF/TBM Avenger Units of World War 2.* Oxford, U.K.: Osprey, 1999.

Tillman, Barrett, and Robert L. Lawson, *U.S. Navy Dive and Torpedo Bombers of WW II.* St. Paul: MBI, 2001.

Trimble, William F. *Jerome C. Hunsaker and the Rise of American Aeronautics.* Washington, D.C.: Smithsonian, 2002.

———. *Wings for the Navy: A History of the Naval Aircraft Factory 1917–1956.* Annapolis, Md.: Naval Institute, 1990.

White, Graham. *Allied Piston Engines of World War II.* Warrendale, Pa.: SAE, 1995.

Winchester, Jim. *Douglas A-4 Skyhawk: Attack and Close-Support Fighter-Bomber.* Barnsley, U.K.: Pen & Sword, 2005.

Wildenberg, Thomas. *Destined for Glory: Dive Bombing, Midway, and the Evolution of Carrier Airpower.* Annapolis, Md.: Naval Institute Press, 1998.

Wolf, William. *U.S. Aerial Armament in World War II: The Ultimate Look.* Vol. 2, *Bombs, Bombsights, and Bombing.* Atglen, Pa.: Schiffer, 2010.

Woogstad, James, and Friddell, Phillip. *Grumman EA-6B Prowler and EA-6A Intruder.* Aerophile Extra Number 2. San Antonio, Tex.: Aerophile, 1985.

Y'Blood, William T. *Hunter-Killer: U.S. Escort Carriers in the Battle of the Atlantic.* Annapolis, Md.: Naval Institute Press, 1983.

Zaloga, Steven J. *Red SAM: The SA-2 Guideline Anti-Aircraft Missile.* Oxford, U.K.: Osprey, 2007.

Zichek, Jared A. *Secret Aerospace Projects of the U.S. Navy: The Incredible Attack Aircraft of the USS United States, 1948–1949.* Atglen, Pa.: Schiffer, 2009.

Index

About the Author

Norman Friedman is a defense analyst and historian specializing in the intersection of policy, strategy, and technology. He has published more than forty books, including an award-winning history of the Cold War, *The Fifty-Year War*; a history of the impact of wargaming at the Naval War College on the successful U.S. Navy strategy against Japan in World War II; a history of naval fighter aircraft; and design histories of many U.S. and British warships. For many years he authored the World Naval Developments column in U.S. Naval Institute *Proceedings*. He has been widely published in other defense magazines.